Radiation Transfer

Radiation Transfer

Statistical and Wave Aspects

L.A. Apresyan
Radio Engineering Institute,
Russian Academy of Sciences, Moscow, Russia

Yu.A. Kravtsov
Space Research Institute,
Russian Academy of Sciences, Moscow, Russia

Translated from the Russian by M.G. Edelev

CRC Press
Taylor & Francis Group
Boca Raton London New York

CRC Press is an imprint of the
Taylor & Francis Group, an **informa** business

CRC Press
Taylor & Francis Group
6000 Broken Sound Parkway NW, Suite 300
Boca Raton, FL 33487-2742

First issued in paperback 2020

ISBN-13: 978-0-367-45589-7 (pbk)
ISBN-13: 978-2-88124-920-4 (hbk)

Visit the Taylor & Francis Web site at
http://www.taylorandfrancis.com

and the CRC Press Web site at
http://www.crcpress.com

British Library Cataloguing in Publication Data
Apresyan, L.A.
Radiation Transfer: Statistical and Wave Aspects
I. Title II. Kravtsov, Yu.A.
III. Edelev, M.G.
539.2

CONTENTS

Foreword

When, ten years ago, the initial Russian version of this book was discussed at the Editorial Board of the Soviet Physics Uspekhi Journal, Ya.B. Zeldovich asked, "wave content of radiation transfer — what could it mean?" The bewilderment of this prominent physicist reflected a widespread opinion that transfer theory exists independently of wave theory and any discussions of the diffraction content of this theory seemed ridiculous.

Recent years have radically changed this view on radiation transfer theory and brought about a new understanding of the place of this theory in the system of physical knowledge. Today radiation transfer theory is treated as a corollary of wave theory (Maxwell's equations) and rather simple statistical hypotheses, the principal assumption being that statistically the field is quasi-uniform.

The wave statistical aspect of radiative transfer theory, seemingly insignificant for thermal optical radiation, comes into the forefront in the theory of propagation and multiple scattering of waves in randomly inhomogeneous media. In scattering media, the equations for the coherence function of the wave field turned out to be equivalent to the equation of radiation transport; therefore the entire arsenal of methods of radiative transfer theory is placed at the disposal of atmospheric optics, radar, underwater acoustics and hydrophysics.

The authors of this book made a substantial contribution to the statistical wave substantiation of transfer theory, so the present treatise is a highly professional delineation of the subject written with certain pedagogical skill. The book introduces the reader to the phenomena associated with multiple scattering. It will be interesting for experts in the field and may serve as a textbook for graduate and postgraduate students. The teaching objectives are assisted by a selection of instructive exercises presented at the end of each chapter.

I have no doubt that this book, updated and revised from its Russian predecessor, will find readership in the West and will be as popular there as its original version has been in Russia.

Wave Propulsion Laboratory
NOAA, Boulder, Colorado

Valerian I. Tatarskii

Preface

In writing this book we had a few objectives: the main objective was to equip optics, acoustics, and radio engineers with a modern statistical wave treatment of radiative transport processes. With an aim to bring the advances of transfer theory as close as possible to the practical needs of experimenters in all areas of wave physics, we wanted to write a textbook that could be an introduction to this interesting though specific field of knowledge. We outlined our approach to the wave fundamentals of transfer theory in the Russian version of this book in 1983 and in a review paper (1984). In this English edition we extend the analysis with an account of the latest achievements in this area but in a form readily accessible to a novice in this field.

Our second aim was to outline the fundamentals of phenomenological transfer theory — primarily for polarized electromagnetic radiation — in a more detailed exposition than had been done before. The major difference of our presentation from other treaties in transfer theory consists in a more general approach. On the basis of common phenomenological considerations we derived a transport equation for polarized radiation in the general case which takes into account the inhomogeneity and time variance of the medium, the effects due to curvature of rays, spatial dispersion and some other complicating factors.

The third objective was to expound on non-traditional phenomena for transfer theory but of considerable interest in wave measurements.

The book consists of five chapters. Following the historic overview of Chapter 1, Chapter 2 is devoted to the phenomenological theory of radiative transport. Formally this chapter is more closely connected with traditional theory than other chapters, though it contains a number of original concepts. This concerns the description of polarization, boundary conditions for radiance, and the behavior of radiance in media with variable parameters. Whenever possible, we trace the connection of radiometric quantities with wave theory.

In Chapter 3, the transport equation is derived from the first principles, that is, from the Dyson and Bethe-Salpeter equations. In multiple scattering theory, these equations may be written for the average field and coherence function. This chapter discusses the diffraction content of the transport equation and non-classical radiometry. This discussion is illustrated by practical examples related to measurements of wave field parameters in various fields of physics.

Chapter 4 presents information about the most popular techniques of solving the transport equation. Materials for this chapter were selected with an eye to the interests of wave physics, that is, the problems considered are oriented to computing the coherence function.

Chapter 5 treats the coherence effects of multiple scattering. These are the effect of backscatter enhancement brought about by coherent Watson channels, and memory effects which characterize internal correlations in a multiply-scattered field. This chapter demonstrates how much information can be extracted from the solution of an intensity transport equation.

Chapter 6 presents allied problems which are no longer covered by the traditional radiative transfer theory. It expounds, for instance, on the properties of radiation in media described by a transport equation with random parameters, nonlinear transport problems, and kinetic equations for waves.

Finally, the reference materials related to the method of geometrical optics are compiled in Appendices A, B and C.

The authors would like to express their sincere gratitude to Professor T.A. Germogenova for her valuable advice concerning the numerical techniques and to Dr. B.I. Klyachin for his assistance in writing Section 6.3. The authors are also indebted to M.G. Edelev for his professional translation of the book into English, as well as to Dr Z.I. Feizulin and Dr A.G. Vinigradov for their invaluable assistance in the process of translation.

1. A Historical Overview

1.1 PHENOMENOLOGICAL APPROACH TO RADIATION TRANSFER THEORY

The theory of radiative transfer evolved almost a century ago in the form of a transport equation for brightness, mainly owing to the efforts of Khvolson (1890) and Schuster (1905). By the 40s and 50s of this century studies of the transport equation had become an independent discipline of mathematical physics, which brought about new efficient techniques of solving integral and integro-differential equations. At present, application areas of transfer equations are many and diverse. These equations describe the scattering of light and radio waves in astrophysics, applied optics, biophysics, geophysics, and plasma physics. An important field is the transfer theory for neutrons. The concepts of transfer theory are widely used also in acoustics and have become popular in oceanology, seismology and other natural sciences.

In the transfer theory, the main characteristic is the radiance or brightness $I(\mathbf{n}, \mathbf{R})$. This is the flux of energy from a unit area of a source or a luminous surface into a unit solid angle in the direction indicated by a unit vector $\mathbf{n}$. The radiance characterizes the angle distribution of radiation, that is, represents an angular spectrum. This spectrum has a local character, for the radiance varies from one point $\mathbf{R}$ to another.

The brightness is an energy variable, therefore it obeys the radiative transport equation (RTE) which is derived from phenomenological considerations and has the form of energy balance equation. By this equation a variation of radiance in the direction $\mathbf{n}$, that is, a change of energy flux in this direction is defined by

(i) a decrement of the energy flux due to the absorption and scattering processes,

(ii) an increment of the energy flux in the direction $\mathbf{n}$ as a result of scattering from other directions $\mathbf{n}'$,

(iii) an increment of the energy flux due to sources of energy.

In general, the balance equation for energy fluxes may be represented in block form as follows

$$\begin{Bmatrix}\text{variation of radiance } I(\mathbf{n},\mathbf{R}),\text{ i.e.,}\\ \text{variation of enegy flux in direction } \mathbf{n}\end{Bmatrix}$$
$$= -\begin{Bmatrix}\text{decrement of energy flux due to absorption}\\ \text{and scattering from direction } \mathbf{n} \text{ into } \mathbf{n}'\end{Bmatrix}$$
$$+\begin{Bmatrix}\text{increment of energy flux in direction } \mathbf{n}\\ \text{due to scattering from other directions } \mathbf{n}'\end{Bmatrix}$$
$$+\{\text{increment of flux due to sources}\} \tag{1.1.1}$$

All the classic radiation transport theory reduces essentially to the analysis of the balance equation (1.1.1).

A more detailed information about radiation can be obtained with the spectral radiance $I_\omega(\mathbf{n},\mathbf{R})$ which represents the rate of change of the radiation not only with angles but also with frequencies. The distribution of radiance over frequencies becomes significant if scattering is accompanied by a change of frequency. In interpreting polarization phenomena, the scalar quantity $I(\mathbf{n},\mathbf{R})$ becomes a matrix function, and the transport equation (1.1.1) acquires terms responsible for the transformation from one polarization into another.

Thus the entire classical transfer theory is based on simple and intuitively appealing considerations of energy balance. Unlike other disciplines of mathematical physics which are known for their complete physical foundations, transfer theory has for long had no reliable substantiation in wave theory and developed as if independently from the latter. A transition from wave fields to an energy balance equation is usually not treated explicitly. In the best case, simple and obviously insufficient considerations are invoked of the type of assumption on incoherence of wave beams. Even best monographs, including books by Chandrasekhar (1960), Born and Wolf (1980), Sobolev (1956, 1975), Bekefi (1966) and Zheleznyakov (1970) introduce the main concepts of transfer theory in a purely phenomenological manner, actually like in the early works on transfer theory, without any attempts to endow these equations with a more rigorous statistical and wave meaning.

In part this can be explained by the simplicity and physical transparency of the geometric optical concepts used by transfer theory. The concept of "ray tube" and considerations of "incoherence of wave beams", which from the very outset allow one to operate with wave intensities, seem to be rather convincing and at first glance do not require any additional explanations. Neutron transfer theory operates with the concepts of "trajectory" and "independent neutron approximation" in place of ray tubes and incoherent beams. The phenomenological approach appears to be justifiable here, at least until the wave properties of neutrons may be neglected (see books of Case and Zweifel, 1967; and Davison, 1957). These seemingly evident heuristic considerations appear to be illusive in close view. This particularly concerns the propagation of laser light and radio waves through a turbid atmosphere or in scattering media when interference and diffraction phenomena cannot be neglected any longer and the applicability of the ray representations becomes questionable.

Indeed, the very terms of transfer theory imply the possibility to operate with plane or quasi-plane waves, which in the general case of an inhomogeneous medium is valid only in the geometrical optics limit of vanishingly small wavelengths $\lambda \to 0$. At final wavelengths ($\lambda \neq 0$) characteristic of modern optical measurements and radio engineering investigations, the concept of "independent and incoherent beams" is frequently inapplicable. This may be true already in a description of free propagated radiation.

If we turn to a more complicated case of scattering media, we see that the phenomenological transfer theory has implicitly implied some averaging over an infinitesimally small (i.e., small compared to the wavelength) volume of the medium and also tacitly implied independence of successive scattering events. As to scattering, the procedure has been reduced to a simple substitution of the term "mean energy" for the term "energy". The hypothesis about independent scattering events has presumed evident for media with "sufficiently random" distribution of scatterers. In scattering with notable variation of frequency, say in Raman scattering or in scattering at movable inhomogeneities, the neglect of interference in successive scattering events may still be justified by the mutual incoherence of the incident and scattered waves. Under the circumstances the applicability limits of the transport equation are indeed turn out to be markedly wider. However, when scattering proceeds without alterations of frequency, such an incoherent summation of intensities is no longer valid.

The validity of the heuristic considerations leading to the transport equation had for long avoided verification by direct comparison with the rigorous results simply because multiple scattering theory lacked any.

Gazaryan (1969) has derived and analyzed an exact expression for the ensemble average intensity of a field for a one-dimensional model of a layer with statistical scatterers. His results turned out to be markedly different from those due to the phenomenological theory of radiative transport. Specifically, the average intensity was found not to fall off linearly across the layer, as this follows from transfer theory, but rather tends – in the limit of strong scattering – to a step distribution with a double intensity of the incident wave in the first half of the layer and a zero intensity in the other half. This finding could not but seem strange, for it testified to an interference formation of the average intensity and to violation of spatial ergodicity with respect to this intensity.

Gazaryan's work have repercussions among the physical researchers involved in the fundamentals of transfer theory. Some even hurried to claim that phenomenological theory is inapplicable altogether, for the new findings doubted the very foundations of this theory operating with incoherent addition of intensities. This readiness to doubt the validity of transfer theory may be referred to the well known isolation of the phenomenological transport theory from electrodynamics. Indeed, optical engineers engaged in applications of transfer theory have not troubled themselves with a substantiation of this theory, whereas electrodynamics researchers have never been satisfied with the substantiation of this theory. The authors of this book still recall the emotional debates aroused by Gazaryan's speech at Rytov's seminar on radio physics in Moscow.

At the time being the status of transfer theory is no longer viewed as uncertain, like it was after Gazaryan's report. The difference between transfer theory and the rigorous statistical approach has been learned to be characteristic only for one dimensional (1-D) problems, and in all probability for 2-D problems. In the final analysis it is due to the phenomenon of strong Anderson localization. Up until the 70s this phenomenon was studied only in quantum solid state physics in connection with the dynamics of electrons in doped systems, and no one thought about possible manifestations of Anderson localization in classical physics, e.g., in optics and acoustics.

In 1-D and 2-D random systems, owing to strong localization the eigenfunctions are localized and vanish at infinity, whereas phenomenological transfer theory is based on travelling quasiplane waves. In 3-D problems, strong localization does not manifest itself up until a certain level of disorder in the system. In contrast to the case of electrons in a solid, in optics this level is not easy to realize: the Anderson transition takes place when the wavelength becomes comparable with the scattering length, $\lambda \sim L_{sc}$ (Ioffe–Regel condition). On the other hand, the ordinary transfer theory is applicable in the region of weak disorder, $\lambda \leq L_{sc}$, characteristic for most classical wave problems.

It is instructive to note that interference phenomena leading to strong localization are essential for static scatterers. Movable scatterers bring about a marked frequency shift in scattering, thus disturbing the temporal coherence of successive acts of scattering and rendering transfer theory applicable only for 1-D and 2-D models.

1.2 REINTERPRETATION OF TRANSFER THEORY

A clear understanding of the fundamentals and applicability conditions of radiative transfer theory can be achieved only by invoking the statistical wave formalism of the propagation of waves in random media. A substantial re-consideration of radiative transfer theory on the base of statistical wave formalism were made in recent two decades. The first attempts of this revision were directed to the substantiation of a condition of incoherent addition of wave beams, which underlies the energy balance in the radiative transport equation. However, much more has come to the surface than one might have expected, namely, that the familiar energy balance conceals an equation for the coherence function of the wave field.

The traditional treatment of the radiative transport equation as a simple condition of energy balance was changed radically only when a direct connection was established between the radiance and the coherence function of the wave field. For an equilibrium thermal radiation such a relation can be readily deduced from the general theory of thermal fluctuations developed by Rytov (1953). In 1956 Rytov and Levin derived from this relation an expression for the temporal correlations of unconstrained thermal radiation. Later similar expressions were obtained by Bourret (1960) and Kano and Wolf (1962) and generalized on the case of spatial correlations by Metha and Wolf (1964).

However, the equilibrium thermal radiation is homogeneous and does not result in a transport of energy from point to point. For inhomogeneous radiation, a relation of radiance with the coherence function was established by Dolin (1964b), who demonstrated, in particular, that in the small angle approximation, the transport equation for radiance is equivalent to a parabolic (truncated wave) equation for the coherence function.

Later a universal relationship between the radiance and coherence function was established for wide-angle beams. We consider this relation in a simple background of a free wave field.

We represent the coherence function $\Gamma(\mathbf{r}_1,\mathbf{r}_2) = \langle u(\mathbf{r}_1)u^*(\mathbf{r}_2)\rangle$ for the amplitude $u(\mathbf{r})$ of a monochromatic wave field $u(\mathbf{r})e^{-i\omega t}$ in the form

$$\Gamma(\rho,\mathbf{R}) = \left\langle u(\mathbf{R}+\frac{\rho}{2})u^*(\mathbf{R}-\frac{\rho}{2})\right\rangle. \tag{1.2.1}$$

Let Γ satisfy the statistical quasi-uniformity condition. This implies that the coherence function $\Gamma(\rho,\mathbf{R})$ depends predominantly on the difference coordinate $\rho = \mathbf{r}_1 - \mathbf{r}_2$ and is weakly dependent on the position of the center of gravity $\mathbf{R} = (\mathbf{r}_1 + \mathbf{r}_2)/2$. For a statistically uniform field, the dependence of Γ on $\mathbf{R}$ disappears completely and $\Gamma(\rho,\mathbf{R}) = \Gamma(\rho)$.

From a modern standpoint, the coherence function $\Gamma(\rho,\mathbf{R})$ of a quasi-uniform wave field is related to the radiance $I(\mathbf{n},\mathbf{R})$ by the "angular" Fourier transformation

$$\Gamma(\rho,\mathbf{R}) = \int I(\mathbf{n},\mathbf{R}) e^{ik_0 \mathbf{n}\rho} d\Omega_{\mathbf{n}} , \tag{1.2.2}$$

where k_0 is the wave number (replaced with the effective wave number k^{eff} in the case of a scattering medium), and $d\Omega_{\mathbf{n}}$ is the elementary solid angle in the direction $\mathbf{n}$. Thus, the radiance $I(\mathbf{n},\mathbf{R})$ may be viewed as an angular spectrum of the wave field. This spectrum may be treated as a *local* spectrum, for it relates to the gravity center $\mathbf{R}$ of the coordinates $\mathbf{r}_1$ and $\mathbf{r}_2$ and varies from one point $\mathbf{R}$ to another.

The inversion of Eq. (1.2.2), i. e., an expression representing $I(\mathbf{n},\mathbf{R})$ in terms of $\Gamma(\rho,\mathbf{R})$, is not single valued – this can be seen already from the fact that the spatial argument ρ contains three components (ρ_x, ρ_y, ρ_z) whereas eqn (1.2.2) is a *two-dimensional* Fourier transformation. Moreover, in the general case, eqn (1.2.2) is an approximate expression which can be obtained by dropping inhomogeneous plane waves from the spectrum $\Gamma(\rho,\mathbf{R})$. These waves commonly decay far from the sources and interfaces, but can be significant in the near field. It turns out that the classical transfer theory is valid only asymptotically far from sources and boundaries and corresponds exactly to the limit of the quasi-uniform wave field when in eqn (1.2.2) the dependence on $\mathbf{R}$ is not significant and the field is formed only by the travelling, rather than decaying, plane waves.

For a quasi-uniform field, the inverse Fourier transform of the coherence function $\Gamma(\rho,\mathbf{R})$ turns out to be proportional to the radiance, and the equation for coherence function deduced in the modern theory of multiple scattering of waves is equivalent to the radiative transport equation.

Despite a more or less evident connection between the coherence function and radiance, the way from the first principles of statistical wave theory to the radiative transport equation was not straightaway. From the first advances in the development of multiple scattering theory in the 60s to the successive substantiation of transfer theory in the 70s and 80s there was a stretch of 15-20 years. Nevertheless a reliable bridge between wave theory and the phenomenological theory of radiative transfer had been built at last.

Important elements of this bridge, whose significance has not been recognized until recent time, have been two plain, indirectly used hypotheses — on the weakness of scattering and on the statistical quasi-homogeneity of a primary wave field. The hypothesis on the relative weakness of a scattering process has led Finkelberg and Barabanenkov to the *one-group approximation* (Finkelberg 1967, Barabanenkov and Finkelberg 1967) which radically simplified the stochastic equation for the coherence function of a wave field. The hypothesis on quasi-homogeneity has enabled a meaningful transition from the coherence function to radiance. Various forms of this hypothesis have been introduced in wave theory from the late 60s (Walther 1968, Barabanenkov, Vinogradov, Kravtsov and Tatarskii 1972, Apresyan 1973, 1975, Ovchinnikov and Tatarskii 1972, Wolf and Carter 1977, Wolf 1978; see also Friberg 1993). A logical path from the foundations of wave theory to the radiative transfer theory is schematically presented in Fig. 1.1.

After considerable efforts it has become clear that radiation transfer theory in general is equivalent to the statistical theory of quasi-uniform wave fields. This is the principal, but not unique, result of the recapitulation of the fundamentals of radiative transfer theory.

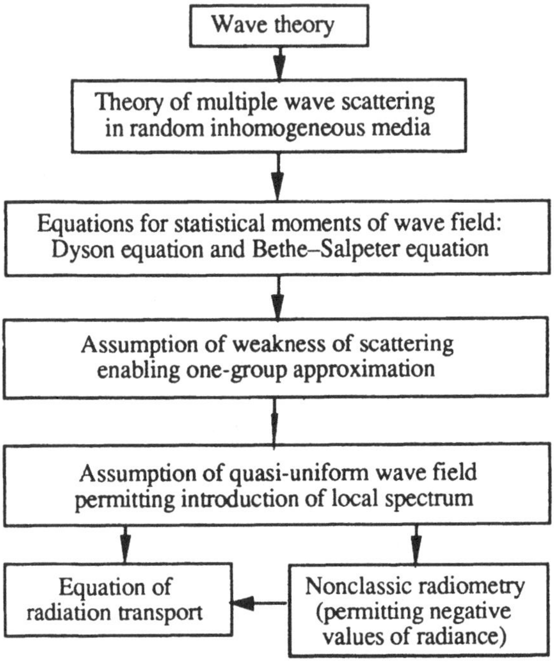

Figure 1.1. Connection of radiative transfer theory with the fundamentals of wave theory.

Once the "wave origin" of radiance has been established it has become possible to infer from the transport equation an information which thus far has been the prerogative of wave theory. This information is listed below.

(i) *Evaluation of the coherence function of the wave field.* When the radiance $I(\mathbf{n},\mathbf{R})$ is measured or calculated from the transport equation, we may determine as a "costless" by-product the coherence function from the radiometry data by Eq. (1.2.2). In other words, purely radiometric, energy data prove sufficient for the recovery of the correlation property of the field. In many a situation this circumstance radically simplifies the computations and measurements.

From the theory standpoint, the perception of the radiance as the spectrum of a random wave field is especially important since it naturally fits the transfer equation in the formalism of a more general statistical wave theory, thus allowing one to directly use the known solutions of this equation in calculations of various coherent effects.

Defining the coherence function in terms of radiance is essential also for the spatial coherent processing of optical, acoustic and radio signals. Most interest in the reconsidered radiometry has been shown by optical researchers involved in laser beam transmission and by experts in underwater acoustics.

(ii) *Incorporation of coherent component in transfer theory.* Attempts to incorporate the primary coherent field, irradiating a scattering medium, in the transport equation have been made earlier by means of boundary conditions. However, these attempts were inconsistent and, to a certain extent, primitive. In the new transfer theory, the coherent component of radiance is combined in a single term and obeys a separate equation, complementary to the transfer equation. This additional equation describes the decay of the coherent component which plays the role of a source of the incoherent scattered field.

(iii) *Applicability conditions for the equation of radiative transport.* In an earlier conceptual framework it seemed that the more disorder is there in the scattering medium the better for the applicability of the transport equation. Re-interpretation of transfer theory from the standpoint of fundamental wave principles provided a solid methodological footing to the applicability limits of the transport equation. In essence the problem focuses on the applicability of one-group approximation in the Dyson and Bethe–Salpeter equations, which requires, in a sense, weakness of scattering, and also on the quasi-homogeneity condition which admits a transition from the coherence function to the radiance. Qualitatively these conditions can be stated in a rather simple form: the wavelength and the characteristic dimension of scatterers must be small compared with the extinction path. However, in this formulation, they are not sufficient.

Thus, the reasons constraining the applicability of the transport equation have been recognized, but no universal applicability conditions have been written so far. As stumbling blocks we see two circumstances: many and varied characteristics of scattering media and certain technical difficulties in recasting the qualitative applicability conditions in the form of inequalities.

(iv) *Consideration of coherent channels in multiple scattering.* Experts in radiometry using the equation of radiative transport have deemed for long time that cooperative, or alternatively, coherent effects are beyond the framework of transfer theory and can only limit the size of this framework (Rozenberg 1970). Whereas the latter has proved unconditionally true the former is true only in part.

Indeed, the transport equation is based on the incoherent (intensity) summation of waves propagating in different directions. In a real scattering medium there always exist coherent pairs of channels responsible for the effect of backscatter enhancement. This effect concerns a maximum in the radiation pattern of a wave scattered backwards by a medium. These pairs of coherent channels, first reported by Watson (1969), seemed to have a purely wave nature and so, in principle, could not be described by transport theory. However, a procedure has been proposed recently that describes the backscattering effect in terms of transport theory. Roughly speaking, one should double the contribution due to the terms responsible for multiple scattering. Thus, transport theory has proved to be a rather flexible formalism to take into account even some coherent effects in multiple scattering.

When the place of transfer theory with respect to statistical wave theory was established, its range of capabilities became clear as well. While it enables one to describe the diffraction of quasi-homogeneous part of a wave field, it says nothing about the behavior of the other part of the spectrum, which is normally small, but does present in actual media. Further on, transfer theory fails to describe the effects of spatial dispersion related to the finite size of the effective inhomogeneities of the medium and other phenomena arising in strongly scattering media. Specifically, the so-called effects of strong (Anderson) localization, which owe their existence entirely to interference phenomena, do not fit in the radiative transport formalism.To describe such phenomena one has to resort to the initial Bethe-Salpeter equation in a more general formulation without reduction to the form of an ordinary transport equation for radiance (some results obtained in this direction will be discussed in Chapter 5).

Many workers were active in deriving a radiative transport equation from the first principles. Running the risk of missing some essential contributions we list some reports of the 60s and 70s devoted to this problem. The earliest efforts are due to Bugnolo (1960) and Gnedin and Dolginov (1963). These followed by the contribution of Borovoi (1966) and a very important series of works by Barabanenkov (1967, 1969b, 1975), Barabanenkov and Finkelberg (1967) and Barabanenkov, Vinogradov, Kravtsov, and Tatarskii (1972). The last report already contains the main features of the modern theory. This achievement, however, had precursors in the papers of Walther (1968), Stott (1968), Watson (1969), Peacher and Watson (1970), and Galinas and Ott (1970). The seventies have seen also the publications of Ovchinnikov and Tatarskii (1972), Howe (1973), Ovchinnikov (1973), Apresyan (1973, 1974, 1975), Furutsu (1975), Carter and Wolf (1975), Aquista and Anderson (1977), Baltes (1978), Wolf and Carter (1977), and Wolf (1978). Some results achieved in the substantiation of transfer theory have already been reflected in textbooks by Ishimaru (1978), Dolginov, Gnedin and Silantyev (1979) and Rytov, Kravtsov and Tatarskii (1989b).

1.3 DIFFRACTION CONTENT OF THE RADIATIVE TRANSPORT EQUATION. NONCLASSICAL RADIOMETRY

Despite the physical transparency of the links between the fundamentals of wave theory and radiative transport theory presented in Fig 1.1 some sequels of wave theory may nonplus a novice to the field. The main sequel of this kind is negative radiance under certain conditions, for example, in a bold computation of the angular spectrum $I(\mathbf{n},\mathbf{R})$ for a field passed through an aperture. From the standpoint of traditional radiometry which treats radiance as energy density, negative values of radiance are nonsense. From the point of view of wave theory, they are a natural consequence of basic principles.

Let a coherent wave pass through an aperture in an opaque screen. If we associate with this wave an angular spectrum $I(\mathbf{n},\mathbf{R})$ in a formal manner, then this spectrum need not necessarily be positive. The reason is that a field passed through an aperture does not meet the quasi-homogeneity condition. On the other hand, radiance becomes positive far from the screen, where the field from the aperture acquires the properties of a quasi-uniform field.

Admission of positive and negative values for radiance brought about a separate branch of transfer theory, called "nonclassical photometry" in the Russian edition of this book (1983) and in a review paper of Apresyan and Kravtsov (1984). Nonclassical radiometry includes in the scope of the transport equation those diffraction objects which would have been unthinkable in the framework of classical radiometry. Quite appropriately this radiometry may be also called "diffraction" transfer theory to contrast it to the classic geometrical optics transfer theory.

Accordingly, In Fig. 1.1 nonclassical radiometry appears as a separate box leading to the transport equation from the side of negative radiance, $I(\mathbf{n},\mathbf{R}) < 0$.

This extended treatment of radiance does not exhaust the diffraction content of transfer theory. There are many other examples illustrating the fact that results of transfer theory contain diffraction effects. Three of them follow.

(i) The transport equation in the small angle approximation proves to be identical to the diffraction parabolic equation for the coherence function (Dolin, 1964b).

(ii) Consideration of weak scattering in the transport equation by perturbation theory leads to exactly the same results as wave scattering theory in the first Born approximation.

(iii) The radiance $I(\mathbf{n},\mathbf{R})$ obeys the wave uncertainty relation which cannot be explained in the framework of purely radiometric representations.

Summing up, nonclassical radiometry enables us to describe and solve some diffraction problems. This is the radical change of the role of the transport equation in physics. It ceases to be a simple analogue of Boltzmann's formula because it follows from the basic principles of the theory and has a diffraction content.

Of course we do not admit the thought of recommending one to solve diffraction problems by solving radiative transport equations, the more so that the capabilities of the transport equation are limited by simplifying idealizations made in its derivation. However, the very possibility of the transport equation to be a carrier of diffraction information is wonderful.

2. Phenomenological Theory of Radiative Transfer

In this chapter we consider the fundamental concepts of the classical theory of radiation transport in free space and in scattering media. The main emphasis will be placed on the description of different complicating factors, such as inhomogeneity and time variability of media, spatial and temporal dispersion, polarization of radiation, anisotropy, etc. The classical transfer theory is built phenomenologically, without invoking rigorous statistical concepts.

A transition from wave theory to radiative transport theory can be effected in two ways. The earlier method resorts to the geometrical optics limit $\lambda \to 0$ and subsequent statistical averaging. This approach corresponds to the classical phenomenological transfer theory, whose main concepts do not involve the wavelength explicitly, but actually treat it as an infinitesimal quantity. The second method consists in the reverse sequence of these operations; it derives first the equations for statistical moments, then allows the wavelength to go to zero, $\lambda \to 0$. This approach has served as a basis for the statistical derivation of a radiative transfer equation, which will be elucidated in Chapter 3. In this chapter, we intend to resort to the first, traditional, approach.

Section 2.1 introduces the basic concept of classic radiometry and radiative transfer theory such as irradiance, radiant intensity, etc. It demonstrates how one can derive the basic equation relating the radiance with the correlation characteristics of the field within the framework of this theory.

Section 2.2 presents a phenomenological derivation of the equation describing in a general form the transport of scalar radiation in scattering media in the presence of various complicating factors. A description of electromagnetic radiation within the framework of classical transfer theory, which allows for polarization, will be the subject of Section 2.3.

2.1 BASIC CONCEPTS OF RADIOMETRY. RADIATION TRANSPORT IN HOMOGENEOUS MEDIA

2.1.1 Incoherent Beams and the Concept of Radiance

To begin with we recall the main concepts of the classical theory of radiation transport. This theory may be applied to the description of waves of any physical nature. To be specific, we shall speak of light radiation using photometric or radiometric terminology. Photometry, as the name implies, is concerned with measurement of light quantities; however, the specific virtues of electromagnetic radiation are not used in the construction of photometric concepts. Therefore, in place of photometry, one may equally speak of "acoustometry" or, more generally, of "energometry" of radiation.

Phenomenological photometry is based on an intuitively appealing "hydrodynamic" pattern of flows of radiant energy along the rays. This image system exploits the representation of a wave field as an incoherent family of bundles of rays or, alternatively, sets of random quasiplane waves which may have any direction at each point and which propagate along the rays in accordance with the laws of geometrical optics. Here incoherence implies, by assumption, simply that the average energy of the field is equal to the sum of energies of different bundles, that is, the expression for energy does not contain interference terms. To make long things short, this is customarily described as the addition of intensities, rather than amplitudes.

In this pattern, radiation sources are completely characterized by a quantity named the *radiance*, or *brightness*, of sources. Some authors use the name specific intensity of radiation, shortened to simple "intensity" in applications.[1]

Consider a radiant surface source. Let $\mathbf{n}_\mathbf{r}$ be the normal to an elementary area $d\Sigma_\mathbf{r}$ at a point $\mathbf{r}$. The radiance of the source is defined as the average flux of energy emitted by this unit area within a unit solid angle around the direction

$$\mathbf{n} = (n_x, n_y, n_y) = (\sin\theta\cos\phi, \sin\theta\sin\phi, \cos\theta)$$

(Fig. 2.1) and denoted by $I = I(\mathbf{n}, \mathbf{r})$. In a nonstationary, i.e. time dependent, case I may also depend on time and so the argument $\mathbf{r}$ must be replaced with $x = (\mathbf{r}, t)$.

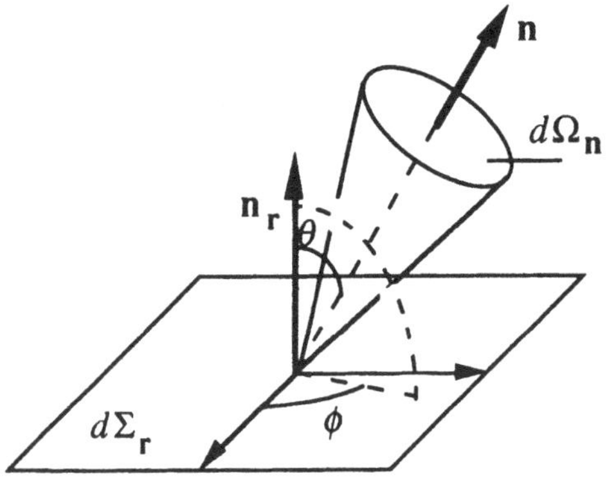

Figure 2.1. Geometry relevant for the definition of the radiance of a radiant surface.

[1] In photometry, the average characteristics of a light fields are commonly defined with allowance for the spectral sensitivity of the eye, the radiant intensity is called luminance and its symbol may be augmented with a subscript v for "visual." In radiometry, which covers all e.m. wavelengths including the optical band, this subscript is e for "energy". In what follows we will be interested in, and tacitly imply, predominantly energy characteristics.

Thus, to a beam wave emanating from a radiant area within the element $d\Omega_{\mathbf{n}}$ of solid angle around the direction **n** we will assign the vector of average energy flux density

$$d\mathbf{S} = d\mathbf{S}(\mathbf{n},\mathbf{r}) = \mathbf{n}I(\mathbf{n},\mathbf{r})d\Omega_{\mathbf{n}}\,, \qquad d\Omega_{\mathbf{n}} = \sin\theta d\theta d\phi\,, \tag{2.1.1}$$

where the average is meant to be a quantity averaged in time (over many periods of emission) and in space (over many wavelengths).

In place of a radiating surface one may consider an imaginary area pierced by radiant beams, but then one should speak of *radiation* or *specific intensity* rather than of source radiance. Further on, in contrast to the approach customary in physical optics (Born and Wolf, 1980) we shall not distinguish this two concepts because they are physically identical, and the context usually suggests whether the radiance relates to sources or to a radiant field.

From eqn (2.1.1) it follows that the total (in all directions) radiant flux at a given point **r** is given by the integral of the radiance over the unit sphere,

$$\mathbf{S}(\mathbf{r}) = \oint_{4\pi} d\mathbf{S}(\mathbf{n},\mathbf{r}) = \oint_{4\pi} \mathbf{n}I(\mathbf{n},\mathbf{r})d\Omega_{\mathbf{n}}\,, \tag{2.1.2}$$

which corresponds to the incoherent addition of the energies of the beams in all directions **n**.

The flux of energy in the direction **n** crossing an oriented element $\mathbf{n}_{\mathbf{r}}d\Sigma_{\mathbf{r}}$ of area with normal $\mathbf{n}_{\mathbf{r}}$ is

$$dP = dP(\mathbf{n},\mathbf{r}) = \cos\theta I(\mathbf{n},\mathbf{r})d\Omega_{\mathbf{n}}d\Sigma_{\mathbf{r}}\,, \tag{2.1.3}$$

where $\cos\theta = \mathbf{n}\cdot\mathbf{n}_{\mathbf{r}}$, and θ is the angle between the normal $\mathbf{n}_{\mathbf{r}}$ to the elementary area and the direction **n** of the radiant flux, as shown in Fig. 2.1.

In agreement with eqn (2.1.3) the radiance I is related to the energy flux dP by

$$I(\mathbf{n},\mathbf{r}) = \frac{dP(\mathbf{n},\mathbf{r})}{\cos\theta d\Omega_{\mathbf{n}}d\Sigma_{\mathbf{r}}} \tag{2.1.4}$$

and measured in $\mathrm{W\,sr^{-1}\,m^{-2}}$ or in $\mathrm{erg\,s^{-1}sr^{-1}\,cm^{-2}}$.

2.1.2 Spectral Radiance

For stationary radiation, we may refer the energy flux to a frequency interval $(\omega, \omega + d\omega)$ and introduce the spectral flux density $d\mathbf{S}_{\omega}$ by the formula

$$d\mathbf{S}(\mathbf{n},\mathbf{r},\omega) = d\mathbf{S}_{\omega}(\mathbf{n},\mathbf{r})d\omega\,. \tag{2.1.5}$$

This allows us to determine, by analogy with eqn (2.1.1), the spectral density of the radiance in the unit frequency interval $I_\omega(\mathbf{n},\mathbf{r})$ by the relation

$$d\mathbf{S}(\mathbf{n},\mathbf{r},\omega) = d\mathbf{S}_\omega(\mathbf{n},\mathbf{r})d\omega = \mathbf{n}I_\omega(\mathbf{n},\mathbf{r})d\Omega_\mathbf{n}d\omega\,. \tag{2.1.6}$$

In analyses involving spectral decompositions, experimenters often confine themselves with positive frequencies. This choice implies transition to the co-called analytic signal (for virtues of this signal, see, e.g., Peřina, 1972; and Rytov, Kravtsov, Tatarskii, 1987b). In this text we shall treat equally both positive and negative frequencies. A spectrum I_ω defined over the entire frequency axis, $-\infty < \omega < \infty$, is connected to the similar spectrum for positive frequencies, $I_{\omega>0}$, by the obvious relation $2I_\omega = I_{\omega>0}$. The quantity I_ω is commonly called the *spectral radiance*, or simply radiance where the dependence on frequency is not significant (the subscript ω is then dropped). The spectral radiance is measured in $\mathrm{W\,sr^{-1}\,m^{-2}\,Hz^{-1}}$ or in $\mathrm{erg\,s^{-1}\,sr^{-1}cm^{-2}\,Hz^{-1}}$.

The total quantities $d\mathbf{S}$ and I are expressed as integrals of the respective spectral densities

$$\begin{aligned} d\mathbf{S}(\mathbf{n},\mathbf{r}) &= \int_{-\infty}^{\infty} d\mathbf{S}_\omega(\mathbf{n},\mathbf{r})d\omega, \\ I(\mathbf{n},\mathbf{r}) &= \int_{-\infty}^{\infty} I_\omega(\mathbf{n},\mathbf{r})d\omega\,. \end{aligned} \tag{2.1.7}$$

2.1.3 Irradiance, Emittance, Radiant Intensity and Energy Density

Radiometry, and especially photometry, make use of a host of close terms and units of measurement. This diversity inspired Hilborn (1984) in the following piece of rhetoric. "After many frustrating battles with jargon of the field I was convinced that radiometry is an acronym for *r*evulsive *a*rchaic *d*iabolical *i*nvidious *o*dious *m*ystifying *e*xotic *t*erminology *r*egenerating *y*awns". We will therefore not indulge into the niceties of the existing terminologies and units of measurement and confine ourselves to the basic concepts necessary for subsequent considerations.

All radiometric characteristics can be expressed in terms of radiance I or spectral radiance I_ω with the aid of intuitively appealing representations about flows of energy propagated along the rays. We recall the most important of these characteristics.[2]

The *irradiance*, E, is defined as the full flux of energy incident on a unit area from all directions at the side of the normal $\mathbf{n_r}$, as shown in Fig. 2.2,

---

[2] For a more detailed review of classical radiometry concepts, the reader is referred to the book of Born and Wolf (1980), and the paper of Wolfe (1980).

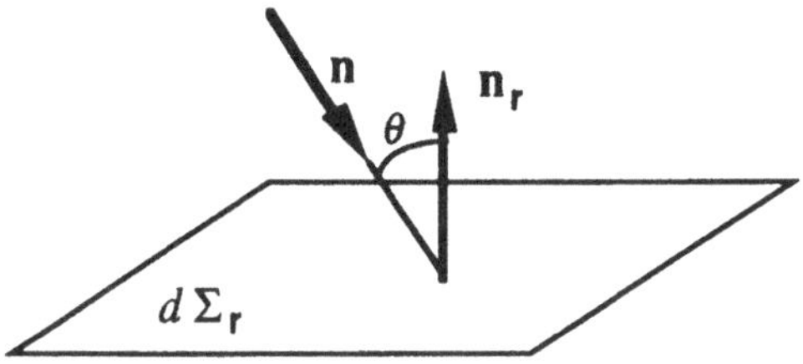

Figure 2.2. Incidence of radiation on an elementary area.

$$E = E(\mathbf{r}) = -\int_{2\pi} \mathbf{n_r} d\mathbf{S} = \int I(\mathbf{n},\mathbf{r}) d\Omega_\mathbf{n} \,. \tag{2.1.8}$$

The minus sign corresponds to the *obtuse* angle $\mathbf{n} \cdot \mathbf{n_r} = -\cos\theta$ formed by the unit vector of wave direction **n** and the normal $\mathbf{n_r}$ to the unit area. A similar quantity to that given by eqn (2.1.8) but defined for the radiating surface is the surface density of the emitted energy flux, known as the *radiant emittance*.

Another radiometric characteristic of radiant sources is the *radiant intensity*. If the point of observation is far from the plane radiating surface Σ, at a distance greater than its size, then Σ may be treated as a point source. The radiant intensity $\mathcal{J}(\mathbf{n})$ of such a source is defined as the flux of energy propagated in unit solid angle and expressed as the integral of the radiance over the surface of the source,

$$\mathcal{J}(\mathbf{n}) = \int_\Sigma \cos\theta I(\mathbf{n},\mathbf{r}) d\Sigma_\mathbf{r} \,. \tag{2.1.9}$$

In classical radiometry it is the radiant intensity $\mathcal{J}(\mathbf{n})$ rather than the radiance $I(\mathbf{n},\mathbf{r})$ that is used as a cornerstone in constructing the conceptual basis. This situation is related to the fact that historically point sources were the first for which quantitative measurements were made.

For an isotropically radiating plane surface of radiance I identical in all directions **n**, the angular dependence of the radiant intensity is seen to be defined by the factor $\cos\theta = \mathbf{n} \cdot \mathbf{n_r}$ (Lambert's law; see also Problem 2.1).

The spatial density of radiant energy $w(\mathbf{r})$ may also be expressed in terms of radiance $I(\mathbf{n},\mathbf{r})$ assuming that the radiation propagates at the velocity c. Since $Id\Omega_\mathbf{n}$ is the amount of energy and c is the volume occupied by the radiation which passes through the unit area $\Sigma = 1$ in a unit of time, each beam contributes to w an amount equal to $c^{-1} I d\Omega_\mathbf{n}$ so that the total energy density is given by the integral

$$w = w(\mathbf{r}) = c^{-1} \int I(\mathbf{n},\mathbf{r}) d\Omega_\mathbf{n} \,. \tag{2.1.10}$$

Most applications of classical photometry understood in a narrow sense of illumination engineering deal predominantly with the evaluation of illumination given by the distribution of point sources with specified characteristics, or else with the selection of characteristics of light sources necessary to provide specified illumination. For the sake of illustration, several worked problems of this kind are appended at the end of this chapter.

2.1.4 Radiometric Quantities in Anisotropic, Weakly Inhomogeneous and Dispersive Media

The main radiometric notion is the beam of radiation. To rightly apply this notion it is required that the medium should afford propagation of locally plane waves. To be more specific, the properties of the medium should afford description of the wave field in the framework of the geometrical optics formalism. This method is applicable not only in the simplest case of homogeneous and stationary isotropic media without absorption and dispersion, but also in a more general case of other media which may be weakly inhomogeneous and slowly nonstationary, anisotropic, weakly absorbing, and showing spatial and frequency dispersion behaviour.[3] According to the geometrical optics approximation, such media propagate beam energy with a group velocity $\mathbf{v}_g$. The magnitude of this velocity may depend on the frequency and direction of the wave. In the general case of an anisotropic medium, the direction of group velocity vector no longer coincides with the wave vector $\mathbf{k}$, and rays, the geometrical lines along which the radiation propagates, can be curved because of refraction at inhomogeneities and can change with time because of the nonstationarity of the medium. Under the definition of group velocity, for a single quasi-plane wave with wave vector $\mathbf{k}$, the energy density $w_\mathbf{k}$ is related to the energy flux density $\mathbf{S}_\mathbf{k}$ by

$$\mathbf{S}_\mathbf{k} = \mathbf{v}_g w_\mathbf{k} . \tag{2.1.11}$$

Whence it follows

$$w_\mathbf{k} = S_\mathbf{k} / v_g . \tag{2.1.12}$$

Consider now a bundle of quasi-plane waves with ray directions $\mathbf{n} = \mathbf{v}_g / v_g$ confined in the element $d\Omega_\mathbf{n}$ of solid angle and distributed over the frequency interval $(\omega, \omega + d\omega)$. The energy flux density $d\mathbf{S}_\mathbf{k}$ corresponding to this bundle is given as

$$d\mathbf{S}_\mathbf{k} = \mathbf{n} I_\omega(\mathbf{n}, \mathbf{r}) d\Omega_\mathbf{n} d\omega , \tag{2.1.13}$$

which corresponds to the "incoherent" addition of the energies due to different directions and frequencies.

[3] For a substantial text on the fundamentals of geometrical optics the reader should consult Kravtsov and Orlov (1990). A minimum background in this field may be acquired from Appendices A, B and C to the present text.

A relation, similar to eqn (2.1.12), for the energy density associated with the beam has the form

$$dw = v_g^{-1} dS_{\mathbf{k}} = v_g^{-1} I_\omega d\Omega_{\mathbf{n}} d\omega . \tag{2.1.14}$$

Integrating this relation we arrive at the *spectral energy density*

$$w_\omega(\mathbf{r}) = \int v_g^{-1} I_\omega d\Omega_{\mathbf{n}} \tag{2.1.15}$$

and the mean energy density

$$w(\mathbf{r}) = \int w_\omega(\mathbf{r}) d\omega = \int v_g^{-1} I_\omega(\mathbf{n},\mathbf{r}) d\Omega_{\mathbf{n}} d\omega . \tag{2.1.16}$$

In agreement with eqn (2.1.13) the average energy flux density is

$$\mathbf{S}(\mathbf{r}) = \int d\mathbf{S}_{\mathbf{k}} = \int \mathbf{n} I_\omega(\mathbf{n},\mathbf{r}) d\Omega_{\mathbf{n}} d\omega . \tag{2.1.17}$$

It should be emphasized that in contrast to the case of isotropic media, for anisotropic media, eqns (2.1.13)–(2.1.17) have $\mathbf{n} = \mathbf{v}_g / v_g$ rather than **k**/k as the unit vector of ray direction. Furthermore, in eqns (2.1.16) and (2.1.17) integration must be carried carried out over the directions and frequencies corresponding to the propagating waves. Profound texts on waves in anisotropic media are those of Ginzburg (1970), Bekefi (1966), and Allis, Buchsbaum and Bers (1963).

2.1.5 Applicability of Radiometric Concepts

Photometry dates as far back as the 15th century and is associated with many famous scientists such as Leonardo da Vinci, Galileo (17th century), and especially Bouger (1729) and Lambert (1760). The first rigorous mathematical formulation of radiative transport theory in turbid media owes its existence mainly to Khvolson (1890) and Schuster (1905) who derived the key equation for this theory – the equation of radiative transport. The history of this development and later advances of the phenomenological theory have been covered in the paper of Rozenberg (1977).

In the early days of photometry, the applicability of its concepts was not doubted though from the very beginning it was obvious that these concepts are only approximate. We intend to formulate the basic heuristic conditions of applicability of classical radiometry remaining within the framework of modern phenomenological theory of radiative energy transfer.

The applicability of these approximations are associated primarily with the methods of measurement of light radiation whose wavelengths λ is exceedingly small compared with the characteristic dimensions L^* of recording devices. The condition

$\lambda << L^*$ corresponds formally to the geometrical optics limit $\lambda \to 0$. In addition, classic photometry has operated only with light sources of natural thermal origin which may be deemed chaotic to a good degree of accuracy owing to the small correlation radius. Now, the basic assumptions of classic photometry follow.

(i) *The ray approach.* It is assumed that the applicability conditions of the geometrical optics approximation are satisfied for every quasi-plane wave and so the *wave field* is treated as the *ray field.*

(ii) *Spatial incoherence of radiation.* It is assumed that rays arriving at a given point from different directions are totally incoherent, which, in a certain sense, corresponds to the statistical independence of the sources of this radiation (see also Section 2.1.6 below).

(iii) *Averaged description.* Measured parameters are assumed to be not local or current values, but rather some time and space averaged, squared field characteristics, and radiative transfer theory operates with these characteristics.

(iv) *The ensemble average approach.* Radiation possesses the properties of stationarity and ergodicity and so the averaged characteristics coincide with the statistical mean.

The condition (i) is the principal one and allows one to speak of radiant fluxes propagating along the rays. In agreement with this conditions, classical transport theory pays no attention to diffraction effects in wave propagation analyses.

The condition (ii) allows one to sum energy quantities rather than fields, thus excluding the possibility of manifestation of interference effects.

The conditions (ii) and (iv) are normally not formulated explicitly in photometric manuals, but they are tacitly assumed. These conditions put into correspondence radiometric quantities and parameters measured in typical experiments.

If we confine this consideration to the description of free, non-scattered radiation, then there will be no difficulties with the substantiation of radiometric concepts. In this case, the sufficiency of the conditions (i) through (iv) is obvious, and radiometry acts as a theory linking the energy relations of the method of geometrical optics with the statistical assumption on incoherence of wave bundles.

Looking a little bit ahead we remark that the sufficiency of the conditions (i) through (iv) does not imply their necessity. It will be demonstrated later that the refusal of the requirement of non-negative radiance enables one in some situations to extend the framework of classical radiometry and partially take into account diffraction effects.

It will be much more difficult to evaluate the applicability limits of the radiometric description of the behavior of radiation in turbid media where scattering is significant. The scattering becomes significant when the medium contains inhomogeneities which are no longer smooth in the wavelength scale. However, near such inhomogeneities, the applicability condition of the method of geometrical optics breaks down and the applicability condition (i) of radiometric concepts is not satisfied. Notwithstanding this fact the radiometric concepts are widely used in turbid media as well. This usage is justified since in turbid media, the applicability of geometrical optics is not required everywhere but also in the mean with respect to the lengths over which averaging is performed. Fine features, such as the behavior of the field near sharp inhomogeneities are excluded from radiometric considerations.

Therefore, in the case of turbid area, applicability of the radiometric description is not an easy matter and can be handled in the framework of the statistical-wave formalism, a more rigorous treatment than the phenomenological approach. We turn to this formalism in Chapter 3.

2.1.6 Sources of Incoherent Radiation

In agreement with the condition (ii) radiometry considers only sources of incoherent ray field. In this section we look into details of this notion.

For simplicity we assume that the monochromatic radiation (we suppress the factor $e^{-i\omega t}$) is produced by random currents $j = j(\mathbf{r})$ with zero mean, $\langle j \rangle = 0$, distributed over a distance surface Σ. (Henceforth angular brackets imply statistical averaging.) In physical optics, saying incoherent for such currents one means their spatial delta-correlation in various points on Σ, *viz.*,

$$\langle j(\mathbf{r}) j(\mathbf{r}_1) \rangle = b(\mathbf{r}) \delta_\Sigma(\mathbf{r} - \mathbf{r}_1), \tag{2.1.18}$$

where $\delta_\Sigma(\mathbf{r} - \mathbf{r}_1)$ is the delta function on the surface Σ. This function satisfies the conditions $\delta_\Sigma(0) = \infty$, $\delta_\Sigma(\mathbf{r}) = 0$ for $\mathbf{r} \neq 0$, and $\int_\Sigma \delta_\Sigma(\mathbf{r} - \mathbf{r}_1) d\Sigma_\mathbf{r} = 1$ for $\mathbf{r}_1 \in \Sigma$.

In the general case, however, the radiometric description of the field does not need incoherence of the field sources $j(\mathbf{r})$ in the sense of eqn (2.1.18). It is required only that the bundles of rays of different directions be incoherent in the range of the formed geometrical optics pattern where the radiometric concepts are applicable. This condition, as will be demonstrated below, does not require that the sources be delta-correlated.

Let us consider a simplest problem of the diffraction of a plane wave by an aperture σ in a plane screen Σ at $z = 0$ with the characteristic dimension $a >> \lambda$, where λ is the wavelength used. In this case, the surface sources j of the field may be deemed given on Σ and we may let in the Kirchhoff approximation $j = 1$ on σ and $j = 0$ beyond it. Then at small z, namely for $z \ll a^2/\lambda$, we obtain a pencil of light which may be described by the ordinary geometrical optics formalism. In the Fresnel (near field) zone, $z \sim a^2/\lambda$, and the diffraction of the beam precludes the applicability of the ray approach. Finally, in the Fraunhofer (far field) zone, for $z \gg a^2/\lambda$, the field is a spherical wave. Its propagation may be described again with the aid of ray theory, taking the aperture as an effective point source. The radiation pattern of such a source depends on the distribution of the field over the entire aperture. It can be obtained as the Fourier transform of the distribution of currents in the source plane. In this case, one may speak about the nonlocal initial condition.

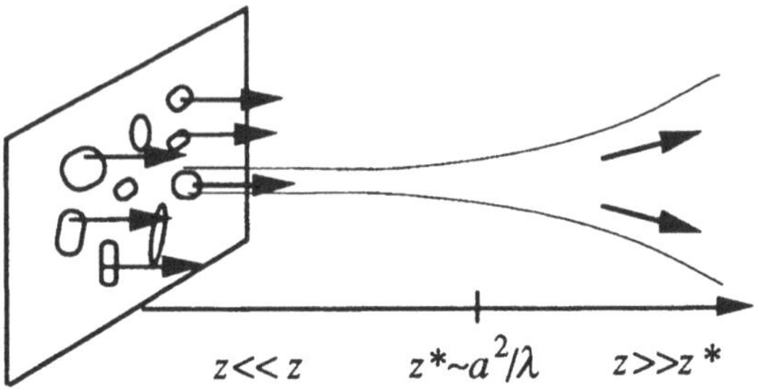

Figure 2.3. Two regions of applicability of the geometrical optics approximation (near field and far field) for the free diffraction of radiation from random sources ($a \sim l_c$ is the correlation radius of the sources).

Assume now that all the plane Σ is covered by approximately identical spots σ of mutually incoherent sources, i.e. with different phase states, so that the size of spots a is the order of the correlation radius of the field on the screen, $a \sim l_c$, as schematically illustrated in Fig. 2.3. Each spot will produce such a field pattern and the average intensity at each point will be equal to the sum of intensities due to different spots.

For the free field, the radiometric formulas are none other than the averaged energy relations of geometrical optics, therefore, in this case, the radiance and other radiometric concepts would be wise to use in both near field ($z \ll a^2 / \lambda$) and far field ($z \gg a^2 / \lambda$). These two fields represent essentially independent problems which differ from one another by boundary conditions.

In the near field, the radiometric description is rather artificial and is not of much interest, since every point is hit by only one ray and so to describe the radiance of the field as a function of direction one would have to introduce the respective delta function. Conversely, in the far field, each point is hit by many rays. Hence the radiometric description of radiance proves to be natural and intuitively appealing. The condition (ii) of incoherence of rays with different directions is fulfilled by virtue of the assumption on the incoherence of the spots. As a result, not delta correlated sources (2.1.18) alone become admissible, but also the sources with finite correlation radius l_c:

$$\langle j(\mathbf{r}) j(\mathbf{r}_1) \rangle = \langle j^2 \rangle K(\mathbf{r}, \mathbf{r}_1) \xrightarrow[|\mathbf{r}-\mathbf{r}_1| \gg l_c]{} 0, \qquad (2.1.19)$$

but the point of observation must be in the Fraunhofer zone with respect to the correlation radius so that $z \gg a^2 / \lambda = l_c^2 / \lambda$.

Thus, the radiometric assumption that the rays with different directions are incoherent does not imply that the sources must be incoherent, that is, spatially uncorrelated in the sense of eqn (2.1.18). All these qualitative considerations can be expressed in a fairly rigorous form by resorting to the known solution of the diffraction of a random field at half-space; we shall do this in Section 3.2. At the moment we note only that in most practical applications the phase of the sources is not piecewise constant on the "coherence spots" but changes continuously. For such sources with continuous correlation functions, the position of the boundary between the near and far fields, z^*, depends not only on the characteristics of inhomogeneities, but also on the size of the radiating surface.

2.1.7 Relation of the Radiance with the Coherence Function of the Field

Classical radiometry deals with the mean energy characteristics of radiation. In the language of statistical theory treating radiation as a random field these quantities correspond to one-point second moments of the field (see Chapter 3). However, already in the framework of the phenomenological theory it is an easy matter to demonstrate that the brightness enables one to express not only one-point moment, but also two-point statistical moment – the coherence function which characterizes the correlation properties of the radiation. Below we intend to obtain such expression for the simplest case of the scalar field in an isotropic medium. Note in passing that classical radiometry deals with measurements of intensity and does not touch upon correlation problems.

The real field $v = v(x)$, $x = (\mathbf{r}, t)$ will be expressed traditionally as the real part of the complex field u, namely, $v = \mathrm{Re}\, u$. By assumption, the radiance $I_\omega(\mathbf{n}, x)$ corresponds to a bundle of locally plane waves propagating at point x in the direction $\mathbf{n}$. Each of the waves is characterized by a (mean) wave vector $\mathbf{k}$ and frequency ω. Therefore, near a point x it has the complex field

$$u_\mathbf{k}(x) = A_\mathbf{k}(x) e^{i(k\mathbf{n}\mathbf{r} - \omega t)}, \tag{2.1.20}$$

where $\mathbf{k} = k\mathbf{n}$ is the wave vector with the wave number $k = k(\omega)$ satisfying the dispersion equation for wave in homogeneous media (in the simplest case of a nondispersive medium, $k = \omega / c = k_0$, where c is the wave velocity), and $A_\mathbf{k}(x)$ is the complex random amplitude which varies slowly over the characteristic wavelengths and periods of oscillation. Under an appropriate normalization, the radiance I_ω will be equal to the average intensity, that is, to the squared amplitude of the quasi-plane wave,

$$I_\omega(\mathbf{n}, x) = \left\langle |u_\mathbf{k}(x)|^2 \right\rangle = \left\langle |A_\mathbf{k}(x)|^2 \right\rangle. \tag{2.1.21}$$

The average energy density W may be deemed proportional to the mean intensity $\langle |u|^2 \rangle$ of the complex field u. Therefore, using eqn (2.1.21) and trivially extending the radiometric relation (2.1.16) we may write

$$\left\langle |u|^2 \right\rangle \equiv \mathrm{Re}\langle u(x)u^*(x)\rangle = \int c_\omega I_\omega(\mathbf{n},x)d\Omega_\mathbf{n}d\omega$$
$$= \mathrm{Re}\int c_\omega \left\langle u_\mathbf{k}(x)u_\mathbf{k}^*(x)\right\rangle d\Omega_\mathbf{n}d\omega \,, \tag{2.1.22}$$

where, in contrast to eqn (2.1.16), we introduce a proportionality coefficient c_ω which depends on the nature of the field, and also – for the sake of more appealing representation – the symbol of real part.

For sufficiently small values of the quantity $\rho = (\boldsymbol{\rho}, \tau)$ when the difference between the amplitudes $A_\mathbf{k}(x)$ and $A_\mathbf{k}(x+\rho)$ may be neglected, in view of eqn (2.1.20) we have

$$\mathrm{Re}\left\langle u_\mathbf{k}(x+\rho)u_\mathbf{k}^*(x)\right\rangle = \mathrm{Re}\left\langle A_\mathbf{k}(x+\rho)A_\mathbf{k}^*(x)\right\rangle e^{i(k\mathbf{n}\boldsymbol{\rho}-\omega\tau)}$$
$$\approx \mathrm{Re}\left\langle |A_\mathbf{k}(x)|^2\right\rangle e^{i(k\mathbf{n}\boldsymbol{\rho}-\omega\tau)}$$
$$= \mathrm{Re}\, I_\omega(\mathbf{n},x)e^{i(k\mathbf{n}\boldsymbol{\rho}-\omega\tau)} \,. \tag{2.1.23}$$

However, in view of the assumption on the incoherence of the bundles, the quasi-plane waves with different wave vectors yield independent contributions into the quantities quadratic with respect to the field. If we take this fact into account, then eqn (2.1.23) leads us to the following generalization of eqn (2.1.22) to the case of two-point moment of the field, commonly called the coherence function,

$$\mathrm{Re}\left\langle u(x+\rho)u^*(x)\right\rangle = \mathrm{Re}\int c_\omega \left\langle u_\mathbf{k}(x+\rho)u_\mathbf{k}^*(x)\right\rangle d\Omega_\mathbf{n}d\omega$$
$$\approx \mathrm{Re}\int c_\omega I_\omega(\mathbf{n},x)e^{i(k\mathbf{n}\boldsymbol{\rho}-\omega\tau)}d\Omega_\mathbf{n}d\omega \,, \tag{2.1.24}$$

or, depressing the Re symbol,

$$\left\langle u(x+\rho)u^*(x)\right\rangle \approx \int c_\omega I_\omega(\mathbf{n},x)e^{i(k\mathbf{n}\boldsymbol{\rho}-\omega\tau)}d\Omega_\mathbf{n}d\omega. \tag{2.1.25}$$

Thus, the coherence function of the complex field is expressed as the Fourier transform (2.1.25) of the radiance I_ω. For a strictly monochromatic field, when $u(x) = u(\mathbf{r})\exp(-i\omega_0 t)$ and the dependence of u on time reduces to the factor $\exp(-i\omega_0 t)$, only spatial correlations are of interest. Clearly, the spectral radiance must be localized at the frequency ω_0, that is, must be of the form $I_\omega = I(\mathbf{n},\mathbf{r})\delta(\omega-\omega_0)$, where $I(\mathbf{n},\mathbf{r})$ is the total radiance (2.1.7). Letting $\tau = 0$ in eqn (2.1.25) we obtain

$$\left\langle u(\mathbf{r}+\boldsymbol{\rho})u^*(\mathbf{r})\right\rangle = c_{\omega 0}\int I(\mathbf{n},\mathbf{r})e^{ik(\omega_0)\mathbf{n}\boldsymbol{\rho}}d\Omega_\mathbf{n} \,. \tag{2.1.26}$$

A more correct derivation of eqns (2.1.25) and (2.1.26) will be presented in Chapter 3 on the base of statistical wave theory. The form of the coefficient $c_{\omega 0}$ entering these equations will be also evaluated there.

It should be clear from the derivation of eqn (2.1.25) that it is valid only for modest separations ρ which must be small compared to the characteristic scale of variation of amplitudes $A_{\mathbf{k}}(x)$, that is, to the scales of variation of $I_\omega(\mathbf{n},x)$ with respect to x, or else to the scales of statistical inhomogeneity and nonstationarity of the field. On the other hand, ρ and τ may be large compared to the characteristic wavelength λ_0 and period of oscillations τ_0, respectively.

2.1.8 Quantum Approach and the Radiance of Thermal Radiation

When the radiometric conditions (i)–(iv) are satisfied, it is natural and useful to treat the wave field as a set of noninteracting quanta each of which propagates along a respective ray and gives an independent contribution $\hbar\omega$ ($\hbar$ is the Planck constant) in the total energy of the field. This approach allows the description of the quantum effects and also proves useful even in the field of applicability of classical physics rendering intuitively appealing geometric image to some physical considerations — this relates primarily to nonlinear problems of plasma theory (Tsytovich, 1971). At the same time the quantum consideration is an important auxiliary tool of mathematical analysis of the radiative transport equation fundamental to radiometric studies. In this connection we should notice results of Sobolev (1963) devoted to probability treatment of the transport equation.

By way of a simple example of physically transparent considerations on the "quantum" parlance we obtain an expression for the radiance of equilibrium thermal radiation in a homogeneous nonabsorbing medium.

The energy dw in a unit volume (per element d^3k) equals the product of the energy of a quantum $\hbar\omega$ by the density of quanta $N_{\mathbf{k}} = [\exp(\hbar\omega / \kappa T) - 1]^{-1}$ (where T is temperature and κ is the Boltzmann constant of Planck's distribution) and by $d^3k / (2\pi)^3$, the number of modes of given polarization in the element d^3k,

$$dw = \hbar\omega N_{\mathbf{k}} \frac{d^3k}{(2\pi)^3} = \theta'(\omega,T)\frac{d^3k}{(2\pi)^3}, \tag{2.1.27}$$

where the function

$$\theta'(\omega,T) = \frac{\hbar\omega}{e^{\hbar\omega/\kappa T} - 1} \equiv \theta(\omega,T) - \frac{\hbar\omega}{2} \tag{2.1.28}$$

is the mean energy of the quantum oscillator $\theta(\omega,T) = (\hbar\omega/2)\coth(\hbar\omega / 2\kappa T)$ diminished by the energy of zero oscillations [4] $\hbar\omega/2$.

[4] For situations when zero oscillations should be taken into account, see Rytov, Kravtsov and Tatarskii (1987b)

From eqn (2.1.27) with the use of eqn (2.1.14) we obtain the expression for the radiance of thermal radiation for one of two independent polarizations in a unit interval of positive frequencies (see, e.g., Bekefi, 1966)

$$I_\omega = v_g \frac{dW}{d\omega d\Omega_{\mathbf{n}}} = v_g \theta'(\omega,T)\frac{1}{(2\pi)^3}\left|\frac{d^3k}{d\omega d\Omega_{\mathbf{n}}}\right| = n_r^2 I_\omega^0 . \qquad (2.1.29)$$

Here

$$\frac{d^3k}{d\omega d\Omega_{\mathbf{n}}}$$

is the Jacobian from k_x, k_y, k_z to ω, $\Omega_{\mathbf{n}}$ ($\Omega_{\mathbf{n}}$ is the solid angle around the direction $\mathbf{n} = \mathbf{v}_g / v_g$), the function

$$n_r^2 = \frac{v_g}{k_0^2}\left|\frac{d^3k}{d\omega d\Omega_{\mathbf{n}}}\right| \qquad (2.1.30)$$

is the square of the so-called ray refractive index n_r, $k_0 = \omega / c$, and

$$I_\omega^0 = \frac{k_0^2}{(2\pi)^3}\theta'(\omega,T) \qquad (2.1.31)$$

is the radiance of equilibrium thermal radiation in a unit interval of positive frequencies per one polarization in vacuum.

Equation (2.1.29) for radiance per one polarization holds in the general case of anisotropic, nonabsorbing media (Rytov, 1953). In a particular case of an isotropic medium with permittivity ε such that $k = k_0\sqrt{\varepsilon}$ we have

$$\left|\frac{d^3k}{d\omega d\Omega_{\mathbf{n}}}\right| = \left|\frac{k^2 dk d\Omega_{\mathbf{n}}}{d\omega d\Omega_{\mathbf{n}}}\right| = \frac{k^2}{v_g},$$

$n_r^2 = \varepsilon$ and so eqn (2.1.29) takes the form

$$I_\omega = \varepsilon I_\omega^0 \qquad (2.1.32)$$

known as the Clausius law.

The classical Rayleigh-Jeans limit $\hbar\omega << \kappa T$ for the expressions (2.1.28)-(2.1.31) (see Problem 2.5) can be obtained in the purely "classical" language, however, the quantum approach proves to be not only a more general formalism, but also a simpler treatment.

The use of quantum concepts proves to be especially useful in describing the scattering in strongly rarified gases, such as plasma of some astrophysical objects (Dolginov, Gnedin, Silant'ev, 1995; Zheleznyakov, 1970). In such rarified media, the field may be deemed free almost everywhere except in a relatively small area near the scatterers. Furthermore, the acts of scattering and the processes of propagation of radiation between the scatterers can be naturally described in the quantum language. Turning to the substantiation of photometry in strongly turbid media, the quantum approach would only complicate the task since in strongly scattering media, even without allowance for quantum effects (in the quasi-classical limit), the concept of a ray or, what is the same, the concept of the trajectory of a quantum is devoid of its meaning.

In what follows we shall not consider quantum effects whatsoever and restrict our attention to a simpler classical random field. It does not mean, however, that we shall completely refuse ourselves the use of the quantum language – it will be used sparingly for description of purely classical phenomena.

2.2 RADIATIVE TRANSFER IN SCATTERING MEDIA. THE SCALAR MODEL

2.2.1 Transport Equation in Nonscattering Media. The Ray Refractive Index

When the notion of radiance I is given the next natural problem concerns the methods for calculation of this quantity in various physical systems. All of them are based on the radiative transport equation.We consider such an equation first for the simplest model of a scalar field. Strictly speaking, this model is applicable only for the description of a one-component, say acoustic, field. For such a field, there will be no question about polarization, that is, about the relationship between the components. Nonetheless, the scalar model is often used — without sufficient grounds — for electromagnetic fields as well. The radiometric description of such a field will be discussed in Section 2.3.

This equation is commonly derived by considering the energy balance in an infinitesimally small volume of the medium and assuming that the properties of the ray field outlined above hold. This derivation of the transport equation is pictorial, physically transparent and therefore well elaborated (see, e.g., Chandrasekhar, 1960; Bekefi, 1966; and Sobolev, 1975). In this section we also resort to energy balance considerations, but we cover the most general case of geometrical optics applicability and bring into the pattern spatial dispersion, inhomogeneity, nonstationarity, anisotropy of the medium, and other aggravating factors.

Consider first a monochromatic radiation in free space, or in a more general case, in a homogeneous nonabsorbing medium. This means that there is no refraction, the energy of plane waves does not change in propagation and so the energy flux of the beam confined in a ray tube is conserved.

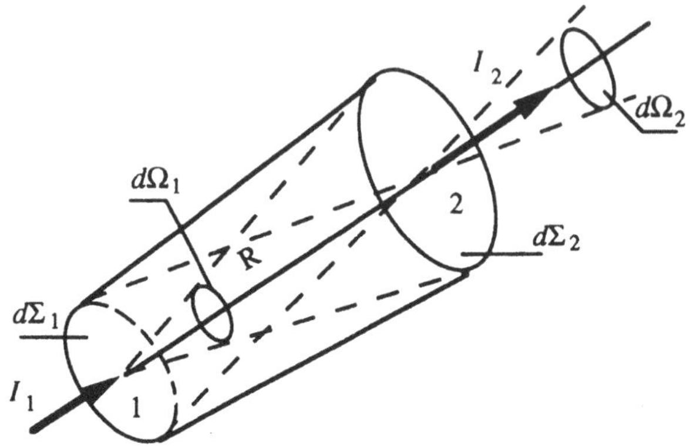

Figure 2.4. Propagation of a radiant beam in a homogeneous medium.

Let two points in a ray tube, *1* and *2*, be separated by a distance R, as shown in Fig. 2.4. If the radiances of the beam at these points are I_1 and I_2, then the flux of energy arriving from the entrance section $d\Sigma_1$ to the exit section $d\Sigma_2$ in a solid angle $d\Omega_1$ is $I_1 d\Sigma_1 d\Omega_1$ and it is equal to the flux $I_2 d\Sigma_2 d\Omega_2$ outgoing from the exit section in a solid angle $d\Omega_2$. As should be clear from Fig. 2.4

$$d\Sigma_1 d\Omega_1 = d\Sigma_1 d\Sigma_2 / R^2 = d\Sigma_2 d\Omega_2 .$$

From this relationship and the equality of the fluxes $I_1 d\Sigma_1 d\Omega_1 = I_2 d\Sigma_2 d\Omega_2$ it follows $I_1 = I_2$. In a homogeneous medium, this implies that the derivative of the radiance I along the ray is zero:

$$d_s I(\mathbf{n},\mathbf{r}) = 0 , \qquad (2.2.1)$$

where $d_s = \mathbf{n}\nabla$ is the differentiation operator along the ray, and $\mathbf{n}$ is the unit vector oriented with the ray.

This relation is the simplest form of transfer equation for radiance in the absence of scattering, refraction and absorption.The spectral radiance I_ω also obeys this equation because the frequency ω does not change when the wave propagates in a stationary medium.

In agreement with this equation, the radiance I remains unchanged along the ray. This important property allows one to treat it as a direct characteristic of the radiative properties of the source surface, which is independent of the distance to the source. This is an essential difference from and advantage over the common intensity which is defined as a quantity proportional to the energy density or the square of the wave amplitude.

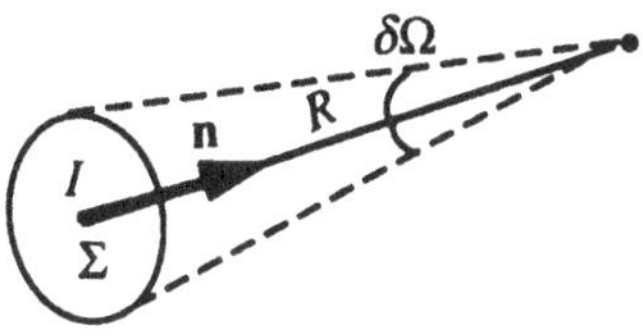

Figure 2.5. Radiation of a small source Σ.

For better insight into this difference we consider a simple example. Let the detector respond to the intensity of radiation incoming from a distant source Σ of radiance I, as shown in Fig. 2.5. Detector readings are proportional to the energy density of the incident radiation, which at a large distance R from the source will fall off as R^{-2}. If we normalize these readings by the solid angle $\delta\Omega$ subtending the source, then this angle will also fall off as R^{-2}. Then the normalized indications of the detector will be proportional to the radiance of the source and independent of the distance to this source.

In an inhomogeneous refracting medium, ray tubes bend and the radiance no longer satisfies eqn (2.2.1). To generalize the transfer equation (2.2.1) for this case we assume first that the medium is nonabsorbing and resort to the law of energy conservation in the form of the continuity equation

$$\partial_t w + \nabla \mathbf{S} = 0, \tag{2.2.2}$$

where w is the energy density and $\mathbf{S}$ is the flux of energy. Substituting the radiometric expressions for w and $\mathbf{S}$ from (2.1.16) and (2.1.17) we obtain

$$\partial_t \int_B v_g^{-1} I_\omega d\omega d\Omega_{\mathbf{n}} + \nabla \int_B \mathbf{n} I_\omega d\omega d\Omega_{\mathbf{n}} = 0 . \tag{2.2.3}$$

Here the integration is over the domain $B = \Delta\omega\Delta\Omega_{\mathbf{n}}$ corresponding to the beam wave.

While in radiometry the variables are normally the frequency and direction of the ray $\mathbf{n} = \mathbf{v}_g / v_g$, in the method of geometrical optics it is more convenient to use the components of the wave vector $\mathbf{k} = (k_x, k_y, k_z)$. If we integrate in eqn (2.2.3) in the components of $\mathbf{k}$ then the elementary volume $d\omega d\Omega_{\mathbf{n}}$ becomes $\left| d\omega d\Omega_{\mathbf{n}} / d^3k \right| d^3k$, where $\left| d\omega d\Omega_{\mathbf{n}} / d^3k \right|$ stands for the Jacobian of the transformation to the new variables, and the frequency ω is related to the wave vector $\mathbf{k}$ by a dispersion equation:

$$\omega = \omega(\mathbf{k}, \mathbf{r}), \tag{2.2.4}$$

which allows for a slow dependence of the medium parameters upon the coordinates **r**.

In the general case, the operator of differentiation along the ray, d_s, entering in the transfer equation (2.2.1), is of the form

$$d_s = \mathbf{n}\nabla + v_g^{-1}\partial_t + \dot{\mathbf{n}}\nabla_{\mathbf{n}} + \dot{\omega}\partial_\omega. \tag{2.2.5}$$

This expression is a full differential operator along the curvilinear space-time ray. The first summand is the spatial derivative along the direction **n**, the second addend allows for the nonstationarity of the radiation, and the last two addends take care of the possible variation of the direction **n** and frequency ω (in homogeneous and stationary media, these terms vanish). In the Appendix A it is shown that

$$\begin{aligned}\dot{\mathbf{n}} \equiv d_s\mathbf{n} &= v_g^{-1}(\hat{\mathbf{1}} - \mathbf{n}\otimes\mathbf{n})\dot{\mathbf{v}}_g \\ &= v_g^{-1}(\hat{\mathbf{1}} - \mathbf{n}\otimes\mathbf{n})(\mathbf{n}\nabla - v_g^{-1}\nabla\omega\nabla_{\mathbf{k}} + v_g^{-1}\partial_t)\dot{\mathbf{v}}_g\end{aligned} \tag{2.2.6}$$

and

$$\dot{\omega} \equiv d_s\omega = v_g^{-1}\partial_t\omega, \tag{2.2.7}$$

where the symbol $\otimes$ implies a tensor product, so that $\mathbf{n}\otimes\mathbf{n}$ is the matrix with components $(\mathbf{n}\otimes\mathbf{n})_{ij} = n_i n_j$, and $\hat{\mathbf{1}}$ is the unit tensor.

If we deem the medium stationary, in contrast to the field, then after some manipulations involving eqns (2.2.4) – (2.2.6) is is not hard to demonstrate that eqns (2.2.2) – (2.2.3) are equivalent to the relation (see Appendix C)

$$\int d\omega d\Omega_{\mathbf{n}}(d_s I + I d_s \ln n_r^{-2}) = 0, \tag{2.2.8}$$

where n_r is the *ray refractive index* (2.1.30). In the particular case of a homogeneous isotropic nondispersive medium with a dispersion law $k = (\omega / c)\sqrt{\varepsilon} = k_0\sqrt{\varepsilon}$, where $\varepsilon = \text{const.}$ is the dielectric permittivity, the group velocity being equal to the phase velocity, $v_g = c / \sqrt{\varepsilon}$, and collinear **n** and **k**, eqn (2.1.30) yields

$$n_r = \frac{c}{\omega}\left[\frac{c}{\sqrt{\varepsilon}}\left|k^2\frac{dkd\Omega_{\mathbf{n}}}{d\omega d\Omega_{\mathbf{n}}}\right|\right]^{1/2} = \sqrt{\varepsilon}, \tag{2.2.9}$$

that is, the ray refractive index coincides with the ordinary refractive index. In the general case of an anisotropic medium, these quantities do not coincide any longer. A more detailed discussion of the ray refractive index may be found in book of Bekefi (1966): see also Problem 2.10.

At each point of nonscattering media, quasi-plane waves with different wave vectors **k** are independent, therefore the radiance I_ω may be let other than zero at arbitrary ω and **n** corresponding to these waves. It follows therefore that the integrand in eqn (2.2.8) must vanish. These condition leads us to the following generalization of the transfer equation (2.2.1) to the case of the radiance in an inhomogeneous medium:

$$d_s I_\omega + I_\omega d_s \ln n_r^{-2} \equiv (d_s + \alpha^{\text{inh}}) I_\omega = 0. \tag{2.2.10}$$

The quantity

$$\alpha^{\text{inh}} = d_s \ln n_r^{-2} = -2 d_s n_r / n_r \tag{2.2.11}$$

may be called the *coefficient of* (spatial) *inhomogeneity*.

The radiative transfer equation (2.2.10) may also be recast to the form

$$\left(d_s + \alpha^{\text{inh}}\right) I_\omega = n_r^2 d_s \left(I_\omega / n_r^2\right) = 0, \tag{2.2.12}$$

which reveals that I_ω / n_r^2 remains constant along the ray, that is, if we allow for the curvature of rays in inhomogeneous stationary media, then what is conserved along the ray is not the radiance I_ω, but rather the ratio I_ω / n_r^2 [Clausius law, see also eqn (2.2.21)].

Under the above assumptions the transfer equation (2.2.10) is applicable for the description of waves of any nature, with an arbitrary dispersion law $\omega = \omega(\mathbf{k})$. In particular, it is applicable for anisotropic media where the direction of the group velocity $\mathbf{n} = \mathbf{v}_g / v_g$ may not coincide with the direction of the wave vector **k**, and the ray refractive index n_r becomes dependent on the direction **n**. Of course, all these conclusion overlook polarization of the waves and birefringence. To take these phenomena into account one has to go beyond the framework of the scalar model.

2.2.2 Relation of the Radiance with the Energy Density of Quasi-Plane Waves

If the medium is filled homogeneously with plane waves of various wave vectors **k**, then by virtue of assumed incoherence the waves with different **k** give independent contributions into the mean energy density w, which may be represented as an integral:

$$w = \int w_{\mathbf{k}} d^3 k, \tag{2.2.13}$$

where $w_{\mathbf{k}}$ has the sense of the energy of a plane wave with wave vector **k**. This relation holds if we replace plane waves with quasi-plane waves which deviate from the former not significantly.

On the other hand, the quantity w in eqn (2.2.13) can be expressed in terms of the radiance in accordance with eqn (2.1.16), viz.,

$$\int w_{\mathbf{k}} d^3k = \int v_g^{-1} I_\omega d\omega d\Omega_{\mathbf{n}} = \int v_g^{-1} I_\omega \left| d\omega d\Omega_{\mathbf{n}} / d^3k \right| d^3k. \tag{2.2.14}$$

This expression makes it clear that the radiance I_ω is related to the energy density of plane waves, $w_{\mathbf{k}}$ by

$$w_{\mathbf{k}} = v_g^{-1} I_\omega \left| \frac{d\omega d\Omega_{\mathbf{n}}}{d^3k} \right| = \frac{I_\omega}{(\omega / c)^2 n_{\mathrm{r}}^2}. \tag{2.2.15}$$

2.2.3 Transformation of the Radiance on the Plane Interface

Let us consider how the radiance varies when the beam wave is incident on the interface between two homogeneous media. This incidence give rise to two secondary beams, reflected and refracted. The radiances of these beams can be found if the law of refraction of the plane wave at this interface is known

$$u^{\mathrm{i}} = u^0 e^{i(\mathbf{k}\mathbf{r} - \omega t)}, \quad \mathbf{k} = k_0 n_1 \mathbf{n}_0 . \tag{2.2.16}$$

For the case of scalar acoustic waves and isotropic media, we write the expressions for the reflected, u^{r}, and refracted (transmitted), u^{t}, waves as follows (Born and Wolf, 1980)

$$u^{\mathrm{r}} = \mathcal{R} u^0 e^{i(\mathbf{k}_1\mathbf{r} - \omega t)}, \quad \mathbf{k}_1 = k_0 n_1 \mathbf{n}_0^*, \tag{2.2.17}$$

$$u^{\mathrm{t}} = \mathcal{T} u^0 e^{i(\mathbf{k}_2\mathbf{r} - \omega t)}, \quad \mathbf{k}_2 = k_0 n_2 \mathbf{n}_2 . \tag{2.2.18}$$

Here $\mathcal{R}$ and $\mathcal{T}$ are the Fresnel coefficients of reflection and refraction, respectively; $\mathbf{n}_0, \mathbf{n}_0^* = (n_{0x}, n_{0y}, -n_{0z})$ and $\mathbf{n}_2$ are the unit vectors in the directions of the incident, reflected and refracted waves, n_1 and n_2 are the refractive indexes of the first and second media and the angle of incidence equals the angle of reflection ($\theta_0 = \theta_1$). The angle of refraction θ_2 is defined by Snell's law $n_1 \sin\theta_0 = n_2 \sin\theta_2$, as shown in Fig. 2.6. For acoustic waves, $\mathcal{R}$ and $\mathcal{T}$ coincide in form with the expressions for $R_\perp$ and $T_\perp$ (2.3.111) if we replace there the refractive indices n for $1/\rho c$, where ρ is the medium's density, and c is the velocity of sound.

Now we come from the case of one plane wave to the case of beam waves, that is, sets of plane waves. We assume that the wave vectors of the incident, reflected and refracted beams are confined, respectively in the solid angles $d\Omega_0$, $d\Omega_1$, and $d\Omega_2$. Then, observing that the elementary solid angle does not change in reflection, $d\Omega_0 = d\Omega_1$, and that the radiance I is proportional to the energy, i.e., to the square of the amplitude of the quasi-plane wave, for radiance of the reflected beam we have

$$I^{\mathrm{r}} = I^{\mathrm{r}}(\mathbf{n}^*) = |\mathcal{R}|^2 I^0(\mathbf{n}), \tag{2.2.19}$$

where $I^0(\mathbf{n})$ is the radiance of the incident beam, and $\mathbf{n}^* = (n_x, n_y, -n_z)$ is the unit vector of the reflected beam.

The radiance of the refracted beam can be found from the condition that the z component of the energy flux

$$dS_z = I^0(\mathbf{n}_0)_z d\Omega_0 + I^{\mathrm{r}}(\mathbf{n}_0^*)_z d\Omega_1 = I^{\mathrm{t}}(\mathbf{n}_2)_z d\Omega_2, \tag{2.2.20}$$

is conserved. With reference to Fig. 2.6, in this expression, $(\mathbf{n}_0)_z = -\cos\theta_0$, $(\mathbf{n}_0^*)_z = \cos\theta_0$, and $(\mathbf{n}_2)_z = -\cos\theta_2$. Recognizing further that $d\Omega_0 = d\Omega_1 = \sin\theta_0 d\theta_0 d\varphi_0$, $d\Omega_2 = \sin\theta_2 d\theta_2 d\varphi_2$, $d\varphi_0 = d\varphi_2$, Snell's law $n_1 \sin\theta_0 = n_2 \sin\theta_2$, and its derivative relation $n_1 \cos\theta_0 d\theta_0 = n_2 \cos\theta_2 d\theta_2$ we obtain from eqn (2.2.20)

$$\frac{I^0}{n_1^2} - \frac{I^{\mathrm{r}}}{n_1^2} = \frac{I^{\mathrm{t}}}{n_2^2}, \tag{2.2.21}$$

which may be viewed as an extension of the Clausius law (2.2.12) to the case of a reflecting boundary. Substituting here I^{r} from eqn (2.2.19) we finally obtain for the radiance of the refracted beam

$$I^{\mathrm{t}} = \left(\frac{n_2}{n_1}\right)^2 (1 - |\mathcal{R}|^2) I^0. \tag{2.2.22}$$

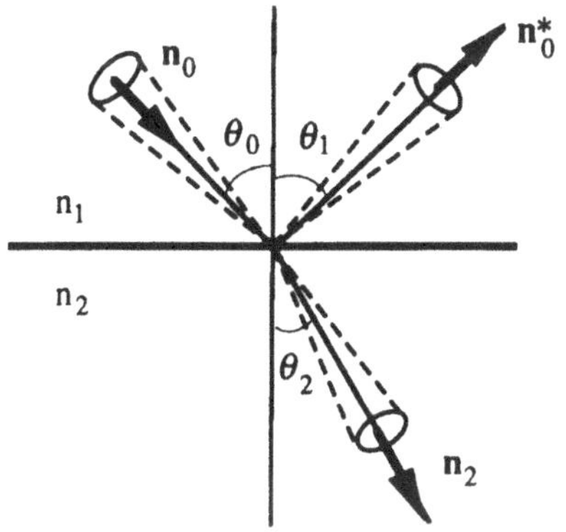

Figure 2.6. Refraction at a plane interface.

The first factor on the right hand side differs this expression from a similar relation for the fluxes of energy and owes its existence to the change of the solid angle $d\Omega$ in refraction. For unmovable boundaries, the same relationships hold for the spectral radiance I_ω as well.

2.2.4 Quasistationary Conservative Media. Conservation of the Adiabatic Invariant

In a medium whose properties vary slowly with time (quaistationary medium),the energy conservation law (2.2.2) is no longer fulfilled because of the parametric interaction of the waves with the medium. From the classical mechanics we know that for an oscillator with slowly variable parameters, the approximately constant parameter is not its energy $W(\omega)$ but rather the so-called adiabatic invariant $W(\omega)/\omega$ (Landau and Lifshitz, 1976). In the quantum mechanical language, $N_\omega = W(\omega)/\hbar\omega$ is the number if quanta with energy $\hbar\omega$, and the conservation of the adiabatic invariant implies the conservation of the number of quanta.

In the framework of phenomenological theory it is natural to expect that a similar quantity N_ω is conserved also when the object of research is quasiplane waves instead of the oscillator. Clearly then in place of N_ω we must consider the quantity $N_{\mathbf{k}} = w_{\mathbf{k}} / \hbar\omega$ equal to the number of quanta in the unit volume of the phase space $(\mathbf{r},\mathbf{k})$. We assume that this quantity remains constant along the ray and express $N_{\mathbf{k}}$ in terms of I_ω using eqn (2.2.15). Then we arrive at the condition of conservation of the adiabatic invariant

$$N_{\mathbf{k}} = w_{\mathbf{k}}/\hbar\omega = c^2 I_\omega/\hbar\omega^3 n_r^2 = \text{const}, \tag{2.2.23}$$

which must be satisfied along the ray. By differentiating this relation along the ray we obtain the equation

$$\left(d_s + \alpha^{\text{inh}}\right)I_\omega \equiv \left\{d_s + \left[d_s \ln\left(\omega^{-3} n_r^{-2}\right)\right]\right\}I_\omega = 0, \tag{2.2.24}$$

which generalizes the transport equation (2.2.10) for the case of a nonstationary medium.

It is instructive to notice that the number of quanta $N_{\mathbf{k}}$ remains constant not in all nonstationary media. Specifically, $N_{\mathbf{k}}$ varies in the process of plasma ionization, though is conserved in recombinations. An analysis of this topic has been given by Kravtsov and Orlov, 1990). In this treatise we hypothesize that the adiabatic invariant exists, but note that in the general case a variability of $N_{\mathbf{k}}$ may be observed by an additional term to α^{inh} in the transport equation (2.2.24).

Now the full coefficient of medium inhomogeneity

$$\alpha^{\text{inh}} = d_s \ln\left(\omega^{-3} n_r^{-2}\right) = d_s \ln n_r^{-2} - 3\left(\omega v_g\right)^{-1} \partial_t \omega \tag{2.2.25}$$

takes into account not only the spatial inhomogeneity, but also the nonstationarity of the medium, or, stated differently, the space-time inhomogeneity.[5]

For a stationary medium, $\partial_t \omega = 0$ so that eqn (2.2.25) carries over to eqn (2.2.11).

2.2.5 Allowance for Absorption and Nonstationarity

In phenomenological theory, weak absorption can be taken into account by adding to the inhomogeneity coefficient α^{inh} the energy absorption coefficient of the plane wave α^{abs}. This addition carries the transfer equation (2.2.24) to the form

$$\left(d_s + \alpha^{\text{inh}} + \alpha^{\text{abs}}\right) I_\omega = 0\,. \tag{2.2.26}$$

Below we restrict our consideration to absorbing media where $\alpha^{\text{abs}} > 0$. If the amplitude of a locally plane wave propagating in such a medium decays along the ray of length s as $e^{-\gamma s}$, then the energy absorption coefficient α^{abs} will be 2γ. In particular, in a homogeneous and isotropic medium, $\gamma = \text{Im}\,k$, where k is the complex wave number characterizing the propagation of plane waves.

2.2.6 Transfer Equation in a Scattering and Radiating Medium

When radiation propagates in a turbid medium, the quasiplane waves with different wave vectors **k** become interrelated because of scattering. Therefore, although in the conservative case the conservation law (2.2.2) holds, we cannot come from this law to the differential transfer equation (2.2.10) at least because the fact that the integral (2.2.8) is zero does not necessarily imply that its integrand is zero.

In radiometry, the radiative transfer equation can be extended to turbid media in a purely phenomenological manner in which the process of propagation is deemed independent upon the process of scattering. By standard reasoning we assume that the beams travel in some qusihomogeneous *effective* medium in agreement with the geometrical optics approximation, and describe the scattering by introducing the scattering cross-section of a unit volume, σ. This method allows us to account for the diffraction (i.e., a change of the direction of propagation) in scattering, while the diffraction spreading will be neglected.

We partition the scattering volume into small elementary volumes (Ivanov, 1969; Rozenberg, 1970 and 1977) each of which is viewed by other such volumes as an effective point scatterer converting the incident radiation into an incoherent manifold of scattered beams. This approach yields the *scattering section per unit volume* $\sigma = \sigma(\omega, \mathbf{n} \leftarrow \omega_0, \mathbf{n}_0)$ as the ratio of the power $dP^{\text{sc}} = dP^{\text{sc}}(\omega, \mathbf{n})$ scattered at a frequency ω by an elementary volume dV into an elemental solid angle $d\Omega_{\mathbf{n}}$ to the density of the incident flux of energy $S_0 = dP_0 / d\Sigma$ corresponding to a frequency ω_0 and direction $\mathbf{n}_0$, as shown in Fig. 2.7, viz.,

[5] This form of the inhomogeneity coefficient has been obtained by Apresyan (1981b), and, for a paticular case, by Pomraning (1968).

$$\sigma(\omega,\mathbf{n} \leftarrow \omega_0,\mathbf{n}) = \frac{dP^{sc}}{dVd\Omega_{\mathbf{n}}} \Big/ S_0 = \frac{dP^{sc}}{dVd\Omega_{\mathbf{n}}} \Big/ \frac{dP_0}{d\Sigma}. \tag{2.2.27}$$

We see that the scattering cross section of an elementary volume has dimension $[\sigma] = \mathrm{m}^{-1}\mathrm{sr}^{-1}$.

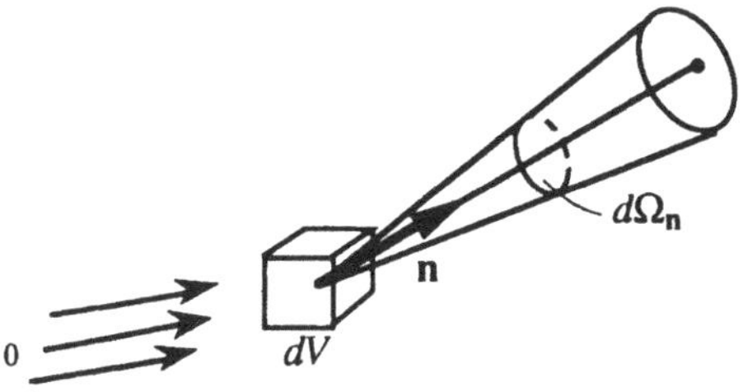

Figure 2.7. Scattering by an element dV of the scattering volume.

If we observe that the incident flux S_0 is proportional to the mean square amplitude of the incident wave $\langle |u^0|^2 \rangle$ and that $dP^{sc}/d\Omega_{\mathbf{n}} \propto R^2 \langle |u^{sc}|^2 \rangle$, where u^{sc} is the amplitude of the scattered wave, and R is the distance to the scatterer, the definition (2.2.27) may be rewritten as

$$\sigma(\omega,\mathbf{n} \leftarrow \omega_0,\mathbf{n}_0) = \frac{1}{dV} \lim_{R\to\infty} R^2 \frac{\langle |u^{sc}|^2 \rangle}{\langle |u^0|^2 \rangle}. \tag{2.2.28}$$

Finally, at large distances R we have

$$\langle |u^{sc}|^2 \rangle = R^{-2} \langle |u^0|^2 \rangle dV \sigma(\omega,\mathbf{n} \leftarrow \omega_0,\mathbf{n}_0). \tag{2.2.29}$$

At first glance the definition (2.2.28) differs from the similar cross section of a single scatterer insignificantly

$$\sigma_0(\omega,\mathbf{n} \leftarrow \omega_0,\mathbf{n}_0) = \frac{dP^{sc}}{d\Omega_{\mathbf{n}}} \Big/ S_0 = \lim_{R\to\infty} R^2 \frac{\langle |u^{sc}|^2 \rangle}{\langle |u^0|^2 \rangle} \tag{2.2.30}$$

and therefore seems to possess the same degree of rigor. Actually, a use of eqn (2.2.28) assumes the possibility of incoherent addition of intensities scattered by different elements of the volume, otherwise it would be senseless to speak of a power scattered by a *unit* volume. This assumption is rather arbitrary and becomes completely justified only for strongly rarefied media. A rigorous justification of this assumption can be given only in the framework of statistical wave theory (to be discussed in Chapter 3 of this book).

If we chose the elementary volume dV in the form of a cylinder of length ds and base Σ so that $dV = \Sigma ds$, as shown in Fig. 2.8, and write the density of the incident flux as $I_{\omega'}(\mathbf{n}')d\omega' d\Omega_{\mathbf{n}'}$, then the flux scattered along ds will be

$$\delta \mathbf{F} \equiv \Sigma dI_{\omega}(\mathbf{n}) = \Sigma ds \sigma(\omega,\mathbf{n} \leftarrow \omega',\mathbf{n}')I_{\omega'}(\mathbf{n})d\omega' d\Omega_{\mathbf{n}'}, \tag{2.2.31}$$

where $dI_{\omega}(\mathbf{n})$ is the increment in the radiance of the scattered radiation corresponding to δF.

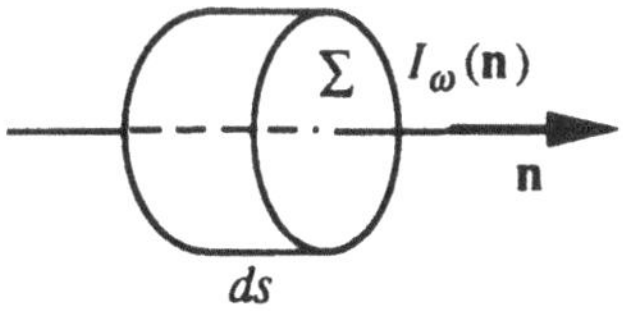

Figure 2.8. Scattering at a cylindrical elementary volume.

The variation rate of the radiance along ds is

$$\begin{aligned} d_s I_{\omega} &\equiv dI_{\omega}(\mathbf{n})/ds \\ &= \sigma(\omega,\mathbf{n} \leftarrow \omega',\mathbf{n}')I_{\omega'}(\mathbf{n}')d\omega' d\Omega_{\mathbf{n}'} . \end{aligned}$$

In order to allow for scattering in the transfer equation this quantity must be integrated with respect to all directions $\mathbf{n}'$ and frequencies ω' of the beams incident on the elementary volume and add the result

$$\int \sigma(\omega,\mathbf{n} \leftarrow \omega',\mathbf{n}')I_{\omega'}(\mathbf{n}')d\omega' d\Omega_{\mathbf{n}'} \equiv \hat{\sigma} I_{\omega} . \tag{2.2.32}$$

to the right-hand side of eqn (2.2.26). If we assume further that the energy is conserved in an elementary scattering act, then in the above setting we can separate the "true scattering" inferred in the propagation of the wave from the "attenuation in an elementary scattering act" associated with the conversion of energy into other directions $\mathbf{n}$ and frequencies ω. To this end we add in eqn (2.2.26) the scattering coefficient

$$\alpha^{sc} = \alpha^{sc}(\omega,\mathbf{n}) = \int d\omega' d\Omega_{\mathbf{n}'} \sigma(\omega',\mathbf{n}' \leftarrow \omega,\mathbf{n}), \tag{2.2.33}$$

so that eqn (2.2.26) eventually becomes

$$(d_s + \alpha^{inh} + \alpha^{abs} + \alpha^{sc})I_{\omega} = \hat{\sigma} I_{\omega} . \tag{2.2.34}$$

In the absence of scattering ($\alpha^{abs} = 0$), the integration of both sides of the resultant equation with respect to all directions $\mathbf{n}$ and frequencies ω causes the terms with α^{sc} and cross-section σ to cancel each other and the transfer equation for a stationary medium goes over to the energy conservation law (2.2.2), viz.,

$$\int d\omega d\Omega_{\mathbf{n}}(d_s + \alpha^{\text{inh}} + \alpha^{\text{sc}} - \hat{\sigma})I_\omega = \int d\omega d\Omega_{\mathbf{n}}(d_s + \alpha^{\text{inh}})I_\omega$$
$$= \partial_t \int v_g^{-1} I_\omega d\omega d\Omega_{\mathbf{n}} + \nabla \int \mathbf{n} I_\omega d\omega d\Omega_{\mathbf{n}}$$
$$= \partial_t w + \nabla \mathbf{S} = 0\,. \tag{2.2.35}$$

The relation (2.2.33), sometimes called the optical theorem, assumes a rigorous sense with respect to energy conservation only in the limit of infinitesimal variation of frequency when the scattering cross-section σ vanishes for $\omega \neq \omega'$, i.e., takes on the form

$$\sigma(\omega, \mathbf{n} \leftarrow \omega', \mathbf{n}') = \sigma_\omega(\mathbf{n} \leftarrow \mathbf{n}')\delta(\omega - \omega')\,, \tag{2.2.36}$$

because from the quantum point of view a variation of frequency would mean that the energy of the quantum is not conserved. Practically eqn (2.2.33) holds provided that in scattering the frequency varies relatively little, $\Delta\omega = |\omega - \omega'| << \omega$.

The sum of the coefficients of absorption and scattering is usually called the *extinction coefficient*:

$$\alpha^{\text{ext}} = \alpha^{\text{abs}} + \alpha^{\text{sc}}. \tag{2.2.37}$$

This coefficient has a simple physical meaning in Bouger's law describing the exponential attenuation of a coherent beam due to absorption and scattering. The inhomogeneity coefficient α^{inh} falls out of this law because allowance for this coefficient reduces to the substitution of $I_\omega / \omega^3 n_{\text{r}}^2$ for the radiance I_ω in the transfer equation [see eqn (2.2.64)].

Suppose now that the medium not only scatters the incident radiation but radiates itself. To describe the intrinsic radiation, transfer theory introduces the *emission coefficient* (which will be also called the *source function*) $\varepsilon_q = \varepsilon_q(\omega, \mathbf{n})$ equal to the flux of energy radiated by a unit volume of the medium into a unit solid angle about the direction $\mathbf{n}$ at a frequency ω. The radiating element of the medium is then thought of as a point source so that the concept of the emission coefficient is applicable only in the far field with respect to the medium's element under consideration.

With reference to the element of volume pictured in Fig. 2.8 we note that the scattered flux (2.2.31) must be augmented with the flux of intrinsic radiation

$$\delta F_1 = \Sigma dI_1 = \Sigma ds\, \varepsilon_q\,, \tag{2.2.38}$$

which corresponds to adding the term

$$d_s I_1 = \varepsilon_q\,. \tag{2.2.39}$$

into the right-hand side of the transfer equation (2.2.34).

Finally, the transfer equation for radiance in a scattering and radiating medium takes the form

$$(d_s + \alpha)I_\omega = \int \sigma(\omega, \mathbf{n} \leftarrow \omega', \mathbf{n}')I_{\omega'}(\mathbf{n}')d\omega' d\Omega_{\mathbf{n}'} + \varepsilon_q$$
$$\equiv \hat{\sigma} I_\omega + \varepsilon_q , \qquad (2.2.40)$$

where

$$\alpha = \alpha^{\text{inh}} + \alpha^{\text{ext}} = \alpha^{\text{inh}} + \alpha^{\text{abs}} + \alpha^{\text{sc}} \qquad (2.2.41)$$

is the full attenuation coefficient, and $\hat{\sigma}$ is the integral operator with nucleus equal to the scattering cross-section.

The general transfer equation (2.2.40) expresses the energy variation law for the radiance I_ω. In agreement with this equation the full variation rate of the radiance along the ray, $d_s I_\omega$, is governed by the following effects:

(a) refraction and parametric interaction of the beams with the quasi-homogeneous and quasi-stationary medium (addend $\alpha^{\text{inh}} I_\omega$),

(b) attenuation (or amplification) associated with the true absorption (or instability), i.e., with the fact that the medium is not conserving (addend $\alpha^{\text{abs}} I_\omega$),

(c) scattering of the beam (addend $\alpha^{\text{sc}} I_\omega$),

(d) scattering of all other beams (addend $\hat{\sigma} I_\omega$) and

(e) intrinsic radiation (emission coefficient ε_q).

It should be emphasized also that the transfer equation (2.2.40) is applicable for a weakly inhomogeneous and nonstationary medium. The respective parameters α, σ, and ε_q will depend on the coordinates $\mathbf{r}$ and time t.

An important feature of the transfer equation (2.2.40) is that it is non-local despite the fact that this equation has been compiled as a local energy balance equation.The non-local property of this equation is hidden behind the dependence of the radiance $I(\mathbf{n}, \mathbf{r})$ not only upon the coordinates but also upon the directions $\mathbf{n}'$ represented by the non-local term $\hat{\sigma} I$ (2.2.32) describing the scattering from the direction $\mathbf{n}'$ into the direction $\mathbf{n}$. This non-local property is illustrated in Fig. 2.9: the radiance $I(\mathbf{n}, \mathbf{r})$ at point $\mathbf{r}$ depends upon the intensities $I(\mathbf{n}', \mathbf{r})$ of waves incoming from all directions $\mathbf{n}'$. In turn, the intensity $I(\mathbf{n}', \mathbf{r})$ depends upon the intensities $I(\mathbf{n}'', \mathbf{r}')$ of waves scattered at other points $\mathbf{r}'$ from the direction $\mathbf{n}''$ into the direction $\mathbf{n}'$ and so on.

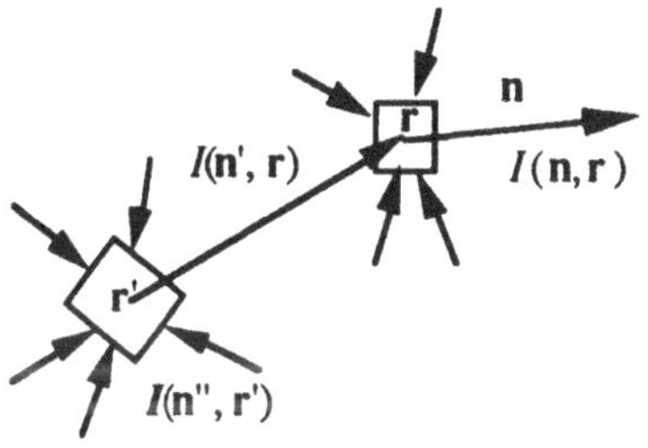

Figure 2.9. Non-local behavior of the radiative transfer in a scattering medium: the radiance $I(\mathbf{n}, \mathbf{r})$ at point $\mathbf{r}$ in direction $\mathbf{n}$ depends upon the values of radiance from other directions $\mathbf{n}'$, which in turn depend upon the values of I at other points and directions.

The said non-local property seriously hinders solution of the transfer equation. Despite its apparent simplicity this is one of the most complex linear equations of mathematical physics.

2.2.7 Measurement of Transfer Equation Parameters. Analytical Models for the Single Scattering Cross Section

For practical application of the transfer equation (2.2.40) we have to specify the relevant parameters for specific turbid media. In principle, this can be done by direct measurement of the phenomenological characteristics of an elementary volume in this medium, namely, the refractive index and the absorption coefficient, and the scattering cross section of the unit volume. Of course the measurement of these characteristics at each point of a real inhomogeneous scattering medium is a task whose practical infeasibility is obvious even to a theoretician. However, if the medium is statistically homogeneous, then it suffices to measure the characteristics of a small, arbitrarily chosen volume of the scattering medium which satisfies certain constraints. To measure the refractive index and the absorption coefficient one may take as such a volume a rather thin but representative layer, that is, one containing many inhomogeneities yet sufficiently thin to neglect the effects of multiple scattering. Then the extinction coefficient can be determined by direct measurements of the intensity of a wave passed through this layer, while the effective refractive index can be obtained from the phase of this wave.

In order to measure the scattering cross-section σ one may resort to analysis of the pattern of scattering from small, elementary volumes of the turbid medium. This type of measurements is conducted with polar nephelometers whose schematic diagram is depicted in Fig. 2.10.

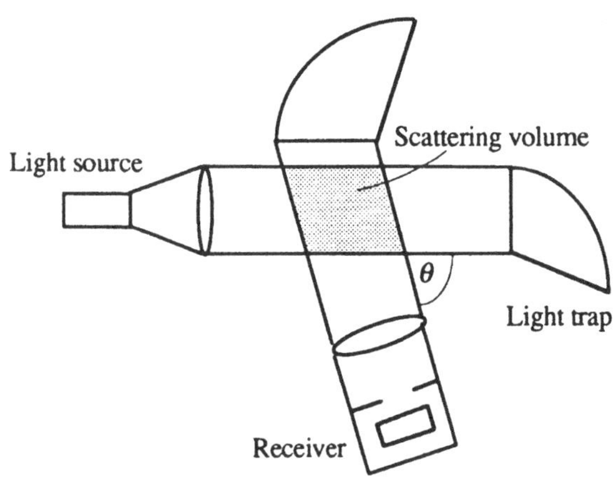

Figure 2.10. Schematic diagram of a polar nephelometer for measurement of the angular characteristics of scattering (after Bohren and Huffman, 1983).

Once the local optical characteristics of the turbid medium have been found by laboratory measurements, we may commence the solution of the transfer equation for certain effects of multiple scattering, which occur in large volumes of the scattering medium. These effects substantially depend upon the geometry of the scattering volume.

Detail measurements of turbid medium characteristics, specifically the scattering cross section of the unit volume, are a difficult task. Therefore, estimations of the effects of multiple scattering with the transfer equation are often based on various analytical approximations, simplifying the mathematical description of the problem. One of the most popular approximations is the *Henyey–Greenstein single scattering cross section* (Henyey and Greenstein, 1941)

$$\sigma_\omega(\mathbf{n} \leftarrow \mathbf{n}_0) = \frac{\alpha^{sc}}{4\pi} \frac{1-g^2}{\left(1+g^2-2g\mathbf{n}\cdot\mathbf{n}_0\right)^{3/2}}, \tag{2.2.42}$$

where g is the asymmetry factor ($-1 \le g \le 1$). This expression is widely used in optics of turbid media (a better physically substantiated approximation has been recently proposed by Cornette and Shanks, 1992)

In the case of strongly scattering media this approach to the transfer equation faces not only experimental complications but also some principal difficulties which go beyond the simplified phenomenological treatment and require that strong statistical theory to be invoked. To be more specific, these difficulties are associated with the matching of the experimental effective characteristics of the medium with the general concept of the transfer equation (2.2.40). The phenomenological theory is based on an intuitive pattern of a strongly rarefied medium. However, a bold application of the phenomenological approach to strongly scattering media leaves without explanation the concepts of effective absorption factor, effective refractive index, and other effective parameters because a simple averaging over the volume of the local characteristics of the medium is justifiable only in the limit of strongly rarefied scatterers.

Discussion of the statistical approach will, however, be left to Chapter 3. Here we dwell on the case of strongly rarefied scattering media.

2.2.8 Sparse Scattering Media

If the scattering medium is always sparse (strongly rarefied), then the above heuristic approach, treating the processes of propagation and scattering independent, may be refined by elementary reasoning without having to resort to a strict statistical analysis. This refinement yields in some zero approximation the parameters of phenomenological theory, namely, the characteristics of the effective medium and the scattering cross section by relating them to more delicate statistical characteristics of the medium.

We shall consider separately the cases of continuous and discrete scattering media and restrict our consideration, for simplicity, to the model of a monochromatic scalar wave field in a steady state medium, where scattering occurs without a change in frequency and the cross-section has the form (2.2.36).

Continuous scattering medium. First of all we note that although scattering is caused by inhomogeneities, nevertheless *in the average* the scattering medium is homogeneous, alternatively speaking it is statistically homogeneous. For continuous scattering fluctuations, it would be natural to assume the mean values of the scattering medium parameters to describe the effective homogeneous medium of phenomenological theory. For example, the case of an electromagnetic field may be handled by assigning to the effective medium a dielectric permittivity ε^{eff} equal to the mean dielectric permittivity of the medium under study $\langle\varepsilon\rangle$. We obtain a somewhat different result if we introduce the effective refractive index $n^{\text{eff}} = \sqrt{\varepsilon^{\text{eff}}}$ as the mean refractive index $n^{\text{eff}} = \langle\varepsilon^{1/2}\rangle$, however, in the limit of weak inhomogeneities corresponding to a strongly rarefied medium the differences of these approaches vanish.

For strongly rarefied media, it is natural to calculate the scattering cross section of a unit volume in the Born approximation. For simplicity we turn to the standard wave equation

$$\left[\Delta + k_0^2(1+\tilde{\varepsilon})\right]u = 0\,, \tag{2.2.43}$$

where $\Delta = \partial_x^2 + \partial_y^2 + \partial_z^2$ is the Laplacian, $k_0 = 2\pi/\lambda = \omega/c$ is the wave number, and $\tilde{\varepsilon} = \varepsilon - 1 = \varepsilon - \langle\varepsilon\rangle$ is the fluctuating part of the real-valued permittivity. In the Born approximation, the scattering cross-section of a unit volume, entering in eqn (2.2.36), is given by

$$\sigma_\omega(\mathbf{n} \leftarrow \mathbf{n}') = (\pi k_0^4/2)\Phi_\varepsilon(\mathbf{q})\,, \tag{2.2.44}$$

where $\mathbf{q} = k_0(\mathbf{n} - \mathbf{n}')$ is the vector of scattering, and $\Phi_\varepsilon(\mathbf{q})$ is the spectral density of permittivity fluctuations related to the fluctuation correlation function $\psi_\varepsilon(\mathbf{r}) = \langle\tilde{\varepsilon}(\mathbf{r}')\tilde{\varepsilon}(\mathbf{r}+\mathbf{r}')\rangle$ by the Fourier transform

$$\Phi_\varepsilon(\mathbf{q}) = \int \psi_\varepsilon(\mathbf{r})e^{-i\mathbf{qr}}(2\pi)^{-3}d^3r\,. \tag{2.2.45}$$

The respective scattering coefficient is

$$\alpha^{\text{sc}} = \int \sigma_\omega(\mathbf{n}' \leftarrow \mathbf{n})d\Omega_{\mathbf{n}'} = \int \frac{\pi k_0^2}{2}\Phi_\varepsilon\big(k_0(\mathbf{n}' - \mathbf{n})\big)d\Omega_{\mathbf{n}'}\,, \tag{2.2.46}$$

while, for a statistically homogeneous and nonabsorbing medium, $\alpha^{\text{inh}} = \alpha^{\text{abs}} = 0$. Eventually the radiative transfer equation becomes

$$(d_s + \alpha^{\text{sc}})I_\omega = \frac{\pi k_0^4}{2}\int \Phi_\varepsilon\big(k_0(\mathbf{n} - \mathbf{n}')\big)I_\omega(\mathbf{n}')d\Omega_{\mathbf{n}'} + \varepsilon_q\,, \tag{2.2.47}$$

In practical applications, to compute the Born scattering cross section (2.2.44) it is customary to adopt a convenient model of the correlation function $\psi_\varepsilon(\mathbf{r})$ so as to match the characteristic dimension of this function with the size of medium's inhomogeneities and simplify the Fourier transformation (2.2.45) as far as possible.

Amidst the most popular models we may point out to the case of a Gaussian correlation $\psi_\varepsilon(\mathbf{r}) \propto \exp-(r/l)^2$ and the case of an exponential correlation $\psi_\varepsilon(\mathbf{r}) \propto \exp-(r/l)$. For a more detailed discussion of the models of $\psi_\varepsilon(\mathbf{r})$, especially for the case of turbulent media, the reader is referred to Rytov, Kravtsov and Tatarskii (1989a).

Discrete scattering media. If the scattering medium may be thought of as a strongly rarefied cloud of discrete scatterers (particles), then in the zero approximation, the effective propagation medium can be described by the parameters of the medium in which the particles are embedded. Then, for the case of particles in free space, the effective medium of propagation will be free space. Adopting this simple assumption we neglect the refraction of waves associated with the presence of particles. Further we may argue as follows: let the density of the particles ν_0, equal to the mean number of particles per unit volume, be constant and so small that the particles may be deemed scattering independently and characterized by some differential scattering cross-section $\sigma_\omega^{(1)}(\mathbf{n} \leftarrow \mathbf{n}')$ and an absorption cross-section σ_1^{abs}. The total scattering cross-section at a single particle is given by the integral

$$\sigma_1^{\text{sc}} = \int \sigma_\omega^{(1)}(\mathbf{n}' \leftarrow \mathbf{n})d\Omega_{\mathbf{n}'}. \tag{2.2.48}$$

In irradiation of a particle by a plane wave the scattering cross-section σ_1^{abs} is defined as the ratio of the power P_{abs} absorbed by the particle to the density of the incident flux of energy $S_0 = dP_0 / d\Sigma$, viz.,

$$\sigma_1^{\text{abs}} = S_0^{-1}P_{\text{abs}} = \left(dP_0 / d\Sigma\right)^{-1} P_{\text{abs}}, \tag{2.2.49}$$

where both σ_1^{abs} and σ_1^{sc} have the dimension of area.

If we assume further that the positions of the particles are uncorrelated and they are uniformly distributed over the volume, then we may let the scattering coefficient α^{sc} (i.e., the total scattering cross-section of a unit volume) be equal simply to $\nu_0\sigma_1^{\text{sc}}$ which corresponds to the coherent addition of the intensities of the waves scattered by different particles. Similarly, the total coefficient of absorption of a unit scattering volume may be deemed equal to $\alpha^{\text{abs}} = \nu_0\sigma_1^{\text{abs}}$. Now the transfer equation (2.2.40) assumes the form

$$d_s I_\omega = \nu_0\left(\sigma_1^{\text{abs}} + \sigma_1^{\text{sc}}\right)I_\omega + \nu_0 \int \sigma_\omega^{(1)}(\mathbf{n} \leftarrow \mathbf{n}')I_\omega(\mathbf{n}')d\Omega_{\mathbf{n}'} + \varepsilon_q. \tag{2.2.50}$$

For the case of discrete scattering particles, this form of the transfer equation along with many specific problem formulations and applications of this equation has been discussed in depth by Ishimaru (1978).

All parameters entering the transfer equation (2.2.50) can be defied by solving the problem of a single scatterer. Let us recapitulate briefly the basic results for this problem.

If a single scatterer, having, generally speaking, random shape, dimension and other internal parameters, is hit by a plane wave $u^0(\mathbf{r}) = \exp(i\mathbf{k}_0\mathbf{r})$, then in the wave zone with respect to the scatterer, that is, for $R >> a$ and $R >> k_0 a^2$, where a is the characteristic dimension of the scatterer, the total field is

$$u = e^{i\mathbf{k}_0\mathbf{r}} + f_\omega e^{ik_0 R}/R\,, \tag{2.2.51}$$

where $f_\omega = f_\omega(\mathbf{n} \leftarrow \mathbf{n}_0)$ is the random amplitude of scattering from the direction $\mathbf{n}_0 = \mathbf{k}_0 / k_0$ to the direction $\mathbf{n} = \mathbf{R}/R$, as shown in Fig. 2.11. The *mean differential scattering cross-section* $\sigma_\omega^{(1)}$ is expressed in terms of the amplitude f_ω as

$$\sigma_\omega^{(1)}(\mathbf{n} \leftarrow \mathbf{n}_0) = \langle |f_\omega(\mathbf{n} \leftarrow \mathbf{n}_0)|^2 \rangle\,. \tag{2.2.52}$$

where the angular brackets imply averaging over the internal parameters of the particle.

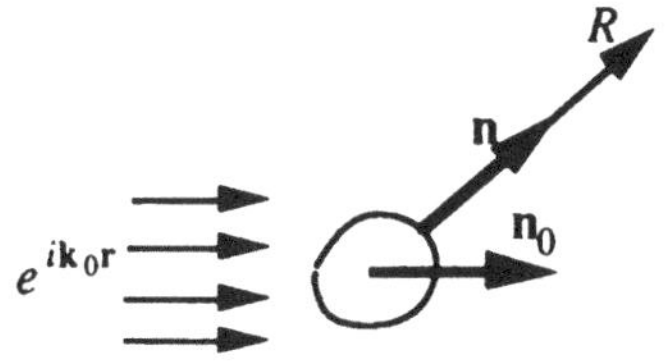

Figure 2.11. Scattering of a plane wave at a single scatterer.

The mean extinction cross-section $\sigma_1^{\text{ext}} \equiv \sigma_1^{\text{abs}} + \sigma_1^{\text{sc}}$ is related to the amplitude of scattering in the forward direction (see Problem 2.13) by the expression

$$\sigma_1^{\text{ext}} = \sigma_1^{\text{abs}} + \sigma_1^{\text{sc}} = (4\pi/k_0)\,\text{Im}\langle f_\omega(\mathbf{n}_0 \leftarrow \mathbf{n}_0)\rangle\,, \tag{2.2.53}$$

which is also called the optical theorem. We have already encountered this name in reference with eqn (2.2.33). This term is often used for various corollaries immediately following from the energy conservation law. A similar discussion of the optical theorem (2.2.53) may be found in the book by Bohren and Huffman (1983).

The relations (2.2.52) and (2.2.53) can be rewritten for the quantities not averaged over the internal parameters of the particle. However here we are interested in the first place in the behavior of the ensemble averages.

Thus, in the model of a strongly rarefied discrete scattering medium, the computation of the transfer equation parameters reduces to the evaluation of the characteristics of single particles, namely, the scattering cross-section $\sigma_\omega^{(1)}$ and the absorption cross-section σ_1^{abs} which can be expressed in terms of the amplitude of scattering f_ω. Evaluation of this amplitude is a cumbersome mathematical problem even for particles of simple geometry. A rigorous solution of this problem for the simple case of a sphere (*Mie solution*) is represented by infinite series. Many approximate techniques for evaluation of f_ω have been devised to simplify the procedure. The more popular solutions include the *electrostatic approximation* applicable to particles of small size a (a must be small compared to the wavelength in a vacuum and inside the particle); the *geometrical optics approximation* describing the opposite limiting case of large particles $k_0a \gg 1$ with $k_0 = \omega / c$; the *Rayleigh–Hans approximation*, widely known as the Born approximation, which requires that the particles should be optically soft, i.e, that the refractive index m of the particle should not differ significantly from the refractive index of the medium (equal to unity), $|m-1| << 1$, and not result in large variations of the incident wave field, $k_0a|m-1| << 1$; and the *van de Hulst approximation*, also known as the straight rays approximation or anomalous diffraction approximation, which requires that $m-1 << 1$ and $k_0a >> 1$. A large number of numerical techniques have been developed for the evaluation of the scattering amplitude. An ample literature is devoted to the scattering at a single particle, including the well known monographs of van de Hulst (1957), Bohren and Huffman (1983), and Ishimaru (1978).

Sparse Medium with Continuous Fluctuations and Discrete Inclusions. In practical applications one often comes across situations when the scattering media contains both continuous fluctuations and discrete inclusions. Typical examples may be a turbulent atmosphere with aerosol particles and water droplets, and a bulk of sea water with plankton and organic inclusions. In general, when we speak about the scattering at discrete scatterers under natural conditions, the medium between these scatterers is, strictly speaking, not always homogeneous and fluctuates itself, although in some situations these fluctuations may be neglected.

To describe such situation in the transfer theory of strongly rarefied media, it is sufficient to sum up the respective coefficients of extinction α^{ext} and the cross-sections of scattering σ after which the transfer equation (2.2.40) becomes

$$\left(d_s + \alpha^{\text{inh}} + \alpha_{\text{cont}}^{\text{ext}} + \alpha_{\text{discr}}^{\text{ext}}\right) I_\omega = \left(\hat{\sigma}^{\text{cont}} + \hat{\sigma}^{\text{discr}}\right) I_\omega + \varepsilon_q \,. \tag{2.2.54}$$

Here $\alpha_{\text{cont}}^{\text{ext}}$ and $\alpha_{\text{discr}}^{\text{ext}}$ are the coefficients of extinction, and $\hat{\sigma}^{\text{cont}}$ and $\hat{\sigma}^{\text{discr}}$ are the scattering operators corresponding to continuous fluctuations and discrete inclusions, respectively; the inhomogeneity coefficient α^{inh}, singled out in an individual term, describes the general inhomogeneity of the background.

It should be kept in mind that the simple addition of the scattering cross section and extinction coefficients is admissible only in the case of weakly scattering media. In denser media where the scattering particles come closer to one another, or when continuous fluctuations of the medium increase, the mutual influence of continuous and discrete inhomogeneities increase appropriately. In the case of strong scattering, this effect cannot be neglected. The coefficients of the transfer equation cease to be additive and their evaluation requires that either a more rigorous theory or direct measurements of the optical characteristics of the scattering medium.

2.2.9 Strongly Rarefied Discrete Media. The Effect of Scattering on Effective Medium Parameters

Thus far we have assumed as an effective medium of propagation the free space between the particles, i.e., deemed the dielectric permittivity of the effective medium ε^{eff} equal to unity. The relation (2.2.51) enables us to refine the transfer equation (2.2.50) already in phenomenological theory by allowing for the refraction due to the presence of the particles. To this end we assume $\varepsilon^{\text{eff}} = 1 + \delta\varepsilon$, where $\delta\varepsilon$ is a small correction to the dielectric permittivity ($|\delta\varepsilon| << 1$) caused by the presence of particles. The the mean field $\langle u \rangle$ corresponding to the medium will satisfy the wave equation $\left[\Delta + k_0^2(1+\delta\varepsilon)\right]\langle u \rangle = 0$, and the mean scattering field, corresponding to the plane incident wave $\exp(i\mathbf{k}_0\mathbf{r})$ far from the scattering volume, in the Born approximation will be given as

$$\langle u^{\text{sc}} \rangle = \int_V \frac{e^{ik_0|\mathbf{R}-\mathbf{r}'|}}{4\pi|\mathbf{R}-\mathbf{r}'|} k_0^2 \delta\varepsilon e^{i\mathbf{k}_0\mathbf{r}'} d^3r' \underset{R>>k_0D^2}{\approx} \frac{e^{ik_0R}}{R} \frac{k_0^2\delta\varepsilon}{4\pi} \int_V e^{-i\mathbf{q}\mathbf{r}'} d^3r' , \tag{2.2.55}$$

where $\mathbf{q} = k_0(\mathbf{n} - \mathbf{n}_0)$ is the vector of scattering, and $D \propto V^{1/3}$. If the characteristic dimension D of the scattering volume V is large compared with the wavelength, then the scattered field differs from zero only near the direction of forward scattering ($\mathbf{n} \approx \mathbf{n}_0$), which corresponds to the zero vector of scattering $\mathbf{q}$. Letting in eqn (2.2.55) $\mathbf{q} = 0$ we find

$$\langle u^{\text{sc}} \rangle \approx \frac{e^{ik_0R}}{R} \frac{k_0^2\delta\varepsilon}{4\pi} V . \tag{2.2.56}$$

On the other hand, in the single scattering approximation, each particle gives a contribution to the scattered field equal to the mean value of the second term in eqn (2.2.51), so that for the forward scattering direction, the mean field is approximately

$$\langle u^{\text{sc}} \rangle \approx V v_0 \langle f_\omega(\mathbf{n}_0 \leftarrow \mathbf{n}_0) \rangle \exp(ik_0R)/R . \tag{2.2.57}$$

Comparing (2.2.57) and (2.2.56) we find the relation of the correction to the dielectric permittivity with the forward scattering amplitude

$$k_0^2\delta\varepsilon/4\pi = v_0 \langle f_\omega(\mathbf{n}_0 \leftarrow \mathbf{n}_0) \rangle , \tag{2.2.58}$$

whence for the effective permittivity we obtain an equation, given, e.g., in the book of van de Hulst (1957),

$$\varepsilon^{\text{eff}} = 1 + \delta\varepsilon = 1 + \left(4\pi/k_0^2\right) v_0 \langle f_\omega(\mathbf{n}_0 \leftarrow \mathbf{n}_0) \rangle . \tag{2.2.59}$$

Here the first addend corresponds to the medium in the absence of particles, and the second gives a small correction associated with the scattering at the particles. Using eqn (2.2.59) we obtain for the wave number in the effective medium

$$\begin{aligned} k^{\text{eff}} &= k_0\sqrt{\varepsilon^{\text{eff}}} \approx k_0(1 + \delta\varepsilon/2) \\ &= k_0\left[1 + \left(2\pi v_0/k_0^2\right)\langle f_\omega(\mathbf{n}_0 \leftarrow \mathbf{n}_0) \rangle\right] . \end{aligned} \tag{2.2.60}$$

Now, it is not hard to derive an expression for the coefficient of effective absorption of the mean field

$$\begin{aligned}\alpha^{abs} &= 2\,\mathrm{Im}\,k^{eff} \approx k_0\,\mathrm{Im}\,\delta\varepsilon \\ &= (4\pi\nu_0/k_0)\,\mathrm{Im}\langle f_\omega(\mathbf{n}_0 \leftarrow \mathbf{n}_0)\rangle \\ &= \nu_0\left(\sigma_1^{abs} + \sigma_1^{sc}\right),\end{aligned} \tag{2.2.61}$$

where we have observed the optical theorem (2.2.53), and the effective refractive index

$$\begin{aligned}n^{eff} &= \mathrm{Re}\sqrt{\varepsilon^{eff}} \approx 1 + \mathrm{Re}(\delta\varepsilon/2) \\ &= 1 + \left(2\pi\nu_0/k_0^2\right)\mathrm{Re}\langle f_\omega(\mathbf{n}_0 \leftarrow \mathbf{n}_0)\rangle.\end{aligned} \tag{2.2.62}$$

It is instructive to note that such an approach takes scattering into account already in the evaluation of the parameters of the effective medium so that one no longer needs to forcibly introduce the scattering coefficient α^{sc} in the transfer equation. Indeed, as can be seen from eqn (2.2.62), in the effective medium, the absorption coefficient of the mean field describes the true absorption, which is described by the absorption cross-section $\nu_0\sigma_1^{abs}$, and automatically observes also the attenuation due to the scattering, which is defined by the total scattering cross-section $\nu_0\sigma_1^{sc}$. In the final analysis we arrive at the same transfer equation (2.2.50) as before. However, in contrast to the above phenomenological approach, we incorporate explicitly the effect of scattering on the refracting properties of the medium. This effect can be significant in the case of an inhomogeneous medium for which the rays cease to be rectilinear.

We see that small corrections to the parameters of the radiative transfer equation caused by the presence of scattering particles can be obtained by an elementary argument. This approach has been used in most applications of radiative transfer theory to strongly rarefied media. However, in phenomenological theory the region of applicability as well as the methods of refining the available equations remain uncertain and call for more rigorous statistical methods, such as those outlined in Chapter 3. Some necessary phenomenological conditions of applicability of transfer theory will be discussed below in Section 2.2.15. These topics become especially important in dense media where scattering inhomogeneities cannot be deemed independent any longer.

2.2.10 Corollaries of the Transfer Equation. Energy Balance Conditions

Let us consider some simple corollaries immediately following from the form of the transfer equation (2.2.40). The absorption coefficient α^{abs} and the function of sources ε_q are, generally speaking, independent quantities. However, if the state of the medium does not change, and the medium-field system has no external sources of energy, then the average power of radiation of the medium must be equal to the average power absorbed by the medium. Hence we obtain the condition of energy balance

$$\int \varepsilon_q(\omega,\mathbf{n})d\omega d\Omega_{\mathbf{n}} = \int \alpha^{\text{abs}}(\omega,\mathbf{n})I_\omega(\mathbf{n})d\omega d\Omega_{\mathbf{n}} \,. \tag{2.2.63}$$

For a homogeneous and stationary medium this condition can be obtained by integrating both sides of the transfer equation (2.2.40) with respect to frequency ω and directions $\mathbf{n}$ and assuming that, despite absorption, for the average energy of the field there holds the integral (in directions and frequencies) condition of energy conservation (2.2.3).

The energy balance condition (2.2.63), equally with the radiative transfer equation, is one of the principal equations of the theory of stellar atmospheres (see, e.g., Sobolev, 1975).

2.2.11 The Generalized Clausius Law

If we rewrite the transfer equation (2.2.40) using eqn (2.2.25) in the form

$$\omega^3 n_r^2\left(d_s + \alpha^{\text{abs}} + \alpha^{\text{sc}}\right)I_\omega / \omega^3 n_r^2 = \hat{\sigma} I_\omega + \varepsilon_q \,, \tag{2.2.64}$$

we readily obtain a few useful corollaries. In the absence of absorption, scattering and sources of radiation ($\alpha^{\text{abs}} = \alpha^{\text{sc}} = 0,\ \sigma = 0,\ \varepsilon_q = 0$) from (2.2.64) it follows that the quantity

$$I_\omega / \omega^3 n_r^2 = \text{const} \,, \tag{2.2.65}$$

proportional to the adiabatic invariant (2.2.23) is constant along the ray. For a stationary medium when the frequency is invariable along the ray, this relation reduces to the condition $I_\omega / n_r^2 = \text{const}$. For an equilibrium thermal radiation, this condition yields the Clausius law saying that the radiance of an equilibrium thermal radiation in a refracting medium is $I_\omega = n_r^2 I_\omega^0$, where I_ω^0 is the radiance of the thermal radiation in free space (2.1.31). Therefore, eqn (2.2.65) may be viewed as a generalization of the Clausius law to the nonequilibrium and nonstationary case.

2.2.12 The Kirchhoff Law for Equilibrium Thermal Radiation and Thermal Sources in Transfer Theory

Suppose that scattering does not change frequency so that the scattering cross-section has the form (2.2.36); in this case it is customary to speak (not apt to our mind) of *coherent* or *Rayleigh* scattering. The first term should be deemed unfortunate because of unnecessary associations with the general concept of coherence, and the second term is not apt because of ambiguity in the use of the name Rayleigh scattering, which more often than not is related to the case of scattering at inhomogeneities whose size is small compared to the wavelength.

For an equilibrium thermal radiation in a homogeneous, isotropic and stationary medium, when $d_s\left(I_\omega/\omega^3 n_r^2\right)=0$, the terms with the scattering coefficient α^{sc} and cross-section σ cancel out in view of (2.2.33) and the assumption that the medium is isotropic. Then from eqn (2.2.64) it follows that the source function is given by

$$\varepsilon_q = \alpha^{\text{abs}} I_\omega = \alpha^{\text{abs}} n_r^2 I_\omega^0 , \tag{2.2.66}$$

where I_ω is the radiance of the equilibrium thermal radiation in the medium (2.1.29), and I_ω^0 is the similar radiance in a vacuum. For the case of the thermal radiation of a plasma, eqn (2.2.66) has ben verified also by direct computations using an equilibrium Maxwellian distribution of frequency (Bekefi, 1966).

The relation (2.2.66) is a particular case of Kirchhoff's law saying that in thermal equilibrium the radiation is proportional to absorption. In its classical form this law relates the radiance of an equilibrium thermal radiation of a body I^{rad} in some direction $\mathbf{n}$ to the absorption factor in this body $\alpha_{\text{tot}}^{\text{abs}}$ when irradiated by a nonpolarized plane wave with backward direction $-\mathbf{n}$ (see, e.g., Rytov, Kravtsov and Tatarskii, 1989a). For a body embedded in a homogeneous medium with ray refractive index n_r, the classical Kirchhoff law has the form

$$I^{\text{rad}} = \alpha_{\text{tot}}^{\text{abs}} n_r^2 I_\omega^0 .$$

We see a full analogy of this expression to eqn (2.2.66), the only difference being that here the role of a source function is played by the radiance of the body I^{rad}, and the role of the coefficient of absorption of a unit volume α^{abs}, having a dimension of L^{-1}, is played by the dimensionless total absorption factor $\alpha_{\text{tot}}^{\text{abs}}$ which, by its physical sense, does not exceed unity.

While eqn (2.2.63) is an integral condition of energy balance, eqn (2.2.66) may be called the condition of *detailed* balance because it relates the power radiated by the medium ε_q with the absorption coefficient α^{abs} at *one* frequency ω and for *one* direction $\mathbf{n}$. Below we show that eqn (2.2.66) results as a corollary of the general fluctuation-dissipation theorem in the statistical wave derivation of the radiative transfer equation (Section 3.5.9); see also Rytov, Kravtsov and Tatarskii (1989a).

The expression (2.2.66) for the source function ε_q is often used in nonequilibrium situations, hypothesizing thereby the so-called "local thermodynamic equilibrium" (Sobolev, 1975) when the radiative properties of the medium are characterized by the local value of temperature which can vary from one point to another.

2.2.13 Approximation of the Total Frequency Redistribution

In astrophysical descriptions of the radiative transfer in stellar atmospheres, a popular factorized model for the scattering cross section is

$$\sigma(\omega,\mathbf{n} \leftarrow \omega_0,\mathbf{n}_0) = \left(4\pi \int \alpha^{\text{abs}}(\omega')d\omega'\right)^{-1} \alpha^{\text{abs}}(\omega)\alpha^{\text{abs}}(\omega_0) , \tag{2.2.67}$$

where α^{abs} is the energy absorption factor. Some authors speak in this case about the *approximation of the total redistribution of frequencies* (Ivanov, 1969) This frequency-factored cross-section markedly simplifies the solution of the transfer equation. It satisfies the optical theorem (2.2.33) if we replace there the scattering coefficient α^{sc} with the absorption coefficient α^{abs}. This substitution is equivalent to assuming that the true absorption, which heats the medium, is absent, therefore, having absorbed a quantum, the medium emits in a while another one at another frequency, in average isotropically, but so that the emission line repeats the profile of the absorption line. We recall, however, that with reference to scattering, quantum language arguments become physically transparent only for strongly rarefied media. It should be noted that a substantial simplification of the solution is retained also for a cross-section equal to the sum of addends of the form (2.2.67).

2.2.14 Useful Auxiliary Parameters

The transfer equation (2.2.40) may be written in the form

$$(d_\tau + 1)\frac{I}{n_r^2} = \frac{1}{4\pi}\hat{X}\left(\frac{I}{n_r^2}\right) + (1 - w_0)S. \tag{2.2.68}$$

Here in place of the length of a ray we use the *optical thickness*

$$\tau = \int_s \alpha^{ext} ds' = \int_s \left(\alpha^{abs} + \alpha^{sc}\right) ds', \tag{2.2.69}$$

and introduce the following new quantities: the *spectral albedo of single scattering*

$$w_0 = \frac{\alpha^{sc}}{\alpha^{ext}} = \frac{\alpha^{sc}}{\alpha^{abc} + \alpha^{sc}}, \tag{2.2.70}$$

and the *source function*

$$S_q = \frac{\varepsilon_q}{n_r{}^2 \alpha^{abs}}. \tag{2.2.71}$$

The operator $\hat{X}$ describing the scattering is given by

$$\hat{X} = \frac{w_0}{n_r{}^2 \alpha^{sc}} \hat{p} n_r^2 \alpha^{sc}, \tag{2.2.72}$$

where the nucleus $p(\omega, \mathbf{n} \leftarrow \omega_0, \mathbf{n}_0)$ of the operator $\hat{p}$ is the *phase function* or *scattering indicatrix*

$$p(\omega, \mathbf{n} \leftarrow \omega_0, \mathbf{n}_0) = 4\pi\sigma(\omega, \mathbf{n} \leftarrow \omega_0, \mathbf{n}_0) / \alpha^{sc}(\omega_0, \mathbf{n}_0). \tag{2.2.73}$$

Normally these parameters are introduced for the case when scattering occurs without a change of frequency [the cross-section is of the form (2.2.36)], the medium is isotropic, α^{abs} and α^{sc} are independent of $\mathbf{n}$, and $\sigma(\omega,\mathbf{n} \leftarrow \omega',\mathbf{n}') = \sigma_\omega(\mathbf{nn}') \times \delta(\omega - \omega')$. Under these assumptions the integral term in eqn (2.2.68) simplifies, $\hat{X} = w_0 \hat{p}$, and so the transfer equation (2.2.67) rewrites as

$$(d_\tau + 1)\frac{I}{n_r^2} = \frac{w_0}{n_r^2}\int p_\omega(\mathbf{nn}')I(\mathbf{n}',\mathbf{r})\frac{d\Omega_{\mathbf{n}'}}{4\pi} + \frac{\varepsilon_q}{\alpha^{ext} n_r^2}, \tag{2.2.74}$$

where $p_\omega(\mathbf{nn}') = 4\pi\sigma_\omega(\mathbf{nn}')/\alpha^{sc}$. This form of transfer equation has been reiterated in the literature (Chandrasekhar, 1960; Bekefi, 1966; Ishimaru, 1978).

The parameters (2.2.70) - (2.2.72) have a simple physical meaning and are often used in applications of transfer theory. The optical thickness τ characterizes extinction, i.e., the overall attenuation of the wave: if $\tau \geq 1$, then the radiation is strongly attenuated due to scattering and absorption, and for a medium with optical thickness $\tau << 1$, the effects of absorption and scattering are small.

The albedo w_0 enables one to assess the relative role of the effects of scattering and absorption: for $w_0 = 1$, absorption is absent, and for $w_0 << 1$, the attenuation due to absorption dominates over the attenuation due to scattering.

As can be seen from (2.2.73) the scattering indicatrix $p(\omega,\mathbf{n} \leftarrow \omega_0,\mathbf{n}_0)$ is normalized to unity

$$\int p(\omega,\mathbf{n} \leftarrow \omega_0,\mathbf{n}_0) d\omega \frac{d\Omega_{\mathbf{n}}}{4\pi} = 1, \tag{2.2.75}$$

and thereby different from the scattering cross-section. Both σ and p are non-negative quantities, therefore the fulfillment of eqn (2.2.75) in the absence of absorption ($w_0 = 1$) allows us to treat $p(\omega,\mathbf{n} \leftarrow \omega_0,\mathbf{n}_0)$ as a conditional transition probability of a quantum from the state $\omega_0,\mathbf{n}_0$ to the state $\omega,\mathbf{n}$ (Sobolev, 1963; Ivanov, 1969).

In thermal equilibrium, the source function (2.2.71) coincides with the radiance of the thermal radiation in free space I_ω^0 [see eqn (2.2.66)] and so is independent of the refractive index and other parameters of the medium which can vary from point to point in the inhomogeneous case. This independence is an advantage of S_q over the source function ε_q (see Problem 2.12).

In order to compare the relative significance of the effects of absorption, scattering and inhomogeneity in qualitative analysis of applications it is often convenient to introduce the characteristic lengths as inverse quantities of the respective coefficients α, namely, the absorption length L_{abs}, the scale of regular inhomogeneity L_{inh}, and the extinction length L_{ext}:

$$L_{abs} = 1/\alpha^{abs}, \quad L_{sc} = 1/\alpha^{sc}, \quad L_{inh} = 1/|\alpha^{inh}|,$$
$$L_{ext} = 1/\alpha^{ext} = (\alpha^{abs} + \alpha^{sc})^{-1} = L_{abs}L_{sc}(L_{abs} + L_{sc})^{-1}. \tag{2.2.76}$$

These lengths characterize the spatial scales at which the respective effects become noticeable. For example, if the absorption length turns out to be small compared with the scattering length, $L_{abs} << L_{sc}$, then the absorption is the governing factor and its significance appears earlier than scattering becomes noticeable.

2.2.15 Necessary Applicability Conditions of the Radiative Transfer Equation in a Scattering Medium

We notice first of all that the general validity conditions of radiometric concepts must hold in order that one be able to apply the transfer equation. At the same time the presence of scattering calls for the introduction of new constraints which must be satisfied otherwise the above phenomenological pattern becomes inapplicable. The principal constrain may be formulated as a requirement that the total attenuation of the beam due to scattering, absorption and inhomogeneous behavior be small over the wavelength, viz., $\lambda|\alpha| = \lambda|\alpha^{inh} + \alpha^{abs} + \alpha^{sc}| << 1$ or

$$|\alpha| \sim L_{inh}^{-1} + L_{abs}^{-1} + L_{sc}^{-1} << \lambda^{-1} . \qquad (2.2.77)$$

In other words, the length of scattering , absorption and inhomogeneity must be large compared to the wavelength. Indeed, if this is not the case the beam has enough time to be scattered, absorbed or to change its amplitude already along a wavelength, i.e., actually before its formation, so that the field loses its running wave behavior.

It should be obvious that for $\alpha \sim 1/\lambda$, the transfer equation become inapplicable. In the absence of absorption in the statistically uniform medium, this condition reduces to the relation $\lambda \sim L_{sc}$, which in solid state physics is known as the *Ioffe–Regel condition*. In this region, coherent effect can occur, such as the strong or Anderson localization of the field, which cannot be described by classical transfer theory. We reiterate these effects in Section 5.4.

The condition (2.2.77) must be satisfied also in the statistical derivation of the transfer equation that will be considered below. Actually the transfer equation evolves as a result of an expansion in the small parameter $\lambda|\alpha|$ and does not observe the effects of higher order with respect to this parameter.

The inequality (2.2.77) is a rough condition of applicability of the transfer equation and must be satisfied in the presence of scattering and without it. This condition is necessary, but, of course, not sufficient to ensure the applicability of transfer theory. Even at small $\lambda|\alpha|$ this condition fails to allow for the coherent effects that occur in backscattering (see Section 5.1). More accurate conditions can be obtained only after refining the structure of the scattering medium and the relevant scattering mechanisms, which in turn calls for more details of the phenomenological pattern developed in Section 2.2.6. By way of example we consider one such condition for a model of a strongly rarefied medium.

In this model, scattering inhomogeneities may be characterized by some size a, having the order of magnitude of the correlation radius of fluctuations for the case of a continuous scattering medium and the order of magnitude of the size of a particle for a medium constituted by independent scatterers. The transfer equation treats the scattering inhomogeneities as point scatterers at which successful scattering acts occur.

The mean distance between such acts, i.e., the mean free path, may be deemed equal to L_{sc} in the order of magnitude. In order that the inhomogeneity may be treated as a point scatterer, each subsequent act of scattering must occur in the wave zone with respect to the inhomogeneity, i.e., not closer than the wavelength of the formation of the spherical wave equal, in the order of magnitude, to ka^2. This leads us to the inequality

$$ka^2 << L_{sc} \tag{2.2.78}$$

which imposes a constraint on the characteristics of the scattering medium. A somewhat more rigorous argument leading to similar inequalities will be outlined in Chapter 3 in the framework of the statistical wave approach.

2.3 PHENOMENOLOGICAL TRANSFER THEORY FOR ELECTROMAGNETIC RADIATION

2.3.1 Polarization of a Plane Electromagnetic Wave

The radiative transfer equation taken in the form considered in the preceding section is, strictly speaking, inapplicable to the description of a multicomponent electromagnetic (EM) field because each EM beam wave is characterized not only by its frequency, direction and intensity, but also by a definite polarization. First generalizations of this transfer equation onto the case of EM radiation have been made by Sobolev (1949), Rozenberg (1955), and Chandrasekhar (1960). We obtain such a generalization first for the case of an EM field in a vacuum, and then take into account the inhomogeneity, nonstationarity and other complicating factors, similar to how we did this earlier for the scalar field. We begin this derivation with a short recollection of the necessary concepts of polarization.

The vector amplitude of a completely polarized plane wave propagating in a homogeneous medium is described by the expression

$$\mathbf{E} = E\mathbf{e}e^{i(\mathbf{kr}-\omega t)} = \mathbf{e}e^{i\psi}, \quad \psi = \mathbf{kr} - \omega t. \tag{2.3.1}$$

For simplicity we set the intensity of this wave equal to unity, $E = |\mathbf{E}| = \sqrt{\mathbf{EE}^*} = 1$, so that the complex unit vector of polarization $\mathbf{e} = \mathbf{a} + i\mathbf{b}$ is normalized according to the condition

$$|\mathbf{e}|^2 = a^2 + b^2 = 1, \quad \mathbf{a} = \mathrm{Re}\,\mathbf{e}, \quad \mathbf{b} = \mathrm{Im}\,\mathbf{e}. \tag{2.3.2}$$

In these operations the polarization of a wave is treated as the direction of the complex vector $\mathbf{e}$ which does not change when this vector is multiplied by an arbitrary factor. The condition (2.3.2) specifies only the magnitude of such a vector. Therefore $\mathbf{e}$ may be deemed specified accurate to an arbitrary phase factor whose variation is equivalent to a change in the origin of the phase ψ (or, alternatively, in the time origin).

In accordance with the condition (2.3.2) we may let

$$a = \cos\beta\,, \quad b = |\sin\beta|\,, \tag{2.3.3}$$

where β is the angle variable from the interval $-\pi/4 \le \beta \le \pi/4$. Later we shall see that β may be chosen such that $\beta > 0$ would imply the right rotation of the polarization vector, and $\beta < 0$ would mean the left rotation.

According to (2.3.1), the real-valued vector of electric field strength is given by

$$\mathbf{E}^{(r)} = \mathrm{Re}\,\mathbf{E} = \mathbf{a}\cos(kr - \omega t) - \mathbf{b}\sin(kr - \omega t)\,. \tag{2.3.4}$$

Multiplying $\mathbf{e}$ by a constant phase factor $\exp(i\psi_0)$ one may always satisfy the condition

$$\mathrm{Im}\,\mathbf{e}^2 \equiv 2\mathbf{ab} = 0\,. \tag{2.3.5}$$

Thus, without any loss of generality the vectors **a** and **b** may be deemed orthogonal, that is, $\mathbf{a}\cdot\mathbf{b} = 0$. For definiteness we assume also that $a > b$, otherwise one can multiply **e** by the factor $i = e^{i\pi/2}$ to transpose **a** and –**b**.

From the condition (2.3.3) we see that the projections $E_{a,b}^{(r)}$ of the strength vector onto the orthogonal principal axes **a** and **b** are

$$E_a^{(r)} = a\cos\psi\,, \quad E_b^{(r)} = -b\sin\psi\,, \tag{2.3.6}$$

so that

$$\left(E_a^{(r)}/a\right)^2 + \left(E_b^{(r)}/b\right)^2 = 1\,. \tag{2.3.7}$$

From these expressions it follows that, when the phase ψ varies, the rotating vector of polarization traces out an ellipse with the semiaxes $a = \cos\beta$ and $b = |\sin\beta|$, as shown in Fig. 2.12. If we consider the field at a fixed point **r**, then the phase $\psi = \mathbf{kr} - \omega t$ will change due to varying time t. If the direction of rotation coincides with the one shown in the figure (counterclockwise looking from the source) and the vectors $\mathbf{e}_1$, $\mathbf{e}_2$, and $\mathbf{e}_3 = \mathbf{k}/k$ form a right-hand basis, then the wave is said to be *right-hand* polarized at azimuth χ, while for the opposite direction of rotation, the wave is *left-hand* polarized. Under this sign convention, the right polarization is associated with the vector product $\mathbf{b}\times\mathbf{a}$ oriented along the direction of propagation **k** which is the same as the rotation of the tip of $\mathbf{E}^{(r)}$ with a left-hand screw. This convention of rotation in polarization is common in chemistry and in some optical literature. In astrophysics and plasma physics, the opposite sign convention is more popular (see Bohren and Huffman, 1983).

In the general case of elliptic polarization, the rotation rate of $\mathbf{E}^{(r)}$ is nonuniform and becomes uniform only at circular polarization (Problem 2.14).

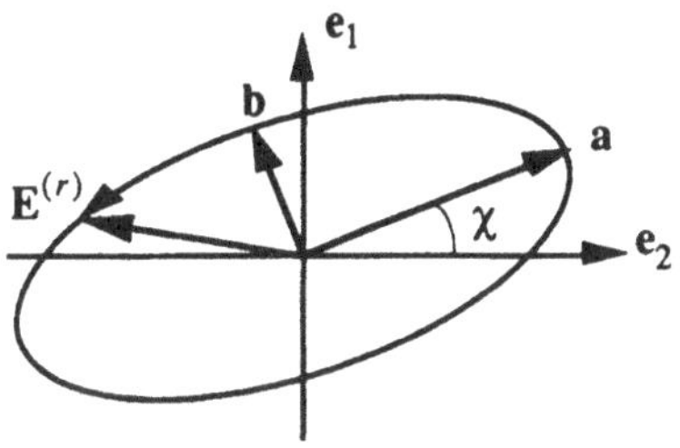

Figure 2.12. Parameters of the polarization ellipse.

The orientation of the polarization ellipse relative to an arbitrary Cartesian basis $(\mathbf{e}_1, \mathbf{e}_2)$ may be specified by the azimuth angle χ (between 0 and π) which its semi-major axis makes with the $\mathbf{e}_2$ axis. Thus, a polarization ellipse is completely specified by two angular variables χ (orientation) and β (form and direction of rotation).

For a partially polarized plane wave, the polarization vector $\mathbf{e}$ in eqn (2.3.1) is random and varies in time. As a consequence, the partially polarized wave is not strongly monochromatic any longer, but it occupies a finite frequency bandwidth $\Delta\omega$. Consequently, the polarization ellipse in Fig. 2.12 does not remain constant, but rotates and deforms with time in a random fashion.

In agreement with the general statistical theory (see, e.g., Rytov, Kravtsov, and Tatarskii, 1987b), a complete statistical description of polarization $\mathbf{e} = \mathbf{e}(t)$ as a random vector function of time is given by an arbitrary set of statistical moments referred to different time instants

$$\left\langle \mathbf{e}(t_1) \otimes \mathbf{e}(t_2) \otimes \ldots \mathbf{e}(t_n) \otimes \mathbf{e}^*(t_{n+1}) \otimes \mathbf{e}^*(t_{n+2}) \otimes \ldots \mathbf{e}^*(t_{n+m}) \right\rangle, \tag{2.3.8}$$

or else, by a set of respective multipoint distribution functions. Here the symbol $\otimes$ implies the direct tensor product because $\hat{\mathbf{d}} = \mathbf{a} \otimes \mathbf{b} \otimes \mathbf{c}\ldots$ is a matrix with elements $d_{ijk\ldots} = a_i b_j c_k \ldots$. Classical radiometry treats only the simple second one-point statistical moment $\langle \mathbf{e}(t) \otimes \mathbf{e}^*(t) \rangle$ whose characteristics can be measured in a rather simple optical experiments using polarizers and receivers detecting average intensities. Clearly the higher moments (2.3.8) may provide additional information about the properties of a partially polarized wave for a non-Gaussian $\mathbf{e}(t)$. Unfortunately, although statistical optics has been intensively developed in recent years, the higher statistical moments of polarization have been studied insufficiently.

If in place of a plane wave (2.3.1) we consider a beam, i.e., a bundle of quasi-plane waves $\mathbf{E}_\mathbf{k}$, then the partial polarization of the beam will not be related formally to its achromaticity. The random behavior of $\mathbf{e}(t)$ is replaced with the randomness of $\mathbf{E}_\mathbf{k}$ as a function of the wave vector $\mathbf{k}$. The difference between these two descriptions disappears if we admit that the vector amplitude $\mathbf{E}_\mathbf{k}$ smoothly depends also upon slow arguments $\mathbf{r}$ and t. Rigorous analysis of these topics falls into the realm of statistical theory which will be presented in Chapter 3.

2.3.2 Phenomenological Description of Electromagnetic Beams

Let us consider changes which should be made in the scalar transfer theory in order to describe polarization of electromagnetic radiation. Normally the radiometric description of a bundle of EM waves boils down to a transition from the scalar radiance $I = I_\omega$ to the radiance matrix $\hat{I}$ taking into account polarization. (In what follows we omit the subscript ω, indicating spectral parameters.) In the scalar theory, the radiance I is proportional to the mean square of the amplitude $u_\mathbf{k}$ of the quasi-plane wave with wave vector $\mathbf{k}$: $I \propto \langle |u_\mathbf{k}|^2 \rangle$. In electromagnetic theory, the matrix $\hat{I}$ is defined as a quantity proportional to the second moment of the vector amplitudes of the electrical field $\mathbf{E}_\mathbf{k}$:

$$\hat{I} \propto \left\langle \mathbf{E}_\mathbf{k} \otimes \mathbf{E}_\mathbf{k}^* \right\rangle = \left[\langle (\mathbf{E}_\mathbf{k})_i (\mathbf{E}_\mathbf{k}^*)_j \rangle \right]. \tag{2.3.9}$$

The matrix $\hat{I}$ is Hermitian,

$$I_{ij} = I_{ji}^* . \tag{2.3.10}$$

In matrix notation this line reads $\hat{I} = \hat{I}^+$, where (+) implies the Hermitian conjugate matrix.

The proportionality coefficient in eqn (2.3.9) may be specified in a natural way by requiring that the trace of the matrix $\hat{I}$ equal the scalar radiance

$$I = \operatorname{tr} \hat{I} = I_{ii} \propto \operatorname{tr} \langle \mathbf{E}_\mathbf{k} \otimes \mathbf{E}_\mathbf{k}^* \rangle = \langle |\mathbf{E}_\mathbf{k}|^2 \rangle . \tag{2.3.11}$$

Henceforth we assume that the summation is over the repeated indices, so that $\operatorname{tr} \mathbf{a} \otimes \mathbf{b} = a_i b_i = \mathbf{a} \cdot \mathbf{b}$. The mean energy density of plane waves

$$w_\mathbf{k} = \frac{\langle |\mathbf{E}_\mathbf{k}|^2 \rangle}{8\pi} \tag{2.3.12}$$

must be connected to this scalar radiance (2.3.11) by the radiometric relation (2.2.14). Writing

$$\hat{I} = c_\mathbf{k} \langle \mathbf{E}_\mathbf{k} \otimes \mathbf{E}_\mathbf{k}^* \rangle , \tag{2.3.13}$$

where $c_\mathbf{k}$ a real-valued proportionality coefficient, and substituting eqn (2.3.12) and $I = \operatorname{tr} \hat{I}$ in eqn (2.2.14) we obtain

$$w = \int \frac{\langle |\mathbf{E}_\mathbf{k}|^2 \rangle}{8\pi} k^2 dk d\Omega_\mathbf{n} = \int c_\mathbf{k} \langle |\mathbf{E}_\mathbf{k}|^2 \rangle dk d\Omega_\mathbf{n} , \tag{2.3.14}$$

where we have observed the dispersion relation $\omega = kc$. We now see that $c_\mathbf{k} = k_0^2 / 8\pi$.

In the general case of a homogeneous nondispersive medium with real-valued permittivity, this relation becomes

$$c_{\mathbf{k}} = \frac{k_0^2 \varepsilon}{8\pi}. \tag{2.3.15}$$

In the general case of an isotropic dispersive medium,

$$w_{\mathbf{k}} = \frac{(\partial_\omega \omega^2 \varepsilon)\langle |\mathbf{E}_{\mathbf{k}}|^2 \rangle}{16\pi\omega}$$

and ε in (2.3.15) gives way to $\left(\partial_\omega \omega^2 \varepsilon\right)/2\omega$. This yields

$$c_{\mathbf{k}} = k_0^2 \left(\partial_\omega \omega^2 \varepsilon\right)/(16\pi\omega).$$

In the literature, the radiance matrix is often defined for a single plane wave with a fixed wave vector $\mathbf{k}_0$ and a random vectorial amplitude $\mathbf{E}_0$ as $(c/8\pi)\langle \mathbf{E}_0 \otimes \mathbf{E}_0^* \rangle$ (Dolginov, Gnedin, and Silantev, 1995). This expression can be obtained from eqn (2.3.13) after integrating the latter with respect to the frequency ω and directions $\mathbf{n}$ if we observe that to a plane wave there corresponds the spectrum

$$\langle \mathbf{E}_{\mathbf{k}} \otimes \mathbf{E}_{\mathbf{k}}^* \rangle = \langle \mathbf{E}_0 \otimes \mathbf{E}_0^* \rangle \delta(\mathbf{k} - \mathbf{k}_0). \tag{2.3.16}$$

2.3.3 Transfer Equations in Anisotropic Media

In deriving the transfer equation for the electromagnetic field, it is convenient to treat separately the cases of anisotropic and isotropic media. In the framework of the radiometric approach, the former case is simpler in a certain sense. Indeed, in an anisotropic medium, in the absence of polarization degeneracy, to each wave vector $\mathbf{k}$ and each type of wave there corresponds a certain polarization vector $\mathbf{e}$ such that in the zero approximation of geometrical optics, the vector amplitude $\mathbf{E}_{\mathbf{k}}$ may be written as (see Appendix A)

$$\mathbf{E}_{\mathbf{k}} = \mathbf{e} A_{\mathbf{k}}.$$

The polarization vector $\mathbf{e} = \mathbf{e}(\mathbf{k})$ is assumed to be normalized to unity: $\mathbf{e} \cdot \mathbf{e}^* = 1$, and specified accurate to a phase multiplier. In agreement with (2.3.13) the matrix $\hat{I}$ has the form

$$\hat{I} = \mathbf{e} \otimes \mathbf{e}^* I,$$

where $I = \operatorname{tr} \hat{I}$ is the scalar radiance. Since the matrix $\mathbf{e} \otimes \mathbf{e}^*$ is known, it suffices to describe how I varies along the ray. However, this quantity must satisfy the scalar transfer equation (2.2.40), because in its derivation we have used only general considerations about energy conservation while the state of polarization has been defined unambiguously. It follows that the scalar transfer theory affords also description of polarized monomodal radiation.

The last remark relates not only to the case of an electromagnetic field, but also to the case of multicomponent waves of arbitrary nature. However in a *multimodal* medium, i.e., one propagating various types of waves, scattering can cause mutual transformation of waves, which is not taken into account in the simple scalar theory. The general form of transfer equation allowing for mutual conversion is discussed in Section 2.3.22.

2.3.4 Electromagnetic Beam in a Homogeneous Isotropic and Weakly Anisotropic Media

In an isotropic medium, transverse waves become polarization degenerate. This implies that the wave vector $\mathbf{k}$ does not describe uniquely the direction of polarization of the wave (see Appendix A). Because the field $\mathbf{E}_\mathbf{k}$ is orthogonal to $\mathbf{k}$ ($\mathbf{k}\cdot\mathbf{E}_\mathbf{k}=0$), the matrix $\hat{I}$ is also transverse with respect to $\mathbf{k}$, i.e.,

$$k_i I_{ij} = I_{ji} k_i = 0\,, \tag{2.3.17}$$

or in matrix form

$$\mathbf{k}\hat{I} = \hat{I}\mathbf{k} = 0\,.$$

From this relation it follows that in a three-dimensional orthonormal basis ($\mathbf{e}_1, \mathbf{e}_2, \mathbf{e}_3 = \mathbf{n} = \mathbf{k}/k$), the matrix $\hat{I}$ has the form

$$\hat{I} = \begin{bmatrix} I_{11} & I_{12} & 0 \\ I_{21} & I_{22} & 0 \\ 0 & 0 & 0 \end{bmatrix} = \begin{bmatrix} \hat{I}_0 & 0 \\ 0 & 0 \end{bmatrix}. \tag{2.3.18}$$

Saying this in other words, in an isotropic medium, the 3×3 matrix $\hat{I}$ reduces to the 2×2 matrix $\hat{I}_0$. This condition may be written also in the invariant form

$$\hat{I} = \sum_{\alpha,\beta=1,2} \mathbf{e}_\alpha \otimes \mathbf{e}^*_\beta I_{\alpha\beta}\,, \tag{2.3.19}$$

where

$$I_{\alpha\beta} = \mathbf{e}^*_\alpha \hat{I} \mathbf{e}_\beta = \operatorname{tr} \mathbf{e}_\beta \otimes \mathbf{e}^*_\alpha \hat{I} = (\hat{I}_0)_{\alpha\beta}\,. \tag{2.3.20}$$

If we use here the quantum-mechanical notations, these relations may be rewritten as $\hat{I} = \sum_{\alpha,\beta} |e_\alpha\rangle I_{\alpha\beta} \langle e_\beta|$ and $I_{\alpha\beta} = \langle e_\alpha | \hat{I} | e_\beta \rangle$. This representation is physically transparent and with some experience in operating tensor products becomes very convenient.

Equations (2.3.18) – (2.3.20) approximately retain their form in weakly anisotropic media where eqns (2.3.17) are valid to a good accuracy. This case often occurs in astrophysical applications in handling strongly rarefied magnetoactive plasmas with a refractive index close to unity.

In eqns (2.3.18) – (2.3.20), the basis vectors $\mathbf{e}_1$ and $\mathbf{e}_2$ may be deemed complex valued to obtain a very simple description of waves with elliptic polarization (Landau and Lifshits, 1975); see Problem 2.15. For a complex basis, the orthonormalization condition is $\mathbf{e}_i\mathbf{e}_j^* = \delta_{ij}$.

If we choose the orthonormalized eigenvectors $\tilde{\mathbf{e}}_1$ and $\tilde{\mathbf{e}}_2$ of the Hermitian matrix $\hat{I}$ to serve as $\mathbf{e}_1$ and $\mathbf{e}_2$, then eqn (2.3.19) becomes

$$\hat{I} = \tilde{\mathbf{e}}_1 \otimes \tilde{\mathbf{e}}_1^* I_{11} + \tilde{\mathbf{e}}_2 \otimes \tilde{\mathbf{e}}_2^* I_{22} . \tag{2.3.21}$$

This representation corresponds to a decomposition of the given beam into a sum of two incoherent beams with mutually orthogonal polarizations $\tilde{\mathbf{e}}_1$ and $\tilde{\mathbf{e}}_2$ and radiances I_{11} and I_{22}; in this case the incoherence is understood as the absence of off-diagonal terms $I_{\alpha\beta}$ with $\alpha \neq \beta$ in eqn (2.3.21). The representation (2.3.21) should not be confused with the decomposition in terms of normal modes, where the basis consists of eigenvectors, which characterize the medium, rather than of the matrices $\hat{I}$. The transfer theory now does not reduce to a description of these two radiances, since in the general case the vectors $\tilde{\mathbf{e}}_1$ and $\tilde{\mathbf{e}}_2$ can vary along the ray. This necessitates consideration of all four nonzero elements of the matrix $\hat{I}$. The transfer equation for them may be derived, like in the scalar case, by separating the processes of propagation and scattering. We look at this procedure in some more detail.

2.3.5 Transfer Equation in an Inhomogeneous Nonstationary and Nonscattering Medium. Allowance for Polarization Vectors

In deriving the transfer equation for the matrix $\hat{I}$ we must observe that even in the absence of scattering and anisotropy the state of polarization of the beam traveling in an inhomogeneous medium can be changed because the polarization vectors rotate in moving along a curvilinear ray. In the method of geometrical optics, such a rotation has been treated first by Rytov (1938) and introduced in transfer theory by Law and Watson (1970). Notice that a weak anisotropy of the medium causes additional rotation of polarization vectors (see Appendix B).

It is not hard to see that if we deem the basis vectors $\mathbf{e}_1$ and $\mathbf{e}_2$ in eqn (2.3.21) rotating by the laws of geometrical optics, then the representation (2.3.21) will remain diagonal in a propagation along the ray, because a rotation of the basis vectors will then coincide with the rotation of the field vectors. It is obvious also that in the absence of absorption and scattering, the radiances I_{11} and I_{22} must vary in accordance with eqn (2.2.10) of scalar theory: $d_s I_{11} = -\alpha^{\mathrm{inh}} I_{11}$, $d_s I_{22} = -\alpha^{\mathrm{inh}} I_{22}$. Since $\hat{I}$ reduces to the 2×2 matrix $\hat{I}_0$ in the plane with normal $\mathbf{n}$, the behavior of $\hat{I}$ will be uniquely defined by the projection of the variation rate of $\hat{I}$ along the ray onto this plane, that is, by the quantity

$$\rho(d_s \hat{I})\hat{\rho} \equiv \hat{\rho}\dot{\hat{I}}\hat{\rho}, \tag{2.3.22}$$

where $\hat{\rho} = \hat{1} - \mathbf{n} \otimes \mathbf{n}$ is the matrix of projection onto the plane with normal $\mathbf{n}$ such that $\hat{\rho}^2 = \hat{\rho}$ and $\mathbf{n}\hat{\rho} = \hat{\rho}\mathbf{n} = 0$.

We observe also that the polarization vector $\mathbf{e}$ obeys the geometrical optics equations of the quasi-isotropic approximation (Kravtsov, 1968)

$$\hat{\rho} d_s \mathbf{e} \equiv \hat{\rho}\dot{\mathbf{e}} = \hat{A}\mathbf{e}. \tag{2.3.23}$$

Here the matrix

$$\hat{A} = \frac{ik_0}{2\sqrt{\varepsilon_0}} \hat{\rho}\hat{\chi}\hat{\rho} \tag{2.3.24}$$

is expressed in terms of the medium's anisotropy tensor $\hat{\chi}$, i.e., in terms of the difference between the full permittivity tensor $\hat{\varepsilon}$ and its isotropic part $\varepsilon_0 = \operatorname{Re} \operatorname{tr} \hat{\varepsilon}/3$:

$$\hat{\chi} = \hat{\varepsilon} - \varepsilon_0 \hat{1}. \tag{2.3.25}$$

When the anisotropy of the medium is small, the tensor components $|\chi_{ij}| << |\varepsilon_0|$.

According to eqn (2.3.23), the matrix $\hat{A}$ defines the rotation of the polarization vector $\hat{\mathbf{e}}$ along a curvilinear ray in a weakly anisotropic medium. Equation (2.3.23) is derived in the Appendix B. Within the limits of this approximation, the components of the anisotropy tensor $\hat{\chi}$ are treated as small perturbations and taken into account in equations describing the evolution of the amplitudes of two mutually orthogonal polarizations.

Substituting now (2.3.21) into (2.3.22) and using eqn (2.3.23) we obtain, after termwise differentiation, the transfer equation in the form

$$\begin{aligned} \hat{\rho}\dot{\hat{I}}\hat{\rho} &= \hat{\rho} \sum_{\alpha=1,2} \left[(\dot{\tilde{\mathbf{e}}}_\alpha \otimes \tilde{\mathbf{e}}^*_\alpha + \tilde{\mathbf{e}}_\alpha \otimes \dot{\tilde{\mathbf{e}}}^*_\alpha) I_{\alpha\alpha} + \tilde{\mathbf{e}}_\alpha \otimes \tilde{\mathbf{e}}^*_\alpha \dot{I}_{\alpha\alpha} \right] \hat{\rho} \\ &= \hat{\rho} \sum_{\alpha=1,2} \left\{ \left[\left(\hat{A}\tilde{\mathbf{e}}_\alpha \right) \otimes \tilde{\mathbf{e}}^*_\alpha + \tilde{\mathbf{e}}_\alpha \otimes \left(\hat{A}\tilde{\mathbf{e}}_\alpha \right)^* \right] I_{\alpha\alpha} - \alpha^{\mathrm{inh}} \tilde{\mathbf{e}}_\alpha \otimes \tilde{\mathbf{e}}^*_\alpha I_{\alpha\alpha} \right\} \hat{\rho}, \end{aligned} \tag{2.3.26}$$

or in matrix form

$$\hat{\rho}\dot{\hat{I}}\hat{\rho} = \hat{\rho}(\hat{A}\hat{I} + \hat{I}\hat{A}^+ - \alpha^{\mathrm{inh}} I)\hat{\rho} = \hat{A}\hat{I} + \hat{I}\hat{A}^+ - \alpha^{\mathrm{inh}} \hat{I}. \tag{2.3.27}$$

Here the inhomogeneity coefficient α^{inh} is given by eqn (2.2.25). In deriving eqn (2.3.27) we have observed the transverse behavior of $\hat{A}$ and $\hat{I}$ with respect to $\mathbf{n}$, which implies $\hat{\rho}\hat{A} = \hat{A}\hat{\rho} = \hat{A}$ and $\hat{\rho}\hat{I} = \hat{I}\hat{\rho} = \hat{I}$.

Although eqns (2.3.27) and (2.3.24) have been derived in the assumption that there is no absorption, these relations must hold in the general case of a weakly absorbing medium. For an isotropic medium with permittivity $\varepsilon = \varepsilon^{h} + i\varepsilon^{a}$, where $\varepsilon^{h} = \mathrm{Re}\,\varepsilon$, the matrix $\hat{A}$ becomes $-k_0\varepsilon^{a}\hat{\rho}/2\sqrt{\varepsilon^{h}}$ and eqn (2.3.27) simplifies to the form

$$\hat{\rho}\dot{\hat{I}}\hat{\rho} = (-k_0\varepsilon^{a}/2\sqrt{\varepsilon^{h}} - \alpha^{\mathrm{inh}})\hat{I}\,. \tag{2.3.28}$$

Here the first term on the right-hand side describes weak absorption, and the second term allows for the variation of the the ray tube volume associated with the refraction in the medium.

The transfer equation (2.3.27) is independent of the choice of a basis in a three-dimensional space and quite appropriately may be termed the *invariant transfer equation*. The transfer equation for components $I_{\beta\gamma}$ in a certain basis follows from eqn (2.3.27) after representing $\hat{I}$ in the form (2.3.19) and applying the operation $\mathrm{tr}\,\mathbf{e}_\gamma \otimes \mathbf{e}^*_\beta(\cdot) = \mathbf{e}^*_\beta(\cdot)\mathbf{e}_\gamma$, i.e., by projecting (2.3.22) onto the vectors $\mathbf{e}_\beta$ and $\mathbf{e}_\gamma$:

$$d_s I_{\beta\gamma} + \sum_{\delta=1,2}\left[\left(\mathbf{e}^*_\beta\dot{\mathbf{e}}_\delta - A_{\beta\delta}\right)I_{\beta\gamma} + I_{\beta\delta}(\mathbf{e}_\beta\dot{\mathbf{e}}^*_\delta - A^*_{\gamma\delta})\right] + \alpha^{\mathrm{inh}} I_{\beta\gamma} = 0. \tag{2.3.29}$$

Here $\beta,\ \gamma = 1, 2,\quad \dot{\mathbf{e}}_\beta = d_s\mathbf{e}_\beta,\quad A_{\beta\gamma} = \mathrm{tr}\,\mathbf{e}_\gamma \otimes \mathbf{e}^*_\beta\hat{A} = \mathbf{e}^*_\beta\hat{A}\mathbf{e}_\gamma$.

Unlike the diagonal representation (2.3.19) where the vectors $\tilde{\mathbf{e}}_1,\ \tilde{\mathbf{e}}_2$ are tied with the state of polarization, now, in eqn (2.3.27), we may take complex basis vectors $\mathbf{e}_1,\ \mathbf{e}_2$ arbitrarily rotating when they shift along the ray, as shown in Fig. 2.13 for real-valued $\mathbf{e}_1,\ \mathbf{e}_2$.

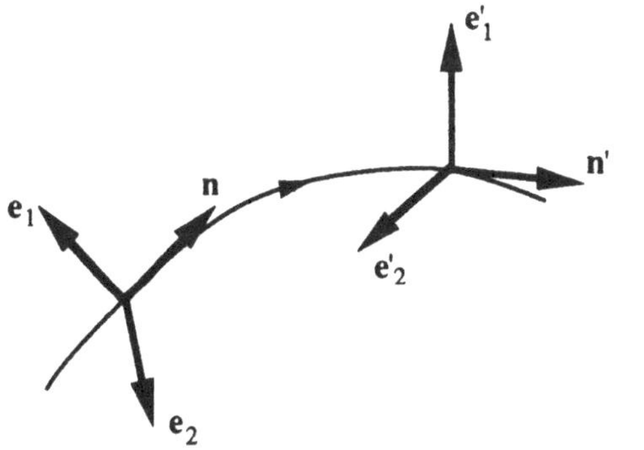

Figure 2.13. Rotation of basis vectors in passing along the ray.

In the case of a curvilinear ray, for which $\mathbf{n} \neq$ const., the basis $\mathbf{e}_1,\ \mathbf{e}_2$ varies from point to point due to the condition of transversality of $\mathbf{e}_1,\ \mathbf{e}_2$ with respect to $\mathbf{n}$. In a particular case when the basis $\mathbf{e}_1,\ \mathbf{e}_2$ rotates in agreement with the equation of quasi-isotropic approximation (2.3.23), eqn (2.3.29) takes on an especially simple form

$$(d_s + \alpha^{\text{inh}})I_{\beta\gamma} = 0 \tag{2.3.30}$$

and coincides in the form with the scalar transport equation (2.2.24). This coincidence should be expected since this choice of the local basis $\mathbf{e}_1$, $\mathbf{e}_2$ completely describes polarization in the absence of scattering.

If we choose basis vectors real-valued, $\mathbf{e}_1 = \mathbf{x}$, $\mathbf{e}_2 = \mathbf{y}$, then from the normalization conditions $\mathbf{e}_i\mathbf{e}_j = \delta_{ij}$ it follows that

$$\mathbf{e}_1\dot{\mathbf{e}}_1 = \mathbf{e}_2\dot{\mathbf{e}}_2 = 0, \quad \mathbf{e}_1\dot{\mathbf{e}}_2 = -\mathbf{e}_2\dot{\mathbf{e}}_1 \equiv -k', \tag{2.3.31}$$

where κ' is some function dependent on the choice of basis $\mathbf{x}$, $\mathbf{y}$. In a particular case when the normal and binormal to the ray are chosen to represent $\mathbf{e}_1$ and $\mathbf{e}_2$, it is easy to demonstrate that the function κ' coincides with the ray torsion κ (see Appendix A). Using the explicit form of the inhomogeneity coefficient (2.2.25) we then find that, for an isotropic nonabsorbing medium ($\hat{A} = 0$), the transfer equation (2.3.27) has the form

$$\omega^3 n_r^2 d_s \frac{1}{\omega^3 n_r^2}\begin{bmatrix} I_{11} & I_{12} \\ I_{12}^* & I_{22} \end{bmatrix} + \begin{bmatrix} 0 & -k' \\ k' & 0 \end{bmatrix}\begin{bmatrix} I_{11} & I_{12} \\ I_{12}^* & I_{22} \end{bmatrix} + \begin{bmatrix} I_{11} & I_{12} \\ I_{12}^* & I_{22} \end{bmatrix}\begin{bmatrix} 0 & k' \\ -k' & 0 \end{bmatrix} = 0. \tag{2.3.32}$$

This equation describes the polarization and intensity of a wave in an inhomogeneous and nonstationary medium in the absence of scattering.

2.3.6 Transfer Equation in a Scattering Medium

We will describe scattering phenomena on the ground of simple considerations similar to that exploited in the preceding section and assume that between acts of scattering the wave propagates in some *effective* medium, being homogeneous (or quasi-homogeneous) despite the presence of scattering inhomogeneities in the real medium. Then we may introduce on the right-hand side of eqn (2.3.27) an integral term proportional to the scattering cross section. Clearly both the scattering cross section and the scattering coefficient must be tensors. In full analogy with the scalar equation (2.2.40) we can write the polarized radiation transfer equation in the radiance matrix $\hat{I} = \hat{I}_\omega(\mathbf{n})$:

$$\hat{\rho}\left(d_s\hat{I}\right)\hat{\rho} - \hat{A}\hat{I} - \hat{I}\hat{A}^+ + \alpha^{\text{inh}}\hat{I} + \hat{\alpha}^{\text{sc}}\hat{I} = \int \hat{\sigma}(\omega, \mathbf{n} \leftarrow \omega', \mathbf{n}')\hat{I}_{\omega'}(\mathbf{n}')d\omega' d\Omega_{\mathbf{n}'}, \tag{2.3.33}$$

which is invariant under the choice of the basis. If we use a real-valued orthonormal basis $(\mathbf{x}, \mathbf{y}, \mathbf{z}) = (\mathbf{e}_1, \mathbf{e}_2, \mathbf{e}_3)$, then individual addends of eqn (2.3.33) may be expressed in terms of components

$$\left(\hat{\alpha}^{sc}\hat{I}\right)_{ij} = \alpha^{sc}_{ij,lm} I_{lm} ,$$
$$\left(\hat{\sigma}\hat{I}\right)_{ij} = \sigma_{ij,lm} I_{lm} , \tag{2.3.34}$$

where all indexes take values 1, 2, and 3. Here $\hat{\alpha}_s = \left[\alpha_{ij,lm}\right]$ is the tensor scattering coefficient, and $\hat{\sigma} = \left[\sigma_{ij,lm}\right]$ is the tensor scattering cross section per unit volume. Like in the scalar case (2.2.29), the tensor scattering cross section of a unit volume may be defined by the relation

$$\left\langle E_i^{sc} E_i^{sc*} \right\rangle = \lim_{R\to\infty} R^{-2} dV \sigma_{ij,lm}(\omega, \mathbf{n} \leftarrow \omega_0, \mathbf{n}_0) \left\langle E_1^0 E_m^{0*} \right\rangle, \tag{2.3.35}$$

where E_l^0 are the components of the vector amplitude of the plane wave incident on the elementary volume dV, and E_l^{sc} are the components of the scattered field (Fig. 2.14 shows the case of real-valued $\mathbf{E}^0$ and $\mathbf{E}^{sc}$).

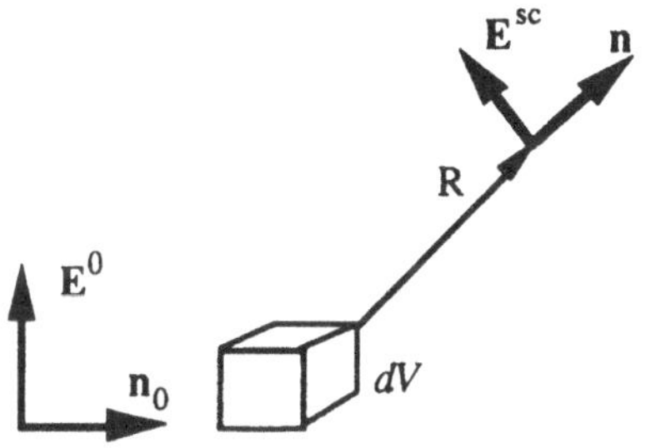

Figure 2.14. Scattering of a polarized wave by an elementary volume *dV*.

With such a definition, the scattering cross section $\sigma_{ij,lm}$ characterizes only the properties of the scattering medium and does not depend on the state of the field. In the literature, other definitions are often used which include the polarization matrix of the incident wave in the scattering cross section. A popular approach is to consider the cross section of the form

$$\sigma' = \sigma_{ii,lm} \frac{\langle E_l^0 E_m^{0*} \rangle}{\langle |E^0|^2 \rangle}, \tag{2.3.36}$$

which depends upon the state of polarization of the incident wave. To avoid confusing the properties of the medium with the properties of the field we prefer using the definition (2.3.35) which ensures the closest consistency of the vectorial and scalar theories.

In contrast to the analogous coefficient α^{sc} in the scalar theory, the matrix coefficient $\hat{\alpha}^{sc}$ describes not only attenuation of the wave, but also the change of its polarization due to the scattering in propagation. If the effect of scattering is not taken into account, then the action of the matrix $\hat{\alpha}^{sc}$ in an isotropic medium will be reduced to the multiplication by the scalar α^{sc}, so that in agreement with the definition (2.3.34) we may write

$$\alpha^{sc}_{ij,lm} = \alpha^{sc}\delta_{il}\delta_{jm} \,. \tag{2.3.37}$$

Because $\hat{I}$ is transverse with respect to $\mathbf{n}$ in eqn (2.3.33), it is sufficient to consider only the tensor components coplanar with $\mathbf{n}$. Therefore, in place of eqn (2.3.37) one may use also the relation

$$\alpha^{sc}_{ij,lm} = \alpha^{sc}\rho_{il}\rho_{jm} \,, \tag{2.3.38}$$

where $\rho_{il} = \delta_{il} - n_i n_l$.

Finally, if this model of effective medium itself allows for scattering, then the coefficient $\hat{\alpha}^{sc}$ in the transfer equation (2.3.33) should be omitted, for in this case the scattering will be described by the matrix $\hat{A}$. This remark will become obvious in considering models of strongly rarefied media in Section 2.3.14. For the scalar problem, a similar result has been obtained in Section 2.2.

2.3.7 Two-dimensional Description and the Choice of a Basis

Let us consider the matrix $\hat{I} = \hat{I}_\omega(\mathbf{n})$ in the basis $\{\mathbf{e}_i\} = (\mathbf{e}_1, \mathbf{e}_2, \mathbf{e}_3 = \mathbf{n})$ and the matrix $\hat{I}_{\omega'}(\mathbf{n}')$ in the basis $\{\mathbf{e}'_i\} = (\mathbf{e}'_1, \mathbf{e}'_2, \mathbf{e}'_3 = \mathbf{n}')$, as shown in Fig. 2.15. If we observe that $\hat{I}$ is transverse with respect to $\mathbf{n}$, the transfer equation (2.3.33) can be written in a form similar to eqn (2.3.26) where all indexes take on two values 1 and 2 rather than three. To this end we substitute eqn (2.3.19) into (2.3.33) and project both sides of the resultant equation onto the unit vectors $\mathbf{e}_1$ and $\mathbf{e}_2$, i.e., apply the operation $\operatorname{tr}\mathbf{e}_\gamma \otimes \mathbf{e}^*_\beta$. We have

$$\begin{aligned} d_s I_{\beta\gamma} + \left(\mathbf{e}^*_\beta \dot{\mathbf{e}}_\delta - A_{\beta\delta}\right) I_{\delta\gamma} + I_{\beta\gamma}\left(\dot{\mathbf{e}}^*_\delta \mathbf{e}_\gamma - A^*_{\gamma\delta}\right) + \alpha^{\text{inh}} I_{\beta\gamma} + \overline{\alpha}_{\beta\gamma,\varepsilon\mu} I_{\varepsilon\mu} \\ = \int \overline{\sigma}_{\beta\gamma,\varepsilon\mu}(\omega, \mathbf{n} \leftarrow \omega', \mathbf{n}') I_{\varepsilon\mu}(\omega', \mathbf{n}') d\omega' d\Omega_{\mathbf{n}'} \,. \end{aligned} \tag{2.3.39}$$

Here

$$\overline{\alpha}_{\beta\gamma,\varepsilon\mu} = e^*_{\beta i} e_{\gamma j} e_{\varepsilon l} e^*_{\mu m} \alpha^{sc}_{ij,lm} \,, \tag{2.3.40}$$

$$\overline{\sigma}_{\beta\gamma,\varepsilon\mu} = e^*_{\beta i} e_{\gamma j} e'_{\varepsilon l} e'^*_{\mu m} \sigma_{ij,lm} \,. \tag{2.3.41}$$

Repeated indexes imply summation where the Greek indexes take on values 1 and 2, the others are 1, 2, and 3, and $e_{\beta i}$ and $e'_{\beta i}$ are the components of the vectors $\mathbf{e}_\beta$ and

$\mathbf{e}'_\beta$, respectively. We shall assume further that $\mathbf{e}_\beta$ and $\mathbf{e}'_\beta$ are defined by one and the same dependence on $\mathbf{n}$ $[\mathbf{e}'_\beta = \mathbf{e}_\beta(\mathbf{n}')]$.

We will call eqn (2.3.39) the *two-dimensional transfer equation* to distinguish it from the invariant transfer equation (2.3.33). The respective description where $\hat{I}_\omega(\mathbf{n})$ is decomposed any time in the basis involving the unit vector $\mathbf{n}$ will be called the two-dimensional representation. This description is widely used in the literature on transfer theory as quite a natural approach, however, it should be emphasized that in the two-dimensional representation, the transfer equation coefficients ($\sigma_{\beta\gamma,\varepsilon\mu}$) are no longer tensor components in a definite basis and therefore in rotations they are not transformed as tensor components.

Let us shed more light on the meaning of the transfer equation (2.3.39). It suggests that the variation of the components of $\hat{I}$ along the ray has been caused by the following effects.

(1) The rotation of the basis and the geometrical optics rotation of the polarization vectors [second and third terms on the left-hand side of eqn (2.3.39)].

(2) The variation of the cross section of the ray tube and the parametric interaction of the waves with the inhomogeneous medium (term $\alpha^{\text{inh}}\hat{I}$).

(3) The scattering and absorption of the given beam (term $\overline{\alpha}_{\beta\gamma,\varepsilon\mu}I_{\varepsilon\mu}$).

(4) The scattering of all other beams [the right-hand side of eqn (2.3.39)].

Notice that effects 1 and 2 are related to the inhomogeneity, nonstationarity or anisotropy of the medium and disappear when these are absent.

We now comment on the bases $\mathbf{e}_1, \mathbf{e}_2$ and $\mathbf{e}'_1, \mathbf{e}'_2$ involved in eqns (2.3.39) – (2.3.41). In constructing these bases it is convenient to take as the datum plane the *plane of scattering* which contains both $\mathbf{n}$ and $\mathbf{n}'$ (it is assumed that $\mathbf{n}\times\mathbf{n}' \neq 0$). Then the bases can be chosen real-valued by taking $\mathbf{e}_1 = \mathbf{e}'_1 = \mathbf{n}\times\mathbf{n}'/|\mathbf{n}\times\mathbf{n}'|$, $\mathbf{e}_2 = \mathbf{n}\times\mathbf{e}_1$, and $\mathbf{e}'_2 = \mathbf{n}'\times\mathbf{e}'_1$, as shown in Fig. 2.15.

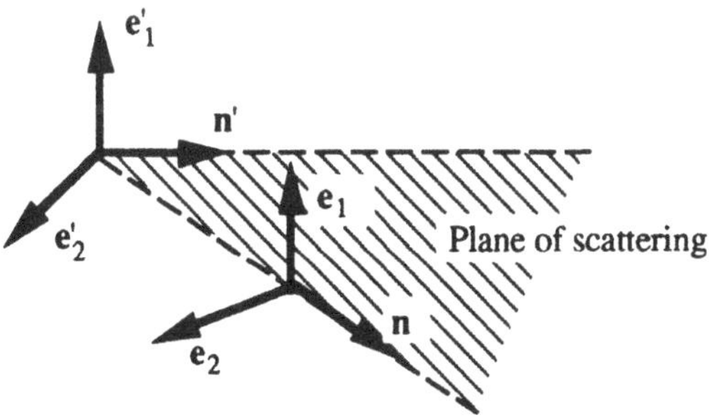

Figure 2.15. Choice of bases in a two-dimensional description of the radiance matrix.

Some authors use in place of the subscripts 1 and 2 other notations, say the symbols "$\parallel$" and "$\perp$" indicating the longitudinal and transverse components relative to the scattering plane (see, e.g., Bohren and Huffman, 1983). Some early papers used the subscripts l and r — the last letters of the words parallel and perpendicular (van de Hulst, 1957). Another approach is to call the components labelled 1 and 2 "vertical" and "horizontal" respectively, in agreement with the geometry of Fig. 2.15, or to use the subscripts V and H. Of course this difference of notations, in addition to opposite definitions for left and right polarizations, should be taken into account in comparison of results presented by different authors.

Sometimes in place of $\mathbf{e}_1$ and $\mathbf{e}_2$ it is more natural to use complex basis vectors of the form

$$\mathbf{e}_+ = \frac{\mathbf{e}_1 + i\mathbf{e}_2}{\sqrt{2}}, \quad \mathbf{e}_- = \frac{\mathbf{e}_1 - i\mathbf{e}_2}{\sqrt{2}}, \tag{2.3.42}$$

and similar vectors $\mathbf{e}'_+, \mathbf{e}'_-$. The vector $\mathbf{e}_+$ corresponds to the right and $\mathbf{e}_-$ to the left circular polarization The reference is readily established by considering the behavior of the vectors

$$\mathrm{Re}\, \mathbf{e}_\pm \exp(-i\omega t) = \frac{\mathbf{e}_1 \cos \omega t \pm \mathbf{e}_2 \sin \omega t}{\sqrt{2}}.$$

2.3.8 The Structure of Coefficients in the Transfer Equation

One can derive an explicit form of the permittivity tensor $\hat{\varepsilon}$ and the parameters $\hat{\alpha}^{sc}$ and $\hat{\sigma}$ entering in the transfer equation only on the basis of some model of the scattering medium. Some obvious constraints on the coefficients of the transfer equation can be obtained from the general considerations. From the fact that $\hat{I}$ is Hermitian we obtain

$$\alpha^{sc}_{ij,lm} = \alpha^{sc*}_{ji,ml}, \quad \sigma_{ij,lm} = \sigma^*_{ji,ml} \tag{2.3.43}$$

and similar conditions for the coefficients $\overline{\alpha}_{\beta\gamma,\varepsilon\mu}$ and $\overline{\sigma}_{\beta\gamma,\varepsilon\mu}$.

From the fact that $\hat{I}$ is transverse [see eqn (2.3.17)] it follows that the components of the tensors $\hat{\alpha}^{sc}$ and $\hat{\sigma}$ may be presented in the form

$$\alpha^{sc}_{ij,lm} = \rho_{ij'}\rho_{jj'}\tilde{\alpha}_{i'j',l'm'}\rho_{l'l}\rho_{m'm}, \tag{2.3.44}$$

$$\sigma_{ij,lm}(\omega, \mathbf{n} \leftarrow \omega', \mathbf{n}') = \rho_{ii'}\rho_{jj'}\tilde{\sigma}_{i'j',l'm'}\rho'_{l'l}\rho'_{m'm}, \tag{2.3.45}$$

where $\tilde{\alpha}_{ij,lm}$ and $\tilde{\sigma}_{ij,lm}$ are some tensors of the form (2.3.45), and $\rho_{ij} = \delta_{ij} - n_i n_j$ and $\rho'_{ij} = \delta_{ij} - n'_i n'_j$ are the matrix components of projection onto the planes with the normals $\mathbf{n}$ and $\mathbf{n}'$ respectively. In symbolic form these relations may be rewritten as

$$\hat{\alpha}^{sc} = (\hat{\rho} \otimes \hat{\rho})\hat{\tilde{\alpha}}(\hat{\rho} \otimes \hat{\rho}), \quad \hat{\sigma} = (\hat{\rho} \otimes \hat{\rho})\hat{\tilde{\sigma}}(\hat{\rho}' \otimes \hat{\rho}'). \tag{2.3.46}$$

The factors $\hat{\rho} \otimes \hat{\rho}$ and $\hat{\rho}' \otimes \hat{\rho}'$ may be omitted since they have been introduced only to emphasize that the matrices $\hat{\alpha}^{sc}$ and $\hat{\sigma}$ act on the matrices $\hat{I}_\omega(\mathbf{n})$ and $\hat{I}_{\omega'}(\mathbf{n}')$ orthogonal with respect to $\mathbf{n}$ and $\mathbf{n}'$. Unlike the earlier used tensor product of *vectors* $(\mathbf{a} \otimes \mathbf{b})_{ij} = a_i b_j$, here $\otimes$ indicates the tensor product of *matrices* $(A \otimes B)_{ij,lm} = A_{il}B_{jm}$. It is useful to note here that $(A \otimes B)C = ABC^{\mathrm{T}}$, where T stands for the transpose.

In the absence of absorption, from the conservation of the adiabatic invariant (or, for a stationary medium, from the conservation of the mean of the full energy) it follows a statement known as the optical theorem for the radiation transfer equation:

$$\sum_{i=1}^{3} \alpha^{sc}_{ii,lm} \equiv \alpha^{sc}_{ii,lm} = \int \sigma_{ii,lm}(\omega',\mathbf{n}' \leftarrow \omega,\mathbf{n}) d\omega' d\Omega_{\mathbf{n}'} \tag{2.3.47}$$

or else in the two-dimensional representation

$$\sum_{\beta=1}^{2} \overline{\alpha}_{\beta\beta,\varepsilon\mu} \equiv \overline{\alpha}_{\beta\beta,\varepsilon\mu} = \int \overline{\sigma}_{\beta\beta,\varepsilon\mu}(\omega',\mathbf{n}' \leftarrow \omega,\mathbf{n}) d\omega' d\Omega_{\mathbf{n}'} . \tag{2.3.48}$$

To derive eqns (2.3.47) and (2.3.48) it suffices to take the trace of both sides in (2.3.33) or (2.3.39) and to integrate these results with respect to the directions $\mathbf{n}$ and frequencies ω, assuming that the effective medium of propagation is nonabsorbing. These relations extend the optical theorem (2.2.33) onto the case of an EM field.

2.3.9 Coefficients of the Transfer Equation in the Isotropic Case

Let us consider in some more detail the case of a statistically isotropic medium where a selected direction is absent in the mean. Under this condition, the effective medium and individual acts of scattering may naturally be deemed also isotropic in view of which the tensor components $\tilde{\alpha}_{ij,lm} = \tilde{\alpha}_{ij,lm}(\omega,\mathbf{n})$ in eqn (2.3.46) may be expressed in terms of n_i and the unit tensor δ_{ij}. By virtue of (2.3.46) the components with n_i drop off. Therefore $\tilde{\alpha}_{ij,lm}$ may be written as

$$\tilde{\alpha}_{ij,lm} = a\delta_{il}\delta_{jm} + b\delta_{ij}\delta_{lm} + c\delta_{im}\delta_{jl} , \tag{2.3.49}$$

where a, b, and c are scalars. Substituting (2.3.49) into (2.3.40) we obtain in the two-dimensional description

$$\overline{\alpha}_{\beta\gamma,\varepsilon\mu} = a(\mathbf{e}^*_\beta \mathbf{e}_\varepsilon)(\mathbf{e}_\gamma \mathbf{e}^*_\mu) + b(\mathbf{e}^*_\beta \mathbf{e}_\gamma)(\mathbf{e}_\varepsilon \mathbf{e}^*_\mu) + c(\mathbf{e}^*_\beta \mathbf{e}^*_\mu)(\mathbf{e}_\gamma \mathbf{e}_\varepsilon) . \tag{2.3.50}$$

If the basis vectors $\mathbf{e}_\alpha$ are real-valued, so that $\mathbf{e}_\alpha \mathbf{e}_\beta = \delta_{\alpha\beta}$, then eqn (2.3.50) takes the form

$$\overline{\alpha}_{\beta\gamma,\varepsilon\mu} = a\delta_{\beta\varepsilon}\delta_{\gamma\mu} + b\delta_{\beta\gamma}\delta_{\varepsilon\mu} + c\delta_{\beta\mu}\delta_{\gamma\varepsilon} . \tag{2.3.51}$$

In this case the optical theorem (2.3.48) becomes

$$(a + 2b + c)\delta_{\varepsilon\mu} = \int \overline{\sigma}_{\beta\beta,\varepsilon\mu}(\omega',\mathbf{n}' \leftarrow \omega,\mathbf{n}) d\omega' d\Omega_{\mathbf{n}'} . \tag{2.3.52}$$

Whence it follows that

$$a+2b+c\equiv L_s^{-1}=\frac{1}{2}\int\bar{\sigma}_{\beta\beta,\gamma\gamma}(\omega',\mathbf{n}'\leftarrow\omega,\mathbf{n})d\omega' d\Omega_{\mathbf{n}'}\,, \tag{2.3.53}$$

where L_s has the meaning of the scattering length. Obviously, one relation (2.3.53) is insufficient to determine all three quantities a, b, and c. However, if we assume that the action of the tensor $\hat{\alpha}_s$ in eqn (2.3.33) reduces to the multiplication by a scalar, i.e. $\hat{\alpha}_s$ has the form (2.3.37), then the parameters b and c become equal to zero and eqn (2.3.53) completely defines the scattering coefficient.

In a similar way, the tensor $\hat{\hat{\sigma}}$ in eqn (2.3.45) must be expressed in terms of n_i, n_i' and δ_{ij}. We make a stronger assumption by letting the quantity $\hat{\hat{\sigma}}$ depends only upon the scalar $\mathbf{nn}'$ rather than upon $\mathbf{n}$ and $\mathbf{n}'$ in separate. This assumption will be advantageous, for example, in the model of a strongly rarefied medium with scatterers small compared to the wavelength. Then, omitting the tensors $\hat{\rho}'$ in eqn (2.3.45) we may write by analogy with eqn (2.3.49)

$$\sigma_{ij,lm}(\omega,\mathbf{n}\leftarrow\omega',\mathbf{n}')=\sigma_1\rho_{il}\rho_{jm}+\sigma_2\rho_{ij}\rho_{lm}+\sigma_3\rho_{im}\rho_{jl}\,, \tag{2.3.54}$$

where $\sigma_i=\sigma_i(\omega\leftarrow\omega',\mathbf{nn}')$ are scalars. According to eqn (2.3.41), in the two-dimensional representation, the cross section (2.3.54) is transformed to the form

$$\bar{\sigma}_{\beta\gamma,\varepsilon\mu}(\omega,\mathbf{n}\leftarrow\omega',\mathbf{n}')=\sigma_1\left(\mathbf{e}_\beta^*\mathbf{e}_\varepsilon'\right)\left(\mathbf{e}_\mu'^*\mathbf{e}_\gamma\right)+\sigma_2\delta_{\beta\gamma}\delta_{\varepsilon\mu}+\sigma_3\left(\mathbf{e}_\beta^*\mathbf{e}_\mu'^*\right)\left(\mathbf{e}_\varepsilon'\mathbf{e}_\gamma\right), \tag{2.3.55}$$

and, for real-valued bases $\mathbf{e}_\beta$ and $\mathbf{e}_\sigma'$, the optical theorem (2.3.53) takes the form

$$L_s^{-1}=\int\left[\frac{(\sigma_1+\sigma_3)[1+(\mathbf{nn}')^2]}{2}+2\sigma_2\right]d\omega' d\Omega_{\mathbf{n}'}\,. \tag{2.3.56}$$

Here we exploited simple symmetrical considerations, and the relationship $\mathbf{e}_1\otimes\mathbf{e}_1^*+\mathbf{e}_2\otimes\mathbf{e}_2^*=\hat{1}-\mathbf{n}\otimes\mathbf{n}$ corresponding to the decomposition of a unit tensor in the basis $(\mathbf{e}_1,\mathbf{e}_2,\mathbf{e}_3=\mathbf{n})$.

In eqn (2.3.55), the first term has a characteristic Rayleigh dependence upon the polarization multipliers. Indeed, if the incident wave is linearly polarized along $\mathbf{e}_\varepsilon'=\mathbf{e}_\mu'=\mathbf{e}$ and one seeks to measure the total intensity of a singly scattered wave then the first term in eqn (2.3.55) after summing over the polarizations of the scattered field yields the Rayleigh factor (Rytov, Kravtsov, and Tatarskii, 1987b)

$$\sum_{\alpha=1,2}\left(\mathbf{e}_\alpha^*\mathbf{e}\right)\left(\mathbf{e}^*\mathbf{e}_\alpha\right)=\mathbf{e}^*\left(\hat{1}-\mathbf{n}\otimes\mathbf{n}\right)\mathbf{e}=1-|\mathbf{ne}|^2$$
$$=\sin^2\chi\,, \tag{2.3.57}$$

where χ is the angle between the polarization vector $\mathbf{e}$ of the primary field and $\mathbf{n}$. A cross section with this angular profile occurs, for example, in the model of strongly rarefied dipole scatterers, specifically, in the model of Thomson scattering by free electrons.

The second summand in eqn (2.3.55) is "isotropic", i.e. independent of polarization. It describes the transformation of an arbitrarily polarized wave into a nonpolarized component of the scattered wave.

The third component differs from Rayleigh type quantities by permuted indexes l and m in eqn (2.3.54). This permutation corresponds to the transposition of the radiance matrix, or else, to the scattering of the right polarized wave into a left polarized wave.

These simple symmetrical considerations substantially refine the form of the local tensor parameters of the medium. Similar considerations often prove useful in describing more complicated nonlocal characteristics related to the problem in the whole (Problem 2.18).

2.3.10 Measurements of Parameters of the Radiation Transfer Equation

The local phenomenological parameters for a homogeneous scattering medium entering in the polarized radiation transfer equation (2.3.39) may in principle be evaluated by direct measurements. A substantial complication as compared to the scalar case is that in such measurements one have to control not only the radiance of the incident and scattering beams, but also the respective states of polarization.

The parameters on the left-hand side of eqn (2.3.39) describe the variation of the radiance matrix in the forward propagation, i.e., without alterations of the frequency and direction of propagation. These parameters may be found by measuring the depolarization of the direct wave, that is, by evaluating the polarization characteristics of a wave passing through a thin layer of the medium. All factors of the elements of the radiance matrix on the left-hand side of eqn (2.3.39) may be included in the coefficients $\overline{\alpha}_{\beta\gamma,\varepsilon\mu}$ because an explicit form of these factors can be justified mainly in a theoretical analysis of depolarization effects.

A direct measurement of the polarization scattering characteristics of an elementary volume, i.e. the measurement of the elements of the scattering matrix $\overline{\sigma}_{\beta\gamma,\varepsilon\mu}$, will be much more complicated than the measurement of depolarization of the direct wave. In the general case this matrix contains sixteen independent elements which are functions of the directions $\mathbf{n}$, $\mathbf{n}'$ and frequencies ω and ω'. Physically the problem of measuring sixteen parameters of the scattering matrix is solved by a nephelometer, shown in Fig. 2.10, whose input and output are equipped with the respective polarizers. Such measurements are rather laborious and has began comparatively recently. A review of pertinent data may be found in the book of Bohren and Huffman (1983).

Of course, everything said in Section 2.2.7 about the known uncertainty of the effective parameters in phenomenological theory is valid for direct measurements of the polarization parameters of the transfer equation. Therefore, for polarized radiation, the simple scattering medium models which enable one to evaluate the explicit form of transfer equation parameters become even more valuable. We consider two such phenomenological models for the case of strongly rarefied media.

2.3.11 Radiative Transfer Equation in a Continuous Weakly Scattering Medium

The parameters entering in the transfer equation for electromagnetic radiation may be computed similar to how this has been done in Section 2.2 for the scalar field. Without repeating the consideration of those section we illustrate briefly those variations which should be introduced to take into account the polarization of radiation separately in the cases of continuous and discrete scattering media. As before we restrict our consideration by the monochromatic case, for which the cross section has the form

$$\sigma_{ij,lm}(\omega,\mathbf{n} \leftarrow \omega',\mathbf{n}') = \sigma_{ij,lm}(\mathbf{n} \leftarrow \mathbf{n}',\omega)\delta(\omega-\omega') \tag{2.3.58}$$

similar to the scalar case (2.2.36).

In a simple model of continuous scattering medium with fluctuations of the scalar dielectric permittivity $\tilde{\varepsilon} = \varepsilon - \langle\varepsilon\rangle$, the scattering cross-section of a unit volume can be readily evaluated in the Born approximation (Rytov, Kravtsov, and Tatarskii, 1989b)

$$\sigma_{ij,lm}(\mathbf{n} \leftarrow \mathbf{n}',\omega) = \rho_{il}\rho_{jm}(\pi k_0^4 / 4)\Phi_\varepsilon(\mathbf{q}), \tag{2.3.59}$$

where $\Phi_\varepsilon(\mathbf{q})$ is the spatial fluctuation spectrum (2.2.45), $k_0 = \omega / c$, $\mathbf{q} = k_0(\mathbf{n}-\mathbf{n}')$ is the scattering vector, and $\rho_{ij} = \delta_{ij} - n_i n_j$. We dropped the factor $\hat{\rho}' \otimes \hat{\rho}'$ in the representation of the cross section (2.3.45). In the two-dimensional representation (2.3.41), from (2.3.59) we obtain

$$\overline{\sigma}_{\beta\gamma,\varepsilon\mu} = (\mathbf{e}_\beta^* \mathbf{e}_\varepsilon')(\mathbf{e}_\mu'^* \mathbf{e}_\gamma)(\pi k_0^4 / 2)\Phi_\varepsilon(\mathbf{q}). \tag{2.3.60}$$

This expression differs from the scalar case considered in Section 2.2 only by a polarization factor. Under the simple assumption (2.3.37) about the form of the scattering coefficient and in view of the optical theorem (2.3.56) this expression completely defines the coefficients of the transfer equation.

2.3.12 Scattering at a Single Scatterer

To take polarization into account in describing the scattering of a plane wave at a single scatterer we have to replace the basic relation (2.2.51) of the scalar theory with a similar vector relation for the electric field strength

$$\mathbf{E} = \mathbf{E}^0 + \mathbf{E}' = \mathbf{e}_0 e^{i\mathbf{k}_0\mathbf{r}} + \mathbf{f}_\omega \frac{e^{ik_0R}}{R}. \tag{2.3.61}$$

Here $\mathbf{e}_0$ is the unit polarization vector of the incident wave, $\mathbf{e}_0\mathbf{e}_0^* = 1$, and $\mathbf{f}_\omega$ is the *vector scattering amplitude*.

Observing the linearity of the field equations, the vector amplitude of eqn (2.3.61) can be written in the form

$$\mathbf{f}_\omega = \mathbf{f}_\omega(\mathbf{n} \leftarrow \mathbf{n}_0) = \hat{f}_\omega(\mathbf{n} \leftarrow \mathbf{n}_0)\mathbf{e}_0 . \tag{2.3.62}$$

The tensor $\hat{f}_\omega(\mathbf{n} \leftarrow \mathbf{n}_0) \equiv \hat{f}$ is the amplitude scattering matrix. The vector scattering amplitude $\mathbf{f}_\omega$ is orthogonal with respect to $\mathbf{n}$. Therefore, the amplitude matrix may be represented as a product of the projection operator $\hat{\rho} = \hat{1} - \mathbf{n} \otimes \mathbf{n}$ by the scattering tensor $\hat{\beta} = \hat{\beta}(\mathbf{n} \leftarrow \mathbf{n}_0)$

$$\hat{f}_\omega(\mathbf{n} \leftarrow \mathbf{n}_0) = \hat{\rho} k_0^2 \hat{\beta}(\mathbf{n} \leftarrow \mathbf{n}_0) , \tag{2.3.63}$$

where the factor k_0^2 is introduced for convenience. In the simplest case, when the scattering particle is a point dipole (this assumption is usually applicable for small particles), the scattering tensor coincides with the particle polarizability tensor.

From eqns (2.3.62), (2.3.61) and the scattering cross section (2.3.35) (where the factor dV may be dropped since we consider a single scatterer) we obtain the following tensor scattering cross section of a single particle

$$\sigma_{ij,lm}^{(1)} = \langle f_{il} f_{jm}^* \rangle = k_0^4 \rho_{ii'} \rho_{jj'} \langle \beta_{i'l} \beta_{j'm}^* \rangle , \tag{2.3.64}$$

where the angular brackets imply averaging over the orientation, form and other scattering parameters which may be random. This expression substitutes the similar relation (2.2.52) of scalar theory and allows for the independent averaging over the ensemble of parameters $\hat{\beta}$ and over the polarizations of the incident wave $\mathbf{E}^0$.

2.3.13 Jones and Müller Scattering Matrices

The vectors $\mathbf{E}^0$ and $\mathbf{E}^{sc}$ are orthogonal to the respective directions of propagation $\mathbf{n}_0$ and $\mathbf{n}$. Therefore it is natural to define them in different bases containing $\mathbf{n}_0$ and $\mathbf{n}$, respectively; this corresponds to the two-dimensional description considered in Section 2.3.7.

Let $\{\mathbf{e}_1(\mathbf{n}), \mathbf{e}_2(\mathbf{n}), \mathbf{e}_3(\mathbf{n}) = \mathbf{n}\}$ be an arbitrary, generally speaking, complex basis involving a unit vector $\mathbf{n}$. For the sake of definiteness we will refer to the basis shown in Fig. 2.15 with $\mathbf{n}' = \mathbf{n}_0'$. Then

$$\mathbf{E}^0 = E_1^0 \mathbf{e}_1(\mathbf{n}_0) + E_2^0 \mathbf{e}_2(\mathbf{n}_0) , \quad \mathbf{f}_\omega = f_1 \mathbf{e}_1(\mathbf{n}) + f_2 \mathbf{e}_2(\mathbf{n}) , \tag{2.3.65}$$

where

$$E_\alpha^0 = \mathbf{e}_\alpha^*(\mathbf{n}_0) \cdot \mathbf{E}^0 , \quad f_\alpha = e_\alpha^*(\mathbf{n}) \cdot \mathbf{f}_\omega , \tag{2.3.66}$$

are the components of $\mathbf{E}^0$ and $\mathbf{f}$.

A representation of the amplitude scattering matrix in the bases $\{\mathbf{e}_\alpha(\mathbf{n})\}$ and $\{\mathbf{e}_\beta(\mathbf{n}_0)\}$ has the form

$$f_{\alpha\beta} = e_\alpha^*(\mathbf{n}) \cdot \hat{f} \cdot \hat{e}_\beta(\mathbf{n}_0), \tag{2.3.67}$$

where α and β take on the values 1 and 2, and

$$f_\alpha = \sum_{\beta=1,2} f_{\alpha\beta} E_\beta^0 . \tag{2.3.68}$$

The matrix $f_{\alpha\beta}$ may be viewed as a particular case of the so-called *Jones matrix* combining the components of the polarization vectors at the input and output of a linear optical system:

Output vector amplitude = [Jones matrix] × Input vector amplitude

Jones matrices are commonly used to describe transformations of the polarization of plane monochromatic waves. In this case a scatterer plays the role of an optical system. It transforms the incident plane wave into a diverging spherical scattered wave.

A two-dimensional representation of the scattering cross section of a single particle, similar to (2.3.41), is given in terms of $f_{\alpha\beta}$ as

$$\overline{\sigma}^{(1)}_{\alpha\beta,\gamma\delta} = \langle f_{\alpha\gamma} f^*_{\beta\delta} \rangle . \tag{2.3.69}$$

This formula may be viewed as a transition from the Jones matrix $f_{\alpha\beta}$ describing the transformation of complex amplitudes to the (generalized) *Müller matrix* $\sigma^{(1)}_{\alpha\beta,\gamma\delta}$ related to the mean square characteristics, or else, to the second moments (or coherence matrix) of a partially polarized radiation:

[Output coherence matrix] = [Müller matrix] × [Input coherence matrix]

Normally the term "Müller matrix" is related to the second moments through the Stokes parameters (to be discussed later), and some authors call the matrix (2.3.69) also the Stokes matrix.

2.3.14 Transfer Equation in a Sparse Discrete Scattering Medium

Let us consider the case of a sparse (rarefied) particulate medium. In the definition of the effective medium parameters we take into account, like in the scalar case, a small variation of dielectric permittivity associated with the presence of scattering particles. Then the tensor of effective dielectric permittivity corresponding to transverse wave propagation and similar to (2.2.59) will have the form

$$\hat{\varepsilon}^{\text{eff}} = \hat{1} + \frac{4\pi\nu_0}{k_0^2} \langle \hat{f}_\omega(\mathbf{n} \leftarrow \mathbf{n}) \rangle = \hat{1} + 4\pi\nu_0 \hat{p} \langle \hat{\beta}(\mathbf{n} \leftarrow \mathbf{n}) \rangle , \tag{2.3.70}$$

where ν_0 is the particle number density. We have assumed here that the medium is strongly rarefied and therefore retained only the terms of the first order of smallness in the density ν_0. This expression relates ε^{eff} to the amplitude forward scattering matrix $\hat{f}_\omega(\mathbf{n} \leftarrow \mathbf{n})$. In this model we no longer need an additional coefficient α^{sc} in the transfer equation, because scattering is taken into account already in computing the effective medium parameters.

Letting $\hat{\varepsilon} = \hat{\varepsilon}^{\text{eff}}$ in eqn (2.3.25) we obtain

$$\begin{aligned} \varepsilon_0 &= \frac{1}{3}\operatorname{Re}\operatorname{tr}\hat{\varepsilon}^{\text{eff}} = 1 + \frac{4\pi}{3}\nu_0 \operatorname{Re}\operatorname{tr}\hat{\rho}\langle\hat{\beta}(\mathbf{n} \leftarrow \mathbf{n})\rangle, \\ \hat{\chi} &= 4\pi\nu_0\hat{\rho}\langle\hat{\beta}(\mathbf{n} \leftarrow \mathbf{n})\rangle - \hat{1}(\varepsilon_0 - 1), \end{aligned} \tag{2.3.71}$$

so that the $\hat{A}$ matrix (2.3.24), entering the transfer equation, takes the form

$$\hat{A} \approx ik_0 2\pi\nu_0\hat{\rho}\left\{\hat{\beta}(\mathbf{n} \leftarrow \mathbf{n}) - \hat{1}\left[\operatorname{Re}\operatorname{tr}\hat{\rho}\hat{\beta}(\mathbf{n} \leftarrow \mathbf{n})\right]/3\right\}\hat{\rho}. \tag{2.3.72}$$

If the contributions of different scatterers in the scattered field are deemed independent, then to obtain the scattering cross section per unit volume it suffices to multiply (2.3.64) by the particle number density ν_0. In the case of a homogeneous effective medium, when the tensor $\langle\hat{\beta}\rangle$ remains constant at each point, the basis vectors $\mathbf{e}_1$ and $\mathbf{e}_2$ may be deemed to be stationary and the transfer equation (2.3.33) takes the form

$$\begin{aligned} \hat{\rho}\left(d_s\hat{I}\right)\hat{\rho} &= \hat{\rho}2\pi i\nu_0 k_0\left[\langle\hat{\beta}(\mathbf{n} \leftarrow \mathbf{n})\rangle\hat{I} - \hat{I}\langle\hat{\beta}^+(\mathbf{n} \leftarrow \mathbf{n})\rangle\right]\hat{\rho} \\ &\quad + \hat{\rho}k_0^4\nu_0\int\langle\hat{\beta}(\mathbf{n} \leftarrow \mathbf{n}')\hat{I}(\mathbf{n}')\hat{\beta}^+(\mathbf{n} \leftarrow \mathbf{n}')\rangle d\Omega_{\mathbf{n}'}\hat{\rho}, \end{aligned} \tag{2.3.73}$$

where it has been recognized that $\hat{I}$ is orthogonal with respect to $\mathbf{n}$ and the angular brackets in the integrand imply averaging over random parameters of the matrices $\hat{\beta}$ and $\hat{\beta}^+$. In a two-dimensional description, this equation may be written as

$$\begin{aligned} d_s I_{\gamma\delta} &= 2\pi i\nu_0 k_0\left[\langle\overline{\beta}_{\gamma\nu}\rangle I_{\nu\delta} - I_{\gamma\nu}\langle\overline{\beta}^*_{\delta\nu}\rangle\right] \\ &\quad + k_0^4\nu_0\int\langle\overline{\beta}_{\gamma\nu}(\mathbf{n} \leftarrow \mathbf{n}')\rangle I_{\nu\mu}(\mathbf{n}')\langle\overline{\beta}^*_{\delta\mu}(\mathbf{n} \leftarrow \mathbf{n}')\rangle d\Omega_{\mathbf{n}'}. \end{aligned} \tag{2.3.74}$$

Here

$$\langle\overline{\beta}_{\mu\nu}\rangle = \langle\overline{\beta}_{\mu\nu}(\mathbf{n} \leftarrow \mathbf{n})\rangle, \text{ and } \langle\overline{\beta}_{\mu\nu}(\mathbf{n} \leftarrow \mathbf{n}')\rangle = \overline{e}^*_{\mu i}(\mathbf{n})\beta_{ij}(\mathbf{n} \leftarrow \mathbf{n}')e_{\nu j}(\mathbf{n}') \tag{2.3.75}$$

are the components of $\langle\hat{\beta}\rangle$ in the two-dimensional representation (which, as has been already noted, are not tensor components in a certain coordinate system).

A transfer equation of the type of eqn (2.3.73) has been examined in depth by Dolginov, Gnedin, and Silant'ev (1995) in connection with astrophysical applications.

To close this topic we point to a relation of our notation with other definitions used in the literature. If the bases $(\mathbf{e}_1', \mathbf{e}_2', \mathbf{n}')$ and $(\mathbf{e}_1, \mathbf{e}_2, \mathbf{n})$ are real-valued and oriented such that $\mathbf{e}_2'$ and $\mathbf{e}_2$ lie in the scattering plane $(\mathbf{n}, \mathbf{n}')$, while $\mathbf{e}_1'$ and $\mathbf{e}_1$ are at right angles with this plane, as depicted in Fig. 2.15, then in a two-dimensional representation, the components of the amplitude scattering matrix $\hat{f}$ will have the form

$$\left[\bar{f}_{\alpha\beta}(\mathbf{n} \leftarrow \mathbf{n}')\right] = \left[\mathbf{e}_\alpha^* \hat{f}(\mathbf{n} \leftarrow \mathbf{n}') e_\beta'\right] = \begin{bmatrix} \bar{f}_{11} & \bar{f}_{12} \\ \bar{f}_{21} & \bar{f}_{22} \end{bmatrix} = -\frac{1}{ik_0} \begin{bmatrix} S_1 & S_4 \\ S_3 & S_2 \end{bmatrix}. \tag{2.3.76}$$

The functions $f_{\alpha\beta}$ are equal to their counterparts in Ishimaru (1978a). The right-hand side of this expression corresponds to the notation used by van de Hulst (1957), where we have observed that this author used the time factor $e^{i\omega t}$ rather than $e^{-i\omega t}$ as we do in this book.

2.3.15 Stokes Parameters

In the case of an isotropic or quasi-isotropic medium, it often proves more convenient to use four real-valued Stokes parameters, than to consider nonzero complex components of the matrix $\hat{I}$, because these parameters are closely related to the quantities directly measured in typical experiments (Bohren and Huffman, 1983). In the literature these parameters are denoted by different letter sets, but a recently popular system is I, Q, U, and V. We will treat these parameters as components of some Stokes four-tuple

$$(I, Q, U, V) = (P_0, P_1, P_2, P_3) = (P_0, \mathbf{P}) = P. \tag{2.3.77}$$

Here P is the *full* Stokes vector and $\mathbf{P}$ is the *polarization* Stokes vector which is zero for a nonpolarized wave and is called a vector merely for convenience. This approach enables us to assign to Stokes parameters transparent geometrical meaning and to substantially simplify many similar relationships, because it reveals latent symmetry of these parameters. The physical meaning and properties of Stokes parameters may be found in optics textbooks, see. e.g., Born and Wolf (1980). We recall the definitions and the basic properties of these parameters below.

Parameters P_i depend upon the choice of a basis. In a real-valued basis they are defined by the relations

$$\hat{I}_0 = \begin{bmatrix} I_{11} & I_{12} \\ I_{21} & I_{22} \end{bmatrix} = c_\mathbf{k} \begin{bmatrix} \langle E_1 E_1^* \rangle & \langle E_1 E_2^* \rangle \\ \langle E_2 E_1^* \rangle & \langle E_2 E_2^* \rangle \end{bmatrix} = \frac{1}{2} \begin{bmatrix} P_0 - P_1 & P_2 + iP_3 \\ P_2 - iP_3 & P_0 + P_1 \end{bmatrix}. \tag{2.3.78}$$

Here $c_\mathbf{k}$ is the real-valued proportionality coefficient (2.3.13), which is normally set to unity (Landau and Lifshitz, 1975), and E_j are the components of the vector $\mathbf{E}_\mathbf{k}$.

From eqn (2.3.78) it follows that the parameter P_0 is equal to the radiance

$$P_0 = I = \operatorname{tr}\hat{I} = I_{11} + I_{22} = c_{\mathbf{k}}\langle |E_1|^2 + |E_2|^2 \rangle, \tag{2.3.79}$$

and the polarization parameters are defined as

$$\begin{aligned} P_1 &= Q = I_{22} - I_{11} = c_{\mathbf{k}}\langle |E_2|^2 - |E_1|^2 \rangle, \\ P_2 &= U = I_{21} + I_{12} = c_{\mathbf{k}} 2\operatorname{Re}\langle E_1 E_2^* \rangle, \\ P_3 &= V = i(I_{21} - I_{12}) = c_{\mathbf{k}} 2\operatorname{Im}\langle E_1 E_2^* \rangle. \end{aligned} \tag{2.3.80}$$

In computing Stokes parameters it is convenient to use Pauli's matrices

$$\hat{\pi}_0 = \begin{bmatrix} 1 & 0 \\ 0 & 1 \end{bmatrix}, \quad \hat{\pi}_1 = \begin{bmatrix} -1 & 0 \\ 0 & 1 \end{bmatrix}, \quad \hat{\pi}_2 = \begin{bmatrix} 0 & 1 \\ 1 & 0 \end{bmatrix}, \quad \hat{\pi}_3 = \begin{bmatrix} 0 & i \\ -i & 0 \end{bmatrix}, \tag{2.3.81}$$

which satisfy the relation

$$\hat{\pi}_l \hat{\pi}_j = i e_{ljk} \hat{\pi}_k + \delta_{lj} \hat{\pi}_0, \tag{2.3.82}$$

where e_{ljk} is the absolutely antisymmetrical tensor, and the indexes $l, j, k = 1, 2,$ and 3. Using this relation it is not hard to prove that the Pauli matrices are orthogonal in the sense that

$$\operatorname{tr} \hat{\pi}_i \hat{\pi}_j = 2\delta_{ij}, \tag{2.3.83}$$

where $i, j = 0, 1, 2, 3$. Using these matrices we can represent the radiance matrix $\hat{I}$ as a linear combination of Stokes parameters

$$\hat{I}_0 = \frac{1}{2} \sum_{i=0}^{3} P_i \hat{\pi}_i. \tag{2.3.84}$$

The Stokes parameters are expressed in terms of $\hat{I}$ by

$$P_i = \operatorname{tr} \hat{\pi}_i \hat{I}_0. \tag{2.3.85}$$

2.3.16 Stokes Parameters of a Completely Polarized Wave and a Poincaré Sphere

For better insight into the physical sense of Stokes parameters we indicate the relation of these parameters with the characteristics of a completely polarized wave. For such a wave we may drop the averaging operations $\langle . \rangle$ in eqn (2.3.78) and see that the elements $I_{\alpha\beta}$ are factored as

$$I_{\alpha\beta} = a_\alpha a_\beta^*. \tag{2.3.86}$$

The determinant $|I_{\alpha\beta}|$ turns out to be zero, and from (2.3.78) it follows that, for a completely polarized wave, the Stokes parameters are related by

$$P_0^2 = P_1^2 + P_2^2 + P_3^2 \equiv \mathbf{P}^2, \tag{2.3.87}$$

where $P_0 = I$ is the intensity of the wave.

It is straightforward to demonstrate (see Problem 2.19) that the polarization parameters **P** are expressed in terms of the polarization ellipse as

$$P_1 = Q = P_0 \cos 2\beta \cos 2\chi\,,$$

$$P_2 = U = P_0 \cos 2\beta \sin 2\chi\,, \tag{2.3.88}$$

$$P_3 = V = P_0 \sin 2\beta\,.$$

The parameters χ and β [$|\tan\beta| = b/a$, see eqn (2.3.3)] define the orientation and form of the polarization ellipse, as depicted in Fig. 2.12.

If we think of P_1, P_2, and P_3 as the components of **P** in a Cartesian system of coordinates, then eqn (2.3.87) is the equation of the Poincaré sphere shown in Fig. 2.16. Therefore, any change in the wave polarization may be represented as a motion of **P** over the Poincaré sphere.

The angles of polar coordinates defining the position of the tip of **P** on the sphere satisfy the conditions following from eqn (2.3.88)

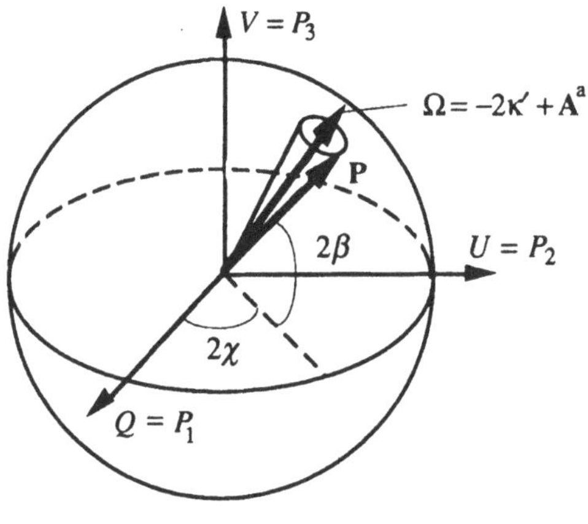

Figure 2.16. Rotation of the polarization Stokes vector on the Poincaré sphere (northern hemisphere $V > 0$ corresponds to the right and the southern hemisphere $V < 0$ to the left elliptic polarization).

$$\tan 2\chi = \frac{P_2}{P_1} = \frac{U}{Q}, \quad \sin 2\beta = \frac{P_3}{P_0} = \frac{V}{I}, \tag{2.3.89}$$

so that the position of **P** on the sphere determines the orientation (χ) and form (β) of the polarization ellipse.

Parameter V characterizes circular polarization. For waves with such polarization, $\beta = \pm\pi/4$, $U = Q = 0$, so that $V = \pm I$, i.e., the whole intensity is due to the component $V = P_3$. The parameter Q describes the linear polarization along the axes $\mathbf{e}_2$ $(\chi = \beta = 0, Q = I, U = V = 0)$ and $\mathbf{e}_1$ $(\beta = 0, \chi = \pi/2, Q = -I, U = V = 0)$, and the parameter U describes the linear polarization along the axes making an angle $\chi = \pm\pi/4$ with the axis $\mathbf{e}_2$ $(\beta = 0, U = \pm I, Q = V = 0)$. Now the right and left circular polarizations correspond, respectively, to the north and south poles on the Poincaré sphere, and the linear polarization corresponds to points on the equator, as shown in Fig. 2.16.

2.3.17 Some Properties of Stokes Parameters

An important characteristic of Stokes parameters P_j is their additive property, namely, if two mutually incoherent beams, propagating in the same direction, overlap, the respective Stokes parameters add together. This property immediately follows from the additivity of the respective radiance matrices. If the beams are mutually coherent, then their radiance matrices are not additive since the matrix of the sum of the beams contains interference terms (Problem 2.20).

In the general case of a partially polarized wave, eqn (2.3.87) gives way to the inequality (see Problem 2.21)

$$\mathbf{P}^2 = P_1^2 + P_2^2 + P_3^2 \le P_0^2 . \tag{2.3.90}$$

If we digress for a moment from the wave intensity and consider the dimensionless Stokes parameters $p_i = P_i / P_0$ (this transition substitutes the dimensionless polarization matrix $\hat{I}_0 / \mathrm{tr}\, \hat{I}_0$ for the radiance matrix $\hat{I}_0$), then these parameter will already describe purely polarization properties of the wave.

In contrast to the full Stokes parameters P_j, the polarization parameters p_i are nonadditive and do not reduce to a simple sum when mutually coherent beams overlap. They are bounded by the condition $|p_i| \le 1$; p_3 characterizes the *degree of circular polarization*, and p_1 and p_2 characterize the degree of linear polarization along the axes $\mathbf{e}_1$ and $\mathbf{e}_2$ at the angles $\pm\pi/4$ with $\mathbf{e}_3$, respectively.

In terms of these nondimensional parameters, the *degree of linear polarization* is

$$l = \sqrt{p_1^2 + p_2^2}, \tag{2.3.91}$$

and the *degree of polarization* is

$$p = \sqrt{p_1^2 + p_2^2 + p_3^2}. \tag{2.3.92}$$

These parameters never exceed unity; for a totally polarized wave $l = 1$, and for a totally linearly polarized wave, we have additionally $p = 1$. It is remarkable that the parameters l, p and p_3 do not change in rotations of the real basis $\{\mathbf{e}_1, \mathbf{e}_2\}$ (Problem 2.24).

These characteristics of Stokes parameters are traditionally outlined in optics textbooks; for more details the reader is referred to the books of Born and Wolf (1980), Shurcliff (1962), and Azzam and Bashara (1977). It is often more convenient to use a less popular approach (Apresyan, 1976) involving the Pauli matrix formalism. It treats the 2×2 Pauli matrices (2.3.81) as projections of 3×3 Hermitian matrices onto the plane $\mathbf{e}_1, \mathbf{e}_2$ (we retained for these matrices the previous notation)

$$\begin{aligned}
\hat{\pi}_0 &= \mathbf{e}_1 \otimes \mathbf{e}_1 + \mathbf{e}_2 \otimes \mathbf{e}_2 = \hat{1} - \mathbf{e}_3 \otimes \mathbf{e}_3, \\
\hat{\pi}_1 &= \mathbf{e}_2 \otimes \mathbf{e}_2 - \mathbf{e}_1 \otimes \mathbf{e}_1, \\
\hat{\pi}_2 &= \mathbf{e}_1 \otimes \mathbf{e}_2 + \mathbf{e}_2 \otimes \mathbf{e}_1 \\
\hat{\pi}_3 &= i(\mathbf{e}_1 \otimes \mathbf{e}_2 - \mathbf{e}_2 \otimes \mathbf{e}_1) = -i\mathbf{e}_3 \times .
\end{aligned} \tag{2.3.93}$$

Here $\hat{1}$ is the 3×3 unit matrix, and $\mathbf{e}_3 \times$ stands for the matrix of the vector product $(\mathbf{z}\times)\mathbf{a} = \mathbf{z} \times \mathbf{a}$. With these matrices, eqns (2.3.84) and (2.3.85) can be rewritten in the three-dimensional invariant form

$$\hat{I} = \frac{1}{2} \sum_{i=0}^{3} P_i \hat{\pi}_i, \quad P_i = \operatorname{tr} \hat{\pi}_i \hat{I}. \tag{2.3.94}$$

The advantages of using this representations will be evident in the transformations of Stokes parameters from one system of coordinates to another and in the evaluation of relations of the Stokes parameters of incident and scattered waves (Problems 2.22 – 2.24).

2.3.18 Transfer Equation for the Stokes Vector

From the transfer equation for the matrix $\hat{I}$ it is not hard to pass over to the transfer equation for parameters P_i (Chandrasekhar, 1960). A simple way to do this involves using the invariant transfer equation (2.3.33) and relations (2.3.94). As a result we have

$$d_s P_i + \sum_{j=0}^{3} \alpha_{ij} P_i = \sum_{j=0}^{3} \int \sigma_{ij}(\omega, \mathbf{n} \leftarrow \omega', \mathbf{n}') P_j(\omega', \mathbf{n}') d\omega' d\Omega_{\mathbf{n}'} . \tag{2.3.95}$$

We intend to analyze this system in more detail and consider separately the cases of nonscattering and scattering media.

2.3.19 Variation of Polarization in the Absence of Scattering. A Geometrical Interpretation

Let the medium be nonscattering so that $\hat{\alpha}^{sc} = 0$ and $\hat{\sigma} = 0$ and eqn (2.3.33) takes the form (2.3.27), viz.,

$$\hat{\rho}\dot{\hat{I}}\hat{\rho} = \hat{A}\hat{I} + \hat{I}\hat{A}^{+} - \alpha^{\text{inh}}\hat{I}. \tag{2.3.96}$$

We substitute in this expression $\hat{I}$ in terms of Stokes parameters from (2.3.94) multiply both sides by $\hat{\pi}_l$ and take the trace. We have

$$\frac{1}{2}\operatorname{tr}\hat{\pi}_l\sum_j\left(\hat{\pi}_j\dot{P}_j + \dot{\hat{\pi}}_j P_j\right) = \dot{P}_l + \frac{1}{2}\sum P_j \operatorname{tr}\hat{\pi}_l\dot{\hat{\pi}}_j$$
$$= \frac{1}{2}\sum_j P_j \operatorname{tr}\hat{\pi}_l\left[\left(\hat{A}\hat{\pi}_j + \hat{\pi}_j\hat{A}^{+}\right) - \alpha^{\text{inh}}\hat{\pi}_j\right]. \tag{2.3.97}$$

Here we observed the relations $\operatorname{tr}\hat{a}\hat{b} = \operatorname{tr}\hat{b}\hat{a}$, $\hat{\pi}_i\hat{\rho} = \hat{\rho}\hat{\pi}_i = \hat{\pi}_i$, and the orthogonality condition (2.3.83). We see that in this case in the system of equations for Stokes parameters (2.3.95) $\sigma_{lj} = 0$ and

$$\alpha_{lj} = \frac{1}{2}\operatorname{tr}\hat{\pi}_l\left(\dot{\hat{\pi}}_j - \hat{A}\hat{\pi}_j - \hat{\pi}_j\hat{A}^{+} + \alpha^{\text{inh}}\hat{\pi}_j\right). \tag{2.3.98}$$

These coefficients are calculated in a straightforward manner by virtue of eqn (2.3.93).

This system of equations can be represented in a more transparent form if we formally treat $\mathbf{P} = (P_1, P_2, P_3)$ as a three-dimensional vector (Apresyan, 1976). Then the system of equations for Stokes parameters becomes

$$\left(d_s + \alpha^{\text{inh}} - A_0^{\text{h}}\right)P_0 - \mathbf{A}^{\text{h}}\mathbf{P} = 0,$$
$$\left(d_s + \alpha^{\text{inh}} - A_0^{\text{h}}\right)\mathbf{P} - \mathbf{A}^{\text{h}}P_0 - \left(\mathbf{A}^{\text{a}} - 2\boldsymbol{\kappa}'\right)\times\mathbf{P} = 0. \tag{2.3.99}$$

Here we denoted $[A_0, \mathbf{A}] = [A_0, A_1, A_2, A_3]$ for the Stokes parameters of an arbitrary 3×3 matrix $\hat{A}$

$$A_i = \operatorname{tr}\hat{\pi}_i\hat{A}. \tag{2.3.100}$$

The Hermitian ($\hat{A}^{\text{h}}$) and anti-Hermitian ($i\hat{A}^{\text{a}}$) parts of $\hat{A}$ may be viewed as generalizations of the concepts of the real and imaginary parts of a complex number onto the case of matrices:

$$\hat{A} = \hat{A}^{\text{h}} + i\hat{A}^{\text{a}}, \quad \hat{A}^{\text{h}} = \frac{1}{2}\left[\hat{A} + \hat{A}^{+}\right], \quad i\hat{A}^{\text{a}} = \frac{1}{2}\left[\hat{A} - \hat{A}^{+}\right], \tag{2.3.101}$$

and $\boldsymbol{\kappa}' = [\kappa_1', \kappa_2', \kappa_3'] = [0, 0, \kappa']$, where $\kappa' = \mathbf{e}_1\dot{\mathbf{e}}_2$ defines the rotation velocity of the basis vectors in moving along the curvilinear ray. In the case of a rectilinear ray, it is natural to choose the basis $\mathbf{e}_1, \mathbf{e}_2$ unmovable, then $\kappa' = 0$. For a curvilinear ray, we can choose as these basis vectors the normal and binormal to the ray, then κ' will be equal to the torsion κ.

Let us examine the physical sense of the parameters in the system (2.3.99). The quantity α^{inh} (2.2.25), entering in this system, takes into account the variations of the cross section of the ray tube and does not cause rotation of the polarization vectors. This quantity can be eliminated by the normalization $\tilde{P}_i = P_i / \omega^3 n_\text{r}^2$. The parameters A_i^a are represented in terms of the anti-Hermitian part of the matrix $\hat{A}$, and correspond to a conserving medium, while the parameters A_i^h are related to the Hermitian part of $\hat{A}$ and describe absorption. In contrast to the case of the permittivity tensor $\hat{\varepsilon}$, here the matrix $\hat{A}^\text{h}$, rather than $\hat{A}^\text{a}$, is responsible for absorption, because $\hat{A} \propto i\hat{\varepsilon}$ [see eqn (2.3.24)]. In the absence of absorption, $A_i^\text{h} = 0$ and the system (2.3.99) can be written as two separable equations for the parameters $P_0'\tilde{P}_0$ and $\tilde{\mathbf{P}}$

$$d_s\tilde{P}_0 = 0, \tag{2.3.102}$$

$$d_s\tilde{\mathbf{P}} = (-2\boldsymbol{\kappa}' + \mathbf{A}^\text{a}) \times \tilde{\mathbf{P}} \equiv \boldsymbol{\Omega} \times \tilde{\mathbf{P}}. \tag{2.3.103}$$

The first line gives the law of conservation of the quantity $\tilde{P}_0 = P_0 / \omega^3 n_\text{r}^2 = I / \omega^3 n_\text{r}^2$ along the ray (derived earlier in scalar theory), and the second line describes the rotation of the Stokes vector in the space of Stokes parameters with a velocity $\boldsymbol{\Omega} = -2\boldsymbol{\kappa}' + \mathbf{A}^\text{a}$.

The use of this geometrical pattern provides an insight into the behavior of polarization Stokes parameters in various specific cases. For example in an inhomogeneous isotropic medium, when $\mathbf{A}^\text{a} = 0$, the angular velocity $\boldsymbol{\Omega} = -2\boldsymbol{\kappa}'$ owes its existence completely to the ray curvature and is directed along the P_3 axis (see Fig. 2.16). Therefore, when the observation point moves along the ray, the vector $\mathbf{P}$ rotates about the P_3 axis so that the circular polarization parameter P_3 does not change, while P_1 and P_2 oscillate. For a completely polarized wave, the conservation of P_3 implies, as can be seen from eqn (2.3.88), that the form of the polarization ellipse remains unchanged ($\beta = \text{const.}$), that is the polarization ellipse rotates as a whole.

In a homogeneous and stationary anisotropic medium, the curvature of rays is absent ($\boldsymbol{\kappa}' = 0$), and the angular velocity vector $\boldsymbol{\Omega}$ coincides with $\mathbf{A}^\text{a}$. Then the Stokes vector rotates around $\mathbf{A}^\text{a}$ so that the projection of $\mathbf{P}$ onto $\mathbf{A}^\text{a}$ is conserved, but, generally speaking, all three Stokes parameters oscillate.

From eqn (2.3.99) it follows that if the wave is completely polarized, i.e. $\mathbf{P}^2 = P_0^2$, then this relation conserves along the ray. Indeed, multiplying the first equation in (2.3.99) by P_0, the second scalar-wise by $\mathbf{P}$, and subtracting the first line from the second we obtain

$$d_s \frac{1}{2}\left(\mathbf{P}^2 - P_0^2\right) + \left(\alpha^{\text{inh}} - A_0^{\text{h}}\right)\left(\mathbf{P}^2 - P_0^2\right) = 0. \tag{2.3.104}$$

Whence it follows that the quantity $\mathbf{P}^2 - P_0^2$ varies along the ray as

$$c_0 \exp\left[-\int_0^s 2(\alpha^{\text{inh}} - A_0^{\text{h}})ds\right]$$

where c_0 is the initial value of $\mathbf{P}^2 - P_0^2$. For a completely polarized wave, $c_0 = 0$, hence, $\mathbf{P}^2 = P_0^2$ everywhere on the ray. Thus, a completely polarized wave, whose degree of polarization is $p = \sqrt{\mathbf{P}^2}/P_0 = 1$, remains completely polarized along the entire path (a specific calculation of A_i^{h} and $\mathbf{A}^{\text{a}}$ is described in Problem 2.25).

2.3.20 Transfer Equation for the Stokes Vector in a Scattering Medium

In the presence of scattering, the transfer equation for the Stokes parameters has the form of eqn (2.3.95) with nonzero right-hand side. We demonstrate how the coefficients of (2.3.95) can be computed, assuming for simplicity that in the initial transfer equation (2.3.33), $\hat{A} = 0$ and $\alpha^{\text{inh}} = 0$, i.e. disregarding the variations of polarization in the propagation between the acts of scattering. This assumption is valid in a homogeneous and isotropic effective medium.

With derivations of Section 2.3.18 it is not hard to write the coefficients (2.3.95) in terms of $\hat{\alpha}^{\text{sc}}$ and $\hat{\sigma}$ entering in eqn (2.3.33)

$$\alpha_{ij} = \frac{1}{2} \text{tr}\, \hat{\pi}_i \hat{\alpha}^{\text{sc}} \hat{\pi}_j, \quad \sigma_{ij} = \frac{1}{2} \text{tr}\, \hat{\pi}_i \hat{\sigma} \hat{\pi}_j', \tag{2.3.105}$$

or in more detail

$$\alpha_{ij} = \frac{1}{2} \pi_{kl}^{(i)} \alpha_{lk,mn} \pi_{mn}^{(j)}, \quad \sigma_{ij} = \frac{1}{2} \pi_{kl}^{(i)} \sigma_{lk,mn} \pi_{mn}'^{(j)}, \tag{2.3.106}$$

where $\pi_{mn}^{(i)} = (\hat{\pi}_i)_{mn}$ are the components of Pauli's matrices, where the matrices $\hat{\pi}_i'$ relate to the basis $(\mathbf{e}_1', \mathbf{e}_2', \mathbf{e}_3' = \mathbf{n}')$.

To illustrate the application of these relations we consider a detailed isotropic situation (2.3.49) – (2.3.54). Substituting (2.3.49) into (2.3.105) we write

$$\alpha_{ij} = \frac{1}{2} \text{tr}\left[a\hat{\pi}_i \hat{\pi}_j + b\hat{\pi}_i\left(\text{tr}\hat{\pi}_j\right) + c\hat{\pi}_i \hat{\pi}_j^{\text{T}}\right],$$

where $\hat{\pi}_j^{\text{T}} = \hat{\pi}_j$, for $j = 0, 1, 2$, and $\hat{\pi}_3^{\text{T}} = -\hat{\pi}_3$ [see eqn (2.3.81)]. Recognizing also that $\text{tr}\, \hat{\pi}_i \hat{\pi}_j = 2\delta_{ij}$ and $\text{tr}\, \hat{\pi}_j = 2\delta_{j0}$ we obtain

$$\alpha_{ij} = (a + 2b\delta_{i0} + c)\delta_{ij} - 2c\delta_{ij}\delta_{i3}\,. \tag{2.3.107}$$

Similarly, substituting (2.3.54) into (2.3.105) we may write

$$\sigma_{ij} = \frac{1}{2}\operatorname{tr}\left[\sigma_1\hat{\pi}_i\hat{\pi}'_j + \sigma_2\pi_i \operatorname{tr}\left(\hat{\pi}_0\hat{\pi}'_j\right) + \sigma_3\hat{\pi}_i\hat{\pi}'^{\mathrm{T}}_j\right]. \tag{2.3.108}$$

Further calculations are also straightforward and eqn (2.3.108) yields an expression for σ_{ij} in a form invariant under the choice of bases $(\mathbf{e}_1, \mathbf{e}_2, \mathbf{n})$ and $(\mathbf{e}'_1, \mathbf{e}'_2, \mathbf{n}')$. By way of example we calculate σ_{00}

$$\begin{aligned}\sigma_{00} &= \frac{1}{2}\operatorname{tr}\left[\sigma_1\hat{\pi}_0\hat{\pi}'_0 + \sigma_2\pi_0 \operatorname{tr}\left(\hat{\pi}_0\hat{\pi}'_0\right) + \sigma_3\hat{\pi}_0\hat{\pi}'_0\right] = \frac{1}{2}(\sigma_1 + 2\sigma_2 + \sigma_3)\operatorname{tr}\left(\hat{\pi}_0\hat{\pi}'_0\right) \\ &= \frac{1}{2}(\sigma_1 + 2\sigma_2 + \sigma_3)\operatorname{tr}\left(\hat{1} - \mathbf{n}\otimes\mathbf{n}\right)\left(\hat{1} - \mathbf{n}'\otimes\mathbf{n}'\right) \\ &= \frac{1}{2}(\sigma_1 + 2\sigma_2 + \sigma_3)\left[1 + (\mathbf{n}\cdot\mathbf{n}')^2\right].\end{aligned}$$

Other coefficients σ_{ij} are calculated in a similar way (see Problem 2.26).

2.3.21 Variation of Polarization in Reflection and Refraction at a Plane Interface

Transmission of a polarized beam of electromagnetic radiation through a plane interface between two homogeneous media is a problem that can be easily solved by analogy with the scalar solution for plane waves (Born and Wolf, 1980). We write this solution without derivation, using the invariant matrix notation.

For the case of isotropic media in the problem geometry shown in Fig. 2.17 we have

$$\begin{aligned}\mathbf{E}^{\mathrm{r}} &= (\mathbf{e}'_1\otimes\mathbf{e}_1\mathcal{R}_{\parallel} + \mathbf{e}'_2\otimes\mathbf{e}_2\mathcal{R}_{\perp})\mathbf{E}^0 \equiv \hat{\mathcal{R}}\,\mathbf{E}^0\,, \\ \mathbf{E}^{\mathrm{t}} &= (\mathbf{e}''_1\otimes\mathbf{e}_1\mathcal{T}_{\parallel} + \mathbf{e}''_2\otimes\mathbf{e}_2\mathcal{T}_{\perp})\mathbf{E}^0 \equiv \hat{\mathcal{T}}\,\mathbf{E}^0\,.\end{aligned} \tag{2.3.109}$$

Here $\mathbf{E}^0$, $\mathbf{E}^{\mathrm{r}}$, and $\mathbf{E}^{\mathrm{t}}$ are the vector amplitudes of the incident, reflected and refracted (transmitted) waves, respectively; $(\mathbf{e}_1, \mathbf{e}_2, \mathbf{n})$, $(\mathbf{e}'_1, \mathbf{e}'_2, \mathbf{n}')$ and $(\mathbf{e}''_1, \mathbf{e}''_2, \mathbf{n}'')$ are orthonormal real-valued bases in which the incident, reflected and refracted waves are decomposed respectively. The vectors $\mathbf{e}_1, \mathbf{e}'_1$, and $\mathbf{e}''_1$ lie in the plane of incidence, whereas the vectors $\mathbf{e}_2 = \mathbf{e}'_2 = \mathbf{e}''_2$ are orthogonal to this plane; $\mathcal{R}_{\parallel}$ and $\mathcal{T}_{\parallel}$ are the Fresnel coefficients of reflection and transmission for the incident wave polarized in the plane of incidence (so-called *p*-polarization)

$$\mathcal{R}_{\parallel} = \frac{n_1\cos\theta_2 - n_2\cos\theta_1}{n_1\cos\theta_2 + n_2\cos\theta_1}, \quad \mathcal{T}_{\parallel} = \frac{2n_1\cos\theta_1}{n_1\cos\theta_2 + n_2\cos\theta_1}, \tag{2.3.110}$$

and $\mathcal{R}_\perp$ and $\mathcal{T}_\perp$ are the analogous quantities for the wave with orthogonal polarization (s-polarization):

$$\mathcal{R}_\perp = \frac{n_1 \cos\theta_1 - n_2 \cos\theta_2}{n_1 \cos\theta_1 + n_2 \cos\theta_2}, \quad \mathcal{T}_\perp = \frac{2n_1 \cos\theta_1}{n_1 \cos\theta_1 + n_2 \cos\theta_2}. \tag{2.3.111}$$

Expressions (2.3.109) take into account that, in reflection, the component of $\mathbf{E}^0$ along $\mathbf{e}_1$ is transformed into the component along $\mathbf{e}_1'$ with the reflection coefficient $\mathcal{R}_\parallel$, while the field component along $\mathbf{e}_2$ is transformed into one along $\mathbf{e}_2'$ with the reflection coefficient $\mathcal{R}_\perp$. The transmitted wave occurs in a similar manner.

In agreement with (2.3.9) the radiance matrix $\hat{I}$ is proportional to the tensor product of the vector amplitudes of quasi-plane waves. In full analogy with the scalar case (2.2.19) we may express the radiance matrix $\hat{I}^r$ of the reflected beam through the radiance matrix $\hat{I}^0$ of the incident beam

$$\hat{I}^r \sim \langle \mathbf{E}^r \otimes \mathbf{E}^{r*} \rangle = \left\langle \left(\hat{\mathcal{R}}\mathbf{E}^0\right) \otimes \left(\hat{\mathcal{R}}\mathbf{E}^0\right)^* \right\rangle = \hat{\mathcal{R}} \langle \mathbf{E}^0 \otimes \mathbf{E}^{0*} \rangle \hat{\mathcal{R}}^+ \sim \hat{\mathcal{R}} \hat{I}^0 \hat{\mathcal{R}}^+, \tag{2.3.112}$$

where the angular brackets imply averaging over the polarizations of the incident beam. We set the proportionality factor in eqn (2.3.112) equal to unity so that in the total reflection with $\mathcal{R}_\perp = \mathcal{R}_\parallel = 1$ we obtain $\hat{I}^r = \hat{I}^0$. Then

$$\hat{I}^r = \hat{\mathcal{R}} \hat{I}^0 \hat{\mathcal{R}}^+. \tag{2.3.113}$$

Recognizing the explicit form of $\hat{\mathcal{R}}$ in eqn (2.3.109), we can represent this expression in the two-dimensional description as

$$\begin{bmatrix} I^r_{11} & I^r_{12} \\ I^r_{21} & I^r_{22} \end{bmatrix} = \begin{bmatrix} \mathcal{R}_\parallel & 0 \\ 0 & \mathcal{R}_\perp \end{bmatrix} \begin{bmatrix} I^0_{11} & I^0_{12} \\ I^0_{21} & I^0_{22} \end{bmatrix} \begin{bmatrix} \mathcal{R}^*_\parallel & 0 \\ 0 & \mathcal{R}^*_\perp \end{bmatrix}$$
$$= \begin{bmatrix} |\mathcal{R}_\parallel|^2 I^0_{11} & \mathcal{R}_\parallel \mathcal{R}^*_\perp I^0_{12} \\ \mathcal{R}_\perp \mathcal{R}^*_\parallel I^0_{21} & |\mathcal{R}_\perp|^2 I^0_{22} \end{bmatrix}, \tag{2.3.114}$$

where the components $I^0_{\alpha\beta}$ refer to the basis $(\mathbf{e}_1, \mathbf{e}_2)$, and the components $I^r_{\alpha\beta}$ refer to the basis $(\mathbf{e}_1', \mathbf{e}_2')$, as shown in Fig. 2.17.

From eqn (2.3.114) it follows a simple relationship of the determinants of the radiance matrices of the incident and reflected beams: $\det I^r = |\mathcal{R}_\parallel \mathcal{R}_\perp|^2 \det \hat{I}^0$. We see that the reflected wave emerges completely polarized ($\det \hat{I}^r = 0$) if the incident wave is completely polarized ($\det \hat{I}^0 = 0$) or if one of the reflection coefficients, $\mathcal{R}_\parallel$ or $\mathcal{R}_\perp$, vanishes.

In the case of a boundary between two dielectric media, only $\mathcal{R}_\parallel$ can vanish when the wave is incident at the Brewster angle $\theta_B = \arctan(n_2 / n_1)$. Alternatively, an arbitrarily polarized wave incident at this angle upon reflection becomes completely polarized along the $\mathbf{e}_2'$ axis. This phenomenon is fundamental to Brewster polarizers.

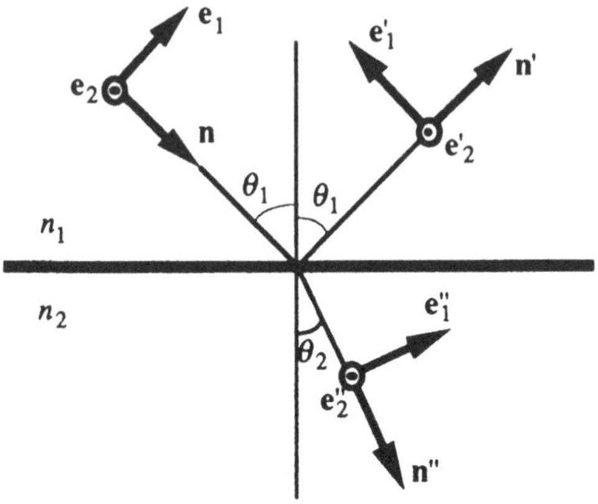

Figure 2.17. Refraction and reflection of an electromagnetic wave at a plane interface.

Similar to eqn (2.3.113), the radiance matrix of the refracted wave has the form

$$\hat{I}^{\mathrm{t}} \propto \overline{\mathbf{E}^{\mathrm{t}} \otimes \mathbf{E}^{\mathrm{t}*}} \propto \hat{\mathcal{T}}\hat{I}^{0}\hat{\mathcal{T}}^{+} \tag{2.3.115}$$

or

$$I^{\mathrm{t}} = c_{\mathrm{t}}\hat{\mathcal{T}}\hat{I}^{0}\hat{\mathcal{T}}^{+}. \tag{2.3.116}$$

The proportionality factor c_{t} may be found from the condition (2.2.21) derived in scalar theory and based upon the polarization state of the wave. Recognizing that the radiance $I = \mathrm{tr}\hat{I}$, from eqn (2.2.21) we obtain

$$\mathrm{tr}\hat{I}^{\mathrm{t}} = c_{\mathrm{t}}\,\mathrm{tr}\,\hat{\mathcal{T}}\hat{I}^{0}\hat{\mathcal{T}}^{+} = (n_2/n_1)^2\left(\mathrm{tr}\,\hat{I}^{0} - \mathrm{tr}\,\hat{I}^{\mathrm{r}}\right).$$

A straightforward evaluation of the traces of matrices from this expression follows in the two-dimensional representation (2.3.114). Accordingly, $\mathrm{tr}\,\hat{I}^{\mathrm{r}} = |\mathcal{R}_{\parallel}|^2 I_{11}^{0} + |\mathcal{R}_{\perp}|^2 I_{22}^{0}$ and from the similar two-dimensional representation for $\hat{I}^{\mathrm{t}}$

$$\hat{I}^{\mathrm{t}} = c_{\mathrm{t}}\begin{bmatrix} |\mathcal{T}_{\parallel}|^2 I_{11}^{0} & \mathcal{T}_{\parallel}\mathcal{T}_{\perp}^{*} I_{12}^{0} \\ \mathcal{T}_{\perp}\mathcal{T}_{\parallel}^{*} I_{21}^{0} & |\mathcal{T}_{\perp}|^2 I_{22}^{0} \end{bmatrix} \tag{2.3.117}$$

it follows that $\mathrm{tr}\hat{I}^{\mathrm{t}} = c_{\mathrm{t}}\left(|\mathcal{T}_{\parallel}|^2 I_{11}^{0} + |\mathcal{T}_{\perp}|^2 I_{22}^{0}\right)$. Finally,

$$c_{\mathrm{t}} = \left(\frac{n_2}{n_1}\right)^2 \frac{\left(1-|\mathcal{R}_{\parallel}|^2\right)I_{11}^{0} + \left(1-|\mathcal{R}_{\perp}|^2\right)I_{22}^{0}}{|\mathcal{T}_{\parallel}|^2 I_{11}^{0} + |\mathcal{T}_{\perp}|^2 I_{22}^{0}}. \tag{2.3.118}$$

In eqn (2.3.117), the components $I^0_{\alpha\beta}$ refer to the basis $(\mathbf{e}_1, \mathbf{e}_2)$ and the components $I^t_{\alpha\beta}$ refer to the basis $(\mathbf{e}_1'', \mathbf{e}_2'')$. If the incident wave is not polarized, $I^0_{\alpha\beta} = \delta_{\alpha\beta} I^0/2$, then according to (2.3.114) and (2.3.117) the reflected and refracted waves will be partially polarized, and their radiances will be

$$I^r = \operatorname{tr} \hat{I}^r = I^0 \frac{|\mathcal{R}_\parallel|^2 + |\mathcal{R}_\perp|^2}{2}, \quad I^t = \left(\frac{n_2}{n_1}\right)^2 \left(1 - \frac{|\mathcal{R}_\parallel|^2 + |\mathcal{R}_\perp|^2}{2}\right). \tag{2.3.119}$$

Using eqns (2.3.114) and (2.3.115) it is not difficult to express the Stokes parameters of the reflected and refracted beams in terms of the Stokes parameters of the incident beam, thus deriving a more transparent description of the polarization in reflection and refraction (Problem 2.27).

2.3.22 Multimode System of Transfer Equations in a Scattering Medium

Thus far we have assumed that a medium can propagate one type of a wave. We now may consider a more general case of a multimodal medium which allows propagation of several types of waves. We will label such waves by subscripts ν. Each wave will be characterized by its own dispersion law $\omega = \omega_\nu(\mathbf{k})$ and its own unit polarization vector $\mathbf{e}_\nu(\mathbf{k})$. The case of polarization degeneracy may be treated as coincident dispersion laws for different types of waves ν. Because transfer theory considers time-averaged energy characteristics of the field, in the framework of this theory the polarization nondegenerate waves need be deemed completely incoherent. This follows already from the description of wave propagation in free space. Indeed, by virtue of the assumption (ii) of Section 2.1.5, waves with different wave vectors $\mathbf{k}$ deemed uncorrelated. Therefore, in multiplying and averaging the results of geometrical optics, we find out that the interference summands become proportional to factors like $\exp\left\{i\left[\omega_\nu(\mathbf{k}) - \omega_\mu(\mathbf{k})\right]t\right\}$ with $\nu \neq \mu$ which vanish upon time averaging. As a result, the radiance matrix (2.3.13) is expressed as a sum of matrices corresponding to each type of propagating waves:

$$\hat{I} = \sum_\nu \mathbf{e}_\nu \otimes \mathbf{e}_\nu^* I_\nu \equiv \sum_\nu \hat{I}_\nu ,$$

where I_ν is the radiance of wave of the type ν.

In the case of polarization degenerate waves of two types, say 1 and 2, the frequencies of these waves are identical, the interference term is equal to unity $\left[\exp i(\omega_1 - \omega_2)t = 1\right]$ and does not change upon time averaging, therefore, we have to observe the correlation of waves 1 and 2 in transfer theory. In place of radiances I_1 and I_2 one must consider the radiance matrix

$$\hat{I} = \sum_{\alpha,\beta=1,2} e_\alpha e_\beta^* I_{\alpha\beta} ,$$

as we did this above for the case of transverse electromagnetic waves.

A phenomenological description of scattering in a multimode medium reduces to the construction of a system of transfer equations that can be deduced heuristically in complete analogy with the above specific cases. It is important to note that, in a multimode medium, the system of transfer equations must take into account not only the scattering but also the mutual transformation of different types of waves. Therefore the scattering cross section will include the summand $\sigma_{\nu\leftarrow\nu}$ corresponding to the conservation of the wave type, and the cross sections of transformation $\sigma_{\nu\leftarrow\mu}$, $\nu \neq \mu$. In addition, even in the conservative case, the energy of different types of waves need not be conserved, because only the total energy of all wave types is conserved. Taking all these conditions into account, it is not hard to repeat the above arguments and write the system of transfer equations that is analogous to the transfer equation (2.3.95) for Stokes parameters and has the form

$$(d_s + \alpha_\nu) I_\nu = \sum_\mu \hat{\sigma}_{\nu\leftarrow\mu} I_\mu \,. \tag{2.3.120}$$

Here I_ν are the radiances or (in the presence of polarization degeneracy) radiance matrices of a wave of type ν, α_ν are the attenuation coefficients, and $\hat{\sigma}_{\nu\leftarrow\mu}$ are the operators which at $\nu = \mu$ describe scattering with the conservation of the type of the wave, and for $\nu \neq \mu$, the transformation of waves in scattering. In the general case, for polarization degenerate waves, both α_ν and the kernels of the operators $\hat{\sigma}_{\nu\leftarrow\mu}$ are some matrices. An example of the system (2.3.120) for the case of longitudinal and transverse waves in plasma has been examined by Apresyan (1973).

PROBLEMS

Problem 2.1. Lambert has arrived at his famous law basing on astronomical observations. According to this law the *radiance* of the solar disc is identical over all its points. This statement contradicted Euler's opinion who believed that each element on the sun's surface provides an identical *radiant intensity*. Actually both laws are approximate in nature. Lambert's law better fits situations with strong absorption, while Euler's law corresponds to weak absorption of light in the solar atmosphere. Both laws are combined in the model law of Blondel (1895):

$$I(\mathbf{n},\mathbf{r}) = I_0 \cos^{m-1}\theta \,, \quad m \geq 0 \,. \tag{1}$$

For $m = 1$, this relationship corresponds to the isotropic radiance (Lambert's law), and for $m = 0$, to the isotropic radiant intensity (Euler's law). Demonstrate that given the radiance distribution (1), the radiant emittance is $I_0 2\pi / (m+1)$.

Solution. The radiant emittance is given by the integral (2.1.8) which is straightforward to take in this case.

Problem 2.2 (Liebenthal) (Fig. 2.18) The maximum irradiance provided by simultaneously operating point sources 1 and 2 at the observation point 3 are S_1 and S_2, respectively. How should the surface Σ be oriented at the observation point to obtain the maximum intensity?

Solution. The fluxes of energy $\mathbf{S}_1$ and $\mathbf{S}_2$ from sources 1 and 2 at point 3 are equal in magnitude to S_1 and S_2 and oriented as shown in Fig. 2.18. Obviously to solve the problem the surface Σ must be oriented at right angles with the vector sum of $\mathbf{S}_1$ and $\mathbf{S}_2$.

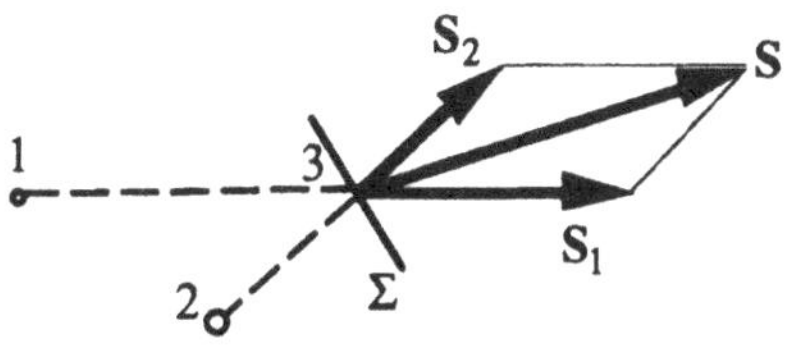

Figure 2.18

Problem 2.3. For a homogeneous isotropic monochromatic radiation, express the energy density in terms of irradiance.

Solution. The irradiance of the area inserted in an isotropic radiation field is

$$E = I_0 \int_0^{\pi/2} d\theta \int_0^{2\pi} d\varphi \sin\theta \cos\theta = \pi I_0, \tag{1}$$

and the energy density (2.1.10) is $w = 4\pi I_0 / c$, whence $w = 4E/c$. These relationships are valid for spectral components of nonmonochromatic thermal radiation. To prove it is sufficient to substitute $2I_\omega^0$ for I_0, where the factor 2 reflects the contributions of two independent polarizations. In the general case of anisotropic radiation, w and E are, in fact, independent quantities.

Problem 2.4. Determine the radiant intensity for a sphere with isotropically radiating surface of radiance I_0.

Solution. For a body with isotropically radiating surface Σ, the radiant intensity $\mathcal{J}$ observed in the direction of the unit vector $\hat{\mathbf{z}}$ is

$$\mathcal{J} = I_0 \int \cos\theta d\Sigma_r = I_0 \int \hat{\mathbf{z}} d\boldsymbol{\Sigma}_r = I_0 \int d\Sigma_z = I_0 \Sigma_z \tag{1}$$

where Σ_z is the projection area of the radiating surface Σ on the plane with normal $\hat{\mathbf{z}}$. Whence, for the radiant sphere we have $\mathcal{J} = I_0 \pi a^2$, where a is the radius of the sphere.

Problem 2.5. For the radiance of an equilibrium thermal radiation I^0, the extreme cases of low, $\hbar\omega > \kappa T$, and high, $\hbar\omega < \kappa T$, temperatures are described by the Rayleigh–Jeans and Wien equations, respectively. Which of these cases corresponds to practical observations of thermal sources?

Solution. The optical range of wavelengths $\lambda \sim 5\times10^{-7}$m corresponds to frequencies $\omega = 2\pi c/\lambda \sim 4\times10^{15}$. Using eqn (2.1.31) and changing from the frequency ω to the wavelength $\lambda = 2\pi c/\omega$, we estimate the position of the bound separating the Rayleigh–Jeans and Wien ranges by considering the position of the maximum of thermal radiation. This estimate leads to the condition $\hbar\omega/\kappa T \sim 5$ ($\hbar = 1.05\times10^{-34}\,\mathrm{J\,s^{-1}}$, $\kappa = 1.38\times10^{-23}\,\mathrm{J\,K^{-1}}$) so that the characteristic temperature $T^* \sim \hbar\omega/5\kappa \sim 6000$ K. Temperatures of common sources lie far below this value, and thus correspond to the Rayleigh–Jeans range. An exclusion is the Sun surface with temperature of 6000 K and for which the visible range lies just near the maximum of thermal radiation where none of the said asymptotic is applicable.

Problem 2.6. Express the full (over all polarizations and frequencies) radiant emittance of a uniformly heated body.

Solution. The spectral radiant emittance at a frequency ω is given by the integral from Problem 2.3, where $2I_\omega^0$ [see eqn (2.1.31)] must be substituted for I_0 to reflect the contributions of two independent polarizations. Taking the integral of the spectral radiant emittance over the frequency range we find

$$\int_0^\infty 2I_\omega^0 d\omega = \sigma T^4\,, \tag{1}$$

where

$$\sigma = \frac{\kappa^4}{(2\pi c)^2\hbar^3}\int_0^\infty \frac{x^3dx}{e^x-1} \tag{2}$$

is Stefan's constant. The last integral equals $\pi^4/15$, thus yielding $\sigma = \pi^2\kappa^4/60\hbar^3c^2$ $= 5.67\times10^{-5}\,\mathrm{g s^{-3} K^{-4}}$.

Problem 2.7. In practical measurements of radiative source characteristics the concept of pseudotemperatures often comes handy. These temperatures are defined by equating some temperature characteristics to the similar characteristics of a blackbody having a given pseudotemperature. The *brightness temperature* T_s of a source of radiance I is determined from the condition $I = I^0(T_s)$, and the *radiation temperature* T_r is determined from the total radiance (integrated over all frequencies, it equals σT_r^4). T_r can be measured by the thermal effect of the radiation, and T_s is determined by comparison with the radiance of a standard emitter. These principles are used in radiation and visual optical pyrometers. The temperature of radiation T_s is commonly defined in the range covered by the Rayleigh–Jeans equation $I_0(T) = (\omega / c)^2 \kappa T / (2\pi)^3$.

Demonstrate that for thermal sources satisfying the Kirchhoff law $I = \alpha I^0$, where α is the absorptance, both T_s and T_r do not exceed the true temperature of the source T.

Solution. The conditions $T_s < T$ and $T_r < T$ are verified directly and follow from the inequality $\alpha < 1$, which reflects the obvious requirement that the power absorbed by the body does not exceed that of the incident radiation.

Problem 2.8. The largest irradiance E_s which the solar power creates on a surface placed beyond the earth's atmosphere at an average distance from the sun to the earth ($R_s \sim 1.5 \times 10^8$ km) is known as the solar constant equal to about 1.4×10^3 W m^{-2}.

Estimate the average radiance and radiant intensity of the solar disc assuming that it is seen from the Earth at the solid angle of 6.8×10^{-5} sr (more accurate data may be found in the handbook of Gutnik, 1965).

Solution. The irradiance E_s is due to all rays incident within the angle $\delta\Omega$ (see Fig. 2.5) and is equal to $I\delta\Omega$, whence $I = E_s / \delta\Omega \approx 2 \times 10^7$ W m^{-2} sr^{-1}. The radiant intensity (see Problem 2.4) is $\mathcal{J} = I\delta\Omega R_s^2 = E_s R_s^2 \approx 3.2 \times 10^{25}$ W.

Problem 2.9. Estimate the average temperature of an Earth's satellite heated by the Sun.

Solution. If we believe that the satellite is a sphere of radius a, then the solar flux incident on its surface is $S_r = \pi a^2 E_s$, where E_s is the solar constant. The flux absorbed by the sphere is $S_\alpha = \alpha S_r$, where α is the average absorption coefficient of the sphere. In accordance with the Stefan and Kirchhoff laws the flux emitted by the sphere is estimated as $S_i \sim \alpha\sigma T^4 4\pi a^2$. From $S_i = S_a$ we find $T \sim (E_s / 4\sigma)^{1/4}$. Letting $E_s \sim 1.4 \times 10^3$ W m^{-1} and $\sigma \sim 5.7 \times 10^{-8}$ W m^{-2} K^{-4} we have $T \sim 280$ K irrespective of the value of the absorption coefficient.

Problem 2.10. Write out the radiance of an equilibrium thermal radiation in a monoaxial crystal (Rytov, 1953).

Solution. For a monoaxial crystal, two out of three eigenvalues of the permittivity tensor $\hat{\varepsilon}$ coincide, so that in the principal axes we have

$$\hat{\varepsilon} = \begin{bmatrix} \varepsilon_\perp & 0 & 0 \\ 0 & \varepsilon_\perp & 0 \\ 0 & 0 & \varepsilon_\parallel \end{bmatrix}, \tag{1}$$

where $\varepsilon_\perp$ and $\varepsilon_\parallel$ are the eigenvalues of $\hat{\varepsilon}$, the subscripts referring to the components perpendicular and parallel to the principal axis. The dispersion equation, defining the relation $\omega = \omega(k)$ (see Appendix B), factors into the product

$$\det\left|k_i k_j - k^2\delta_{ij} + k_0^2\varepsilon_{ij}\right| = \left(k^2 - k_0^2\varepsilon_\perp\right)\left(\varepsilon_\parallel k_\parallel^2 + \varepsilon_\perp k_\perp^2 - k_0^2\varepsilon_\parallel\varepsilon_\perp\right)k_0^2 = 0, \tag{2}$$

where the first factor corresponds to the ordinary (o) and the second to the extraordinary (e) wave. This equation describes two Fresnel ellipsoids in the space of wave vectors $\mathbf{k} = \left[k_x, k_y, k_z\right] = \left[\mathbf{k}_\perp, k_\parallel\right]$. The ellipsoid corresponding to the o-wave degenerates into a circle $k^2 = k_0^2\varepsilon_\perp$. With respect to an ordinary wave the medium behaves as isotropic with dielectric permittivity $\varepsilon = \varepsilon_\perp$, so that for this wave the ray refractive index coincides with the ordinary refractive index, $n_r^o = \sqrt{\varepsilon_\perp}$, and the respective contribution in the radiance is given by eqn (2.1.32).

It is convenient to write the equation for the Fresnel ellipsoid of an e-wave in the form

$$\omega^2 = c^2\left(\frac{k_\parallel^2}{\varepsilon_\perp} + \frac{k_\perp^2}{\varepsilon_\parallel}\right). \tag{3}$$

Differentiating this expression with respect to $\mathbf{k}$ we find for the group velocity

$$\mathbf{v}_g = \frac{d\omega}{d\mathbf{k}} = \frac{c}{k_0}\left(\frac{\mathbf{k}_\parallel}{\varepsilon_\perp} + \frac{\mathbf{k}_\perp}{\varepsilon_\parallel}\right), \tag{4}$$

then

$$v_g = \frac{c}{k_0}\sqrt{\left(\frac{k_\parallel}{\varepsilon_\perp}\right)^2 + \left(\frac{k_\perp}{\varepsilon_\parallel}\right)^2}\,. \tag{5}$$

To find the ray refractive index for an e-wave, n_r^e, it is convenient to use a spherical system of coordinates and specify the vector $\mathbf{k}$ by its magnitude k and the angles θ, φ, and the vector $\mathbf{v}_g$ by its modulus v_g and the angles θ_v, φ_v. From the symmetry considerations it should be clear that $\varphi_v = \varphi$. Then

$$\left|\frac{d^3k}{d\omega d\Omega_\mathbf{n}}\right| = \left|\frac{k^2 dk \sin\theta d\theta d\varphi}{d\omega \sin\theta_v d\theta_v d\varphi_v}\right| = \left|k^2 \frac{\partial k}{\partial\omega}\frac{d\cos\theta}{d\cos\theta_v}\right|. \tag{6}$$

In addition, from eqn (3) it follows that $\partial\kappa/\partial\omega = k/\omega$, and from eqn (4) it follows that $v_{g,\parallel} = v_g\cos\theta_v = ck\cos\theta/\left(k_0\varepsilon_\perp\right)$, whence it is not hard to see that

$$\frac{d\cos\theta_v}{d\cos\theta} = \frac{\varepsilon_\parallel\varepsilon_\perp^2}{\left(\varepsilon_\perp^2\sin^2\theta + \varepsilon_\parallel^2\cos^2\theta\right)^{3/2}}. \tag{7}$$

Using these relations we obtain

$$n_r^e = \frac{1}{\sqrt{\varepsilon_\perp}}\left(\frac{\varepsilon_\perp^2 \sin^2\theta + \varepsilon_\parallel^2 \cos^2\theta}{\varepsilon_\perp \sin^2\theta + \varepsilon_\parallel \cos^2\theta}\right). \tag{8}$$

Finally, for the radiance of the equilibrium thermal radiation in a monoaxial crystal we have the following sum of contributions due to the ordinary and extraordinary waves

$$I_\omega = I_\omega^0\left[\varepsilon_\perp + \left(n_r^e\right)^2\right], \tag{9}$$

where n_r^e is from eqn (8) and I_ω^0 is given by eqn (2.1.31).

Problem 2.11. Estimate the variation in the radiance of an acoustical beam normally incident on the air-water interface.

Solution. The velocity of sound in air of density $\rho_1 = 1.3\times 10^{-3}\ \mathrm{g/cm^3}$ is $c_1 = 330\ \mathrm{m/s}$, and in water of density $\rho_2 = 1\ \mathrm{g/cm^3}$ is $c_2 = 1500\ \mathrm{m/s}$. Keeping in mind the remark to eqn (2.2.18) we express the reflectance R of the acoustical wave and substitute it in eqn (2.2.22). Recognizing that n_2/n_1 is equal to c_1/c_2,.we obtain from this equation the radiance in water

$$\frac{I_t}{I^0} = \left(\frac{c_1}{c_2}\right)^2\left(1 - \left|\frac{\rho_2 c_2 - \rho_1 c_1}{\rho_2 c_2 + \rho_1 c_1}\right|^2\right). \tag{1}$$

We see that, for the air–water interface, the radiance is attenuated by 1.8×10^4 times.

Problem 2.12. Estimate the variation of the brightness temperature for a beam crossing a layer of heated absorbing medium.

Solution. Neglecting scattering in the transfer equation (2.2.74), i.e., letting $w_0 = 0$ and replacing S with I_ω^0 to reflect the thermal radiation of the medium we have

$$(d_\tau + 1)\frac{I}{n_r^2} = I_\omega^0. \tag{1}$$

For the geometry shown in Fig. 2.19, the solution of this equation has the form

$$\left(\frac{I}{n_r^2}\right)_\tau = e^{-\tau}\left(\frac{I}{n_r^2}\right)_0 + \int_0^\tau e^{-\tau'} I_\omega^0(\tau - \tau')\,d\tau'. \tag{2}$$

Here $\tau = \alpha l$ is the optical path length, the inlet quantities are labeled 0 and the outlet quantities are labeled τ. The integration is over the beam path. The brightness temperature computed for the Rayleigh–Jeans region differs from the radiance I by only a constant multiplier $T = (2\pi)^3 I / k_0^2 \kappa$ (see Problem 2.7). Therefore in place of the radiance I and the source function for the heated layer I_ω^0 we can consider directly the temperatures of radiation T and of the layer T_s, respectively. For the simplest case of uniformly heated layer with a constant refractive index $n_{\mathbf{r}}$, from eqn (2) we have

$$T = e^{-\tau} T_0 + n_{\mathbf{r}}^2 T_s (1 - e^{-\tau}). \tag{3}$$

The first term in this expression describes the attenuation of the observed temperature due to absorption in the layer, and the second gives the contribution of the thermal radiation of the layer. For $\tau \to 0$ we have $T = T_0$, whereas for $\tau \to \infty$ the brightness temperature is determined by T_s.

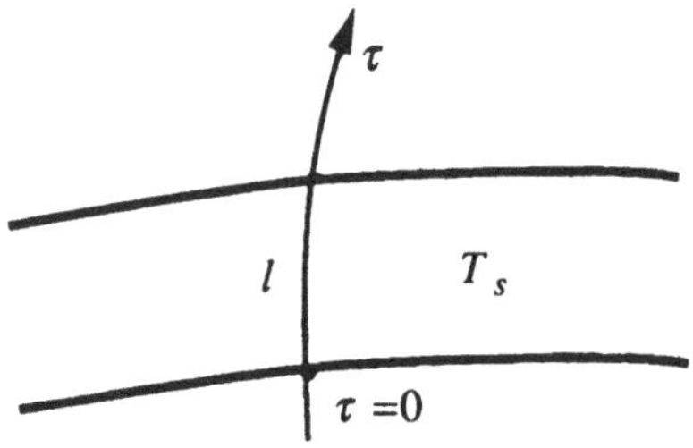

Figure 2.19

Problem 2.13. Prove the optical theorem (2.2.52) for a single scatterer using the expression for the energy flux density $S = (2ik_0)^{-1}\left(u^* \nabla u - u \nabla u^*\right)$ where u is the field amplitude satisfying the wave equation $\left(\Delta + k_0^2 \varepsilon(r)\right) u = 0$, $\varepsilon(\mathbf{r}) \to 1$ as $r \to \infty$.

Solution. To prove eqn (2.2.52) it suffices to write the definition of the absorption cross section (2.2.49) in the form

$$\sigma_a = -\lim_{R\to\infty} S_0^{-1} \oint S d\Sigma = -\lim_{R\to\infty} S_0^{-1} R^2 \oint \mathbf{S n} d\Omega_{\mathbf{n}}\,, \tag{1}$$

where the integration is over the sphere of infinite radius embracing the scatterer. In view of eqns (2.2.52), (2.2.48) and (2.2.51) it is not hard to compute the right-hand integral (1) to get eqn (2.2.52) (Ishimaru, 1978; Bohren and Huffman, 1983).

Problem 2.14. Compute the angular velocity of the polarization vector for an elliptically polarized wave.

Solution. The polarization ellipse in the main axes is described by eqn (2.3.6). Substituting these relations in the expression for the angle $\gamma = \arctan\left(E_b^{(\mathrm{r})} / E_a^{(\mathrm{r})}\right)$ between the polarization vector and the main axis a and differentiating we obtain for the angular velocity

$$\omega_E = \dot{\gamma} = \frac{ab\omega}{a^2 \cos^2 \psi + b^2 \sin^2 \psi}, \tag{1}$$

where $\psi = \omega t$. Therefore, in the general case, ω_E varies with time, and it is only for the circular polarization, when $a = b$, that the angular velocity is constant: $\omega_E = \omega = \text{const.}$

Equation (1) gives a clue to the probability distribution of the phase if we assume that the probability of finding the angle γ in the interval $d\gamma$ is proportional to the time dt the vector $E^{(r)}$resides in this interval. Then $P(\gamma)d\gamma = cdt$ and so $P(\gamma) = c / \dot{\gamma}$, where $\dot{\gamma}$ should be expressed in terms of γ by eliminating t from eqn (1). The coefficient c can be readily derived from the normalization condition

$$\int_0^{2\pi} P(\gamma)d\gamma = \int \frac{c}{\dot{\gamma}} d\gamma = c\int_0^T dt = cT = 1, \tag{2}$$

where $T = 2\pi / \omega$ is the period. In the general case of elliptic polarization, the probability distribution $P(\gamma)$, like the rotation velocity (1), is nonuniform.

Problem 2.15. Calculate the time-average intensity of a plane elliptically polarized wave passing through a linear polarizer.

Solution. Let the incident wave be described by the vector

$$\mathbf{E} = \text{Re}(\mathbf{a} + i\mathbf{b})e^{-i\omega t}, \tag{1}$$

where **a** and **b** are the vectors of the semimajor and semiminor axes of the polarization ellipse, $\mathbf{a} \cdot \mathbf{b} = 0$. The linear polarizer picks up from **E** the projection $E_x = \mathbf{xE}$ on the direction of the main axis x of the polarizer:

$$E_x = \text{Re}(a_x + ib_x)e^{-i\omega t}, \tag{2}$$

whence for the time average intensity we have

$$\overline{I_x} = \overline{|E_x|^2} = \frac{1}{2}\left(a_x^2 + b_x^2\right) = \frac{1}{2}\left(a^2 \cos^2 \theta + b^2 \sin^2 \theta\right), \tag{3}$$

where θ is the angle between the vector **a** and the x axis. We see that in a slow rotation of the polarizer, the polar diagram of its response (3) does not follow an ellipse

$$E^2 = \left[\left(\frac{\cos\theta}{a}\right)^2 + \left(\frac{\sin\theta}{b}\right)^2\right]^{-1}, \tag{4}$$

described by the tip of **E** from eqn (1), but rather a more intricate curve (3). This conclusion has been drawn by Mayes (1976). However, this author has misidentified the time average (3) with the averaging over the *uniform* phase of the rotating vector (1). In reality, owing to the nonuniform rotation of E the distribution of this phase is also nonuniform (see Problem 2.14).

Problem 2.16. Derive the necessary conditions for the 4×4 matrix $\hat{M} = [M_{ij}]$ with real-valued elements to be the Müller matrix of a linear system (Kumar and Simon, 1992).

Solution. The Müller matrix $\hat{M}$ carries the Stokes vector P at the input in a linear system to the output Stokes vector $P' = \hat{M}P$; the components of P and P' satisfy the inequalities $P_0 \geq 0$, $P_0' \geq 0$, and (2.3.90). If we choose P in the form

$$P = \begin{bmatrix} 1 \\ \mathbf{n} \end{bmatrix}, \tag{1}$$

where $\mathbf{n}$ is the unit vector (which corresponds to a fully polarized incident wave), then

$$P_0' = \sum_{j=0}^{3} M_{0j}P_j = M_{00} + \sum_{j=1}^{3} M_{0j}n_j \equiv M_{00} + \mathbf{M}_0 \cdot \mathbf{n} \geq 0, \tag{2}$$

$$(P_0')^2 = \left(M_{00} + \sum_{j=1}^{3} M_{0j}n_j \right)^2 \geq \sum_{j=1}^{3} (P_j')^2 = \sum_{j=1}^{3} \left(M_{j0} + \sum_{j=1}^{3} M_{ji}n_i \right)^2, \tag{3}$$

where $\mathbf{M}_0 = [M_{01}, M_{02}, M_{03}]$. Recognizing that these inequalities must be satisfied for arbitrary directions of $\mathbf{n}$, from eqns (2) and (3) we obtain

$$M_{00} \geq |\mathbf{M}_0| = \sqrt{M_{01}^2 + M_{02}^2 + M_{03}^2}, \tag{4}$$

$$\left\{ \sum_{j=1}^{3} \left(M_{j0} + \sum_{i=1}^{3} M_{ji}n_i \right)^2 \Bigg/ \left(M_{00} + \sum_{j=1}^{3} M_{0j}n_j \right)^2 \right\}_{\max} \leq 1, \tag{5}$$

where $\{\cdot\}_{\max}$ implies the maximum over all directions of $\mathbf{n}$.

Problem 2.17. Derive the Jones and Müller matrices describing the passage of a plane wave through an ideal linear polarizer.

Solution. Let a plane wave $\mathbf{E}$ propagate along the z direction and the polarizer axis is set along the x axis, so that $\mathbf{E}$ travels along x unchanged and attenuates along y. This process is described by the projector $\hat{\Lambda} = \mathbf{x} \otimes \mathbf{x}$ such that $\mathbf{E}' = \hat{\Lambda} \cdot \mathbf{E} = \mathbf{x}(\mathbf{x} \cdot \mathbf{E})$, where $\mathbf{E}'$ and $\mathbf{E}$ are the vector amplitudes of the field before and after the polarizer, respectively, and $\mathbf{x}$ is the unit vector. In the (x,y) basis, the Jones matrix $\hat{\Lambda}$ has the form

$$\begin{bmatrix} \Lambda_{xx} & \Lambda_{xy} \\ \Lambda_{yx} & \Lambda_{yy} \end{bmatrix} = \begin{bmatrix} 1 & 0 \\ 0 & 0 \end{bmatrix}.$$

The respective Müller matrix is expressed by the tensor product $\hat{\Lambda} \otimes \hat{\Lambda}^{+}$ taken in the basis of Stokes vectors

$$M_{ij} = \frac{1}{2} \operatorname{tr} \hat{\pi}_i \left(\hat{\Lambda} \otimes \hat{\Lambda}^{*} \right) \hat{\pi}_j = \frac{1}{2} \operatorname{tr} \hat{\pi}_i \hat{\Lambda} \hat{\pi}_j \hat{\Lambda}^{+} = \frac{1}{2} \operatorname{tr} \hat{\pi}_i \cdot \mathbf{x} \otimes \mathbf{x} \cdot \hat{\pi}_j \cdot \mathbf{x} \otimes \mathbf{x}$$
$$= \frac{1}{2} \left(\hat{\pi}_i \right)_{xx} \left(\hat{\pi}_j \right)_{xx} .$$

According to eqn (2.3.81), $(\hat{\pi}_0)_{xx} = -(\hat{\pi}_1)_{xx} = 1$ and $(\hat{\pi}_2)_{xx} = (\hat{\pi}_3)_{xx} = 0$ so that $-M_{01} = -M_{10} = M_{11} = 1/2$ and the rest $M_{ij} = 0$. A similar reasoning may be applied to a polarizer with an arbitrarily oriented axis.

Problem 2.18. A plane wave incidence on a scattering layer gives rise to, generally speaking, diffuse reflected and transmitted waves with a wide angular spectrum, as shown in Fig. 2.20. Given the layer is *statistically isotropic* and the incident and scattered waves are normal to the layer, describe the variation of the ensemble-average Stokes parameters for the transmitted wave.

Solution. Simple symmetrical considerations, similar to those used in deriving eqn (2.3.49) indicate that in this case, the radiance matrix of the transmitted wave $\hat{I} = \langle E \otimes E^{*} \rangle$ is expressed in terms of the radiance matrix of the incident wave $\hat{I}^{0} = \langle E^{0} \otimes E^{0*} \rangle$ by the relation

$$\langle E_i E_j^{*} \rangle = \left(\alpha \delta_{il} \delta_{jm} + \beta \delta_{ij} \delta_{lm} + \gamma \delta_{im} \delta_{jl} \right) \langle E_l^{0} E_m^{0*} \rangle , \tag{1}$$

where α, β and γ are constants describing the properties of the scattering layer, and the indices run over the values of x and y. Passing over in eqn (1) from the radiance matrices to Stokes parameters with the aid of eqns (2.3.84) and (2.3.85) we obtain

$$\begin{aligned} P_0 &= (\alpha + 2\beta + \gamma) P_0^{0} , \\ P_1 &= (\alpha + \gamma) P_1^{0} , \\ P_2 &= (\alpha + \gamma) P_2^{0} , \\ P_3 &= (\alpha - \gamma) P_3^{0} , \end{aligned} \tag{2}$$

where P_j and P_j^{0} refer to the incident and transmitted wave, respectively. These relations allow one to determine the layer parameters α, β and γ from measurements of the polarization parameters of the forward scattered wave.

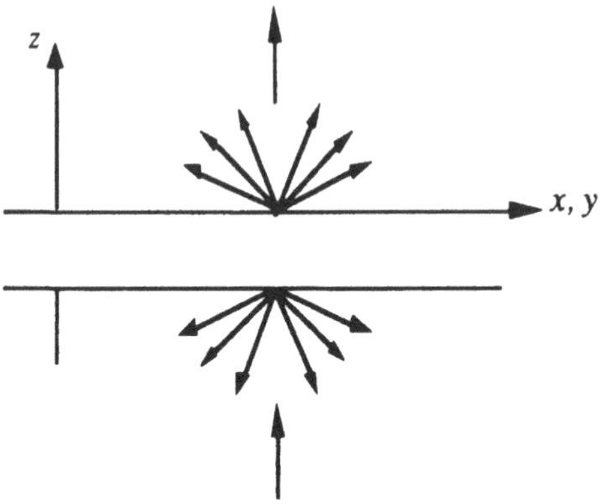

Figure 2.20

Results equivalent to (2) have been obtained by Freund (1991), who used in place of α,β and γ these parameters normalized to their sum.

From eqn (2) it follows that α,β and γ are real-valued parameters. Their physical meaning becomes more transparent if we rewrite eqn (1) in the form

$$\hat{I} = \alpha\hat{I}^0 + \beta\hat{1}\,\mathrm{tr}\,\hat{I}^0 + \gamma\hat{I}^{0T}. \tag{3}$$

Now we see that the parameter α describes the propagation of the wave without a change in its polarization state, β describes the transformation of an arbitrary polarization into a nonpolarized component, and γ describes the variation in the direction of rotation for the component with a circular polarization.

Because multiple scattering depolarizes the wave, and the probability of changing the sense of rotation of the polarization in forward scattering is obviously small we may conclude that in thicker layers the parameter β must become dominant, whereas α and γ must vanish and $\alpha \gg \gamma$. This physically obvious conclusion has been borne out by direct computations for a model of point scatterer set forth by Freund (1991).

Problem 2.19. Derive the relations (2.3.88).

Solution. With reference to eqn (2.3.3) and Fig. 2.12 we see that the components of **a** and **b** are $a_1 = \cos\beta\sin\chi$, $a_2 = \cos\beta\cos\chi$, $b_1 = \sin\beta\cos\chi$, and $b_2 = -\sin\beta\sin\chi$. Substituting these expressions in (2.3.80), where $E_l = a_l + ib_l$, we arrive at (2.3.88).

Problem 2.20. In crystal optics and in optics of anisotropic media one is often faced with the problem of decomposition of the polarization vector in two independent coherent polarizations, alternatively, the problem of a polarization formed in a coherent summation of two polarized beams. A natural follow-up is the question of whether or not the phases of beams with nonidentical polarizations could be compared. Studying the interference phenomena in anisotropic crystals, Pancharatnam (1956) called two mutually coherent beams cophased if the intensity of their superposition is maximal with respect to the phase variations of the beams. He defined the phase difference of two mutually coherent beams with different polarizations as an additional phase, which must be introduced in the polarization vector of one of the beams to attain the maximum intensity of their superposition.

Demonstrate that the natural definition of the phase coincidence due to Pancharatnam leads to an unexpected consequence — it lacks transitivity; indeed, if beam 1 is in phase with beam 2, and beam 2 is in phase with beam 3, then beam 3 need not necessarily be in phase with beam 1.

Solution. Let beams 1 and 2 have a linear polarization and beam 3 has a circular polarization. It is not difficult to demonstrate that the coincident phases in the above sense for beams with linear and circular polarization imply the coincidence of their polarization vectors at the instant when the linear polarization vector is a maximum, and for beams with linear polarizations at an acute angle, this coincidence implies a simultaneous attainment of their maxima. Therefore, all three beams cannot be simultaneously cophased,unless their directions of linear polarizations coincide with one another.

After some three decades the studies of Pancharatnam (summarized in *Current Science*, 1990) have found their continuation in the theory of the geometrical phase introduced in quantum mechanics by Berry in 1984 and later extended to many other physical systems (see Shapere and Wilczek, 1989, appended to Pancharatnam; Berry, 1990).

Problem 2.21. Consider the invariants of the radiance matrix (2.3.78), that is, the quantities which do not change in rotations of the basis $(\mathbf{e}_1,\mathbf{e}_2)$. What inequalities follow from the positive definiteness of $\mathbf{a}^*\hat{I}_0\mathbf{a} = c_k\langle|\mathbf{a}E_k|^2\rangle \geq 0$, where $\mathbf{a} = (a_1,a_2)$ is an arbitrary vector?

Solution. From matrix theory we know that in the space of complex vectors, the invariants of $\hat{I}_0$ under the rotations of the basis $(\mathbf{e}_1,\mathbf{e}_2)$ are the trace $\mathrm{tr}\hat{I}_0 = P_0$ and the determinant $\det\hat{I}_0 = \left(P_0^2 - \mathbf{P}^2\right)/4$. If we restrict consideration by usual rotations of the real-valued basis $(\mathbf{e}_1,\mathbf{e}_2)$ in the space of real vectors, then there occurs a new invariant P_3. Since $\hat{I}_0$ is positive definite, then

$$\lambda_{1,2} = \frac{1}{2}\left(P_0 \pm \sqrt{P_0 - 4\det\hat{I}_0}\right) = \frac{1}{2}\left(P_0 \pm \sqrt{\mathbf{P}^2}\right) \geq 0,$$

and we arrive at the inequality $P_0^2 \geq \mathbf{P}^2$.

Problem 2.22. Demonstrate that the orthogonal eigenvectors $\mathbf{e}_1$ and $\mathbf{e}_2$ of the radiance matrix (2.3.21) are associated with Stokes vectors pointing to the opposite ends of a diameter on the Poincaré sphere.

Solution. The orthogonality condition $\mathbf{e}_1^*\mathbf{e}_2 = 0$ may be expressed in terms of the radiance matrices $\hat{I}_{l1} = \mathbf{e}_1 \otimes \mathbf{e}_1^*$ and $\hat{I}_{l2} = \mathbf{e}_2 \otimes \mathbf{e}_2^*$ as $\mathrm{tr}\hat{I}_{l1}\hat{I}_{l2} = |\mathbf{e}_1 \cdot \mathbf{e}_2^*|^2 = 0$. Substituting in this condition $\hat{I}_{lj}$ expressed by Stokes parameters $P_l^{(j)}$ [normalized to unity

$(\mathbf{P}_l^{(j)})^2 = 1$, because $P_0^{(j)} = \operatorname{tr}\hat{I}_{lj} = \mathbf{e}_j^*\mathbf{e}_j = 1$], and observing the orthogonality condition (2.3.83) we have

$$\left|\mathbf{e}_1^* \cdot \mathbf{e}_2\right|^2 = \operatorname{tr}\sum_{l=0}^{3}\sum_{m=0}^{3}\frac{1}{4}\hat{\pi}_l\hat{\pi}_m P_l^{(1)}P_m^{(2)} = \frac{1}{2}\sum P_l^{(1)}P_l^{(2)}$$
$$= \frac{1}{2}\left(1 + \mathbf{P}^{(1)}\mathbf{P}^{(2)}\right) = 0. \quad (1)$$

Consequently, the unit polarization vectors $\mathbf{P}^{(1)}$ and $\mathbf{P}^{(2)}$ are oppositely directed, because $\mathbf{P}^{(1)}\mathbf{P}^{(2)} = \cos\gamma = -1$, where γ is the angle between them.

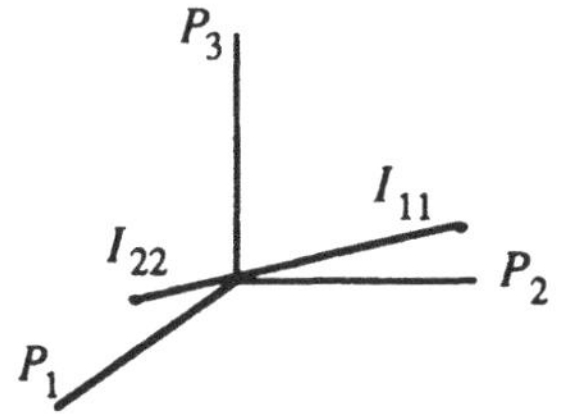

Figure 2.21. Representing a full Stokes vector as two orthogonal polarizations.

The proved property of orthogonal polarizations allows us to give the following simple interpretation to the expansion of the radiance matrix of partially polarized radiation into the sum of incoherent orthogonal contributions (2.3.21). Let us lay off the Stokes vectors corresponding to the eigenpolarizations $\tilde{\mathbf{e}}_1$ and $\tilde{\mathbf{e}}_2$. Multiply each of these vectors by the respective eigenvalue I_{11} or I_{22}. Now, the radiance matrix of partially coherent radiation will be represented by two points in the space of Stokes parameters corresponding to two opposite Stokes vectors, as shown in Fig. 2.21.

Problem 2.23. In addition to the decomposition of the radiance matrix into two mutually incoherent orthogonal polarizations, there exists one more convenient representation

$$I = \frac{1}{2}\mathcal{I}_1\hat{1} + \mathbf{E}\otimes\mathbf{E}^*, \quad (1)$$

where the first term corresponds to the unpolarized fraction and the second to the completely polarized fraction of I. For the first term, the polarization Stokes vector $\mathbf{P}$ is obviously equal to zero. Demonstrate that the similar vector for the second addend coincides with the polarization vector for the initial matrix I.

Solution. The said property immediately follows from the fact that the full Stokes vector $[P_0, \mathbf{P}]$ on the left-hand side of eqn (1) is equal to the sum of analogous vector addends from the right-hand side of eqn (1). In this sense, the said addends are mutually incoherent.

The expansion (1) is similar to one of the Problem 2.22 and also allows a vivid geometrical interpretation in the space of Stokes parameters, namely, in this space, the unpolarized fraction of (1) corresponds to the origin. Therefore, for a partially polarized oscillation, the radiance matrix $\hat{I}$ may be represented by a polarization vector $\mathbf{P}$ corresponding to the completely polarized fraction $E \otimes E^*$ and the radiance of the unpolarized component $\mathcal{J}_1 = P_0 - |\mathbf{P}|$ after referring the latter to the origin as shown in Fig. 2.22.

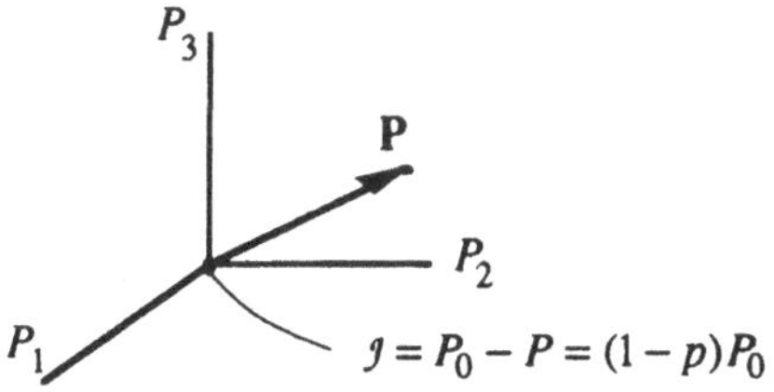

Figure 2.22. Representing the full Stokes vector as unpolarized and completely polarized parts.

Thus, in the case of a partially polarized beam we have two physically transparent representations of the radiance matrix in the Stokes parameter space (Figs. 2.21 and 2.22). These representations differ actually just by a shift of the vector $\mathbf{P}$ by $-I_{22}\mathbf{P}$.

Problem 2.24. Using eqn (2.3.94) find the transformation of Stokes parameter in rotation of the two-dimensional coordinate frame.

Solution. Let P_i and P_i' be the Stokes parameters in the initial basis $(\mathbf{e}_1, \mathbf{e}_2, \mathbf{z})$ and in the basis $(\mathbf{e}_1', \mathbf{e}_2', \mathbf{z})$ rotated from the former about the z axis through an angle θ. Then, in view of (2.3.94) we have

$$\begin{aligned}
P_1' &= \frac{1}{2}\left(P_1 \operatorname{tr} \hat{\pi}_1' \hat{\pi}_1 + P_2 \operatorname{tr} \hat{\pi}_1' \hat{\pi}_2\right) \\
&= \frac{1}{2} P_1 \operatorname{tr}\left(\mathbf{e}_2' \otimes \mathbf{e}_2' - \mathbf{e}_1' \otimes \mathbf{e}_1'\right)\left(\mathbf{e}_2 \otimes \mathbf{e}_2 - \mathbf{e}_1 \otimes \mathbf{e}_1\right) \\
&+ \frac{1}{2} P_2 \operatorname{tr}\left(\mathbf{e}_2' \otimes \mathbf{e}_2' - \mathbf{e}_1' \otimes \mathbf{e}_1'\right)\left(\mathbf{e}_1 \otimes \mathbf{e}_2 + \mathbf{e}_2 \otimes \mathbf{e}_1\right) \\
&= \frac{1}{2} P_1\left[(\mathbf{e}_2'\mathbf{e}_2)^2 + (\mathbf{e}_1'\mathbf{e}_1)^2 - (\mathbf{e}_2'\mathbf{e}_1)^2 - (\mathbf{e}_1'\mathbf{e}_2)^2\right] + P_2\left[(\mathbf{e}_2'\mathbf{e}_2)(\mathbf{e}_2'\mathbf{e}_1) - (\mathbf{e}_1'\mathbf{e}_2)(\mathbf{e}_1'\mathbf{e}_1)\right] \\
&= P_1 \cos 2\theta + P_2 \sin 2\theta \,, \qquad (2)
\end{aligned}$$

and similarly

$$P_2' = -P_1 \sin 2\theta + P_2 \cos 2\theta \,. \qquad (3)$$

Here, $\cos\theta = \mathbf{e}_2'\mathbf{e}_2 = \mathbf{e}_1'\mathbf{e}_1$, $\sin\theta = \mathbf{e}_2'\mathbf{e}_1 = -\mathbf{e}_1'\mathbf{e}_2$, and we made good use of the relation $\operatorname{tr}(\mathbf{a} \otimes \mathbf{b})(\mathbf{c} \otimes \mathbf{d}) = \mathbf{bc} \operatorname{tr} \mathbf{a} \otimes \mathbf{d} = (\mathbf{b} \cdot \mathbf{c})(\mathbf{a} \cdot \mathbf{d})$.

Problem 2.25. The permittivity tensor of a weakly anisotropic rarefied magnetoactive plasma is given by the expression

$$\hat{\varepsilon} = \hat{1} - \frac{\omega_p^2}{\omega\omega'}\left[1 - \left(\frac{\omega_H}{\omega'}\right)^2\right]^{-1}\left(\hat{1} - i\frac{\boldsymbol{\omega}_H \times}{\omega} - \frac{\boldsymbol{\omega}_H \otimes \boldsymbol{\omega}_H}{\omega^2}\right), \tag{1}$$

where $\omega_p = \left(4\pi\rho_0 e^2 / m\right)^{1/2}$ is the electron plasma frequency, $\boldsymbol{\omega}_H = (e/mc)\mathbf{H}$ is the electron gyrofrequency vector, $\mathbf{H}$ is the external magnetic field, $\omega' = \omega + i\nu_{\text{eff}}$, ν_{eff} is the effective collision frequency, and $\boldsymbol{\omega}_H \times$ implies a matrix acting by the rule $(\boldsymbol{\omega}_H \times)\mathbf{a} = \boldsymbol{\omega}_H \times \mathbf{a}$.

Find the parameters A_i^{h} and A_i^{a} (2.3.100) for a high-frequency plasma.

Solution. At high frequencies, i.e., for $\omega \gg \omega_H, \omega_p, \nu_{\text{eff}}$, eqn (1) can be replaced with the approximate expression

$$\hat{\varepsilon} = \hat{1} - \left(\frac{\omega_p}{\omega}\right)^2\left[\left(1 - i\frac{\nu_{\text{eff}}}{\omega}\right)\hat{1} - i\frac{\boldsymbol{\omega}_H \times}{\omega} - \frac{\boldsymbol{\omega}_H \otimes \boldsymbol{\omega}_H}{\omega^2}\right]. \tag{2}$$

Using eqns (2.3.24), (2.3.25), (2.3.100), and (2.3.101), it is not hard to calculate the parameters A_i^{h} and A_i^{a}:

$$A_i^{\text{h}} = \frac{-\nu_{\text{eff}}\delta_{i0}k_0\omega_p^2}{\omega^3\sqrt{\varepsilon_0}}, \tag{3}$$

$$\mathbf{A}^{\text{a}} = \left(A_1^{\text{a}}, A_2^{\text{a}}, A_3^{\text{a}}\right)$$
$$= \frac{k_0\omega_p^2}{2\omega^2\sqrt{\varepsilon_0}}\left[\left(\frac{\omega_H}{\omega}\right)^2 \sin^2\theta_H \cos 2\varphi_H, \left(\frac{\omega_H}{\omega}\right)^2 \sin^2\theta_H \sin 2\varphi_H, 2\frac{\omega_H}{\omega}\cos_H\right], \tag{4}$$

where $\varepsilon_0 \approx 1 - \omega_p^2/\omega^2$, and the angles θ_H and φ_H define the orientation of the external magnetic field $\mathbf{H}$ in the basis $(\mathbf{e}_1, \mathbf{e}_2, \mathbf{e}_3 = \mathbf{n})$, as shown in Fig. 2.23.

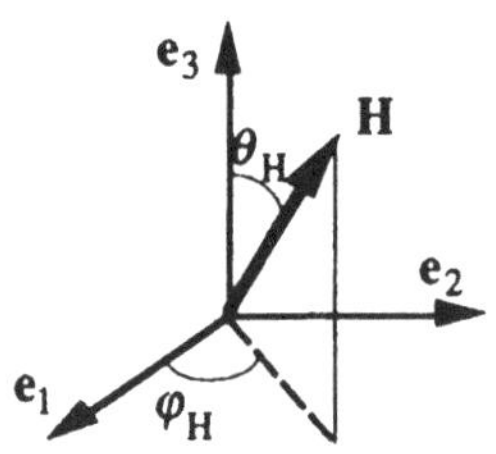

Figure 2.23. Orientation of the basis in a magnetoactive plasma.

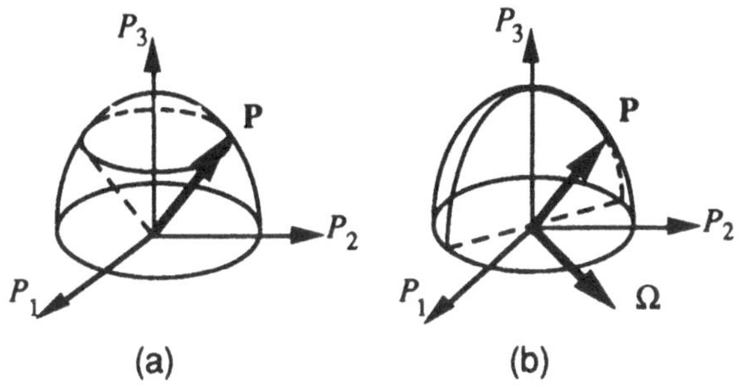

Figure 2.24. Rotation of Stokes vector **P**: (a) Faraday effect, (b) Cotton–Mouton effect.

In the case of *longitudinal* propagation, when the wave vector **k** is directed along the field **H** and $\theta_H = 0$, both vectors $\mathbf{A}^a$ and $\Omega = -2\kappa' + \mathbf{A}^a$ are directed along the axis P_3. In the absence of absorption ($\nu_{eff} = 0$), the Stokes polarization vector **P** rotates around P_3, which corresponds to the rotation of the polarization ellipse as a whole ($\beta = \text{const}$, $\chi \neq \text{const}$), that is, to the Faraday effect (Fig. 2.24a).

In the case of *transverse* propagation $\theta_H = \pi / 2$, the vector **A** is in the plane (P_1, P_2) and for rectilinear rays, when we can pick up $\kappa = 0$, the angular velocity $\Omega = \mathbf{A}^a$ also lies in the plane (P_1, P_2). If in addition **P** is orthogonal to $\mathbf{A}^a$, then we are faced with the Cotton-Mouton effect (Fig. 2.24b) in which the orientation of the polarization ellipse remains intact ($\chi = \text{const}$) but its form varies ($\beta \neq \text{const}$).

Problem 2.26. Write out the Stokes matrix σ_{ij} for the scattering cross section (2.3.54) of the Rayleigh type: $\sigma_2 = \sigma_3 = 0$.

Solution. Calculating the elements of σ_{ij} in agreement with eqn (2.3.105), by analogy with the case considered in Section 2.3.18, we find

$$\sigma_{ij} = \frac{1}{2}\sigma_3(\omega \leftarrow \omega', \mathbf{nn}')M_{ij}\,, \tag{1}$$

where the matrix M_{ij} has the form

$$\begin{bmatrix} 1+(\mathbf{nn}')^2 & (\mathbf{n}'\mathbf{y})^2-(\mathbf{n}'\mathbf{x})^2 & -2(\mathbf{nx})(\mathbf{nx}') & 0 \\ (\mathbf{n}'\mathbf{y})^2-(\mathbf{n}'\mathbf{x})^2 & (\mathbf{xx}')^2+(\mathbf{yy}')^2-(\mathbf{xy}')^2-(\mathbf{yx}')^2 & 2(\mathbf{xx}')(\mathbf{xy}')-2(\mathbf{yx}')(\mathbf{yy}') & 0 \\ -2(\mathbf{n}'\mathbf{x})(\mathbf{n}'\mathbf{y}) & 2(\mathbf{xx}')(\mathbf{yx}')-2(\mathbf{xy}')(\mathbf{yy}') & 2(\mathbf{xx}')(\mathbf{yy}')+2(\mathbf{xy}')(\mathbf{yx}') & 0 \\ 0 & 0 & 0 & 2(\mathbf{nn}') \end{bmatrix}$$

The real-valued bases $(\mathbf{e}_1, \mathbf{e}_2, \mathbf{e}_3) = (\mathbf{x}, \mathbf{y}, \mathbf{n})$ and $(\mathbf{e}_1', \mathbf{e}_2', \mathbf{e}_3') = (\mathbf{x}', \mathbf{y}', \mathbf{n}')$ can be chosen arbitrarily.

Problem 2.27. Derive the transformation of Stokes parameters of a partially polarized wave in reflection and refraction at a plane interface.

Solution. Using eqns (2.3.114) and (2.3.117) it is not difficult to express the Stokes parameters of the reflected and refracted beams in terms of the Stokes parameters of the incident beam. For the reflected beam, from eqn (2.3.114) and (2.3.80) we obtain

$$\begin{bmatrix} P_0^r \\ P_3^r \end{bmatrix} = \frac{1}{2}\begin{bmatrix} |\mathcal{R}_\parallel|^2 + |\mathcal{R}_\perp|^2 & |\mathcal{R}_\parallel|^2 - |\mathcal{R}_\perp|^2 \\ |\mathcal{R}_\parallel|^2 - |\mathcal{R}_\perp|^2 & |\mathcal{R}_\parallel|^2 + |\mathcal{R}_\perp|^2 \end{bmatrix}\begin{bmatrix} P_0^0 \\ P_3^0 \end{bmatrix},$$

$$\begin{bmatrix} P_1^r \\ P_2^r \end{bmatrix} = \frac{1}{2}\begin{bmatrix} \mathrm{Re}\,\mathcal{R}_\parallel\mathcal{R}_\perp^* & \mathrm{Im}\,\mathcal{R}_\parallel\mathcal{R}_\perp^* \\ -\mathrm{Im}\,\mathcal{R}_\parallel\mathcal{R}_\perp^* & \mathrm{Re}\,\mathcal{R}_\parallel\mathcal{R}_\perp^* \end{bmatrix}\begin{bmatrix} P_1^0 \\ P_2^0 \end{bmatrix},$$

where the Stokes parameters $P_1^r = \mathrm{tr}\hat{\pi}_i'\hat{I}^r$ are calculated in the basis $(\mathbf{e}_1', \mathbf{e}_2')$, and $P_r^0 = \mathrm{tr}\hat{\pi}_i\hat{I}^0$ in the basis $(\mathbf{e}_1, \mathbf{e}_2)$. Similar expressions with $\mathcal{R}$ replaced with $\mathcal{T}$ are valid for the refracted beam.

In the particular case of a completely polarized incident beam $\left(P_0^2 = \mathbf{P}^2\right)$, the reflected and refracted beams will also be completely polarized. By the Stokes parameters one can easily determine the orientation and shape of the polarization ellipses of the reflected and refracted beam waves.

3. Statistical Wave Content of Radiative Transfer Theory

This chapter treats the phenomenological radiative transfer theory as a corollary of statistical wave theory. The salient features of this approach vividly manifest themselves already in the relatively simple example of free radiation which will be our point of departure in discussing this topic. The reader is assumed to be familiar with the main concepts of random field theory in an amount approximately corresponding to the expositions in the books of Ishimaru (1978) and Rytov, Kravtsov, and Tatarskii (1989b).

Since the radiance is a particular case of the field spectrum, in Section 3.1 we consider the general properties of spectra of nonuniform fluctuations. Such spectra can be defined in different ways, thus obtaining different generalizations of a regular spectrum of uniform fluctuations. The more general approach is associated with the so-called Wigner function that is used in Section 3.2 to derive a transfer equation for radiation in free space. The transfer equation in a scattering medium is described in Section 3.5, preceded by Sections 3.3 and 3.4 presenting the necessary background of multiple scattering theory and derivations of the key equations of this theory (Section 3.3) and highlighting some features of the description of discrete scattering media in electrodynamics problems (Section 3.4). Finally, Section 3.6 deals with the basic corollaries of the wave approach that are associated with the correlation content of radiative transfer theory.

3.1 RADIANCE AND CORRELATION FUNCTIONS OF THE FREE WAVE FIELD

3.1.1 Statistical Approach versus Phenomenological Approach

Unlike phenomenological transfer theory that considers the geometrical optics asymptotic expansion where wave equations are not involved directly, the statistical approach hinges directly on full stochastic wave equations. Since we have no room here to discuss in some detail mathematical statistics and the fundamental problem of sense and validity of the probabilistic approach, we confine ourselves to some simple considerations concerning the statistical description of random quantities.

In practice, random quantities differ from deterministic functions primarily in complexity and unpredictability of their behavior. Characterizing wave fields, one usually speaks of randomness if such fields are formed by unknown sources with unpredictable but statistically stable behavior, or if they propagate in a complex inhomogeneous medium whose characteristics are not known exactly. Examples of random sources may be all thermal sources of radiation, and examples of complex propagation media may be almost all natural media such as the atmosphere, ionosphere, ocean surface, and the interior of the earth. In nature, as a rule, any wave field is random to a certain extent, and this randomness can be neglected only in some outstanding situations.

Formally treated, the statistical approach differs from the dynamic approach in that the former does not involve predictions of the behavior of single realizations and considers only the properties of the statistical ensemble as a whole. Of course, this does not imply that this approach completely denies a description of single realizations, but the quantitative conclusions refer precisely to the ensemble of realizations.

Ensemble averaging can often be reduced to the averaging over a realization by taking advantage of one or another form of ergodicity (see Sections 3.2.16 and 3.2.18). However, in the general case, a comparison of the results of statistical theory with experimental data usually requires multiple repetition of experiments. This is the main difference of the statistical approach from the phenomenological treatment outlined in the preceding chapter: while the phenomenological theories are based mainly on intuition and simple but not rigorous considerations, the statistical approach relies on the first principles of wave theory with considerable attention to the formalization of the properties of the statistical ensemble.

3.1.2 Statistical Wave Radiative Correlation Theory: Problem Formulation

In order to substantiate phenomenological radiative transfer theory we look at the classical random field (multicomponent in the general case) which is a random function of four variables $x=(\mathbf{r},t)$. For monochromatic radiation, the time dependence is taken proportional to the factor $e^{-i\omega t}$. This gives the real valued field u in the form $u=\mathrm{Re}\,u(x)$, where $u(x)=u(\mathbf{r})e^{-i\omega t}$ is the complex wave field. Now, the role of the random field u will be played by the complex wave field $u(x)$, or, if we drop the factor $e^{-i\omega t}$, the amplitude $u(\mathbf{r})$ depending on the spatial arguments. In the case of quasi-monochromatic radiation, the latter may also slowly depend upon time. In the subsequent exposition we will consider in parallel the cases of monochromatic and quasi-monochromatic fields without writing out the explicit form of the argument x. If x is treated as one scalar variable, then it is assumed that the respective considerations refer to an arbitrary component of the vector $x=(\mathbf{r},t)$ or $x=\mathbf{r}$. The specific form is usually obvious from the context.

In the linear theory, the field u obeys the linear wave equation

$$\hat{L}u=q, \tag{3.1.1}$$

where $\hat{L}$ is the wave operator describing the properties of the medium propagating the radiation, and q is the source function. Both $\hat{L}$ and q may be stochastic: while the operator $\hat{L}$ corresponds to the presence of random inhomogeneities in the medium, the function q enables the stochastic sources to be described, specifically, the random incident radiation corresponding to sources at infinity. All the necessary statistical characteristics of the medium ($\hat{L}$) and the sources (q) are assumed to be given. Specific examples of eqn (3.3.1) will be considered below.

Correlation theory analyzes two first statistical moments of the field

$$M = \langle u_1 \rangle, \tag{3.1.2}$$

$$\Gamma = \langle u_1 u_2^* \rangle, \tag{3.1.3}$$

where $u_1 = u(x_1)$, $u_2 = u(x_2)$. The first moment M is the *mean field*, and in the case of complex wave fields the second moment Γ is usually called the *coherence function*. When the number of arguments of these moments must be highlighted we write M_1 or Γ_{12} in place of M or Γ.

For a real-valued physical field, one could omit in (3.1.3) the complex conjugation symbol. For a quasi-monochromatic complex field, equation (3.1.3) corresponds to the time-averaged second moment of the physical field, in contrast to the moment $\langle u(x_1)u(x_2)\rangle$, a *second coherence function*, that does not contain complex conjugation and is a rapidly oscillating function of time. In measuring average energy characteristics of radiation one usually averages over high frequencies and the contribution due to the second coherence function $\langle u(x_1)u(x_2)\rangle$ vanishes (Problem 3.6). We shall not consider this function below and confine our consideration almost entirely to the coherence function (3.1.3).

In describing the statistical moments, it is convenient to represent the field u as a sum of its mean value $\langle u \rangle$ and the fluctuating part $\tilde{u}$

$$u = \langle u \rangle + \tilde{u}, \tag{3.1.4}$$

where $\langle \tilde{u} \rangle = 0$. This representation is commonly called the decomposition of the field u into the *coherent* $\langle u \rangle$ and *incoherent* $\tilde{u}$ components. These components are independent in terms of their contributions to the second moment Γ. To demonstrate this fact we substitute (3.1.4) in the second moment (3.1.3) and obtain

$$\begin{aligned}\Gamma &= \Big\langle \big(\langle u_1 \rangle + \tilde{u}_1\big)\big(\langle \tilde{u}_2^* \rangle + \tilde{u}_2^*\big)\Big\rangle \\ &= \langle u_1 \rangle \langle u_2^* \rangle + \langle \tilde{u}_1 \tilde{u}_2^* \rangle \equiv \langle u_1 \rangle \langle u_2^* \rangle + \psi_{12}.\end{aligned} \tag{3.1.5}$$

The first addend on the right-hand side of this expression corresponds to the mean field and the *correlation function*

$$\psi_{12} = \Psi = \langle \tilde{u}_1 \tilde{u}_2^* \rangle = \langle u_1 u_2^* \rangle - \langle u_1 \rangle \langle u_2^* \rangle. \tag{3.1.6}$$

describes fluctuations. In agreement with eqn (3.1.5), one can consider $\langle u \rangle$ and the correlation function Ψ instead of the first moment — the mean field $\langle u \rangle$ and coherence function Γ.

In problems of scattering theory, the decomposition of the field into coherent and incoherent components can be endowed with geometrical (interference) meaning. Let the full field u consists of a nonrandom (coherent) incident wave u^0 and the field of the scattered wave u^{sc}, so that $u = u^0 + u^{sc}$, and we need to measure the mean intensity $\langle |u|^2 \rangle = \Gamma(x,x)$. Clearly, the incident wave can interfere in the mean only with the coherent component of the scattered field $\langle u^{sc} \rangle$, because the mean intensity $\langle |u|^2 \rangle = |u^0|^2 + 2\,\mathrm{Re}\, u^{0*} \langle u^{sc} \rangle + \langle |u^{sc}|^2 \rangle$ does not contain the incohrent component $\tilde{u}^{sc} = u^{sc} + \langle u^{sc} \rangle$.

3.1.3 Statistically Uniform and Stationary Fields. Spectra of Uniform Fluctuations

A random field u is called *statistically uniform* or *statistically homogeneous* in x if all its statistical characteristics are invariant under arbitrary translations in x, that is, are independent of such translations. Fields statistically uniform in time are called *stationary*, while those statistically uniform in space are referred to as simply uniform. In addition , one may speak of statistical uniformity of individual characteristics of a random field without having to require the whole statistical homogeneity of the latter. In such cases, we shall, as a rule, refer the concept of homogeneity to an argument of the function under consideration; this may be $x = (\mathbf{r}, t)$, or $x = \mathbf{r}$, or simply t (in the last case, homogeneity coincides with stationarity).

For a uniform and stationary field, the mean value is constant, and the coherence function Γ and the correlation function Ψ depend only on the difference $\rho = x_1 - x_2$. This almost immediately implies that Γ and Ψ are delta-correlated in the spectral-conjugated argument. Indeed, the time-space spectrum of u is defined as

$$u(K) = \int u(x) e^{-iKx} \frac{d^4x}{(2\pi)^4}. \tag{3.1.7}$$

Here we denote $x = (\mathbf{r}, t)$, $K = (\mathbf{k}, \omega)$, $Kx = \mathbf{kr} - \omega t$, $d^4x = d^3r dt$ and $d^4K = d^3k d\omega$.

For a uniform and stationary field, the coherence function is

$$\begin{aligned} \Gamma &= \langle u(x_1) u^*(x_2) \rangle \\ &= \int \langle u(K') u^*(K'') \rangle \exp i(K'x_1 - K''x_2) d^4K' d^4K''. \end{aligned} \tag{3.1.8}$$

In order that this function depend only on the difference $x_1 - x_2$ we need that that amplitude $u(K')$ and $u^*(K'')$ be delta-correlated:

$$\langle u(K') u^*(K'') \rangle = J_{K'} \delta(K' - K''). \tag{3.1.9}$$

Then, from (3.1.8) we have the expression

$$\Gamma(\rho) = \Gamma(x_1 - x_2) = \int J_K e^{iK\rho} d^4 K, \tag{3.1.10}$$

which represents the Wiener-Khinchin theorem.

According to (3.1.10) the quantity J_K, that will be called the *spectral density* (or simply the *spectrum*) is given as the Fourier transform of the second moment Γ over the difference variable $\rho = x_1 - x_2$:

$$J_K = \int \Gamma(\rho) e^{-iK\rho} \frac{d^4\rho}{(2\pi)^4}. \tag{3.1.11}$$

From (3.1.9) it follows formally that J_K is proportional to $\langle |u(K)|^2 \rangle$:

$$J_K \propto \left\langle |u(K)|^2 \right\rangle, \tag{3.1.12}$$

which implies that this quantity is nonnegative, $J_K \geq 0$.

Strictly speaking, the proportionality coefficient in eqn (3.1.12) is an indefinitely large quantity $\delta(0) = \infty$. Difficulties in handling such quantities may be overcome by assuming that the field is bounded in space and time by a large but finite volume $V^{(4)}$. Then the unbounded quantity

$$\delta(0) = \int e^{i0\rho} (2\pi)^{-4} d^4\rho = \infty$$

recasts into the finite expression $\int (2\pi)^{-4} d^4\rho = (2\pi)^{-4} V^{(4)} < \infty$. A similar assumption may be applied to other cases where expressions like $\delta(0)$ occur, see, for example, Problem 5.5. It is important to note only that the proportionality coefficient in eqn (3.1.12) is positive. The statement of non-negativity of J_K constitutes, essentially, the key point of the Wiener-Khinchin theorem. For $K \neq 0$, this quantity may be treated as the intensity of fluctuations of a field with wave vector K.

It should be noted that the property of statistical homogeneity of coherence functions does not imply, generally speaking, the statistical homogeneity of other field characteristics, say, the probability density. For example, in the cased of a deterministic plane wave $u = \exp(iKx)$, the coherence function $\Gamma = \exp(iK(x_1 - x_2))$ is statistically uniform, whereas the mean field $\langle u \rangle = \exp(iKx)$ is not statistically uniform. If we consider the superposition of two plane waves

$$u = a_1 e^{iK_1 x} + a_2 e^{iK_2 x}, \tag{3.1.13}$$

where $K_1 \neq K_2$, and both a_1 and a_2 are random amplitudes, then the coherence function

$$\Gamma = \sum \langle a_\alpha a_\beta^* \rangle \exp i\left(K_\alpha x_1 - K_\beta x_2\right) \tag{3.1.14}$$

is also not statistically uniform in the general case because of the presence of interference terms with $\alpha \neq \beta$.

Clearly, to introduce the spectrum (3.1.11) it suffices to require that only the coherence function Γ should be statistically uniform (rather than all characteristics of the field).

3.1.4 Spectra of Nonuniform Fluctuations. The Wigner Function

The spectral density J_K (3.1.11) is the most important statistical characteristic of a uniform field. Since the spectral density J_K is nonnegative, we may assign to this quantity an energy meaning, and combine it with the main concept of radiometry — the radiance. However, the definition (3.1.11) is applicable only to the statistically uniform coherence function, whereas all the statistical characteristics of real physical fields are nonuniform even in view of the boundedness of these fields in space and time. It is desirable, therefore, to introduce the *local spectrum* $J_K(R)$ that would characterize the intensity of fluctuations with wave vector K in a neighborhood of a point R rather than in the entire space.

For the function Γ one may formally write the general spectral expansion in terms of eigenfunctions $\Psi_n(x)$, that is related to the Karhunen–Loeve theorem (see Problems 3.1 and 3.2). However, this expansion is of low use for practical purposes because of the absence of information about the eigenfunctions $\Psi_n(x)$ which are also not related to the concept of the wave vector K. It is expedient, therefore, to consider another approach.

From physical considerations it should be clear that the local spectrum $J_K(R)$ must be given by some linear transformation $\hat{Q}$ of the second moment Γ, which, in the case of uniform fluctuations, when Γ_{12} depends only upon the difference $\rho = x_1 - x_2$, would coincide with (3.1.11). In the general form, such a transformation may be written as

$$\begin{aligned} J_K(R) &= \hat{Q}\Gamma \\ &\equiv \int Q(R,K;R',\rho')\langle u(R'+\rho'/2)u^*(R'-\rho'/2)\rangle d^4R'd^4\rho'. \end{aligned} \tag{3.1.15}$$

Here, the function $Q = Q(R,K;R',\rho')$ must satisfy the condition

$$\int Q(R,K;R',\rho')d^4R' = (2\pi)^{-4}\exp(-iK\rho'), \tag{3.1.16}$$

which converts eqn (3.1.15) to the ordinary spectrum (3.1.11) in the case of a statistically uniform field (see also Problem 3.3). In addition, it would be natural to require that the function Q depend only upon the difference $R - R'$: $Q = Q(K, R - R', \rho')$, to reflect the homogeneity and stationarity of $\hat{Q}$. Indeed, from eqn (3.1.15) we see that, in this case, to the second moment $\Gamma' = \Gamma(x_1 + a, x_2 + a)$, obtained from Γ by a translation of the the origin by a, there corresponds the spectrum $J'_K(R) = J_K(R + a)$ obtained from the spectrum $J_K(R)$ by the same shift in the argument R; this implies the homogeneity of the map Q, i.e., the absence of singular points R in this transformation.

The reader may found different definitions of the spectra of statistically nonuniform fluctuations corresponding to different functions Q. Walther (1973) used the spectrum $\langle u(x)e^{iKx}u^*(K)\rangle$. Page (1952) and Lampard (1954) proposed their own versions for nonstationary random processes — the instantaneous Page–Lampard spectrum. A widely used definition is the dynamic spectrum based on a finite segment of the process (for this approach, the reader is referred to the book by Rytov, Kravtsov, and Tatarskii, 1989b). All these definitions are covered by the general formalism (3.1.15) – (3.1.16).

We note other interesting attempts of introducing the so-called physical spectrum (Marc, 1970) directly related with the method of detection of physical fields, specifically, with the rate of count of photons (Eberly and and Wodkiewicz, 1977), or with the blackening of the photographic plate in optical measurement (Bartelt, Brenner, and Lohmann, 1980). By refusing to strictly follow the condition like (3.1.16), one may choose the function Q such that the physical spectrum be a nonnegative quantity. Such attempts are justified when particular experiments are to be described, but they are hardly applicable for the construction of a general theory of quasi-uniform field since they lead to the definition of the spectrum that characterizes not only the field but also the method of measurements.

The more popular version of the local fluctuation spectrum is the *Wigner function* which, for the space-time fields under consideration, may be defined as

$$W(K,R) = \int \langle u(R+\rho/2)u^*(R-\rho/2)\rangle e^{-iK\rho}\frac{d^4\rho}{(2\pi)^4}$$
$$\equiv \int \Gamma(R,\rho)e^{iK\rho}\frac{d^4\rho}{(2\pi)^4}, \tag{3.1.17}$$

where $K\rho = \mathbf{k\rho} - \omega\tau$. This definition can be readily applied to the case of spatial fields $[\rho = (\boldsymbol{\rho}, \tau)$ becomes $\rho = (\boldsymbol{\rho})]$ or temporal signals [see eqn (3.1.23)].

We define direct and inverse Fourier transformations by the relations

$$\hat{F}_{K\leftarrow x}f(x) = \int f(x')e^{-iKx'}\frac{d^4x'}{(2\pi)^4}, \quad \hat{F}^{-1}_{x\leftarrow K}f(K) = \int f(K')e^{iK'x}d^4K', \tag{3.1.18}$$

where $f(x)$ is an arbitrary function. Then the Wigner function (3.1.17) can be written as the Fourier transform with respect to the difference variable $\rho = x_1 - x_2$ with the fixed "center of gravity" argument $R = (x_1 + x_2)/2$:

$$W(K,R) = \hat{F}_{K\leftarrow\rho}\Gamma(R,\rho). \tag{3.1.19}$$

Thus, the wave vector K, that emerges in the definition of the Wigner function, is related to the difference argument ρ (see Problem 3.4).

The Wigner function can be expressed also in terms of the spectral amplitude (3.1.7):

$$W(K,R) = \int \langle u(K+K'/2)u^*(K-K'/2)\rangle e^{iK'R}d^4K'. \tag{3.1.20}$$

Using the Wigner function $W(K,R)$ (3.1.17) as the local spectrum $J_K(R)$ (3.1.15) yields the function Q in the form

$$Q = \delta(R - R')(2\pi)^{-4} e^{-iK\rho'}, \tag{3.1.21}$$

which, as can be readily seen satisfies the normalizing condition (3.1.16) and the condition of uniformity in R. It is not hard to verify that the quantity $W(K,R)$ (3.1.17) is real-valued, but need not be positive; that is, with this definition the spectrum can take on negative values as well.

Taking the inverse Fourier transform in eqn (3.1.17) we obtain the expression for the second moment Γ in terms of the Wigner function

$$\Gamma = \langle u(R+\rho/2)u^*(R-\rho/2)\rangle = \int W(K,R)e^{iK\rho}d^4K. \tag{3.1.22}$$

For a random process $\xi(t)$, the local instantaneous frequency spectrum may also be defined in the form of the Wigner function as

$$W(\omega,T) = \int \langle \xi(T+\tau/2)\xi^*(T-\tau/2)\rangle e^{i\omega\tau}(2\pi)^{-1}d\tau. \tag{3.1.23}$$

Both (3.1.17) and (3.1.23) are applicable not only to random but also to deterministic (non-fluctuating) quantities $u(x)$ and $\xi(t)$. Specifically, these representations can be used also to describe pulse signals bounded in time and space.

3.1.5 Wigner Function as a Local Spectral Density: Advantages and Disadvantages

Although different definitions of local spectra of nonuniform fluctuations can be constructed by choosing different functions Q in eqn (3.1.15), the most popular choice is the Wigner function. It has been first introduced in quantum mechanics (Wigner, 1932) in a somewhat different context — as quasi-probability, and has been treated by many authors (see, e.g., Mori et al., 1962; Balescu, 1963; Tatarskii, 1983). At present, in addition to quantum mechanics and statistical optics, Wigner's function is widely used in radar signal processing, in pattern recognition systems, and in many other applications (see, e.g., the review of Cohen, 1989).

Apart from relative simplicity, Wigner's function has many other advantages in the role of a local spectrum of quasi-uniform fields in statistical optics, namely,

- it is directly related with the mean square energy characteristics of radiation that are measured in common optical experiments,
- it naturally includes the case of partially coherent fields,
- it is convenient for a rigorous mathematical description in terms of wave equations,
- it ensures the conversion to the case of the ordinary spectrum for a statistically uniform field,
- for free radiation satisfying the d'Alembert wave equation, it strictly obeys the

transfer equation in free space and remains constant along the ray (although in the general case of nonuniform fields, the concept of a ray loses its informal physical sense), and

- it remains formally useful in the case of nonergodic fields when statistical means do not coincide with the mean over the segment of one realization.

All these advantages by far outperform the disadvantages of this function, namely,

- Wigner's function can assume negative values (for a quasi-uniform field, thought, this function is positive), and

- the function $W(K,R)$ may be nonzero where the field $u(R)$ vanishes.

The last property evolves from the meaning of the non-local argument R as a center of gravity of the observation points x_1 and x_2. We will discuss this property in some more detail in Section 3.1.7 below, and also in the ensuing section when discussing the generalized radiance (see Fig. 3.6).

3.1.6 Properties of the Wigner Function

Here we highlight some more (readily verifiable) properties of Wigner's function; these are also transferred to the case of the frequency spectrum (3.1.23) of the process $\xi(t)$ or the spatial spectrum of the field $u(\mathbf{r})$.

(i) The integral of $W(K,x)$ with respect to K gives the average intensity $\mathcal{J}(x)$ of the field $u(x)$:

$$\int W(K,x)d^4K = \langle |u(x)|^2 \rangle \equiv \mathcal{J}(x).$$

(ii) The integral of $W(K,x)$ with respect to x yields the average intensity $\mathcal{J}_K$ of the spectrum of the field $u(K)$

$$\int W(K,x)d^4x = (2\pi)^4 \langle |u(K)|^2 \rangle \equiv \mathcal{J}_K .$$

(iii) The integral of $W(K,x)$ with respect to K and x yields the total intensity of the field

$$\int W(K,x)d^4Kd^4x = \int \langle |u(x)|^2 \rangle d^4x = (2\pi)^4 \int \langle |u(K)|^2 \rangle d^4K .$$

(iv) As we shall see below, the relation

$$\int |W(K,x)|^2 \, d^4Kd^4x = (2\pi)^{-4} \int |\langle u(x')u^*(x'')\rangle|^2 \, d^4x'd^4x''$$

is intimately connected with the uncertainty principle.

(v) For a plane wave $u(x) = A\exp(ik_0x)$, the Wigner function is

$$W(K,x) = |A|^2 \, \delta(k-k_0).$$

(vi) For the field $u(x) = A\delta(x - x_0)$, localized at a point x_0, the spectrum will also be localized at x_0:

$$W(K,x) = (2\pi)^{-4}|A|^2\,\delta(x - x_0).$$

(vii) If $u(x)$ is zero outside some simply connected domain x with convex boundary, then $W(K,x)$ also vanishes outside this area. A similar property holds with respect to the spectrum $u(K)$ (3.1.7).

(viii) Substituting $u(x + x_0)$ for $u(x)$ or $u(K)\exp(iKx_0)$ for $u(K)$ carries $W(K,x)$ to $W(K,x + x_0)$.

Relations (ii) - (iv) are applicable only to bounded pulse signals rapidly decaying at infinity. This condition is, alas, not satisfied by the statistically uniform field when $W(K,x)$ is independent of x and all integrals in (ii) - (iv) turn to infinity.

3.1.7 Wigner Function and Uncertainty Relations

The Wigner function $W(K,x)$ is given by the Fourier transform with respect to the variable ρ, where the field u behaves as a function of the arguments $R \pm \rho/2$. This is, so to say, an incomplete Fourier transform, characterizing in the mean square the local properties of the Fourier transform $u(K)$ near the point R. This treatment of the Wigner function is appropriate in many, though not in all situations, as we shall see below.

For Fourier conjugated arguments x and K_x, we have the well known uncertainty relationship $\Delta x \Delta K_x \geq 1$ where Δx and Δk_x are the "widths" of the function and its Fourier transform, respectively. We demonstrate that, while R and K in eqn (3.1.17) are not Fourier conjugated, a similar uncertainty relationship holds for them as well.

First of all we note that we may assign the meaning of the uncertainty relation to the above relation (iv). Indeed, let the field u be non-fluctuating and square integrable, so that relations (iii) and (iv) are finite. Denote the full energy of the pulse by

$$\mathcal{E} \equiv \int |u(x)|^2\, d^4x. \tag{3.1.24}$$

Then, in view of (iv) we have

$$(2\pi)^4 \int |W(K,x)|^2\, d^4K d^4x = \int |u(x)|^2\, |u(x')|^2\, d^4x d^4x' = \mathcal{E}^2. \tag{3.1.25}$$

This relation may be called the condition of the constant volume of the Wigner function for fields with fixed total energy $\mathcal{E}$. From this condition it follows that if we change only the shape of the pulse u(x) keeping its total energy (3.1.24) constant, then a reduction of the width of W(K,x) in one of the arguments either K or x will have to increase the width of this function in the other argument, because the integral (3.1.25) must remain constant.

Qualitatively, the uncertainty relationship may be represented in different forms by considering different definitions of the width of W(k,R) as a function of its arguments. We express first the uncertainty relationship in a similar way to its quantum mechanical representation for the coordinate and momentum. To this end we consider an example of a random pulse with finite mean energy $\langle \mathcal{E} \rangle < \infty$. If one treats the quantity $\langle |u(x)|^2 \rangle$ as the mean energy density in the x space, then $(2\pi)^4 \langle |u(K)|^2 \rangle$ has the sense of the mean energy density in the K space. For simplicity we assume that the mean total pulse energy equals unity: $\langle \mathcal{E} \rangle = 1$. Then the volume average value of an arbitrary function of coordinates $f(x)$ may be defined as

$$\tilde{f}(x) \equiv \int f(x) \langle |u(x)|^2 \rangle d^4x = \int f(x) \mathcal{I}(x) d^4x = \int f(x) W(K,x) d^4x\, d^4K, \qquad (3.1.26)$$

and the averaging of an arbitrary function of wave vector K may be defined as

$$\begin{aligned} \tilde{f}(K) &\equiv \int f(K)(2\pi)^4 \langle |u(K)|^2 \rangle d^4K \\ &= \int f(K) \mathcal{I}_K d^4K \\ &= \int f(K) W(K,x) d^4x\, d^4K . \end{aligned} \qquad (3.1.27)$$

Let us assume for simplicity that the pulse average values of the coordinate x and wave vector K are zero:

$$\tilde{x} = \int x W(K,x) d^4x\, d^4K = 0, \qquad (3.1.28)$$

$$\tilde{K} = \int K W(K,x) d^4x\, d^4K = 0. \qquad (3.1.29)$$

Then the mean square pulse width in one of the component $R = (R_x, R_y, R_z, T)$, say,

$$(\Delta R_z)^2 \equiv \tilde{R}_z^2 = \int R_z^2 \mathcal{I}(R) d^4R = \int R_z^2 W(K_\rho, R) d^4R\, d^4K_\rho, \qquad (3.1.30)$$

and the similar width in the K space

$$(\Delta K_{\rho z})^2 \equiv \tilde{K}_{\rho z}^2 = \int K_{\rho z}^2 \mathcal{I}_{K_\rho} d^4K_\rho = \int K_{\rho z}^2 W(K_\rho, R) d^4R\, d^4K_\rho \qquad (3.1.31)$$

satisfy the uncertainty relationship (Problem 3.5)

$$2\Delta R_z \Delta K_{\rho z} \geq 1. \qquad (3.1.32)$$

We replaced here throughout the wave vector notation K with K_ρ to emphasize that this vector is Fourier conjugated with the difference variable ρ.

Similar relations hold true also for other components x and K.

For qualitative estimations, we will write the uncertainty relation [omitting the numerical factor in eqn (3.1.32)] as

$$\Delta R_z \Delta K_{\rho z} \geq 1. \tag{3.1.33}$$

This inequality relates ΔR_z, the width of the amplitude $u(x)$ in the variable z, with $\Delta K_{\rho z}$, the width of the Fourier spectrum of the field $u(K)$in the conjugated variable K_z. This relationship is applicable both for a fluctuating field and for a deterministic field devoid of fluctuations.

Inequality (3.1.33) is not a unique possible form of uncertainty relationship. According to (3.1.30) and (3.1.31), the quantities $\Delta R_z$and $\Delta K_{\rho z}$ in this relation characterize only the mean square width of the intensity distribution $\mathcal{I}(x) = \langle |u(x)|^2 \rangle$ in the x space, and $\mathcal{I}_K = (2\pi)^4 \langle |u(K)|^2 \rangle$ in the K space, but actually spell nothing about the behavior of the correlation of these amplitudes $\langle u(x_1) u^*(x_2) \rangle$ for $x_1 \neq x_2$ and $\langle u(K_1) u^*(K_2) \rangle$for $K_1 \neq K_2$. However, such information is contained in the initial Wigner function. We now demonstrate how additional uncertainty relationships, that are directly related with the correlation scales, can be obtained.

For simplicity we assume that the mean field value is zero, <u> = 0 (otherwise in place of the coherence function it suffices to consider the correlation function). Taking Fourier transforms of relations (i) and (ii) in Section 3.1.6 we have respectively

$$\begin{aligned} \hat{F}_{K_R \leftarrow x} \mathcal{I}(x) &= \int W(K_\rho, R) \frac{\exp(-iK_R R)}{(2\pi)^4} d^4 K_\rho d^4 R \\ &= \int \left\langle u\left(K_\rho + \frac{K_R}{2}\right) u^* \left(K_\rho - \frac{K_R}{2}\right) \right\rangle d^4 K_\rho \\ &\equiv b(K_R), \end{aligned} \tag{3.1.34}$$

and

$$\begin{aligned} F^{-1}_{\rho \leftarrow K_\rho} \mathcal{I}_{K_\rho} &= \int W(K_\rho, R) \exp(iK_\rho \rho) d^4 K_\rho d^4 R \\ &= \int \left\langle u\left(R + \frac{\rho}{2}\right) u^* \left(R - \frac{\rho}{2}\right) \right\rangle d^4 R \\ &\equiv \tilde{b}(\rho). \end{aligned} \tag{3.1.35}$$

The function $b(K_R)$ defined by eqn (3.1.34) describes the rate of decay of the correlations of the K amplitudes $u(K)$, and $b(\rho)$ in eqn (3.1.35) characterizes the similar rate of decay of the correlation of the x amplitudes $u(x)$. Indeed, in view of the assumption on the unit mean energy $\langle \mathcal{E} \rangle = 1$ we have $b(0) = \tilde{b}(0) = 1$. In addition, $b(K_r) \to 0$ for $K_R > \Delta K_R$, and $\tilde{b}(\rho) \to 0$ for $\rho > \Delta\rho$, where ΔK_R and $\Delta\rho$ are the scales of the correlations of the respective amplitudes in the K and x spaces. However, according to eqn (3.1.34), $b(K_R)$is the Fourier transform of the intensity $\mathcal{I}(x) = \langle |u(x)|^2 \rangle$, whose width in the z coordinate is estimated, according to (3.1.30), as ΔR_z. Therefore, in agreement with the well known property of the Fourier conjugated functions, for the pair $b(K_R)$ and $\mathcal{I}(x)$, we may write a new uncertainty relationship:

$$\Delta R_z \Delta K_{Rz} \geq 1\,, \tag{3.1.36}$$

and, similarly, for the pair of functions $\tilde{b}(\rho)$ and $\mathcal{I}_K = (2\pi)^4 \langle |u(k)|^2 \rangle$,

$$\Delta \rho_z \Delta K_{\rho z} \geq 1. \tag{3.1.37}$$

Let us consider the sense of these relationships. The quantity ΔR_z characterizes the width of the mean intensity $\mathcal{I}(R) = \langle |u(R)|^2 \rangle$ in the coordinate R_z in the x space and the quantity ΔK_{Rz} is the rate of decay of the amplitude correlations $\langle u(K_1) u^*(K_2) \rangle$ in the K space. According to eqn (3.1.36), the narrower is the intensity distribution, the wider is the correlation scale of the respective K amplitudes. Similarly, from eqn (3.1.37) it follows that the smaller $\Delta K_{\rho z}$ and, hence, the width of the intensity distribution $\mathcal{I}_K$ in the K space, the wider the respective correlation scale of the field $u(x)$ in the x scale. In contrast to inequality (3.1.32) valid also in the absence of fluctuations, eqns (3.1.36) and (3.1.37) refer only to fluctuating fields.

Obviously, the uncertainty relationships (3.1.36) and (3.1.37) are mutually symmetric under a permutation of x and K spaces. These relationships are directly related to the van Cittert–Zernike theorem which describes the far-field correlations by the Fourier transform of the source intensity. Similar relationships have been discussed in the textbook of Rytov, Kravtsov and Tatarskii (1989) and in a paper of Friberg and Wolf (1983).

The elucidated properties of Wigner's function $W(K,R)$ in many aspects meet intuitive anticipations of the local expansion of the spectrum in terms of plane waves. Therefore one could try to assign to $W(K,R)$ the meaning of a distribution of field mean energy over wave vectors K at a point R. However, in the general case, this cannot be done because the very concept of the wave vector is non-local and cannot be referred to *one* point R. Accordingly, the definition (3.1.17), like, perhaps, other possible definitions of local spectra, fails to meet *all* the requirements that can be imposed on the concept of the local spectrum.

Thus the non-local property of the wave vector is reflected in that $W(K,R)$ is a *two-point* statistical characteristic whose argument R, emerging in the coherence function Γ, refers to the center of gravity between two observation points rather than to a single point of observation. Therefore, if the field $u(R)$ is zero outside the region with a nonconvex boundary, then the function $W(K,R)$ may be other than zero outside this region.

Consider, for example, a system of two parallel slits in a screen. On the opaque screen $0 < x < a$ separating the slits the field $u(x)$ is zero, but the Wigner function is nonzero since the center-of-gravity coordinates $0 < R < a$ are associated with points x_1 and x_2 lying in the lit slits. This possibility seems to disagree with the common concept of the local spectrum and may lead to various paradoxes, if we overlook the nonlocal nature of the argument R. We revisit this topic in the ensuing section.

As will be recalled, for a statistically nonuniform second moment Γ, Wigner's function $W(K,R)$ is not positive; therefore, in the general case one cannot assign to it an energy meaning. The nonnegativity of Γ is guaranteed only for quasi-uniform field, that is, for fields whose deviations from uniform ones are not very large.

3.1.8 Quasi-Uniform Fields and Their Spectra

We call the coherence function quasi-uniform if the derivatives with respect to the coordinates of the center of gravity $R = (x_1 + x_2)/2$ are far smaller than the derivatives with respect to the difference argument $\rho = x_1 - x_2$, that is, if

$$|\partial_R \Gamma| << |\partial_\rho \Gamma|. \tag{3.1.38}$$

We note that this definition of quasi-uniformity does not require factorization of Γ as do the functions R and ρ and as has been assumed in the literature (see, e.g., Carter and Wolf, 1977). Simple consideration demonstrate that, in real situations, the quasi-uniformity condition (3.1.38) cannot be valid for all values of R and ρ because real fields are bounded in space and time. The conditions (3.1.38) should be understood as follows.

In the general case the coherence function Γ may be represented as a sum of the quasi-uniform part $\Gamma^{(1)}$ and the nonuniform part $\Gamma^{(2)}$:

$$\Gamma = \Gamma^{(1)} + \Gamma^{(2)}, \tag{3.1.39}$$

where $\Gamma^{(1)}$ obeys (3.1.38) whereas $\Gamma^{(2)}$ does not. The nonuniform part $\Gamma^{(2)}$ becomes essential near the interfaces, at the surfaces of inhomogeneities, in the presence of regular field nonuniformities near focal points, and at the time the field is switched on and off (nonstationarity). Conversely, far from the interfaces and in the absence of regular nonuniformities, the quasi-uniform part $\Gamma^{(1)}$ provides the predominant contribution.

Even when both terms in (3.1.39) are of the same order of magnitude, $\Gamma^{(2)}$, in contrast to $\Gamma^{(1)}$, is frequently a rapidly oscillating function of the center-of-gravity coordinate R. This behavior permits us to get rid of $\Gamma^{(2)}$ by smoothing in R at distances of several initial characteristic wavelengths.

It is important that, using a certain asymptotic procedure, one can obtain closed equations for the quasi-uniform part $\Gamma^{(1)}$, so that in a certain sense $\Gamma^{(1)}$ is a quantity independent of $\Gamma^{(2)}$. We will not examine specific procedures of splitting Γ into $\Gamma^{(1)}$ and $\Gamma^{(2)}$. Speaking about the quasi-uniform coherence function Γ below we will keep in mind the possibility to isolate the quasi-uniform part $\Gamma^{(1)}$, thus digressing from the nonuniform part $\Gamma^{(2)}$. Thus, the forthcoming exposition will be based on discarding the nonuniform term $\Gamma^{(2)}$ in eqn (3.1.39) and on the approximation of Γ by $\Gamma^{(1)}$, keeping in mind, of course, that actually every coherency function contains a nonuniform part, though small or rapidly oscillating.

The requirement of quasi-uniformity (3.1.38) may be rewritten as

$$|\partial_R \Gamma| \sim \mu |\partial_\rho \Gamma|, \tag{3.1.40}$$

where

$$\mu \sim L_\rho / L_R << 1 \tag{3.1.41}$$

is a small auxiliary parameter equal to the ratio of the scales L_ρ and L_R characterizing the variation rates of Γ in the arguments ρ and R, respectively. Since ρ is the difference variable, we will require that inequality (3.1.38) hold locally, or in the small, only for moderate values of ρ. In order of magnitude estimations we shall assume simply $\rho = 0$.

The quantity L_R entering in eqn (3.1.41) has the meaning of the scale of the statistical nonuniformity of the field. For a statistically uniform field, when $\Gamma(R,\rho) = \Gamma(\rho)$, L_R is infinite. The scale L_ρ is usually within the field correlation radius, but need not coincide with this radius. For example, for a plane wave $u = A\exp(iKx)$ with random amplitude A, we have $\Gamma = \langle |A|^2 \rangle \exp(iK\rho)$. Then, L_ρ is of the order of the characteristic wavelength or period of the field, $L_\rho \sim K^{-1}$, whereas the correlation radius, that depends on the behavior of the magnitude of Γ, is infinite in this case.

The conditions (3.1.38) and (3.1.41) imply that Γ varies along the center-of-gravity coordinate $R = (x_1 + x_2)/2$ slowly as compared with the fast variations in the difference variable $\rho = x_1 - x_2$. For simplicity we assume that the nonuniformity and nonstationarity are described by one small parameter μ, so that eqn (3.1.40) means the identical relative smallness of the spatial and temporal statistical nonuniformity of the coherence function Γ. In the limit as $\mu \to 0$ we arrive at the condition $\partial_R \Gamma = 0$, i.e., we obtain the coherence function Γ that depends only upon the difference variable. For this function, the spectrum (3.1.17) is a non-negative quantity. Therefore, it is natural to expect that , for small but nonzero values of μ, i.e., for a quasi-uniform field, the spectrum (3.1.17) will be non-negative too.

For brevity we will refer to the limit transition when $\mu \to 0$ as the *quasi-uniform limit*. In this limit, one might expect that all definitions of the spectra of nonuniform fluctuations that satisfy the conditions (3.1.38) and (3.1.41) will be asymptotically equivalent, i.e., will tend to a common limit — the spectrum of statistically uniform fluctuations. In the quasi-uniform limit, this behavior allows us to treat the spectrum (3.1.17) as a local energy characteristics of the field, viewing values of K as values of the local wave vector near point R. Actually, the transition from the spectrum of uniform fluctuations to the spectrum of quasi-uniform fluctuations is in many aspects analogous to the geometrical optics transition from the plane wave to the quasi-plane wave that is also characterized by some local value of the wave vector K. Therefore, in the case of eqn (3.1.17) we may speak of the local value of the wave vector, implying usual requirements of slow dependence of R, i.e., only for the quasi-uniform field.[1]

The quasi-uniformity conditions (3.1.41) may be formulated also in the spectral parlance, considering the second moment $\Gamma(K_1, K_2) = \langle u(K_1) u^*(K_2) \rangle$ of the spectral amplitudes $u(K)$, i.e., the coherence function Γ in the K representation. It is

[1] An excellent discussion of similar topics related with the concept of instantaneous frequency of quasi-harmonic oscillations has been given in the classical lectures of Mandelshtam (1972) that, unfortunately, have not been translated into English. About Mandelshtam, see the paper of Gaponov-Grekhov and Rabinovich (1979).

convenient to use in place of wave vectors K_1 and K_2 the vectors $K_\rho = (K_1 + K_2)/2$ and $K_R = K_1 - K_2$ that occur in the Fourier transformation of the second moment $\Gamma = \langle u(R+\rho/2)u^*(R-\rho/2)\rangle$ with respect to variables ρ and R and are the Fourier conjugates of these variables. This transition is schematically depicted in Fig. 3.1.

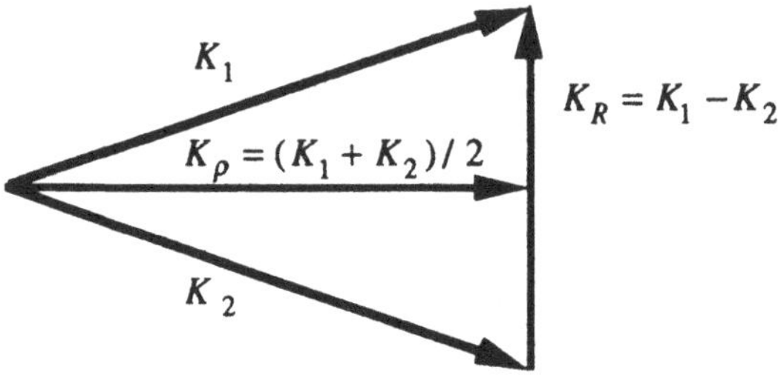

Figure 3.1. Wave vectors K_R and K_ρ corresponding to the sum and difference coordinates

In the statistically uniform case, the function $\Gamma(K_\rho, K_R)$ is strictly localized for $K_R = 0$, i.e., on the diagonal $K_1 = K_2$ on the (K_1, K_2) plane:

$$\Gamma(K_\rho, K_R) = \langle u(K_\rho + K_R/2)u^*(K_\rho - K_R/2)\rangle$$
$$= J_{k_\rho}\delta(K_R) = J_{k_\rho}\delta(K_1 - K_2). \tag{3.1.42}$$

We call the region $K_R \sim 0$ of the (K_1, K_2) plane near the diagonal $K_1 = K_2$ the *quasi-uniformity region*. The quasi-uniform part $\Gamma^{(1)}$ of the coherence function Γ differs from zero exactly in the quasi-uniformity region for $K_1 \approx K_2$. The width of this region ΔK_R may be estimated from the condition of quasi-uniformity (3.1.38) if we observe that, in agreement with the known properties of Fourier transformation, $\partial_R\Gamma \sim L_R^{-1}\Gamma \sim \Delta K_R\Gamma$, $\partial_\rho\Gamma \sim K_\rho\Gamma$, where K_ρ and ΔK_R are the characteristic wave vector and the width of the quasi-uniform part of the spectrum. Using these approximations, we may write eqn (3.1.38) as

$$L_R^{-1}\Gamma \sim \Delta K_R\Gamma << K_\rho\Gamma, \tag{3.1.43}$$

or

$$\Delta K_R << K_\rho \tag{3.1.44}$$

According to eqn (3.1.44), the quasi-uniformity region is associated with the spectral width ΔK_R small compared to the characteristic length of the wave vector K_ρ. From the definition of $W(K,R)$ in eqn (3.1.17) it is not hard to see that to the parameters L_R and K_ρ we may assign the sense of characteristics scales of variation of Wigner's function $W(K,R)$ in the arguments R and K, respectively.

3.1.9 Local Quasi-Uniform Coherence Function in the Geometrical Optics Approximation. The Field of Quasi-Uniform Sources in a Homogeneous Medium

We now demonstrate that, in the absence of multipath propagation, when only one ray passes through a point, an ordinary geometrical optics asymptotic representation of the field satisfies the quasi-uniformity conditions (3.1.38) and (3.1.41). Indeed, geometrical optics takes the field u in the form

$$u(x) = A(\mu x)\exp[i\Psi(\mu x)/\mu], \tag{3.1.45}$$

where A is the amplitude, Ψ is the phase, and μ is a small parameter comparable with the ratio of the wavelength to the characteristic length over which the properties of the medium or wave vary appreciably: $\mu \sim \lambda / L_R$ (see Appendix A). The coherence function for this field has the form

$$\Gamma = A\left(\mu\left(R+\frac{\rho}{2}\right)\right)A^*\left(\mu\left(R-\frac{\rho}{2}\right)\right)$$
$$\times\exp\left\{\frac{i}{\mu}\left[\Psi\left(\mu\left(R+\frac{\rho}{2}\right)\right)-\Psi\left(\mu\left(R-\frac{\rho}{2}\right)\right)\right]\right\}. \tag{3.1.46}$$

Assuming that the gradients of the amplitude and phase are quantities of one order of magnitude, $\nabla A / A \sim \nabla\Psi \sim \mu$, and letting $\mu \to 0$ we obtain

$$\partial_R\Gamma|_{\rho=0} \sim \mu\partial_\rho\Gamma|_{\rho=0} << \partial_\rho\Gamma|_{\rho=0}. \tag{3.1.47}$$

In this case, the small geometrical optics parameter μ plays the role of the small quasi-uniformity parameter (3.1.41).

This reasoning leads us to an important conclusion that the geometrical optics asymptotic of the coherence function Γ may be viewed as a particular case of the quasi-uniform limit. Therefore, in a theory of the quasi-uniform field below we will cover also the results that can be obtained for the second moment Γ using the method of geometrical optics to evaluate the field u. In other words, for the coherence function Γ, the quasi-uniform field theory is a definite generalization of the ordinary geometrical optics method.

In the limiting case of coherent radiation, for relatively small ρ, the quasi-uniformity (*local* quasi-uniformity) does not mean the *global* quasi-uniformity, that is, the fulfillment of the quasi-uniformity conditions (3.1.38) for all ρ. To illustrate this statement we consider the coherence function $\Gamma(\mathbf{R},\rho)$ for a monochromatic geometrical optics field in the form of two mutually coherent beams (Fig. 3.2). One can expect that the respective coherence function may be localized at small ρ and values of $\mathbf{R}$ near each beam, $\mathbf{R} \sim \mathbf{r}_1$ or $\mathbf{R} \sim \mathbf{r}_2$, where $\mathbf{r}_{1,2}$ are two points on the beams. However, in the case of coherent beams, the coherence function Γ acquires an additional term with values $\mathbf{R} \sim (\mathbf{r}_1 + \mathbf{r}_2)/2$ and $\rho = \mathbf{r}_1 - \mathbf{r}_2$. If we forget for a moment that the argument $\mathbf{R}$ is the radius vector of the center of gravity between two observation point, then this addend may be associated with an imaginary interference beam laying in between $\mathbf{r}_1$ and $\mathbf{r}_2$. In Fig. 3.2 this beam is shown by dashed line. Of course, a measurement instrument put in the way of this interference beam could not detect it. Interference rays occur only in interpretations of interference experiments where the field has been detected at distant points (in this case, at points $\mathbf{r}_1$ and $\mathbf{r}_2$) and they reflect the nonlocal character of the auxiliary argument **R**.

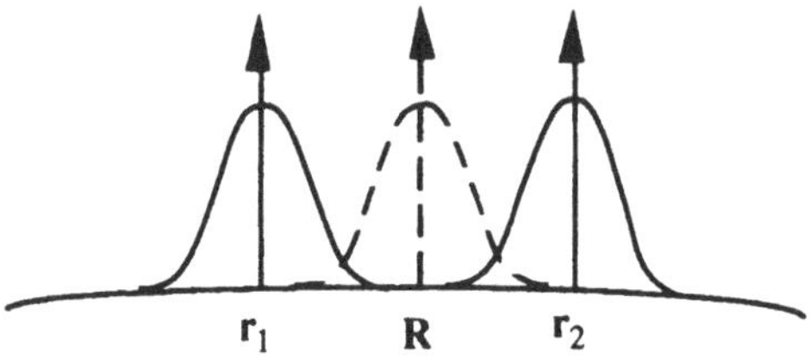

Figure. 3.2. Illustrating the mechanism of virtual interference beams. Two real beams (solid lines) give rise to an interference addend (dashed line) in the coherence function that correspond to $\mathbf{R} = (\mathbf{r}_1 + \mathbf{r}_2)/2$ and $\rho = \mathbf{r}_1 - \mathbf{r}_2$.

The addend correspondent to point $\mathbf{R} \sim (\mathbf{r}_1 + \mathbf{r}_2)/2$ is, generally speaking, nonuniform and rapidly oscillates in **R**, so that its description would go beyond the scope of the quasi-uniform field. This type of addends naturally occurs in descriptions of interference phenomena (see Problem 3.14 and Section 3.2.12).

In multipath propagation, the field is represented as a sum of terms like (3.1.45). In this case, the coherence function will contain rapidly oscillation interference terms that do not obey the quasi-stationarity conditions. Description of such terms is beyond the applicability limits of the theory of the quasi-uniform field, however, often they are of low significance, because they vanish on space or time averaging over lengths exceeding the characteristic scales of interference fringes.

In the forthcoming discussion we intent to use the following simple assertion: in a homogeneous medium, the field of quasi-uniform sources is quasi-uniform. The proof of this statement is straightforward. The field u induced by the sources q in a homogeneous medium has the form

$$u(x) = \int G(x - x')q(x')dx', \tag{3.1.48}$$

where $G(x)$ is the Green's function. For the coherence function we have

$$\begin{aligned}\Gamma &= \left\langle u\left(R + \frac{\rho}{2}\right)u^*\left(R - \frac{\rho}{2}\right)\right\rangle \\ &= \int G\left(R' + \frac{\rho'}{2}\right)G^*\left(R - \frac{\rho'}{2}\right)b(\mu(R - R'), \rho - \rho')d^4R'd^4\rho',\end{aligned} \tag{3.1.49}$$

where

$$b(\mu R,\rho) = \left\langle q\left(R+\frac{\rho}{2}\right)q^*\left(R-\frac{\rho}{2}\right)\right\rangle \tag{3.1.50}$$

is the second moment of the quasi-uniform sources q. The small parameter u owes its existence here to the assumption of quasi-uniform sources. From eqn (3.1.49) we seen that the coherence function Γ_{12} depends only upon μR, i.e., is also quasi-uniform.

Digressing from the possibility of focusing, we may formulate the above assertion as the conservation of quasi-uniformity in propagation of the wave field in a free space, and this formulation proves important in a substantiation of radiometry. Indeed, the propagation of a wave field in agreement with the Huygens-Fresnel principle may be described as the process of successive formation of virtual secondary sources. The total field of such sources gives the field value at a point of interest. If the propagating field turns out to be quasi-uniform at some surface, than this will correspond to the quasi-uniformity of the secondary sources on this surface that, according to eqn (3.1.49), will excite the quasi-uniform wave field. Thus, once the quasi-uniformity occurs, its is conserved in subsequent propagation of the wave field.

3.1.10 Fluctuation Spectrum of the Wave Field Far from the Sources

The statistical moments (3.1.2) and (3.1.3) may refer to arbitrary random fields, specifically, to *wave* random fields. The latter are the solutions of wave equations and therefore impose familiar constraints on feasible field spectra. The spectra of free fields are outstanding in that they are concentrated on a so-called dispersion surface. We elucidate this in some detail.

Consider first the radiation of random sources in a free space, that is, in a homogeneous, stationary, non-scattering and nonabsorbing medium. We assume that this field satisfies eqn (3.1.1); we rewrite this equation in integral form

$$\hat{L}u = \int L(x-x')u(x')d^4x' = q, \tag{3.1.51}$$

where $\hat{L}$ is a deterministic operator having a difference kernel $L(x-x')$, and q is the random function of the sources. For a singular kernel L, eqn (3.1.51) includes also differential wave operators. For simplicity we assume that $\langle q\rangle = 0$ and $\langle u\rangle = 0$ so that the coherence function Γ coincides with the correlation function Ψ.

The principal model of the wave equation (3.1.51) will be d'Alembert's equation

$$\hat{L}u = (\Delta - c^{-2}\partial_t^2)u = 0, \tag{3.1.52}$$

that is, the case when $\hat{L}$ is the d'Alembert operator. This equation describes the propagation of waves of different physical nature in a homogeneous and stationary non-dispersive medium.

Since we seek the field far from the sources we let $q = 0$ in (3.1.51). Then, expanding the field u in the Fourier integral in terms of plane monochromatic waves, i.e., writing the expression

$$u(x) = \int u(K) e^{iKx} d^4K, \tag{3.1.53}$$

inverse to (3.1.7) we obtain for $u(K)$

$$L(K)u(K) = 0, \tag{3.1.54}$$

where

$$L(K) = \int L(x') e^{-iKx'} d^4x' = e^{-iKx} \hat{L} e^{iKx}. \tag{3.1.55}$$

In the general case of a multicomponent field, $L(K)$ is a matrix dependent upon K. In the case of d'Alembert's operator, from eqns (3.1.52) and (3.1.55) we obtain

$$L(K) = e^{-iKx}(\Delta - c^{-2}\partial_t^2)e^{iKx} = e^{-i(\mathbf{kr}-\omega r)}(\Delta - c^{-2}\partial_t^2)e^{i(\mathbf{kr}-\omega r)}$$
$$= -k^2 + (\omega / c)^2 . \tag{3.1.56}$$

From eqn (3.1.54) we see that far from the sources, the random spectrum $u(K)$ is not arbitrary and can be nonzero only for wave numbers K satisfying the dispersion equation

$$\det L(K) = 0. \tag{3.1.57}$$

For one-component field, this determinant coincides with $L(K)$. In the particular case of d'Alembert's equation (3.1.52), the dispersion equation

$$L(K) = -k^2 + (\omega / c)^2 = 0 \tag{3.1.58}$$

gives $k = k(\omega) = \omega / c$, or $\omega = \omega(\mathbf{k}) = kc$.
Thus, we may write

$$u(K) = A'(\mathbf{k})\delta(L(K)) = A_{\mathbf{k}}\delta(\omega - \omega(\mathbf{k})), \tag{3.1.59}$$

where $\omega(\mathbf{k})$ is the root of the dispersion equation (3.1.57), and $A_{\mathbf{k}} = A'(\mathbf{k})|\partial_\omega L(K)|^{-1}$ has the meaning of the amplitude of a plane wave with wave vector $\mathbf{k}$. We used here the known formula

$$\delta(f(x)) = \delta(x - x_0)/|\partial_x f(x)|,$$

where x_0 is the simple root of the equation $f(x) = 0$. The presence of the delta function $\delta(L(K))$ in (3.1.59) implies that the random amplitude $u(K)$ may be other than zero only for wave vectors satisfying the dispersion equation $L(K) = 0$.

In eqn (3.1.59) we assume that the dispersion equation (3.1.57) has a unique and real valued solution $\omega = \omega(\mathbf{k})$, which corresponds to a single mode radiation in a conservative medium. If the medium can propagate several types of waves, then eqn (3.1.59) must be replaced with

$$u(K) = \sum_{\nu} A_{\mathbf{k}}^{\nu} \delta(\omega - \omega_{\nu}(k)), \tag{3.1.60}$$

where the summation is over all types of waves.

According to eqns (3.1.53) and (3.1.59), far from the sources, the field is given as a set of plane traveling waves

$$u(x) = \int A_{\mathbf{k}} e^{i(\mathbf{kr} - \omega(\mathbf{k})t)} d^3k. \tag{3.1.61}$$

Inverting this equation yields

$$A_{\mathbf{k}} e^{-i\omega(\mathbf{k})t} = \int u(x) e^{-i\mathbf{kr}} (2\pi)^{-3} d^3r. \tag{3.1.62}$$

This implies that the spatial spectrum of a free wave field, that is, the right-hand side of (3.1.62), becomes automatically dependent on time as $\exp(-i\omega(\mathbf{k})t)$ in view of the fact that the field satisfies the dispersion equation (3.1.57).

Now we turn to statistical characteristics of the field in a free space. If this field is statistically uniform and stationary, then by the Wiener-Khinchin theorem, plane waves with different wave vectors $K_1 \neq K_2$ are uncorrelated. The fluctuation spectrum J_K, as the random spectrum of the field (3.1.59) must be localized on the dispersion surface (3.1.57), that is,

$$J_K = I'_K \delta(L(K)) = I_{\mathbf{k}} \delta(\omega - \omega(\mathbf{k})), \tag{3.1.63}$$

where $I_{\mathbf{k}} = I'_{\mathbf{k}} |\partial_{\omega} L(K)|^{-1}$. The coherence function of such a field is representable in the form similar to (3.1.61)

$$\Gamma(x_1 - x_2) = \langle u(x_1) u^*(x_2) \rangle = \int I_{\mathbf{k}} e^{i(\mathbf{k}\boldsymbol{\rho} - \omega(\mathbf{k})\tau)} d^3k. \tag{3.1.64}$$

From this expression we obtain a relation like (3.1.62), but this time not for the field u but for the coherence function Γ:

$$I_{\mathbf{k}} e^{-i\omega(\mathbf{k})\tau} = \int \Gamma(\rho) e^{-i\mathbf{k}\boldsymbol{\rho}} (2\pi)^{-3} d^3\boldsymbol{\rho}. \tag{3.1.65}$$

Here, $\rho = x_1 - x_2 = (\boldsymbol{\rho}, \tau)$. This relation indicates that the spatial fluctuation spectrum of a statistically uniform and stationary free wave field depends on the time interval $\tau = t_1 - t_2$ between the observation instants t_1 and t_2 as $\exp(-i\omega(\mathbf{k})\tau)$, that is, by a harmonic law.

In view of the delta-correlation property $\langle u(K) u^*(K_1) \rangle = J_K \delta(K - K_1)$, for stationarity it suffices to assume only that the field is statistically uniform

$$\langle A_{\mathbf{k}} A^*_{\mathbf{k}_1} \rangle = I_{\mathbf{k}} \delta(\mathbf{k} - \mathbf{k}_1). \tag{3.1.66}$$

Indeed, the property of stationarity, i.e. the delta correlated behavior of the second moment $\langle u(K)u^*(K_1)\rangle$ in frequencies ω and ω_1, follows from the strict relation between the wave vector and frequency $\omega = \omega(\mathbf{k})$ imposed by the dispersive equation (3.1.57). In other words, in a nonabsorbing stationary and homogeneous medium, the spatially uniform free wave field is also stationary.

In the multimode case when there is a few solutions of the dispersion equation (3.1.57) $[\omega = \omega_\nu(\mathbf{k}),\ \nu = 1,2,\ldots]$, the spectrum of statistically uniform fluctuations will contain a sum of terms like (3.1.63) corresponding to each type of wave

$$J_K = \sum_\nu I_{\mathbf{k}}^{(\nu)} \delta(\omega - \omega_\nu(\mathbf{k})). \tag{3.1.67}$$

In addition, a correlation term can occur that is other than zero at the intersection of dispersion surfaces $\omega_\nu(\mathbf{k}) = \omega_\mu(\mathbf{k}),\ \nu \neq \mu$. This type of terms have to be taken into account only in describing the polarization degenerate waves.

3.1.11 Extension to Quasi-Uniform Fields

When one speaks about the uniformity and stationarity of physical fields, he in fact implies the quasi-uniformity and quasi-stationarity, because real fields are always to some or other degree nonuniform and non-stationary. For a rigorous description of quasi-uniform fields, the uniform fluctuation spectrum J_K (3.1.11) must be replaced with the nonuniform fluctuation spectrum $J_K(R)$ for which we have assumed to use the Wigner function $W(K,R)$. Substituting in (3.1.20) the spectral amplitude (3.1.59) and thus taking into account the wave nature of the field, after simple algebra we have

$$W(K,R) = \int \left\langle A_{\mathbf{k}+\mathbf{k}'/2} A^*_{\mathbf{k}-\mathbf{k}'/2} \right\rangle \delta\left(\omega - \frac{\omega_+ + \omega_-}{2}\right) \times \exp\left[i\mathbf{k}'\mathbf{R} - i(\omega_+ - \omega_-)T\right] d^3k', \tag{3.1.68}$$

where $\omega_\pm = \omega(\mathbf{k} \pm \mathbf{k}'/2)$, and $R = (\mathbf{R},T)$.

In the case of a spatially homogeneous field, $W(K,R)$ does not depend on the vector $\mathbf{R}$ and the integrand in (3.1.68) may be other than zero only at $\mathbf{k}' = 0$. In the quasi-uniform limit, $\mathbf{k}'$ remains far smaller than $\mathbf{k}$ in magnitude: $k' << k$. This allows us to expend $\omega_\pm$ in powers of $\mathbf{k}'$ letting $\omega_\pm \approx \omega(\mathbf{k}) \pm \mathbf{v}_g \mathbf{k}'/2$, where $\mathbf{v}_g = d_{\mathbf{k}}\omega(\mathbf{k})$ is the group velocity. As a result, eqn (3.1.68) becomes

$$W(K,R) = I_{\mathbf{k}}(\mathbf{R} - \mathbf{v}_g T)\delta(\omega - \omega(\mathbf{k})), \tag{3.1.69}$$

where

$$I_{\mathbf{k}}(\mathbf{R}) = \int \left\langle A_{\mathbf{k}+\mathbf{k}'/2} A^*_{\mathbf{k}-\mathbf{k}'/2} \right\rangle \exp(i\mathbf{k}'\mathbf{R}) d^3k'. \tag{3.1.70}$$

This relation extends the expression

$$I_{\mathbf{k}}(\mathbf{R}) = \int \left\langle A_{\mathbf{k}+\mathbf{k}'/2} A^{*}_{\mathbf{k}-\mathbf{k}'/2} \right\rangle d^3k' \tag{3.1.71}$$

following from (3.1.66) and valid for the uniform field, to the quasi-uniform case.

The dependence of $W(K,R)$ on R describes the statistical uniformity of the field, and the components of R enter eqn (3.1.69) only in the form of $\mathbf{R} - \mathbf{v}_g T$; therefore, from (3.1.72) it follows that the nonuniformities of wave field fluctuations travel with the group velocity $\mathbf{v}_g$. Having represented the coherence function Γ in terms of Wigner's function $W(K,R)$ in agreement with (3.1.22), we have

$$\Gamma(R,\rho) = \int I_{\mathbf{k}}\left(\mathbf{R} - \mathbf{v}_g T\right) \exp\left[i(\mathbf{k}\boldsymbol{\rho} - \omega(\mathbf{k})\tau)\right] d^3k. \tag{3.1.72}$$

This relation can be inverted with the Fourier transformation thus relating the spectral density with the field coherence function

$$I_{\mathbf{k}}\left(\mathbf{R} - \mathbf{v}_g T\right) e^{-i\omega(\mathbf{k})\tau} = \int \Gamma(R,\rho) e^{-i\mathbf{k}\boldsymbol{\rho}} \frac{d^3\boldsymbol{\rho}}{(2\pi)^3}. \tag{3.1.73}$$

We recall that here $\Gamma(R,\rho) = \langle u(R+\rho/2) u^*(R-\rho/2) \rangle$ with $R = (\mathbf{R},T)$ and $\rho = (\boldsymbol{\rho}, \tau)$.

Equation (3.1.73) generalizing eqn (3.1.65) onto the quasi-uniform case allows us to write the spectral density $I_{\mathbf{k}}$ in the form

$$I_{\mathbf{k}}\left(\mathbf{R} - \mathbf{v}_g T\right) = e^{i\omega(\mathbf{k})\tau} \int \Gamma(R,\rho) e^{-i\mathbf{k}\boldsymbol{\rho}} \frac{d^3\boldsymbol{\rho}}{(2\pi)^3}. \tag{3.1.74}$$

In particular, letting here $\tau = 0$ yields

$$I_{\mathbf{k}}\left(\mathbf{R} - \mathbf{v}_g T\right) = \int \Gamma(R,\rho)\big|_{\tau=0} e^{-i\mathbf{k}\boldsymbol{\rho}} \frac{d^3\boldsymbol{\rho}}{(2\pi)^3}. \tag{3.1.75}$$

Finally, calculating the temporal spectrum of both sides of (3.1.73) we obtain

$$\int \Gamma(R,\rho) e^{-i\mathbf{k}\boldsymbol{\rho}} \frac{d^4\rho}{(2\pi)^4} = I_{\mathbf{k}}\left(\mathbf{R} - \mathbf{v}_g T\right) \delta(\omega - \omega(\mathbf{k})). \tag{3.1.76}$$

We note that the inversion of eqn (3.1.72), i.e. the representation of the spectrum $I_{\mathbf{k}}$ in terms of the second moment Γ is not one-one. For example, multiplying both parts of (3.1.76) by an arbitrary function $f(\omega)$ and integrating with respect to frequency ω we can obtain another expression for $I_{\mathbf{k}}$ in terms of Γ different from (3.1.75). This lack of uniqueness is associated with the fact that (3.1.73) is valid not for an arbitrary field, but only for the quasi-uniform wave field, for which the Fourier transform on the right hand side of (3.1.73) automatically yields a harmonic dependence on the difference time like $\exp[-i\omega(\mathbf{k})\tau]$. In the general case of an arbitrary nonuniform wave field, eqn (3.1.73) does not hold.

3.1.12 Frequency-Angular Spectrum and Radiance

In order to pass over from integration with respect to $\mathbf{k}$ to the integration with respect to ω and wave vector directions $\mathbf{n} = \mathbf{k} / k$ we change the normalization of $I_{\mathbf{k}}$. To this end it suffices to use in place of $I_{\mathbf{k}}$ the frequency-angular spectrum $I_{\omega\mathbf{n}}$ defined by

$$I_{\mathbf{k}} d^3k = I_{\omega\mathbf{n}} d\omega d\Omega_{\mathbf{n}}, \tag{3.1.77}$$

where ω and $\mathbf{k}$ are related by the dispersion equation (3.1.57). Substituting this expression in (3.1.72) we evaluate the relation of $I_{\omega\mathbf{n}}$ with the coherence function Γ

$$\Gamma(R,\rho) = \int I_{\omega\mathbf{n}}(\mathbf{R} - \mathbf{v}_g T) e^{i(\mathbf{k}\rho - \omega\tau)} d\omega d\Omega_{\mathbf{n}}. \tag{3.1.78}$$

For a non-dispersive medium with phase velocity $c = \omega / k = const$, (3.1.77) yields

$$I_{\omega\mathbf{n}} = I_{\mathbf{k}} d^3k / d\omega d\Omega_{\mathbf{n}} = I_{\mathbf{k}} k^2 d_\omega k = I_{\mathbf{k}} k^2 / c, \tag{3.1.79}$$

then eqn (3.1.76) may be rewritten as

$$\int \Gamma(R,\rho) e^{-ik\rho} \frac{d^4\rho}{(2\pi)^4} = I_{\omega\mathbf{n}}(\mathbf{R} - \mathbf{n}cT) k^2 \delta(k - k(\omega)), \tag{3.1.80}$$

where $k(\omega) = \omega / c$ is the wave number and $\mathbf{n}c$ plays the role of group velocity $\mathbf{v}_g$.

We consider in some detail the case of monochromatic field when in place of $u(x)$ we take the function $u(\mathbf{r})\exp(-i\omega_0 t)$, where the amplitude $u(\mathbf{r})$ describes the spatial structure of the field. Substituting $u(x) = u(\mathbf{r})\exp(-i\omega_0 t)$ on the left-hand side of (3.1.80) and taking the integral with respect to the difference time τ we find

$$\begin{aligned} & I_{\omega\mathbf{n}}(\mathbf{R} - \mathbf{n}cT) k^{-2} \delta(k - k(\omega)) \\ & = \delta(\omega - \omega_0) \int \left\langle u\left(\mathbf{R} + \frac{\rho}{2}\right) u^*\left(\mathbf{R} - \frac{\rho}{2}\right) \right\rangle \frac{\exp(-ik\rho)}{(2\pi)^3} d^3\rho. \end{aligned} \tag{3.1.81}$$

In the case of monochromatic field, the frequency-angular spectrum $I_{\omega\mathbf{n}}$ is concentrated at a frequency ω_0, that is,

$$I_{\omega\mathbf{n}} = I_{\mathbf{n}} \delta(\omega - \omega_0), \tag{3.1.82}$$

where $I_{\mathbf{n}} = I_{\mathbf{n}}(\mathbf{R})$ is the angular spectrum of this radiation. Substituting this expression in (3.1.81) and equating the factors at $\delta(\omega - \omega_0)$ we obtain

$$\begin{aligned} & I_{\mathbf{n}}(\mathbf{R}) k^{-2} \delta(k - k(\omega_0)) \\ & = \int \left\langle u\left(\mathbf{R} + \frac{\rho}{2}\right) u^*\left(\mathbf{R} + \frac{\rho}{2}\right) \right\rangle \frac{\exp(-i\mathbf{k}\rho)}{(2\pi)^3} d^3\rho \equiv W(\mathbf{k},\mathbf{R}), \end{aligned} \tag{3.1.83}$$

where $W(\mathbf{k},\mathbf{R})$ is the Wigner distribution for the spatial amplitude $u(\mathbf{R})$. We also let here $T=0$ because the right- and, hence, the left-hand side of (3.1.81) does not depend on T so that the radiance $I_{\omega\mathbf{n}}$ is stationary. Taking the Fourier transform of both parts of this relation we express the second moment of the spatial amplitude $u(\mathbf{r})$ in terms of the angular spectrum $I_{\mathbf{n}}$:

$$\Gamma(\mathbf{R},\rho) \equiv \left\langle u\left(\mathbf{R}+\frac{\rho}{2}\right)u^*\left(\mathbf{R}-\frac{\rho}{2}\right)\right\rangle = \int I_{\mathbf{n}}(\mathbf{R})\exp\left(ik(\omega_0)\mathbf{n}\rho\right)d\Omega_{\mathbf{n}}\,. \tag{3.1.84}$$

The inverse expression for the angular spectrum $I_{\mathbf{n}}$ in terms of the second moment $\Gamma(\mathbf{R},\rho)$ may be obtained by integrating both sides of (3.1.83) with respect to k (Ovchinnikov and Tatarskii, 1972)

$$I_{\mathbf{n}}(\mathbf{R}) = \int_0^\infty k^2dk\int\left\langle u\left(\mathbf{R}+\frac{\rho}{2}\right)u^*\left(\mathbf{R}-\frac{\rho}{2}\right)\right\rangle\frac{\exp(-i\mathbf{k}\rho)}{(2\pi)^3}d^3\rho$$

$$\equiv \int_0^\infty k^2dkW(\mathbf{k},\mathbf{R})\,. \tag{3.1.85}$$

We see that in a free space the spectral densities $I_{\mathbf{k}}$ and $I_{\omega\mathbf{n}}$ differ from the radiometric radiance $I = I_\omega$ only by constant multipliers, since in agreement with eqns (2.1.1), (3.1.66) and (3.1.79), all three quantities are proportional to the mean square amplitudes of quasi-plane waves. Therefore, we may write

$$I = C(k)I_{\mathbf{k}} = C(k)(k^2d_\omega k)^{-1}I_{\omega\mathbf{n}}, \tag{3.1.86}$$

where $C(k)$ is a proportionality coefficient depending upon the nature of the field. The explicit form of this coefficient can be readily evaluated at each particular case (see an example of thermal radiation in Section 3.1.14 and Problem 3.6). In what follows we will use the term radiance with respect to all three quantities I, $I_{\mathbf{k}}$, and $I_{\omega\mathbf{n}}$. More often than not, we shall consider the frequency-angular spectrum $I_{\omega\mathbf{n}}$ as radiance. Thus, the radiance allows us to find not only the flux and energy density from the classical radiometric relations like (2.1.16) and (2.1.17), but also the coherence function Γ of the wave field. In Section 2.1, from radiometric considerations we conjectured a similar relation (2.1.25), however, within the phenomenological theory, the proportionality coefficient entering in this relation remained undetermined.

According to (3.1.78), for the coherence function of the free wave field, the dependence on delay time τ is achieved by taking the Fourier transform of the radiance $I_{\omega\mathbf{n}}$ with respect to ω and the dependence on on the spatial separation ρ is achieved by taking the Fourier transform with respect to the angular variables $\mathbf{n}$. Essentially, eqn (3.1.78) is nothing other than a special form of the Wiener–Khinchin theorem, which takes into account the wave nature of the object under study.

While the relation of the frequency spectrum with the temporal correlation of the field, i.e. with the dependence of Γ on delay time τ is well known for a long time, the meaning of radiance as the angular spectrum of the spatial correlation of radiation has attracted attention only after the paper of Dolin (1964) considered the connection of the radiative transfer theory in a free space with the rigorous statistical wave theory in the small angle approximation (Section 4.1). We mention that for a equilibrium thermal radiation, this type of connection has been used before, for example, by Bourret (1960), however, this case is considerably simpler than eqn (3.1.78), because it corresponds to a uniform and isotropic radiation (Section 3.1.15). Later, expressions of the type (3.1.78) have been considered from various aspects by a number of authors (Walther, 1968; Apresyan, 1975; Carter and Wolf, 1975; Baltes, 1978; Wolf, 1978).

3.1.13 Uncertainty Relations for the Radiance

The uncertainty relations for the Wigner function from Section 3.1.7 are valid for an arbitrary wave field u. They can be extended to the case of a free wave field by writing them for the radiance. The principal difference from the general case consists here in that the wave vector of the free wave field is not arbitrary, but must satisfy the dispersion equation $k = \omega / c$.

Consider the case of monochromatic stationary radiation. Specifying a frequency defines the wave vector magnitude k, thus allowing one to cross from the wave vector $\mathbf{k}$ to the respective angular variable. To demonstrate, we take the vector $\mathbf{k}$ in the form $\mathbf{k} = k(\omega)\mathbf{n}$, where $k(\omega) = \omega / c = 2\pi / \lambda$, and $\mathbf{n}$ is the unit vector. For a beam propagating around a direction $\mathbf{z}$, the uncertainty of the wave vector in the transverse direction $\mathbf{x}$ is given by $\Delta k_x = k(\omega)\Delta n_x \equiv k(\omega)\Delta\theta$, where $\Delta\theta$ is the uncertainty of the orientation angle of $\mathbf{k}$ about the z axis (Fig. 3.3)

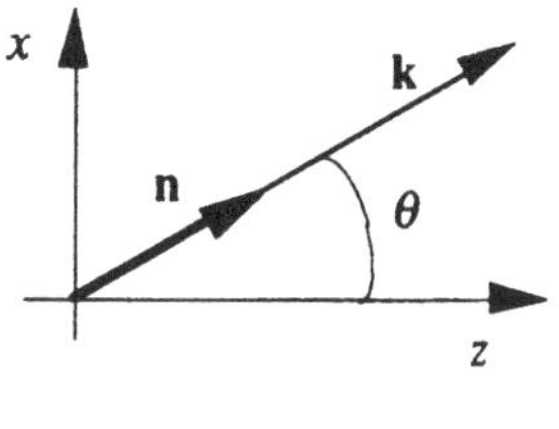

Figure 3.3

This expression can be substituted into the uncertainty relations (3.1.33), (3.1.36) and (3.1.37) thus passing to the uncertainty of the respective angle $\Delta\theta$. As a result, accurate to multipliers of order unity we have

$$\Delta R_x \Delta\theta_\rho \geq \lambda, \tag{3.1.87}$$

$$\Delta R_x \Delta\theta_R \geq \lambda, \tag{3.1.88}$$

$$\Delta \rho_x \Delta\theta_\rho \geq \lambda. \tag{3.1.89}$$

In agreement with the results of Section 3.1.7, ΔR_x and $\Delta\theta_\rho$ are the uncertainties of the space-angle beam intensity distribution that characterize the width of the Wigner function $W(\mathbf{k}, \mathbf{R})$, or else the radiance $I_{\mathbf{n}}(\mathbf{R})$ from (3.1.85), in the argument $\mathbf{R}$ and directions $\mathbf{n}$. These quantities are applicable both to coherent and to partially coherent fields.

By contrast, the uncertainties $\Delta\rho_x$ and $\Delta\theta_R$ characterize, respectively, the scales of the spatial and angular correlations and become meaningful only for the partially coherent, i.e. fluctuating, radiation. These uncertainties describe the width of the correlation functions $b(\mathbf{k}_R)$ (3.1.34) and $\tilde{b}(\rho)$ (3.1.35),that in view of eqns (3.1.83) and (3.1.84) take the form

$$b(\mathbf{k}_R) = \int \left\langle u\left(\mathbf{k}_\rho + \frac{\mathbf{k}_R}{2}\right) u^*\left(\mathbf{k}_\rho - \frac{\mathbf{k}_R}{2}\right) \right\rangle d^3 k_\rho$$

$$= \int I_\mathbf{n}(\mathbf{R}) \exp(-i\mathbf{k}_R \mathbf{R}) \frac{d^3 R}{(2\pi)^3} d\Omega_\mathbf{n} \tag{3.1.90}$$

and

$$\tilde{b}(\rho) = \int \left\langle u\left(\mathbf{R} + \frac{\rho}{2}\right) u^*\left(R - \frac{\rho}{2}\right) \right\rangle d^3 R$$

$$= \int I_\mathbf{n}(\mathbf{R}) \exp(i\mathbf{k}_{(\omega)} \mathbf{n}\rho) d^3 R \, d\Omega_\mathbf{n} \,. \tag{3.1.91}$$

In our subsequent consideration, we shall refer to the uncertainty relations (3.1.87) through (3.1.89).

3.1.14 Equilibrium Thermal Electromagnetic Radiation

This case differs from that of the ordinary scalar field by the need to take into account polarization. As an initial equation, we may take here the wave equation for the electrical field $\mathbf{E}$ in vacuum:

$$L\mathbf{E} = -(\nabla \times \nabla \times + c^{-2} \partial_t^2)\mathbf{E} = 0, \tag{3.1.92}$$

where c is the velocity of light, and $\nabla\times = \mathrm{curl}$. Here $\mathbf{E}$ will be understood as the complex field, so that the physical field is $\mathrm{Re}\,\mathbf{E}$. Then

$$L(K) = \mathbf{k} \times \mathbf{k} \times + (\omega / c)^2 = [(\omega / c)^2 - \mathbf{k}^2]\hat{\rho} + (\omega / c)^2 \hat{\pi}. \tag{3.1.93}$$

The term with $\hat{\rho} = \hat{1} - \hat{\pi} = 1 - \mathbf{k} \otimes \mathbf{k} / k^2$ corresponds to transverse waves, and the term with $\hat{\pi} = \mathbf{k} \otimes \mathbf{k} / k^2$ corresponds to the non-propagating longitudinal field that can be disregarded for $\omega \neq 0$. The solution of the dispersion equation

$$|L(K)| = \det L(K) = [(\omega / c)^2 - k^2]^2 (\omega / c)^2 = 0 \tag{3.1.94}$$

for transverse waves yields $\omega(\mathbf{k}) = kc$, and eqn (3.1.59) becomes

$$\mathbf{E}(K) = \mathbf{A_k}\delta(\omega - kc) = A_\mathbf{k}\mathbf{e_k}\delta(\omega - kc), \tag{3.1.95}$$

where $\mathbf{A_k}$ is the amplitude, and $\mathbf{e_k}$ is the unit vector of polarization that must be orthogonal to $\mathbf{k}$: $\mathbf{k}\mathbf{e_k} = 0$, $\mathbf{e_k}\mathbf{e_k^*} = 1$. Substituting $\mathbf{A_k} = A_\mathbf{k}\mathbf{e_k}$ in the spatial uniformity condition (3.1.66) of the field we obtain

$$\left\langle \mathbf{A_k} \otimes \mathbf{A^*_{k_1}} \right\rangle = \hat{I}_\mathbf{k}\delta(\mathbf{k} - \mathbf{k}_1) = \frac{1}{2} I_\mathbf{k}\hat{\rho}\delta(\mathbf{k} - \mathbf{k}_1). \tag{3.1.96}$$

Since the equilibrium radiation is isotropic, the matrix $I_\mathbf{k}$ may be written as $\hat{I}_\mathbf{k} = (1/2)I_\mathbf{k}\hat{\rho}$, where $I_\mathbf{k} = \mathrm{tr}\hat{I}_\mathbf{k}$. Equation (3.1.96) lead us to an analog of eqn (3.1.72) for the correlation function associated with transverse waves

$$\Gamma_{12} = \langle \mathbf{E}(x_1) \otimes \mathbf{E}^*(x_2) \rangle = \int \frac{1}{2} I_\mathbf{k}\hat{\rho} e^{i(\mathbf{k\rho} - \omega(\mathbf{k})\tau)} d^3k, \tag{3.1.97}$$

where $\omega(\mathbf{k}) = kc$.

The spectral density $I_\mathbf{k}$ may be expressed in terms of the radiance of equilibrium thermal radiation I^0_ω. To this end, in similarity with the previous example and in view of (3.1.97) we write the mean energy density as

$$\langle w \rangle = (8\pi)^{-1}\langle |E|^2 \rangle = (8\pi)^{-1}\int I_\mathbf{k} k^2 dk d\Omega_\mathbf{n} = c^{-1}\int I \, d\omega d\Omega_\mathbf{n}. \tag{3.1.98}$$

This equation suggests that the spectrum $I_\mathbf{k}$ is related with the radiance I by

$$I_\mathbf{k} = 8\pi I / ck^2 d_\omega k = 8\pi I / k^2, \tag{3.1.99}$$

or $I = (k^2/8\pi)I_\mathbf{k} = (c/8\pi)I_{\omega\mathbf{n}}$.

We now substitute (3.1.99) in (3.1.97) and observe that the radiance of equilibrium thermal radiation is given by eqn (2.1.31) with the coefficient two reflecting the number of independent polarizations. This implies that $I = 2I^0_\omega = 2\hbar k^3 c(2\pi)^{-3}[\exp(\hbar\omega/kT) - 1]^{-1}$. Then

$$\begin{aligned}\Gamma_{12} &= \int \frac{4\pi}{k^2} 2I^0_\omega \hat{\rho} \exp[ik(\mathbf{n\rho} - c\tau)] d^3k \\ &= \frac{\hbar c}{\pi^2} \int \frac{k\left(k^2\hat{1} - \mathbf{k}\otimes\mathbf{k}\right)}{\exp(\alpha k) - 1} e^{ik(\mathbf{n\rho} - c\tau)} dk d\Omega_\mathbf{n} \\ &= -(\hat{1}\Delta - \nabla\otimes\nabla)\frac{4\hbar c}{\pi\rho}\int_0^\infty \frac{\sin k\rho e^{-ikc\tau}}{\exp(\alpha k) - 1} dk,\end{aligned} \tag{3.1.100}$$

where $\alpha = \hbar c / kT$

Expanding the integrand in (3.1.100) in powers of $\exp(\alpha k)$ one may express eqn (3.1.100) in terms of the generalized Riemann ς-function (Kano and Wolf, 1962;

Metha and Wolf, 1964). For the sake of illustration, we limit our consideration to the simple particular case of calculating the temporal correlation ($\rho = 0$) for the components of the physical field $\mathbf{E}' = \mathrm{Re}\,\mathbf{E}$ (Bourret, 1960). The correlation of the components of this field is given by the matrix

$$\langle \mathbf{E}'(x_1) \otimes \mathbf{E}'(x_2) \rangle = \frac{1}{2}\mathrm{Re}\langle \mathbf{E}(x_1) \otimes \mathbf{E}^*(x_2) + \mathbf{E}(x_1) \otimes \mathbf{E}(x_2) \rangle. \qquad (3.1.101)$$

After time averaging, the second term here will be negligibly small and may be omitted, then from (3.1.100) and (3.1.101) we get

$$\begin{aligned} \Gamma(\tau) &\equiv \langle \mathbf{E}(\rho = 0, \tau) \otimes \mathbf{E}(0,0) \rangle \\ &= \frac{1}{2}\mathrm{Re}\frac{\hbar c}{\pi^2}\int \frac{k^3 e^{-ikc\tau}(\hat{1} - \mathbf{n} \otimes \mathbf{n})}{\exp(\alpha k) - 1} dk d\Omega_{\mathbf{n}} \\ &= -\frac{2}{3}\hbar c \frac{\pi^3}{\alpha^4} \hat{I} L'''(\pi c \tau / \alpha), \end{aligned} \qquad (3.1.102)$$

where $L'''(x) = d_x^3 L(x)$ and

$$L(x) = \coth x - \frac{1}{x} \qquad (3.1.103)$$

is the Langevin function. We see that the characteristic correlation time of thermal radiation is in the order of the inverse Wien frequency

$$\tau_k \sim \frac{\alpha}{\pi c} \sim \frac{\hbar}{\pi k T} \sim \frac{1}{\omega_W}, \qquad (3.1.104)$$

where $\omega_w = 2{,}82kT / \hbar$ is the frequency at which the radiance of equilibrium thermal radiation (2.1.28) is a maximum.

From (3.1.100) it is not hard to establish that the components of the electrical field vector for a frequency ω have correlation radii in the order of the wavelength λ regardless of temperature T. This result has no bearing on the nature of electromagnetic radiation and is common for arbitrary isotropic random wave fields in homogeneous and isotropic media.

Similar expressions can be derived also for correlations of the form $\langle \mathbf{E}_1 \otimes \mathbf{H}_2^* \rangle$ and $\langle \mathbf{H}_1 \otimes \mathbf{H}_2^* \rangle$, and these equations will retain their form upon allowance for the quantum nature of radiation (Metha and Wolf, 1964).

3.1.15 Statistical Definition of Radiance for an Arbitrary Wave Field in a Free Space

We noted above that a rigorous substantiation of the radiometric pattern where the wave field is treated as a bundle of incoherent beams can be managed and the wave content of radiance can be elucidated only on the assumption that the second moments of the field are quasi-uniform. If we abandon the concept of beams and confine ourselves to considering only *part* of radiometric relationships, then we may conjecture several definitions of radiance each of which will satisfy some, but not all, radiometric relations for an arbitrary nonuniform wave field. However, these definitions will be of doubtful value, since the radiance introduced in the absence of quasi-uniformity does not permit a close mathematical description and, generally speaking, does not satisfy the transfer equation. We demonstrate this point with an example of an electromagnetic field.

The flux density **S** and the volume density w of EM field energy are given by the expressions

$$\mathbf{S} = \frac{c}{4\pi}\mathbf{E} \times \mathbf{H}\,,$$
$$w = \frac{1}{8\pi}(E^2 + H^2)\,, \tag{3.1.105}$$

where **E** and **H** are real-valued physical fields. Taking the statistical means and using familiar averaging rules we may write

$$\langle \mathbf{S} \rangle = \langle |S|\mathbf{n} \rangle = \langle \langle S \rangle_{\mathbf{n}} \mathbf{n} \rangle = \int P(\mathbf{n}) \langle S \rangle_{\mathbf{n}} \mathbf{n}\, d\Omega_{\mathbf{n}}\,,$$
$$\langle w \rangle = \langle \langle w \rangle_{\mathbf{n}} \rangle = \int P(\mathbf{n}) \langle w \rangle_{\mathbf{n}}\, d\Omega_{\mathbf{n}}\,. \tag{3.1.106}$$

Here, $\mathbf{n} = \mathbf{S}/S, \langle \cdot \rangle_{\mathbf{n}}$ implies the conditional averaging at a fixed **n**, and $P(\mathbf{n})$ is the probability distribution function of the random vector **n**. These expressions suggest two definitions of radiance, viz.,

$$I_1 = P(\mathbf{n}) \langle S \rangle_{\mathbf{n}}\,,$$
$$I_2 = P(\mathbf{n}) \langle w \rangle_{\mathbf{n}} c\,. \tag{3.1.107}$$

Now I_1 is used in the expression for the average flux

$$\langle S \rangle = \int \mathbf{n} I_1 d\Omega_{\mathbf{n}}\,, \tag{3.1.108}$$

and I_2 in the expression for the energy density

$$w = c^{-1} \int I_2 d\Omega_{\mathbf{n}}\,. \tag{3.1.109}$$

These relations coincide with the basic formulas of radiometry (2.1.2) and (2.1.10) and remain valid in the general case of an appreciably nonuniform and nonstationary field. However, these radiances I_1 and I_2 need not coincide and both of them are consistent with the physically transparent pattern of the flow of energy along the rays.

These definitions prove to be mutually consistent only in the case of the quasi-uniform and quasi-stationary coherence function, when the energy flux and density are related by the same expression $\langle S\rangle_{\mathbf{n}} = c\langle w\rangle_{\mathbf{n}}$ that is used for the plane wave.

3.2 TRANSFER EQUATION FOR RADIATION IN A FREE SPACE

3.2.1 Derivation of the Transfer Equation for Free Radiation

In the preceding section we demonstrated that far from the sources, the Wigner function $W(K,R)$ (3.1.17) for a quasi-uniform field has a special form (3.1.69) that depends on R only in the combination $\mathbf{R}-\mathbf{v}_g T$. From this reasoning it follows immediately that $W(K,R)$ satisfies the transfer equation

$$(\mathbf{v}_g\nabla + \partial_T)W(K,R) = 0. \tag{3.2.1}$$

The same equation is satisfied by the radiance $I_k(R) = I_{\mathbf{k}}(\mathbf{R}-\mathbf{v}_g T)$ entering in (3.1.69).

The transfer equation (3.2.1) can be readily derived directly from the initial wave equation without resorting to the expansion of the field in terms of plane waves. We prove this statement, confining the proof for simplicity to the case of scalar wave field u satisfying the wave equation

$$\hat{L}u = (\Delta - c^{-2}\partial_t^2)u = 0. \tag{3.2.2}$$

Multiplying both sides of this equation, taken at a point x_1, by the complex conjugated field $u_2^* = u^*(x_2)$, averaging the result, and using the permutability of the operations of differentiation and averaging we obtain the wave equation for the coherence function: $\hat{L}_1\Gamma = 0$. In a similar way we derive the equation $\hat{L}_2\Gamma = 0$. Note thot the operator $\hat{L}_1$ acts in the argument x_1, and $\hat{L}_2$ in the argument x_2. Following the procedure of Dolin (1964) we take

$$(\hat{L}_1 \pm \hat{L}_2)\Gamma = 0. \tag{3.2.3}$$

It is not hard to see that the correlation function $\Psi = \Gamma - \langle u_1\rangle\langle u_2^*\rangle$ satisfies the same equations, because the averaging of eqn (3.2.2) yields $\hat{L}_1\langle u_1\rangle = \hat{L}_2\langle u_2\rangle = 0$.

For simplicity we assume that the mean field is zero, $\langle u\rangle = 0$, and so $\Gamma = \Psi$. We change the variables x_1 and x_2 for $R = (x_1 + x_2)/2$ and $\rho = x_1 - x_2$, substitute in (3.2.3) the coherence function Γ expressed in terms of the Wigner function by eqn (3.1.22) and replace everywhere R with μR to consider also the quasi-uniform limit by letting $\mu \to 0$. Then for $W(K,\mu R)$ we obtain two exact equations corresponding to the choice of plus or minus sign in eqn (3.2.3):

$$\left[k^2 - \left(\frac{\omega}{c}\right)^2 - \frac{\mu^2}{4}\left(\Delta_{\mu R} - c^{-2}\partial_{\mu T}^2\right)\right]W(K,\mu R) = 0, \tag{3.2.4}$$

$$\mu\left[\mathbf{k}\nabla_{\mu\mathbf{R}} + \frac{\omega}{c^2}\partial_{\mu T}\right]W(K,\mu R) = 0. \tag{3.2.5}$$

The last formulation is the transfer equation that, in this case, is an exact corollary of the wave equation $(\hat{L}_1 - \hat{L}_2)\Gamma = 0$ irrelevant of the quasi-uniformity conditions. This fact is associated with the special form of the initial wave equation (3.2.2) and does not hold for more complex wave operators.

In the general case, the wave vector $\mathbf{k}$ and frequency ω entering in (3.2.5) need not satisfy the dispersion equation for waves in a free space, i.e. $\omega \neq kc$, because the Wigner function may be other than zero outside the dispersion surface $\omega = kc$. However, in the quasi-uniform limit attained as $\mu \to 0$, we may neglect small terms of order $0(\mu^2)$ in eqn (3.2.4) and obtain

$$\left[k^2 - (\omega / c)^2\right]W(K,\mu R) = 0. \tag{3.2.6}$$

This equation shows that $W \neq 0$ only when $\omega = kc$ that is,

$$W(K,\mu R) = I_{\mathbf{k}}(\mu R)\delta(\omega - kc). \tag{3.2.7}$$

where $I_{\mathbf{k}}(\mu R)$ satisfies eqn (3.2.5), i.e., has the form $I_{\mathbf{k}} = I_{\mathbf{k}}(\mu\mathbf{R} - \mathbf{v}_g\mu T)$, $\mathbf{v}_g = c\mathbf{k} / k$. Thus, we returned to the Wigner function (3.1.69) of the quasi-uniform field that has been derived above from other considerations for the general case of a dispersive medium.

Thus, the transfer equation (3.2.1) follows from the wave equation (3.2.2) on the assumption that the coherence function is quasi-uniform. Expanding perturbation theory in μ, we may deduce the transfer equation also for the general type of wave operator L (Apresyan, 1973). We shall do this in Section 3.5 for the general case of a weakly absorbing, scattering medium.

3.2.2 Conditions of Field Quasi-Uniformity

We now abandon using the formal auxiliary parameter μ by letting $\mu = 1$ and estimate qualitatively the conditions that preserve the expansions considered in Section 3.2.1. With $\mu = 1$, to pass from (3.2.4) to (3.2.6) we obviously need the inequalities

$$|\Delta_R W(K,R)| << k^2 W(K,R), \tag{3.2.8a}$$

$$|\partial_T^2 W(K,R)| << \omega^2 W(K,R). \tag{3.2.8b}$$

Assuming that $|\Delta_R W| \sim L_R^{-2} W$ and $|\partial_T^2 W| \sim T_R^{-2} W$, where L_R and T_R are the characteristic variation scales of $W(K,R) = W(K;\mathbf{R},T)$ in the arguments $\mathbf{R}$ and T,.we obtain

$$L_R >> \frac{1}{k} = \frac{\lambda}{2\pi}, \quad T_R >> \frac{1}{\omega} = \frac{\tau}{2\pi}. \tag{3.2.8c}$$

These inequalities represent the conditions of quasi-uniformity of the wave field second moment over the characteristic wavelength λ and period τ and coincide with the earlier condition (3.1.41), since, in this case, the characteristic variation length of the second moment Γ in the difference variable ρ are of the order of λ and τ: $L \sim \lambda$, $L_\tau \sim \tau$.

In the considered case, the variation scales of free radiation $W(K,R)$ in the center-of-gravity coordinates L_R and T_R are defined only by the boundary conditions imposed on $W(K,R)$. By contrast, in the scattering medium, these scales will depend also upon the characteristic of the medium.

3.2.3 Radiometric Description of the Radiation of Plane Sources

Consider a simple problem on the radiation of stationary monochromatic sources. This problems admits an exact solution in the framework of both transfer theory and wave theory — a convenient example to compare both treatments. The subsequent consideration will be based on the treaties of Ovchinnikov and Tatarskii (1972) and Apresyan (1975).

In radiometry, radiation sources are described by the distribution of radiance I^0 over the plane $z = 0$. For $z > 0$, the radiance of a radiation I is defined by the transfer equation (3.2.1) that for the stationary monochromatic case, when $\partial_T I = 0$, may be described in the form

$$\mathbf{n}\nabla I = 0. \tag{3.2.9}$$

Here, $\mathbf{n} = \mathbf{v}_g / v_g = \mathbf{k} / k$, and we assume that only one monochromatic component of the radiation is considered. The dependence on frequency ω, that enters in all spectral characteristics as a factor $\delta(\omega - \omega_0)$ in the monochromatic case, is tacitly assumed. We intend to describe the properties of the spatial coherence of radiation treating the radiance I as a function of $\mathbf{r}$ and $\mathbf{n}$: $I = I(\mathbf{n},\mathbf{r})$.

The transfer equation (3.2.9) suggests that the radiance is invariable along the ray with direction $\mathbf{n}$ so that $I(\mathbf{n},\mathbf{r}) = I(\mathbf{n},\mathbf{r}_{0\perp}) = I^0(\mathbf{n},\mathbf{r}_{0\perp})$, where $\mathbf{r}_{0\perp}$ and $\mathbf{r}$ correspond to the initial and final points on the ray, as shown in Fig. 3.4. Eliminating $\mathbf{r}_{0\perp}$ between this expression and the ray equation $\mathbf{r} = \mathbf{r}_{0\perp} + \mathbf{n}s$ (s is the ray path) we find for $z > 0$ the radiance I as a function of the source radiance

$$I(\mathbf{n},\mathbf{r}) = I^0\left(\mathbf{r}_\perp - \frac{\mathbf{n}_\perp}{n_z} r_z\right). \tag{3.2.10}$$

Here and below we denote $\mathbf{r} = (\mathbf{r}_\perp, r_z = z)$ and $\mathbf{n} = (\mathbf{n}_\perp, n_z)$; the subscript $\perp$ labels vector components transverse to the z axis.

In the far zone where the distance from the source is large compared to the source size we may treat it as a point source and introduce the radiant intensity (2.1.9)

$$J(\mathbf{n}) = \cos\theta \int I^0(\mathbf{n},\mathbf{r}_\perp) d^2 r_\perp . \tag{3.2.11}$$

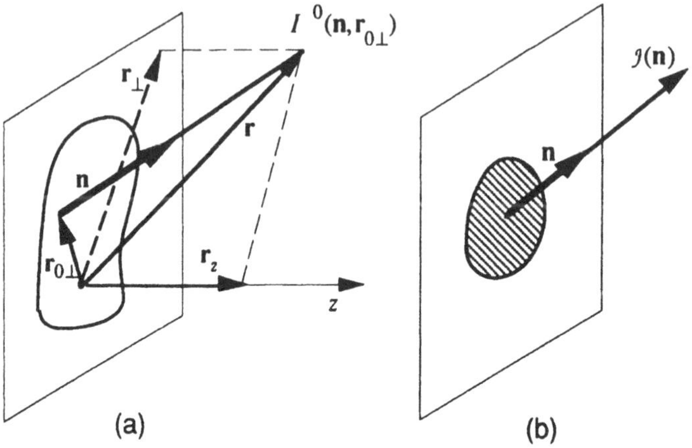

Figure 3.4. (a) Radiance and (b) the radiant intensity of a plane source.

The quantity $J(\mathbf{n})$ describes the angular distribution of the radiant flux from the small source (Fig. 3.4b) and is related to the distribution of source intensity by the asymptotic expression

$$J(\mathbf{n}) = \lim_{r\to\infty} r^2\langle |u(r)|^2\rangle, \tag{3.2.12}$$

which is deduced in Problem 3.7.

In transfer theory, the radiance I^0 is the main characteristic of the sources; it must be non-negative and arbitrary specified in other respects. For example, in the case of plane sources, we may specify an arbitrary angular distribution of radiance $I^0(\mathbf{n},\mathbf{r}_\perp)$ at any point of the plane $z = 0$. By contrast, in wave theory, the sources are described by a distribution of currents or by a distribution of the field u^0 over the surface $z = 0$ equivalent to the surface currents. It turned out that radiometry may be treated as the quasi-uniform limiting case of wave theory where the source radiance I^0 is expressed in terms of the surface field u^0. We deduce this expression by comparing the exact diffraction solution with eqn (3.2.10) derived with the formalism of transfer theory.

3.2.4 Exact Solution for the Diffraction in a Half-Space

Wave theory gives an exact expression for the amplitude of diffracted field u in terms of its initial value $u^0 = u(\mathbf{r}_\perp, 0)$ at the screen $z = 0$. This solution may be expressed either by the Rayleigh method (expansion in terms of plane waves, see Problem 3.9) or using the Green's function known for this application (see, e.g., Rytov, Kravtsov, and Tatarskii, 1989b). The last method yields

$$u(\mathbf{r}) = \int G(\mathbf{r}-\mathbf{r}'_\perp)u^0(\mathbf{r}'_\perp)d^2\mathbf{r}'_\perp, \tag{3.2.13}$$

where

$$G(\mathbf{r}) = \frac{k_0}{2\pi i}\frac{z}{r}\frac{e^{ik_0 r}}{r} \tag{3.2.14}$$

is the Green's function, and $k_0 = \omega / c$.

From eqn (3.2.13) for the coherence function $\Gamma = \langle u(\mathbf{R}+\rho/2)u^*(\mathbf{R}-\rho/2)\rangle$ in the half-space $z \geq 0$ we have

$$\Gamma = \int G\left(\mathbf{R}-\mathbf{R}'_\perp + \frac{\rho-\rho'_\perp}{2}\right)G^*\left(\mathbf{R}-\mathbf{R}'_\perp - \frac{\rho-\rho'_\perp}{2}\right)\Gamma^0(\mathbf{R}'_\perp,\rho'_\perp)d^2R'_\perp d^2\rho'_\perp, \tag{3.2.15}$$

where

$$\Gamma^0(\mathbf{R}_\perp,\rho_\perp) = \left\langle u^0\left(\mathbf{R}_\perp + \frac{\rho_\perp}{2}\right)u^{0*}\left(\mathbf{R}_\perp - \frac{\rho_\perp}{2}\right)\right\rangle \tag{3.2.16}$$

is the coherence function of the field at the screen.

3.2.5 Relation of Transfer Theory with Exact Wave Theory

Let us compare the diffraction relation (3.2.15) with the expression for the coherence function following from transfer theory

$$\begin{aligned}\Gamma &= \int I(\mathbf{n},\mathbf{R})\exp(ik_0\mathbf{n}\rho)d\Omega_\mathbf{n} \\ &= \int I^0\left(\mathbf{n},\mathbf{R}_\perp - \mathbf{n}_\perp\frac{R_z}{n_z}\right)\exp(ik_0\mathbf{n}\rho)d\Omega_\mathbf{n} .\end{aligned} \tag{3.2.17}$$

This expression results from the substitution of (3.2.10) into (2.1.26) where in this case we may put $c_{\omega_0} = 1$.

The product of Green's functions in the integrand of eqn (3.2.15) may be approximated by the quantity

$$GG^* \approx |G(\mathbf{R}-\mathbf{R}'_\perp)|^2\exp(ik_0\mathbf{n}(\rho-\rho'_\perp)), \tag{3.2.18}$$

thus separating the dependencies on the variables $\rho-\rho'_\perp$ and $\mathbf{R}-\mathbf{R}'_\perp$ [here, $\mathbf{n} = (\mathbf{R}-\mathbf{R}'_\perp)/|\mathbf{R}-\mathbf{R}'_\perp|$]. Then, eqn (3.2.15) assumes the radiometric form of (3.2.17). Calculations indicate that the source radiance I^0 may be expressed in terms of the coherence function of the field on the screen:

$$I^0(\mathbf{n},\mathbf{R}_\perp) = \left(\frac{k_0}{2\pi}\right)^2\cos\theta\int\Gamma^0(\mathbf{R}_\perp,\rho_\perp)\exp(-ik_0\mathbf{n}_\perp\rho_\perp)d^2\rho_\perp \tag{3.2.19a}$$

or alternatively

$$I^0(\mathbf{n},\mathbf{R}) = k_0^2 n_z W^0(k_0\mathbf{n}_\perp,\mathbf{R}_\perp), \tag{3.2.19b}$$

where $n_z = \cos\theta$, and

$$W^0(\mathbf{K}_\perp,\mathbf{R}_\perp) = \int \Gamma^0(\mathbf{R}_\perp,\rho_\perp)\exp(-i\mathbf{k}_\perp\rho_\perp)\frac{d^2\rho_\perp}{(2\pi)^2} \tag{3.2.20}$$

is the Wiener function for the field distribution on the screen. This expression for I^0 may be obtained by resorting to the expansion in terms of plane waves (this instructive alternative approach is treated in depth in Problem 3.9).

The approximation (3.2.18) is applicable provided that in the region essential for integration in (3.2.15)

$$|\mathbf{R} - \mathbf{R}'_\perp| >> \left|\frac{\rho - \rho'_\perp}{2}\right|, \tag{3.2.21}$$

$$|\mathbf{R} - \mathbf{R}'_\perp| >> \sqrt{k_0\mathbf{n}\left(\frac{\rho - \rho'_\perp}{2}\right)\left|\frac{\rho - \rho'_\perp}{2}\right|}. \tag{3.2.22}$$

The former of these inequalities allows us to let $(\rho - \rho'_\perp)/2 \approx 0$ in the amplitude factors of Green's functions, and the latter allows us to keep in the phase factors only the terms linear in $\rho - \rho'_\perp$. The quadratic terms in the exponents cancel out, therefore, the condition (3.2.22) actually implies the smallness of third-order terms.

The source radiance expression (3.2.19) can be deduced directly from eqn (3.1.84) for the coherence function of the quasi-uniform field if we let there $R_z = \rho_z = 0$ and take into account that $d\Omega_\mathbf{n} = d^2 n_\perp / n_z$, where $n_z = \sqrt{1 - n_\perp^2}$. Taking the inverse Fourier transform we immediately arrive at (3.2.19). However, this approach does not give us the constraints that would allow us to neglect the nonuniform part of the coherence function and assume the field to be quasi-uniform.

Thus the source radiance (3.2.19) and radiation radiance (3.1.85) are mutually consistent, except that the former is clearly dependent on the choice of the source plane, whereas the latter is invariant under the choice of coordinate system. This apparent contradiction can be explained by viewing eqn (3.2.19) as a particular case of (3.1.85) tied to a fixed source plane and valid only subject to the inequalities (3.2.21) and (3.2.22) constraining the positions of observation points relative to this plane.

Expression (3.2.19) for radiance of monochromatic sources was first derived by Walther (1968). Subsequently it was discussed by many authors from different points of view (see, e.g., the reviews of Baltes (1978) and Wolf (1978) and the literature cited therein). Walther (1973) revealed some drawbacks of eqn (3.2.19) and set forth an alternative definition of I^0. The possibility of various definitions for radiance is associated with the above lack of uniqueness in the local spectrum concept.

3.2.6 Generalized Radiance of Plane Sources and Nonclassic Radiometry

In the final analysis, the source radiance (3.2.19) has a direct bearing on the choice of Wigner's function as a local spectrum of nonuniform fluctuations. Alternative definitions of local spectra may lead to different expressions for the radiance of the sources; however, in the quasi-uniform limit, all these definitions must give identical results and must reduce to the common limit in terms of the radiometric radiance.

Thus, the radiance of plane sources I^0 is expressed through the boundary value of the correlation function of the field in agreement with (3.2.19). We consider some corollaries of this expression.

First of all we note that the radiance (3.2.19) need not be positive and, in principle, can take on negative values. Therefore, in the general case it can be treated as some *generalized radiance of the sources*. This conclusion does not seem strange if we observe that the field quasi-uniformity conditions generally do not require that the sources be quasi-regular. Near appreciably irregular sources, the field is also appreciably nonuniform, thus invalidating a radiometric description in this region. At the same time, the quasi-uniformity conditions could be met far from the sources, where a radiometric description becomes valid. It is in this area that eqn (3.2.19) should be understood as the generalized source radiance. In the region of inapplicability of classical radiometry, eqn (3.2.19) does not possess all the properties of common radiometric radiance. Thus, the generalized radiance allow one to derive a correct description of the field in the quasi-uniformity region. Moreover, in conjunction with the radiative transfer equation, this formula covers the effects of diffraction — an unattainable goal in the framework of classical radiometry.

Thus we extended the applicability range of radiative transfer theory at the expense of the abandoned heuristic condition of non-negative radiance. We will call the approach *nonclassic radiometry* or *diffraction radiative transfer theory*. Below we consider some applications of such a nonclassic approach.

3.2.7 Diffraction and Interference of Coherent Radiation

Consider the diffraction of spatially coherent (non-fluctuating) plane wave

$$u^0 = e^{i\mathbf{kr}}, \tag{3.2.23}$$

normally incident on an aperture Σ with the characteristic size $a >> \lambda$ in the plane $z = 0$, as shown in Fig. 3.5. While in this problem fluctuations are absent, all above considerations concerning the coherence function $\Gamma = \langle u_1 u_2^* \rangle = u_1 u_2^*$ remain in force and the aperture may be treated as a *coherent source* of radiation.

In the physical optics approximation, the field within the plane $z = 0$ may be let equal to u^0 in the aperture and zero elsewhere. We write

$$u^0(\mathbf{r}_\perp) = u^0 \theta_\Sigma(\mathbf{r}) = \theta_\Sigma(\mathbf{r}), \tag{3.2.24}$$

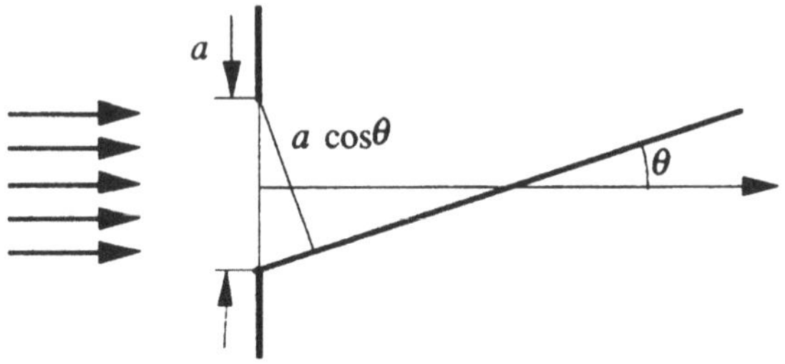

Figure 3.5. Diffraction at an aperture.

where $\theta_\Sigma(\mathbf{r})$ is the transmission function of the aperture which is unity for $\mathbf{r}$ in Σ and zero outside Σ so that $\theta_\Sigma^2 = \theta_\Sigma$. Substituting (3.2.24) in (3.2.19) yields the generalized radiance of the coherent source in the form

$$I^0(\mathbf{n},\mathbf{R}_\perp) = \left(\frac{k_0}{2\pi}\right)^2 n_z \int \theta_\Sigma\left(\mathbf{R}_\perp + \frac{\rho_\perp}{2}\right)\theta_\Sigma\left(\mathbf{R}_\perp - \frac{\rho_\perp}{2}\right)\exp(-ik_0\mathbf{n}_\perp\rho_\perp)d^2\rho. \quad (3.2.25)$$

Let Σ be a y-unbounded slit of width $2a$. Taking the integral on the right-hand side of (3.2.25) yields

$$I^0(\mathbf{n},\mathbf{R}_\perp) = \frac{n_z}{\pi n_x}\delta(n_y)\sin\big(2k_0 n_x(a-|R_x|)\big)\theta(a-|R_x|), \quad (3.2.26)$$

where $\theta(x)$ is the Heaviside step function equal to unity for $x \geq 0$ and zero for $x < 0$.

The function (3.2.26) does not obey the requirements imposed on the source radiance in phenomenological theory because it is a negative quantity. Nonetheless, substituting (3.2.25) in the radiant intensity (3.2.11) gives a positive quantity

$$J(\mathbf{n}) = \left|\frac{k_0\cos\theta}{2\pi}\int \theta_\Sigma(\mathbf{r})\exp(-ik_0\mathbf{rn})d^2r\right|^2, \quad (3.2.27)$$

coincident with the exact result of diffraction theory. Substituting the generalized radiance (3.2.25) in eqn (3.2.17) we obtain a correct description of Γ in the quasi-uniformity region. To demonstrate, we calculate the coherence function Γ for observation points in the Fraunhofer zone for the aperture, i.e., in the region where $R >> a^2/\lambda$. The aperture plays the role of a point source of a divergent spherical wave, so that the field is given by

$$u(\mathbf{R}) = \left(\frac{k_0 n_z}{2\pi iR}\right)\exp(ik_0R)\int u^0(\mathbf{r}_\perp')\exp(-ik_0\mathbf{nr}_\perp')d^2r_\perp', \quad (3.2.28)$$

where $\mathbf{n} = \mathbf{R}/R$ is the unit vector directed to the point of observation $\mathbf{R}$ from the center of the aperture. If comparatively mild inequalities $R >> \rho, \lambda, \sqrt{\rho^3/\lambda}$ hold and we may let $(\mathbf{R}\pm\rho/2)/|\mathbf{R}\pm\rho/2|\approx\mathbf{R}/R=\mathbf{n}$, then the second moment of the field is

$$\Gamma\approx\left(\frac{k_0n_z}{2\pi R}\right)^2\int u^0\left(\mathbf{R}'_\perp+\frac{\rho'_\perp}{2}\right)u^{0*}\left(\mathbf{R}'_\perp-\frac{\rho'_\perp}{2}\right)$$
$$\times\exp(ik_0\mathbf{n}(\rho-\rho'_\perp))d^2R'_\perp d^2\rho'_\perp. \qquad (3.2.29)$$

The Fraunhofer zone $R>>k_0a^2$ is the domain of quasi-uniformity of the moment Γ_{12}, because here $\partial_R\Gamma_{12}\sim R^{-1}\Gamma_{12}$, $\partial_\rho\Gamma_{12}\sim\lambda^{-1}\Gamma_{12}>>\partial_R\Gamma_{12}$. The condition of quasi-uniformity may be somewhat relaxed if we turn to the above quasi-uniformity conditions (3.2.21) and (3.2.22) derived in Problem 3.10.

Thus, the generalized radiance has allowed us to describe the diffraction of a coherent radiation far from the aperture. The generalized radiance (3.2.25) differs from the phenomenological radiance of the sources also by its possibly nonzero value at a point $\mathbf{R}_\perp$ even when the initial field $u^0(\mathbf{R}_\perp)$ at the same point is zero. This behavior is feasible if the boundary of Σ is irregular or has concave segments, as exemplified in Fig. 3.6. In the final analysis, this property of the generalized radiance is associated with he nonlocal behavior of the argument $\mathbf{R}=(\mathbf{r}_1+\mathbf{r}_2)/2$.

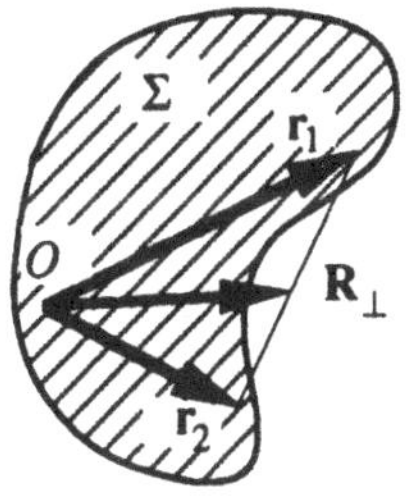

Figure. 3.6. For an aperture with convex edge, the radius vector $\mathbf{R}_\perp$ of the center of gravity may find itself in the region where the field is zero. However the respective generalized radiance does not vanish.

One more example of nonlocal behavior of the generalized radiance is given by a Young interferometer consisting of two apertures. The transmission function of the system of two apertures is given by the sum of the transmission functions of individual apertures

$$\theta_\Sigma=\theta_\Sigma^{(1)}+\theta_\Sigma^{(2)}.$$

The respective generalized radiance $I^0(\mathbf{n},\mathbf{R}_\perp)$ will contain an interferometric term proportional to $\theta_\Sigma^{(1)}\theta_\Sigma^{(2)}$ and nonzero between the apertures. Therefore, the ordinary radiometric description based on the formula (3.2.25) is inapplicable near the apertures, but becomes valid far from them, where both slits may be treated as one source of a directed cylindrical wave.

3.2.8 Spatially Incoherent Random Source

When the field on the screen $z=0$ is short-correlated so that the correlation radius of this field is less than any one characteristic parameter of the problem, then we may apply the model of random (or delta correlated)sources by letting

$$\left\langle u^0\left(\mathbf{R}_\perp+\frac{\rho_\perp}{2}\right)u^{0*}\left(\mathbf{R}_\perp-\frac{\rho_\perp}{2}\right)\right\rangle=\sigma^2(\mathbf{R}_\perp)\delta(\rho), \tag{3.2.30}$$

where $\sigma^2(\mathbf{R}_\perp)=\langle|u^0(\mathbf{R}_\perp)|^2\rangle$ is the intensity of field fluctuations on the screen. Such sources are sometimes referred to as *spatially incoherent*. Substituting (3.2.30) in (3.2.19) we obtain the radiance of random sources in the form

$$I^0(\mathbf{n},\mathbf{R}_\perp)=\left(\frac{k_0}{2\pi}\right)^2 n_z\sigma^2(\mathbf{R}_\perp)=\left(\frac{k_0\sigma(\mathbf{R}_\perp)}{2\pi}\right)^2\cos\theta. \tag{3.2.31}$$

According to this expression, the angular dependence of random sources is proportional to $n_z=\cos\theta$, i.e., is not constant as with a radiation corresponding to Lambert's law (Problem 3.11). Thus, the approximation of delta-correlated sources leads to an anisotropic distribution of radiance.

Expression (3.2.31) represents small-scale sources producing a short-correlated field on the plane $z=0$. However, a knowledge that the sources are small is insufficient to describe these sources by the radiance with an angular dependence of the form (3.2.31). For example, if we characterize the sources by the normal derivative $\partial_z u^0=\partial_z u|_{z=0}$ on the surface $z=0$ rather than by the surface field $u^0=u(r_\perp,\ z=0)$, alternatively, if we formulate a Neumann problem rather than Dirichlet's, then, in the limit of small correlation radii, these sources may be associated with the radiance function $I^0\propto\sec\theta$ rather than $I^0\propto\cos\theta$ as in the case of eqn (3.2.31). Indeed, an expression for the field similar to (3.2.28) but given in terms of its normal derivative on the boundary of the surface $z=0$ has the form

$$u(\mathbf{R})=(2\pi R)^{-1}\exp(ik_0R)\int\partial_z u^0(\mathbf{r}'_\perp)\exp(-ik_0\mathbf{n}\mathbf{r}'_\perp)d^2r'_\perp.$$

Clearly, to pass to this case we need to replace u^0 in (3.2.19) with $\partial_z u^0/(-ik_0n_z)$, then

$$I^0(\mathbf{n},\mathbf{R}_\perp)=\frac{1}{\cos\theta}\int\left\langle\partial_z u^0\left(\mathbf{R}_\perp+\frac{\rho'_\perp}{2}\right)\partial_z u^{0*}\left(\mathbf{R}_\perp-\frac{\rho'_\perp}{2}\right)\right\rangle\frac{\exp(-ik_0\mathbf{n}\rho'_\perp)}{(2\pi)^2}d^2\rho'_\perp$$

For small-scale fluctuations with $|\mathbf{k}_0\rho'|<<1$, the radiance I^0 is proportional to $\sec\theta$. For small angles of diffraction, $\theta<<1$ when $\cos\theta\approx1$, this dependence does not differ from the case of eqn (3.2.31): $I^0\propto\cos\theta$, whereas at higher θ the radiances for two cases under consideration are no longer coincident.

Therefore, the angular dependence of source radiance (3.2.31) is not versatile and is valid not for all short-correlated sources. The final decision on which source model should be chosen for a particular application needs additional physical arguments.

From eqn (3.2.31) we see that the radiance of random sources is non-negative and other than zero only in the region where the source intensity $\sigma^2(R_\perp)$ is nonzero; that is, the quantity (3.2.31) possesses all the properties of ordinary radiometric radiance, therefore, in this case, we deal with classical radiometry. This stems from the fact that the random sources (3.2.30) are the limiting case of the so-called quasi-uniform sources which, essentially, were the only ones treated in phenomenological theory. Below we consider the properties of such sources in some more detail.

3.2.9 Quasi-Uniform Sources

For quasi-uniform sources, the coherence function

$$\Gamma^0 = \left\langle u^0\left(\mathbf{R}_\perp + \frac{\rho_\perp}{2}\right) u^{0*}\left(\mathbf{R}_\perp - \frac{\rho_\perp}{2}\right)\right\rangle$$

smoothly varies in its argument $\mathbf{R}_\perp$ as compared to the fast variation in $\rho_\perp$. For simplicity only we assume that Γ^0 can be factored:

$$\Gamma^0 = \sigma_0^2(\mathbf{R}_\perp)\gamma^0(\rho_\perp), \tag{3.2.32}$$

the variance $\sigma_0^2(\mathbf{R}_\perp) = \langle |u^0(\mathbf{R}_\perp)|^2 \rangle$ as a function of $\mathbf{R}_\perp$ describes a smooth statistical nonuniformity with characteristic scale L_R, and the correlation coefficient $\gamma^0(\rho_\perp)$ tends to zero for $\rho \geq l_c << L_R$. To clarify the meaning of the approximation (3.2.32) we write

$$\Gamma^0 = \sigma_0^2(\mathbf{R}_\perp)\gamma^0(\mathbf{R}_\perp, \rho_\perp), \tag{3.2.33}$$

where $\gamma^0(\mathbf{R}_\perp, \rho_\perp) = \Gamma^0 / \sigma_0^2(\mathbf{R}_\perp)$ is normalized: $\gamma^0(\mathbf{R}_\perp, 0) = 1$. In order to pass from the general expression (3.2.33) to the approximation (3.2.32) we need to substitute $\gamma^0(\mathbf{R}_\perp, \rho_\perp)$ for the normalized correlation function $\gamma^0(\rho_\perp)$. This substitution is possible if we neglect the nonuniformity of the correlation coefficient at the source. For sources with sharp boundaries, this approximation will not be applicable in the rims with width in the order of correlation radius l_c. However, these rims may be neglected if the source is sufficiently long.

Substituting (3.2.32) in eqn (3.2.19) for the radiance of quasi-uniform sources we obtain

$$I^0(\mathbf{n}, \mathbf{R}_\perp) = k_0^2 n_z \sigma_0^2(\mathbf{R}_\perp)\tilde{\gamma}^0(k_0 \mathbf{n}_\perp), \tag{3.2.34}$$

where the quantity

$$\tilde{\gamma}^0(\mathbf{k}_\perp) = \int \gamma^0(\rho_\perp)\exp(-i\mathbf{k}_\perp\rho)(2\pi)^{-2}d^2\rho$$

has the sense of the normalized angular spectrum of plane sources

$$\int \tilde{\gamma}^0(\mathbf{k}_\perp)d^2k_\perp = \gamma^0(0) = 1.$$

For statistically uniform sources, the function $\tilde{\gamma}^0(\mathbf{k}_\perp)$ is nonnegative. For quasi-uniform sources, this property must hold approximately. Additionally, if we disregard the boundary rims of width in the order of l_c, then the quantity (3.2.34) may be deemed equal to zero outside the domain of the sources, that is, $I_\mathbf{n}^0$ possesses all the properties of radiometric radiance. This result should be expected, for the field of quasi-uniform sources is quasi-uniform. Therefore, in this case, the radiometric description of radiance is applicable (in the wave zone) everywhere, so that the sources as well as the field may be characterized by traditional radiometric quantities.

Substituting the radiance (3.2.34) in the coherence function (3.2.17) we obtain

$$\Gamma = \int k_0^2 n_z \sigma_0^2\left(\mathbf{R}_\perp - \mathbf{n}_\perp \frac{R_z}{n_z}\right)\tilde{\gamma}^0(k\mathbf{n}_\perp)\exp(ik_0\mathbf{n}\rho)d\Omega_\mathbf{n}. \tag{3.2.35}$$

For a field statistically uniform on the screen $z = 0$, the variance σ_0^2 is constant, and since $d\Omega_\mathbf{n} = d^2n_\perp / n_z = d^2k_\perp / n_z k_0^2$, this expression becomes

$$\Gamma = \sigma_0^2 k_0^2 \int \tilde{\gamma}^0(k_0\mathbf{n}_\perp)\exp(ik_0\mathbf{n}\rho)n_z d\Omega_\mathbf{n}$$
$$= \sigma_0^2 \int_{k_\perp^2 < k_0^2} \tilde{\gamma}^0(\mathbf{k}_\perp)\exp\left(ik_\perp\rho_\perp + i\rho_z\sqrt{k_0^2 - k_\perp^2}\right)d^2k_\perp,$$

that coincides with the well known result for diffraction at a random screen (Rytov et al., 1989a). For $\sigma_0^2 = \text{const}$, the scale L_R is infinite, therefore, in this case, nothing limits separation of observation points ρ. Of course, as before, these points must be separated from the plane $z = 0$ at a distance of at least a few wavelengths.

3.2.10 Diffraction of a Random Field at an Aperture. Van Cittert-Zernike Theorem

In Section 3.2.7 we considered the diffraction of a coherent (nonfluctating) field at an aperture Σ. Now with the aid of eqn (3.2.32) we describe the diffraction of a random (partially coherent) quasi-uniform field.

We suppose that the aperture size $a \sim \sqrt{\Sigma}$ are large compared to the correlation radius of the incident field, $a >> l_c$. If we neglect, as before, the rims of width about l_c near the aperture edge and treat $\mathbf{R}$ as the radius vector of the center of gravity of two observation points, then the model of quasi-uniform sources (3.2.32) is applicable. For such sources, $\sigma_0^2(\mathbf{R}_\perp) = 0$ beyond the aperture.

For simplicity we confine out consideration to the small angle observation range where we may let $n_z = \cos\theta \approx 1$ and replace in the limit $R_z \to \infty$ the integration with respect to the elementary solid angle $d\Omega_{\mathbf{n}} = d^2 n_\perp / n_z$ with the integration with respect to $n_\perp$ in infinite limits. Then eqn (3.2.35) may be recast as

$$\Gamma = k_0^2 \int F(\mathbf{k}'_\perp)\tilde{\gamma}^0(k_0 \mathbf{n}_\perp)\exp[i\mathbf{k}'_\perp(\mathbf{R}_\perp - R_z \mathbf{n}_\perp) + ik_0(\mathbf{n}_\perp \rho_\perp + \rho_z)]d^2 n_\perp d^2 k'_\perp , \tag{3.2.36}$$

where

$$F(\mathbf{k}_\perp) = \int \sigma_0^2(\mathbf{R}_\perp)\exp(-i\mathbf{k}_\perp \mathbf{R}_\perp)(2\pi)^{-2} d^2 R_\perp$$

is the Fourier transform of variance fluctuations.

As $R \to \infty$, the integrand in (3.2.36) contains a rapidly oscillating phase factor. Therefore, preexponential factors may be factored out of the integral at the point of stationary phase $\mathbf{k}'_\perp = k_0 \rho_z / R_z$, $\mathbf{n}_\perp = \mathbf{R}_\perp / R_z$. The remaining integral is readily taken to give

$$\Gamma(\mathbf{R},\rho) = \Gamma(\mathbf{R},0)\gamma(\mathbf{R},\rho), \tag{3.2.37}$$

where

$$\Gamma(\mathbf{R},0) = \left(\frac{k_0}{R_z}\right)^2 \int \sigma_0^2(\mathbf{R}'_\perp) d^2 R'_\perp \tilde{\gamma}^0\left(k_0 \frac{\mathbf{R}}{R_z}\right),$$

$$\gamma(\mathbf{R},\rho) = F\left(\frac{k_0 \rho_\perp}{R_z}\right)\exp\left[ik_0\left(\rho_z + \frac{\rho_\perp \mathbf{R}_\perp}{R_z}\right)\right](2\pi)^2\left[\int \sigma^2(\mathbf{R}'_\perp) d^2 R'_\perp\right]^{-1}$$

with $\gamma^0(\mathbf{R},0) = 1$.

From this expression we see that the intensity distribution of the diffracted field $\Gamma(\mathbf{R},0) = \langle |u(\mathbf{R})|^2 \rangle$ depends on the type of correlation function of the source field u^0 (in terms of γ^0 and has an angular dependence identical with the source radiance I^0 given by (3.2.34). This result should be obvious already from classical transfer theory; however, the expression for source radiance (3.2.19) remained unknown until recent time. Indeed, if we reformulate the problem under consideration in the language of transfer theory by specifying the radiance distribution I^0 over the aperture Σ, then the intensity $\langle |u|^2 \rangle$ will be proportional to the energy density w at the point of observation. Recognizing further that the radiance is conserved along the ray, and the energy density is represented by the integral (2.1.10) of the radiance with respect to directions, for distant observation points in the limit $R \sim R_z \to \infty$ we obtain

$$w(\mathbf{R}) = c^{-1} \int \mathrm{I} d\Omega_{\mathbf{n}} \approx c^{-1} I^0(\mathbf{n})\delta\Omega,$$

where $\delta\Omega = \Sigma / R^2$ is the solid angle at which the aperture is seen from the point of observation (Fig. 3.7). This expression has the same angular dependence as the source radiance $I^0(\mathbf{n})$.

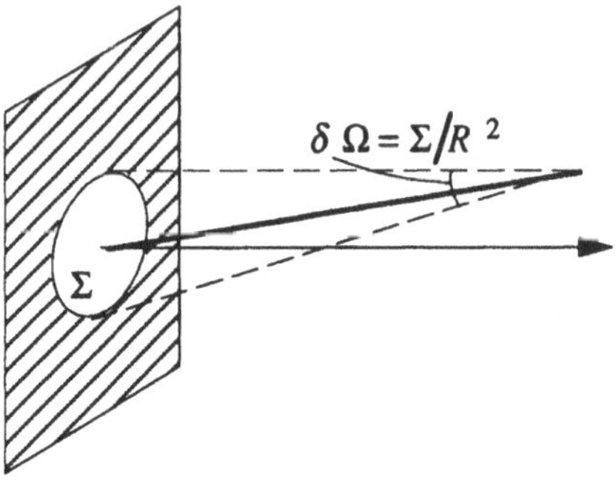

Figure 3.7. Geometry depicting diffraction at an aperture.

The correlation properties of the diffracted field, that is, the behavior of Γ as a function of ρ, are governed mainly by the form of the aperture which controls the function γ entering eqn (3.2.37).

The dependence of Γ on the aperture shape represents *the van Cittert–Zernike theorem* relating the coherence function of the radiation of spatially incoherent sources with the diffraction of a plane wave at an aperture. In agreement with eqn and (3.2.37), the dependence of $|\Gamma|$ on $\rho_\perp$ is governed by the Fourier transform $\mathbf{F}$ of the source *intensity distribution* $\sigma^2(\mathbf{R}_\perp)$. For a plane wave diffracted at an aperture with transmission function proportional to $\sigma^2(\mathbf{R}_\perp)$, this same Fourier transform defines the diffracted field $u^d(\mathbf{R}_\perp)$ as a function of $\mathbf{R}_\perp$ in the Fraunhofer zone. Therefore, if one of the observation points ($\mathbf{r}_1$ or $\mathbf{r}_2$) satisfying the applicability conditions (3.2.37) is arbitrary fixed in the plane $r_{1z} = r_{2z} = R_z$ and the other point is moved, the the dependence of $|\Gamma|$ on the difference $\rho_\perp = \mathbf{r}_{1\perp} - \mathbf{r}_{2\perp}$ will repeat the dependence of $|u^d|$ on $\mathbf{R}_\perp$. Domains of coherence about a given point for which $|\Gamma|$ appreciably differs from zero will alternate with domains of incoherence where $|\Gamma| \approx 0$. For example, for a field diffracted at a circular aperture of radius a, we have the Airy pattern

$$\left|u^d(\mathbf{R})\right| \propto \left|2J_1(\xi)/\xi\right|, \quad \xi = k_0 a\left|\mathbf{R}_\perp\right|/R_z, \tag{3.2.38}$$

where J_1 is the Bessel function, and $\mathbf{R} = (\mathbf{R}_\perp, R_z)$ (Born and Wolf, 1980). Accordingly, the dependence of $|\Gamma_{12}|$ on $\rho_\perp$ is defined by the same function $|2J_1(X)/X|$, where $X = k_0 a|\rho_\perp|/R_z)|$. Each observation point is surrounded by a coherence spot which is in turn encircled by concentric circles of full coherence where $J_1 = 0$. These circles correspond to Airy rings in diffraction theory.

To satisfy the applicability conditions (3.2.37) one need not place the observation point in the Fraunhofer zone relative to the aperture ($R_z >> k_0 a^2$), instead it suffices to move it at a distance $R_z >> z = k_0 a l_c$ (see Rytov et al., 1989a). This allowance implies that eqn (3.2.37) will be valid already in the near field for the aperture (for $k_0 a l_c << R_z \leq k_0 a^2$) and does not require $R_z >> k_0 a^2$ any longer. This fact is important for experimental radiometry with its typical dimensions of $a \geq 1$. For coherent illumination, the Fraunhofer zone $z >> k_0 a^2 \sim 10^5$ cm = 1 km is beyond practical distances, whereas with incoherent irradiation (assuming the correlation radius $l_c \sim \lambda$) the condition $z >> k_0 a l_c \sim a$ is met at any distance exceeding the aperture size $a \sim 1$ cm.

3.2.11 Radiant Intensity and the Inverse Problem for Correlation of Sources

In this section we examine the expression for radiant intensity. When the point of observation is far from the aperture (z >> a), then it can be treated as some point source. The respective radiant intensity (3.2.11) is obtained by integration of the radiance (3.2.19) over the plane of the source $z = 0$:

$$
\begin{aligned}
J(\mathbf{n}) &= n_z \int I^0(\mathbf{R}_\perp, \mathbf{n}) d^2 R_\perp \\
&= \left(\frac{k_0 n_z}{2\pi}\right)^2 \int \left\langle u^0\left(\mathbf{R}_\perp + \frac{\rho_\perp}{2}\right) u^{0*}\left(\mathbf{R}_\perp - \frac{\rho_\perp}{2}\right)\right\rangle \exp(-ik_0 \mathbf{n} \rho_\perp) d^2\rho_\perp d^2 R_\perp \\
&= \left(\frac{k_0 n_z}{2\pi}\right)^2 \left\langle \left| u^0(\mathbf{k}_\perp = k_0 \mathbf{n}_\perp)\right|^2 \right\rangle,
\end{aligned} \tag{3.2.39}
$$

where

$$
u^0(\mathbf{k}_\perp) = \int u^0(\mathbf{R}_\perp) \exp(i\mathbf{k}_\perp \mathbf{R}_\perp) d^2 R_\perp
$$

is the spatial spectrum of the source field taken over the plane $z = 0$.

We see that the radiant intensity $J(\mathbf{n})$, corresponding to the generalized source radiance (3.2.19), is a nonnegative quantity depending only upon the low-frequency, i.e. large-scale, part of the field spectrum $u^0(\mathbf{k}_\perp)$ that corresponds to a rather small wave numbers $|\mathbf{k}_\perp| = k_0 |\mathbf{n}_\perp| \leq k_0$. The high-frequency part of the spectrum $|\mathbf{k}_\perp| > k_0$ describes exponentially decaying nonuniform waves and does not taken into account in (3.2.39).

When the angular distribution of radiant intensity $J(\mathbf{n})$ is known, one may solve the inverse problem by determining the contribution of the low-frequency part of the spectrum in the field correlation function at the aperture [this problem was solved by Carter and Wolf (1975)]. Suppose that the incident field $u^0(\mathbf{r}_\perp)$ is statistically uniform. In the physical optics approximation, the field in the plane $z = 0$ is $u^0 = u^0(\mathbf{r}_\perp)\theta_\Sigma(\mathbf{r}_\perp)$, where $\theta_\Sigma(\mathbf{r}_\perp)$ is the transmission function of the aperture. Substituting this expression in (3.2.39) and assuming that the aperture is large compared to the correlation radius of the field at the aperture ($a >> l_c$) we obtain

$$J(\mathbf{n})=\left(\frac{k_0 n_z}{2\pi}\right)^2\int\theta_\Sigma\left(\mathbf{R}_\perp+\frac{\rho_\perp}{2}\right)\theta_\Sigma\left(\mathbf{R}_\perp-\frac{\rho_\perp}{2}\right)\Gamma^0(\rho_\perp)$$
$$\times\exp(-ik_0\mathbf{n}\rho)d^2R_\perp d^2\rho_\perp$$
$$\approx\left(\frac{k_0 n_z}{2\pi}\right)^2\int\theta_\Sigma(\mathbf{R}_\perp)\Gamma^0(\rho_\perp)\exp(-ik_0\mathbf{n}\rho_\perp)d^2R_\perp d^2\rho_\perp$$
$$=\left(\frac{k_0 n_z}{2\pi}\right)^2\Sigma\int\Gamma^0(\rho_\perp)\exp(-ik_0\mathbf{n}\rho_\perp)d^2\rho_\perp\ , \tag{3.2.40}$$

where Σ is the aperture area, and $\Gamma^0(\rho_\perp)=\langle u^0(\rho_\perp)u^{0*}(0)\rangle$ is the coherence function of the field at the aperture.

Denote the contributions of the low-frequency ($|\mathbf{k}_\perp|<k_0$) and high-frequency ($|\mathbf{k}_\perp|>k_0$) parts of the spectrum $u^0(\mathbf{k}_\perp)$ in the correlation function Γ^0 as Γ^0_{lf} and Γ^0_{hf} so that $\Gamma^0=\Gamma^0_{\text{lf}}+\Gamma^0_{\text{hf}}$. This representation corresponds to the above decomposition of Γ in (3.1.39), where Γ^0_{lf} corresponds to the quasi-uniform and Γ^0_{hf} to the nonuniform part of Γ localized near the plane $z=0$

Taking the inverse Fourier transform in (3.2.40) we easily obtain

$$\Gamma^0_{\text{lf}}(\rho_\perp)=\frac{1}{\Sigma}\int\limits_{n_\perp\le 1}\frac{J(\mathbf{n})\exp(ik_0\mathbf{n}\rho_\perp)}{1-n_\perp^2}d^2n_\perp . \tag{3.2.41}$$

Here, we observed that $n_z=(1-n_\perp^2)^{-1/2}$.

The uniform and isotropic radiance of the source $I^0(\mathbf{r}_\perp,\mathbf{n})=I^0=\text{const}$ is associated with the radiant intensity $J(\mathbf{n})=n_z\Sigma I^0$ whose angular dependence is proportional to $n_z=\cos\theta$ and is thus consistent with Lambert's law. In this case, from (3.2.41) we find

$$\Gamma^0_{\text{lf}}(\rho)=I^0\int\limits_{n_\perp\le 1}(1-n_z^2)^{-1/2}\exp(ik_0\mathbf{n}\rho_\perp)d^2n_\perp .$$

Taking this integral we get (Carter and Wolf, 1975)

$$\Gamma^0_{\text{lf}}(\rho_\perp)=I^0 2\pi\frac{\sin(k_0|\rho_\perp|)}{k_0|\rho_\perp|}.$$

In contrast to the example on the isotropic radiance of radiation elucidated in Problem 3.10, this expression corresponds to a radiation propagating in the halved solid angle $\cos\theta>0$. This situation realizes for a thermal source of large aperture $a\gg\lambda$. For such a source, as in general for the coherence function with the characteristic dependence on ρ in the form $\sin(k_0\rho_\perp)/(k_0\rho_\perp)$, the contribution from the high-frequency part of the spectrum Γ vanishes and so $\Gamma^0=\Gamma^0_{\text{lf}}$.

In some cases, using the analytical properties of the function (3.2.40), Γ^0_{hf} may be represented in terms of Γ_{lf}^{0} with the aid of the analytical continuation of this function to the high-frequency spectrum (Baltes, 1978). Thus, under certain conditions, one can completely recover the correlation function of a quasi-uniform source from the measured angular distribution of radiant intensity.

The above examples demonstrates that, for diffraction of partially coherent radiation at an aperture, many results of wave theory can be obtained by radiation transport theory if the radiance of the sources $I^0(\mathbf{n},\mathbf{r}_\perp)$ is specified not arbitrarily but with the relation (3.2.19). It should be kept in mind that the common radiometric conceptual system becomes applicable only in domains where the wave field is quasi-uniform.

3.2.12 Generalized Radiance for Paraxial Optical Systems

The concept of generalized radiance comes handy in the description of paraxial propagation of pencil beams in optical systems. It is common to describe diffraction effects in such systems with the aid of the so-called small angle or parabolic approximation. An alternative approach is the use of the generalized radiance, that is directly related with the measured intensities, naturally covers the case of partially coherent radiation, and is conserved along the ray. The last property lends the physical transparency of geometrical optics to many diffraction analyses (Problem 3.12). Below we take an example of a monochromatic source to briefly consider the procedure of using the generalized radiance in paraxial optics.

For pencil beams propagating within a narrow sector of angles close to the optical axis, the generalized radiance is concentrated near $n_z \sim 1$, that is, at $|\mathbf{n}_\perp| \le \delta\theta << 1$. Therefore, we may let in (3.1.84) $\rho_z = R_z = 0$ and $d\Omega_\mathbf{n} \approx d^2 n_\perp$ and pass over to the expression for the transverse coherence function (see also Problem 3.13)

$$\Gamma(\mathbf{R}_\perp,\rho_\perp) = \int I(\mathbf{n}_\perp,\mathbf{R}_\perp)\exp(ik_0\mathbf{n}_\perp\rho_\perp)d^2n_\perp\,. \tag{3.2.42}$$

Whence, we obtain the generalized radiance as

$$I(\mathbf{n}_\perp,\mathbf{R}_\perp) = \left(\frac{k_0}{2\pi}\right)^2 \int \Gamma(\mathbf{R}_\perp,\rho_\perp)\exp(-ik_0\mathbf{n}_\perp\rho_\perp)d^2\rho_\perp\,. \tag{3.2.43}$$

This expression differs from the source radiance (3.2.19) by $\cos\theta$ replaced with unity.

In free space, the solution of the transfer equation (3.2.10) permits one to recalculate the generalized radiance from one plane to another. In the paraxial approximation, this recalculation rule takes the form

$$I(\mathbf{n}_\perp,\mathbf{R}_\perp,z) = I(\mathbf{n}_\perp,\mathbf{R}_\perp - \mathbf{n}_\perp z,0). \tag{3.2.44}$$

Relations (3.2.42) – (3.2.44), describing the propagation of narrow beams in free space, may be viewed as a particular case of transformations effected by paraxial optical systems. For such systems, the amplitude distribution u in the output plane is related with the amplitude distribution u^0 in the input plane by the relation

$$u(\mathbf{r}_\perp) = \int P(\mathbf{r}_\perp, \mathbf{r}'_\perp) u^0(\mathbf{r}'_\perp) d^2 r'_\perp, \tag{3.2.45}$$

where $P(\mathbf{r}_\perp, \mathbf{r}'_\perp)$ is the impulse response or point spread function. For simplicity we will assume that fluctuations are absent in the system, therefore the function $P(\mathbf{r}_\perp, \mathbf{r}'_\perp)$ is deterministic.

If we recall that the transverse coherence function is

$$\Gamma(\mathbf{R}_\perp, \rho_\perp) = \left\langle u\left(\mathbf{R}_\perp + \frac{\rho_\perp}{2}\right) u^*\left(\mathbf{R}_\perp - \frac{\rho_\perp}{2}\right)\right\rangle$$

we readily obtain from (3.2.45) that the generalized radiance $I(\mathbf{n}_\perp, \mathbf{R}_\perp)$at the output to the system is related to the radiance at the input by the formula

$$I(\mathbf{n}_\perp, \mathbf{R}_\perp) = \int \mathcal{P}(\mathbf{n}_\perp, \mathbf{R}_\perp; \mathbf{n}'_\perp, \mathbf{R}'_\perp) I(\mathbf{n}'_\perp \mathbf{R}'_\perp) d^2 n'_\perp d^2 R'_\perp. \tag{3.2.46}$$

The function

$$\mathcal{P}(\mathbf{n}_\perp, \mathbf{R}_\perp; \mathbf{n}'_\perp, \mathbf{R}'_\perp) = \left(\frac{k_0}{2\pi}\right)^2$$
$$\times \int d^2\rho'_\perp d^2\rho''_\perp P\left(\mathbf{R}_\perp + \frac{\rho'_\perp}{2}, \mathbf{R}'_\perp + \frac{\rho''_\perp}{2}\right) P^*\left(\mathbf{R}_\perp - \frac{\rho'_\perp}{2}, \mathbf{R}'_\perp - \frac{\rho''_\perp}{2}\right) \exp[ik_0(\mathbf{n}'_\perp \rho''_\perp - \mathbf{n}_\perp \rho'_\perp)] \tag{3.2.47}$$

is usually called the *ray spread function*.

In the particular case of an ideal optical system with lateral magnification $M = r_\perp / r'_\perp$ we have

$$P(\mathbf{r}_\perp, \mathbf{r}'_\perp) = \delta(\mathbf{r}_\perp - M\mathbf{r}'_\perp), \tag{3.2.48}$$

$$\mathcal{P}(\mathbf{n}_\perp, \mathbf{R}_\perp, \mathbf{n}'_\perp, \mathbf{R}'_\perp) = \delta(\mathbf{R} - M\mathbf{R}'_\perp)\delta(\mathbf{n}M - \mathbf{n}'). \tag{3.2.49}$$

Therefore eqn (3.2.46) leads to

$$I(\mathbf{n}_\perp, \mathbf{R}_\perp) = \frac{1}{M^2} I^0\left(M\mathbf{n}_\perp, \frac{\mathbf{R}_\perp}{M}\right), \tag{3.2.50}$$

which corresponds to the treatment of the generalized radiance $I(\mathbf{n}_\perp, \mathbf{R}_\perp)$ as the distribution density of "generalized" rays.

Indeed, if we treat $I^0(\mathbf{n}_\perp, \mathbf{R}_\perp)$ as the intensity distribution of generalized rays over the initial "pattern" at the optical system input, then (3.2.50) will give a pattern expanded M times (replacement of $\mathbf{R}_\perp$ with $\mathbf{R}_\perp / M$) with the intensity reduced M^2

times and with the angular distribution narrowed M times (replacement of $\mathbf{n}_\perp$ with $M\mathbf{n}_\perp$).

However, the potential of such a treatment should not be overestimated, especially when the object of description is coherent radiation. Indeed, as noted in Section 3.2.7, the argument $\mathbf{R}$ of the generalized radiance is nonlocal, the transition from diffraction transfer theory to ordinary radiometric treatment is possible only for partially coherent radiation with a rather small coherence radius, and in the general case of coherent radiation, $\mathbf{R}$ does not refer to values of the field $u(\mathbf{R})$, but only points to the position of the center of gravity between two points where this field is observed.

This property of generalized radiance is retained in the geometrical optics limit of exceedingly small wavelengths. Therefore, if we view the generalized radiance as the density of generalized rays, than these rays may include "false" or "interference" rays corresponding to the position of $\mathbf{R}$ at the center of gravity between two "physical" rays (see Problem 3.14; we already mentioned these rays in Section 3.1.9). Interference rays are not associated with the transport of field energy. Such rays are the price that has to be paid for the possibility to describe the interference of coherent radiation in terms of the generalized radiance while the ordinary radiance prevent us from carrying out such an analysis.

3.2.13 Uncertainty Relations and the Resolving Capacity in Radiometric Measurements

In Section 3.1.13 we obtained the uncertainty relations (3.1.87) – (3.1.89) that express the principal limitations imposed by the wave nature of radiation on the accuracy of radiometric measurements. These limitations relate the uncertainties in the spatial distribution (ΔR_x) and the angular distribution ($\Delta\theta_\rho$) of beam intensity on the one hand and the the scales of the spatial correlations ($\Delta\rho_x$) and angular correlations ($\Delta\theta_R$) of the field in the beam on the other.

The inequality (3.1.87) $\Delta R_x \Delta\theta_\rho \geq \lambda$ combines the uncertainties in two arguments of the radiance $I(\mathbf{n},\mathbf{R})$: the direction of propagation $\mathbf{n}$ and the radius vector $\mathbf{R}$. This implies that in any measurements of radiance I, the values of its arguments cannot be obtained with accuracy better than allowed by this inequality. Thus, an improvement in the angular resolution will have to entail a reduction in the resolution in coordinates. We trace out how this inequality evolves by an example of an aperture of diameter D with the angular distribution of radiation measured in its Fraunhofer zone, as shown in Fig. 3.8a.

The angular resolution of such an instrument $\Delta\theta = \lambda / D$ is limited by the diffraction at the aperture and can be increased by expanding D. However, larger aperture reduce the localization of radiance: the measured radiance must be assigned to a spatial domain with of size D rather than to point $\mathbf{R}$, that is the uncertainty in the coordinate ΔR_x is approximately D. Thus, the product of uncertainties $\Delta\theta\Delta R_x$ is approximately equal to the wavelength: $(\lambda / D)D \sim \lambda$.

This result may be expressed alternatively by saying that every attempt to localize the radiance in space leads to a delocalization (broadening) of the angular distribution of radiance, and, conversely, a narrowing of the angular distribution broadens the radiance distribution in space. This treatment of the uncertainty relation is close to the quantum-mechanical perception of this relationship for coordinate and momentum.

The above example, as well as inequality (3.1.87) itself, is applicable to the case of partially coherent radiation and for the case of a coherent plane wave when the intensity distribution in the Fraunhofer zone for a circular aperture is given by the Airy formula (3.2.38).

Two other uncertainty relations (3.1.88) and (3.1.89) describe field correlations, and lose their meaning for a completely coherent wave. These two relations may be also illustrated in the screen aperture context, but only for the case of partially coherent incident radiation (Friberg and Wolf, 1983).

In order to illustrate eqn (3.1.89) we consider the distribution of the average intensity in the far field for the diffraction of a statistically uniform field at the aperture (Fig. 3.8b). If the aperture size D is sufficiently large compared to the field correlation scale $\rho_c \sim \Delta\rho_x$, then the mean intensity distribution in the Fraunhofer zone will not be defined by the full size D, but by the scale of field nonuniformity $\Delta\rho_x$, and will have the angular width $\Delta\theta \geq \lambda / \Delta\rho_x$. This agrees with the uncertainty relation (3.1.89).

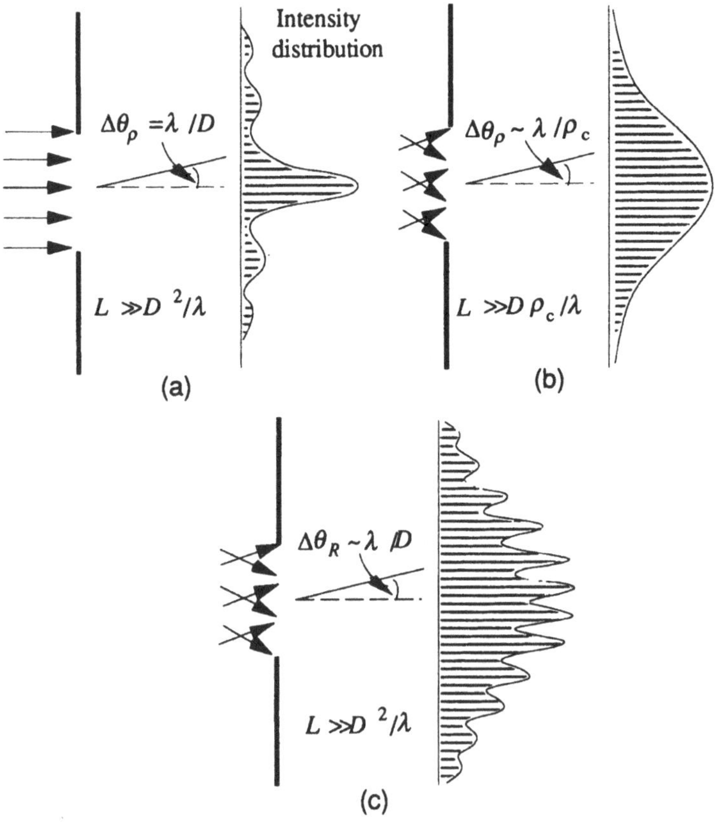

Figure 3.8. Diffraction at an aperture of diameter D: (a) intensity distribution in the Fraunhofer zone; (b) diffraction of a statistically uniform wave for $D \gg \rho_c$, where ρ_c is the coherence radius; (c) angular correlations in diffraction of radiation of random sources with $\rho_c \to 0$.

It is important to emphasize that, in this case, the intensity distribution for individual realizations may appreciably differ from the mean intensity distribution that is obtained by ensemble averaging.

In order to illustrate eqn (3.1.88) it is convenient to consider the case of a short-correlated incident field, or else, the case of random sources on the aperture surface (Fig. 3.8c). In this case, diffraction at the aperture affects not only the angular distribution of mean intensity, but, by van Cittert–Zernike theorem (Section 3.2.10), also the scale of angular correlations $\Delta\theta_R$. The presence of such correlations may be revealed by inserting a Young interferometer in the observation plane and recording interference fringes in the intensity distribution behind the interferometer screen. In this case, for uncertainties, we have $\Delta R_x \sim D$ and $\Delta\theta_R \geq \lambda / D$ so that $\Delta R_x \Delta\theta_R \geq \lambda$.

Thus, the uncertainty relations (3.1.87) – (3.1.89) combine absolutely different parameters of partially coherent sources and have different physical meaning.

3.2.14 Coherent (Antenna) Measurements and Optical Heterodyning

Let us consider the case of monochromatic radiation, omitting the time factor $\exp(-i\omega t)$. In radar and radio astronomy (Fig. 3.9), signals fed from the antenna to the amplifier are proportional to

$$v_{out} = \int_{\Sigma} u(\mathbf{r}_\perp) f^*(\mathbf{r}_\perp) d^2 r_\perp, \tag{3.2.51}$$

where the integral is taken over the antenna aperture Σ, u is the amplitude of the incident wave, and the weighting function $f(\mathbf{r}_\perp)$ describes the distribution of sources (currents) over the aperture in the radiation mode. This relation is written in the scalar approximation (a description of polarization effects may be found in Problem 3.15). This type of measurement are commonly terms coherent, for the output signal v_{out} depends upon the phase distribution of u over the aperture.

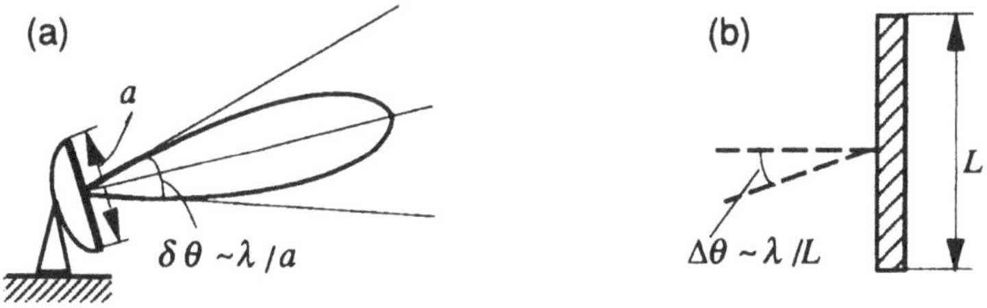

Figure 3.9. Antenna measurements: (a) receiving antenna, (b) source of radiation.

In optics, the coherent measurement scheme cannot be realized directly because available instruments are unable to record phases of rapidly oscillating optical signals. However, a relation for the photodetector current, very similar to eqn (3.2.51), occurs when the so-called optical heterodyning is used. This method feeds the signal wave with amplitude u and frequency ω_1 to the input of a quadrature photodetector simultaneously with the local oscillator wave with amplitude u_0 and frequency ω_0

(Fig. 3.10). The photo current is measured at the difference frequency $\omega_0 - \omega_1$, and is given by

$$i(t) = \int \eta(\mathbf{r}_\perp)|u_0 \exp(-i\omega_0 t) + u_1 \exp(-i\omega_1 t)|^2 \, d^2 r_\perp, \tag{3.2.52}$$

where the weighting factor $\eta(\mathbf{r}_\perp)$ is called the quantum efficiency.

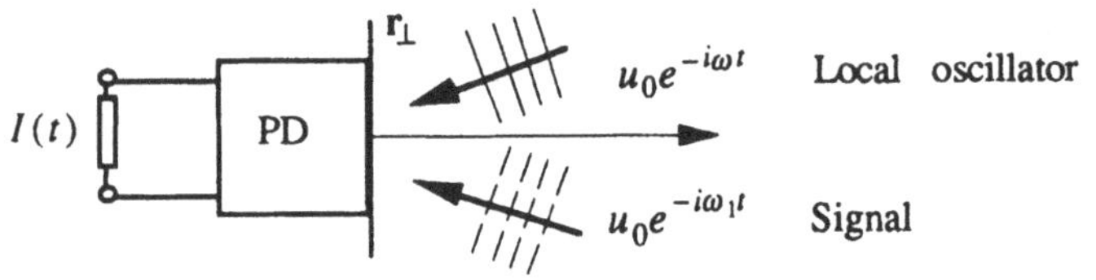

Figure 3.10. Optical heterodyne receiver (after Siegman, 1966)

For the current at the difference frequency $\Delta\omega = \omega_0 - \omega_1$, from (3.2.52) it follows

$$i_{\Delta\omega} = 2\,\mathrm{Re}\int u(\mathbf{r}_\perp) u_0^*(\mathbf{r}_\perp)\eta(\mathbf{r}_\perp) d^2 r_\perp . \tag{3.2.53}$$

This expression is analogous to eqn (3.2.51) combining the output signal with the field distribution over the antenna aperture. The weighting function here is the product $u_0^*\eta$. Similarity of both expressions allows us to carry results obtained for coherent measurements onto the case of optical radiation (Siegman, 1966).

Returning, for definiteness sake, to the antenna measurement scheme (3.2.51), we consider the field far from the sources. This field may be represented as an expansion in terms of traveling waves satisfying the dispersion equation $k = \omega / c = k_0$, namely,

$$u(\mathbf{r}) = \int A_\mathbf{n} \exp(ik_0 \mathbf{n}\mathbf{r}) d\Omega_\mathbf{n} . \tag{3.2.54}$$

Substituting this expression in (3.2.51) we get

$$v_{out} = \int A_\mathbf{n} f(\mathbf{n}_\perp) d\Omega_\mathbf{n} , \tag{3.2.55}$$

where

$$f(\mathbf{n}_\perp) = \int_\Sigma f(\mathbf{r}_\perp) \exp(ik_0 \mathbf{n}_\perp \mathbf{r}_\perp) d^2 r_\perp \tag{3.2.56}$$

is the directional response pattern of the antenna.

In the general case of quasi-monochromatic random sources, the amplitude of the incident wave, $A_\mathbf{n}$, slowly varies in time in a random manner. After detection and time averaging, or, alternatively, narrow band filtering, the receiver output is proportional to the quantity

$$\langle |\mathbf{v}_{out}|^2 \rangle = \int \langle A_{\mathbf{n}'} A^*_{\mathbf{n}''} \rangle f^*(\mathbf{n}'_\perp) f(\mathbf{n}''_\perp) d\Omega_{\mathbf{n}'} d\Omega_{\mathbf{n}''}. \tag{3.2.57}$$

Radio astronomy and many other applications detect radiation coming from very distant sources that may be deemed independent from each other. In eqn (3.2.57), different sources are associated with different $\mathbf{n}'_\perp \neq \mathbf{n}''_\perp$, and we may assume the correlation $\langle A_{\mathbf{n}'} A^*_{\mathbf{n}''} \rangle$ to be other than zero only at $\mathbf{n}'_\perp = \mathbf{n}''_\perp$, viz.,

$$\langle A_{\mathbf{n}'} A^*_{\mathbf{n}''} \rangle \approx I(\mathbf{n}'_\perp) \delta_2(\mathbf{n}'_\perp \mathbf{n}''_\perp), \tag{3.2.58}$$

where $\delta_2(\mathbf{n}'\mathbf{n}'')$ is the delta function on a unit sphere, and

$$I(\mathbf{n}') = \int \langle A_{\mathbf{n}'} A^*_{\mathbf{n}''} \rangle d\Omega_{\mathbf{n}''} \tag{3.2.59}$$

is the radiance of radiation. Equation (3.2.58) implies that the radiation incident on the antenna is spatially quasi-uniform.

Substituting (3.2.58) in (3.2.57) we obtain the output intensity as

$$\langle |\mathbf{v}_{out}|^2 \rangle = \int I(\mathbf{n}'_\perp) |f(\mathbf{n}'_\perp)|^2 d\Omega_{\mathbf{n}'}. \tag{3.2.60}$$

In radio engineering, this expression is usually written as (Pacholczyk,1970)

$$P = S_e \int I(\mathbf{n}'_\perp) P_e(\mathbf{n}'_\perp) d\Omega_{\mathbf{n}'}, \tag{3.2.61}$$

where P is the power of the output signal, S_e is the effective surface of the antenna, and

$$P_e(\mathbf{n}_\perp) = |f(\mathbf{n}_\perp)|^2 / |f(0)|^2 \tag{3.2.62}$$

is the power directional pattern normalized by the condition $P_e(\hat{\mathbf{z}}) = 1$, where $\hat{\mathbf{z}}$ is the normal to the antenna surface.

We see that the standard radio astronomy relation (3.2.61) is valid for a quasi-uniform beam obeying the quasi-uniformity condition (3.2.58). When the radiation is not quasi-uniform, the correlation $\langle A_{\mathbf{n}'} A^*_{\mathbf{n}''} \rangle$ may be other than zero also at $\mathbf{n}' \neq \mathbf{n}''$. In this case, eqn (3.2.58) does not hold and the output power cannot be expressed in terms of the intensity $\langle A_{\mathbf{n}} A^*_{\mathbf{n}} \rangle$ alone. Below we estimate conditions which enable the approximation (3.2.58).

Let the correlation $\langle A_{\mathbf{n}'} A^*_{\mathbf{n}''} \rangle$ be different from zero in the interval $|\mathbf{n}'_\perp - \mathbf{n}''_\perp| = \Delta n_\perp \leq \Delta\theta$. For a distant thermal (or, alternatively, spatially incoherent) source of size L (Fig. 3.9b), the van Cittert–Zernike theorem gives $\Delta\theta \sim \lambda / L$. The characteristic variation dimension of the function $f(\mathbf{n}_\perp)$ in (3.2.57) has the order of magnitude of the angular width of the antenna directional pattern $\delta n_\perp \sim \delta\theta \sim \lambda / a$, where a is the antenna aperture. In order to let $f(n'_\perp) f^*(n''_\perp) \approx |f(n'_\perp)|^2$ in taking the integral (3.2.57), which is equivalent to using the approximation (3.2.58), the angular separation of the correlated waves, $\Delta\theta$, must be small compared to the characteristic scale $\delta\theta$ of variation of $f(\mathbf{n}_\perp)$:

$$\Delta\theta \sim \frac{\lambda}{L} << \delta\theta \sim \frac{\lambda}{a}. \tag{3.2.63}$$

This condition reduces to the inequality $L >> a$ and implies that the antenna dimensions are small compared to the size of the source. In radio astronomy, this inequality is naturally always true.

Consider now the opposite limiting case of a wide-angle coherent beam and a narrow directional response pattern: $\Delta\theta >> \delta\theta \sim \lambda / a$. We assume that $f(\mathbf{n}_\perp)$ differs from zero near the direction $\mathbf{n}_{0\perp}$, that is, for $|\mathbf{n}_\perp - \mathbf{n}_{0\perp}| < \delta\theta$. Then, we may factor $\langle A_{\mathbf{n}_0} A^*_{n_0} \rangle$ outside the integral sign in (3.2.57), to get

$$\langle |\mathbf{v}_{\text{out}}|^2 \rangle \approx \langle |A_{\mathbf{n}_0}|^2 \rangle \left| \int f(\mathbf{n}'_\perp) d\Omega_{\mathbf{n}'} \right|^2. \tag{3.2.64}$$

The intensity of the received signal turns out to be proportional to the intensity of the wave directed in $\mathbf{n}_0$. It is natural that even a set of such measurements obtained by scanning over all feasible directions $\mathbf{n}_0$ is incapable of describing the correlation properties of the field that, in general, depend on $\langle A_{\mathbf{n}'} A_{\mathbf{n}''} \rangle$ for $\mathbf{n}'_\perp \neq \mathbf{n}''_\perp$.

It is customary to characterize the radiation pattern of an antenna by the *antenna solid angle*

$$\Omega_a = \int P_e(\mathbf{n}) d\Omega_{\mathbf{n}} \tag{3.2.65}$$

or the *directivity* $D = 4\pi / \Omega_a$ which indicates by how many times the antenna pattern is narrower than the full solid angle 4π corresponding to the isotropic pattern. We note that in antenna theory, in place of $P(n)$, one often uses of the concept of *gain*

$$G(\mathbf{n}) = DP(\mathbf{n}),$$

which differs from $P(\mathbf{n})$ by normalization: in the direction of the normal to the antenna $G(\mathbf{n})$ must be equal to D.

The directivity D is related to the effective antenna area S_e by

$$D = \frac{4\pi}{\lambda^2} S_e. \tag{3.2.66}$$

Recalling the definition of D we immediately obtain

$$\Omega_a S_e = \lambda^2. \tag{3.2.67}$$

This inequality agrees with the uncertainty relation (3.1.87) for antenna measurements: the square of the uncertainty in the coordinate ΔR_x has the order of magnitude of the antenna area S_e, and the square of the angular uncertainty $\Delta\theta_p$ has the order of magnitude of the antenna solid angle Ω_a.

3.2.15 Incoherent Radiometry

In radiometric measurements of optical radiation, phases are normally not recorded, the detection being performed locally, at each point of the recording device. The incoherent device may be, for example, a photographic plate, photo counter, or the eye.

In contrast to the coherent receiver — the antenna, an incoherent detector by itself is almost insensitive to the direction of the incident radiation. In this case, selection by directions is achieved only with the aid of additional optical systems, say telescopes, used in conjunction with the sensor.

Consider the response of an incoherent device, for example, the output current of a photodetector. This response is proportional to the integral of the intensity $J(\mathbf{r}_\perp) = \langle |u(\mathbf{r}_\perp)|^2 \rangle$ over the plane area of the surface Σ:

$$P = \int_\Sigma \langle |u(\mathbf{r}_\perp)|^2 \rangle d^2 r_\perp = \int \theta_\Sigma(\mathbf{r}_\perp) \langle |u(\mathbf{r}_\perp)|^2 \rangle d^2 r_\perp. \tag{3.2.68}$$

Here, θ_Σ is the transmission function of the aperture. Equation (3.2.68) is not sensitive to the phase of the field u, and this differs the incoherent measurement scheme from the coherent scheme considered above.

Substituting (3.2.54) in (3.2.68) yields

$$P = \int \langle A_{\mathbf{n}'} A^*_{\mathbf{n}''} \rangle \theta_\Sigma(\mathbf{n}' - \mathbf{n}'') d\Omega_{\mathbf{n}'} d\Omega_{\mathbf{n}''}. \tag{3.2.69}$$

The function

$$\theta_\Sigma(\mathbf{n}' - \mathbf{n}'') = \int \theta_\Sigma(\mathbf{r}_\perp) \exp(ik_0(\mathbf{n}' - \mathbf{n}'')\mathbf{r}) d^2 r_\perp,$$

differs from zero when $|\mathbf{n}' - \mathbf{n}''| \le \lambda / a$, where $a \sim \Sigma^{1/2}$ is the typical dimension of the aperture Σ.

If u is quasi-uniform, that is, the approximation (3.2.58) is applicable, then eqn (3.2.69) becomes

$$P = \int I(\mathbf{n}) \theta_\Sigma(0) d\Omega_{\mathbf{n}} = \Sigma \int I(\mathbf{n}) d\Omega_{\mathbf{n}}, \tag{3.2.70}$$

where Σ is the illuminated area. In order to make this approximation applicable, as in the case of the coherent measurement configuration, we require, by analogy with eqn (3.2.63), that $\Delta\theta << \delta\theta \sim \lambda / a$. From eqn (3.2.69) we see that the considered measurement configuration is isotropic, that is, identically sensitive to radiance $I(\mathbf{n})$ incident from arbitrary direction $\mathbf{n}$.

In the opposite limiting case of a wide-angle coherent beam with $\Delta\theta >> \delta\theta \sim \lambda / a$, in eqn (3.2.63), we may use the approximation

$$\theta_\Sigma(\mathbf{n}'-\mathbf{n}'') \approx \theta_\infty(\mathbf{n}'-\mathbf{n}'') \equiv \int \exp\left(ik_0(\mathbf{n}'-\mathbf{n}'')\mathbf{r}_\perp\right)d^2\mathbf{r}_\perp = \left(\frac{2\pi}{k_0}\right)^2 \delta(\mathbf{n}'_\perp - \mathbf{n}''_\perp).$$

Substituting this expression in (3.2.69) yields the received power as

$$P \approx \left(\frac{2\pi}{k_0}\right)^2 \int \langle A_{\mathbf{n}'} A^*_{\mathbf{n}''}\rangle \delta(\mathbf{n}'_\perp - \mathbf{n}''_\perp) d\Omega_{\mathbf{n}'} d\Omega_{\mathbf{n}''}$$

$$= \left(\frac{2\pi}{k_0}\right)^2 \int \left\langle |A_{\mathbf{n}'}|\right\rangle^2 \frac{1}{n'_z} d\Omega_{\mathbf{n}'} ,$$

This expression is similar to the approximation (3.2.64) for the coherent measurement scheme. However, if we detect the intensity $\langle |A_{\mathbf{n}_0}|^2\rangle$ with the coherent measurement scheme for one direction $\mathbf{n}_0$ that is cut out by the antenna directional response, then the incoherent scheme remains almost isotropic — it integrates the intensities $\langle |A_{\mathbf{n}'}|^2\rangle$ over all directions $\mathbf{n}'$ with a weighting factor $1/n'_z$.

Thus, in both coherent and incoherent measurement scheme, depending on the size of the receiving aperture, one and the same incident beam with a maximum angular spreading of correlated waves $\Delta\theta$ may be detected as coherent (for $\Delta\theta << \delta\theta$) and as incoherent (for $\Delta\theta >> \delta\theta$).

3.2.16 Time Averaging and Temporal Ergodicity

Processes or fields are called ergodic if their time and/or spatial averages over one realization are close to their ensemble averages. A guess on ergodicity is usually presented as an *ergodicity hypothesis*. The objective of this hypothesis is to assign to the statistical moments, arising in the theory, some specific values following from the experiment. Without such a hypothesis, the main objects of statistical theory prove to be pointless.

The temporal ergodicity of the random process $\xi(t)$ assumes that the time average

$$\overline{\xi}^T = \frac{1}{T}\int_t^{t+T} \xi(t')dt', \tag{3.2.71}$$

is close to the ensemble average $\langle\xi\rangle$. It is believed normally that the values $\overline{\xi}^T$ and $\langle\xi\rangle$ will be close provided that the averaging interval T is much larger than the correlation time τ_c of $\xi(t)$. Intuitively it is assumed that "much larger" means 10 times as large, or at best 100 times as large. However, even 100-fold excess of T over τ_c need not guarantee the expected closeness of $\overline{\xi}^T$ and $\langle\xi\rangle$. In what follows we present simple considerations that are helpful in choosing the appropriate averaging interval.

Consider the difference $\eta \equiv \overline{\xi}^T - \langle\xi\rangle$. For an overwhelming majority of realizations, the fluctuating quantity η does not exceed a few, say three, standard deviations $\sigma_\eta = \langle \eta^2 \rangle^{1/2}$. As a rule, this implies

$$|\eta| \equiv |\bar{\xi}^T - \langle \xi \rangle| \le C\sigma_\eta = C\left(T^{-2} \int_0^T \int_0^T \langle \tilde{\xi}(t')\tilde{\xi}(t'') \rangle dt dt' \right)^{1/2}, \tag{3.2.72}$$

where $C \approx 3$.To ensure that η be small, we require that the right-hand side of eqn (3.2.72) vanish as $T \to \infty$. This is the so-called *necessary and sufficient condition of temporal ergodicity* of $\xi(t)$ (Rytov, Kravtsov, and Tatarskii, 1987a, Section 2.5).

Determine the integral correlation time by

$$\tau_c = \sigma_\xi^{-2} \int_{-\infty}^{\infty} \langle \xi(t)\xi(t+\tau) \rangle d\tau, \quad \sigma_\xi^2 \equiv \langle \tilde{\xi}^2 \rangle. \tag{3.2.73}$$

Then the inequality (3.2.72) may be rewrite as

$$|\eta| \le C_\eta \sigma_\eta \approx C_\eta \sigma_\xi (\tau_c/T)^{1/2}. \tag{3.2.74}$$

From this relation we see that when $T \to \infty$, η tends to zero as $\sqrt{(\tau_c/T)}$, therefore, for a sufficiently large averaging interval T, the residual η becomes small. Having specified a tolerable $|\eta|_{\max}$ and assumed $C_\eta = 3$ common in engineering calculations, we can estimate the necessary averaging interval T. For $T = 10\tau_c$, the expected residual $|\eta|_{\max}$ will be around $3\sigma_\xi \sqrt{0.1} \approx \sigma_\xi$; and, for $T = 100\tau_c$, the accuracy is improved only to $0.3\sigma_\xi$, i.e., is considerably smaller than might be expected intuitively.

Estimators of the type of (3.2.74) reflect the common statistical regularity saying that the relative fluctuations of a sum of a large number N of independent (and equivalent) addends decrease as $N^{1/2}$. In this case, the number of correlation intervals fitted within the averaging interval plays the role of $N \approx T/\tau_c$. For slowly decaying correlation functions, $|\eta|_{\max}$ will decrease even slower than $N^{-1/2} = (\tau_c/T)^{1/2}$ (Picibono, 1977).

By way of example, we examine conditions enabling us to interpret the readings of the photodetector in terms of radiometry, replacing the averaging over the realization with the averaging over the statistical ensemble.

The basic photometric instrument is the square-law photodetector. Its output i is proportional to the intensity of the light field averaged over a time T

$$\overline{i(t)}^T = K\frac{1}{T}\int_0^T v^2(t+t')dt' \equiv K\overline{v^2(t)}^T \tag{3.2.75}$$

For v we can substitute here the full amplitude of the light field $E(t)$, or one of components of vector $\mathbf{E}$, if a polarizer is installed in front of the photodetector. The geometrical dimensions of the photodetector are deemed to be small compared with the spatial correlation radius of the light field, i.e., we consider a point detector. Finite dimensions of the detector will be accounted for later.

The averaging interval T in (3.2.75) is governed by the inertia of the photodetector. In optical measurements, it by far exceeds the characteristic period of light oscillations $2\pi/\omega$. For a quasi-monochromatic field, we may write $v(t) = \mathrm{Re}[u(t)\exp(-i\omega t)]$, where $u(t)$ is the slow complex amplitude of the field. Then

$$v^2(t) = \frac{1}{2}|u(t)|^2 + \frac{1}{4}u^2(t)e^{-2i\omega t} + \frac{1}{4}[u^*(t)]^2 e^{2i\omega t}.$$

On averaging, the rapidly varying addends vanish and the final result is

$$\overline{i(t)}^T = \frac{K}{2}\overline{|u(t)|^2}^T \tag{3.2.76}$$

Thus, the output current is proportional to the squared complex field amplitude averaged over time T.

Let the inertia of the photodetector be small, and the slow amplitude $u(t)$ cannot vary significantly over time T. Then the quantity detected is nothing other than *local* (at the location of the detector) *instantaneous* intensity $J(r,t) \equiv |u(r,t)|^2$. In this case, the photodetector output does not allow interpretation in terms of ensemble average quantities, since the photocurrent fluctuates in time from one point to another.

If the time of averaging is large compared to the correlation time, $T >> \tau_c$, and the process $u(t)$ is stationary, then we may make a good use of the ergodicity hypothesis by assuming that the measured value of intensity $\overline{J}^T = \overline{|u|^2}^T$ approaches its ensemble average $\langle J\rangle = \langle |u|^2\rangle$. For sufficiently large values of the ratio T/τ_c, the photocurrent $\overline{i}^T$ is almost not fluctuating and describes the average field intensity $\langle J\rangle$ at point r at frequency ω. This implies that the photodetector output is proportional to the integral of the radiometric radiance $I_\omega(n)$ over a cone of directions depending on the photodetector design (it is assumed that the photodetector includes some optical system that takes care of direction selectivity). If we manage to make the field of view of the photodetector sufficiently small, for example by hooking up the detector to a telescope, we can measure the radiance $I_\omega(\mathbf{n})$ in a certain direction. Radiance measurements will be reiterated in Section 3.2.20 below.

3.2.17 Fourier Spectroscopy

Measurement of the width and form of spectral lines calls for sophisticated high-resolution spectral instruments. Therefore, a spectral analysis of light fields has its technical and cost limitations. However, under certain conditions, such measurements can be carried out without spectral instruments by replacing measurements of I_ω with measurements of the correlation function of photocurrent

$$\Psi_i(\tau) = \overline{i(t)i(t+\tau)}^T. \tag{3.2.77}$$

In the case of Gaussian fields, such a correlation function is proportional to the square of the temporal coherence function:

$$\Psi_i(\tau) = |\Gamma(\tau)|^2, \quad \Gamma(\tau) = \langle u(t)u^*(t+\tau)\rangle. \tag{3.2.78}$$

Now, measuring $\Psi_i(\tau)$ yields an information on the modulus of the field coherence function. If the function $\Gamma(\tau)$ is positive, then we will be able to determine the frequency spectrum I_ω by Fourier transforming $\Gamma(\tau)$, viz.,

$$I_\omega \sim \frac{1}{2\pi}\int \Gamma_{12}(\tau)e^{i\omega\tau}d\tau. \tag{3.2.79}$$

This method of spectral analysis has come to be known as Fourier spectroscopy. It is widely used in applied optics (Wells, 1988).

We notice the possibility of the inverse operation — evaluation of the temporal coherence function $\Gamma(\tau)$ by Fourier transforming the spectral radiance I_ω. For thermal electromagnetic radiation, we considered the relation of $\Gamma(\tau)$ with I_ω in Section 3.1.14.

3.2.18 Spatial Averaging and Spatial Ergodicity

Up to this point we have considered only temporal averaging and temporal ergodicity. However, the spatial averaging action of the instrument reduces also to the ensemble averaging if radiation possesses the property of *spatial ergodicity*.

As an example of the ergodic field we above all should consider the wave intensity $\mathcal{J}(r,t) = |u(r,t)|^2$. Ergodicity manifests itself in measurements of energy fluxes. For pencil beam waves, the flux of energy is proportional to the integral of intensity taken over the aperture S of the energy receiver:

$$P(t) = \int_S \mathcal{J}(r,t)d^2r. \tag{3.2.80}$$

The aperture-averaged intensity is defined in terms of the flux P:

$$\overline{\mathcal{J}(t)}^S = \frac{1}{S}P(t) = \frac{1}{S}\int_S \mathcal{J}(r,t)d^2r. \tag{3.2.81}$$

The larger the aperture area S compared to the squared intensity correlation radius $r_{\mathcal{J}}^2$ the closer the aperture average intensity $\overline{\mathcal{J}(t)}^S$ will be to the ensemble average $\langle\mathcal{J}\rangle$. We will characterize this closeness by the difference $\Delta\mathcal{J} = \overline{\mathcal{J}}^S - \langle\mathcal{J}\rangle$. The relative fluctuations $\Delta\mathcal{J}/\langle\mathcal{J}\rangle$ are defined by the number N_c of correlation spots fitted within the detector area. If $r_{\mathcal{J}}^2$ is the "correlation area" of intensity in the receiver plane, then $N_c \sim S/r_{\mathcal{J}}^2$ and, as shown in Problem 3.17,

$$\frac{\langle(\Delta \jmath)^2\rangle}{\langle \jmath\rangle^2} \approx \frac{\langle \tilde{\jmath}^2\rangle}{\langle \jmath\rangle^2}, \qquad \frac{r_J^2}{S} = \frac{1}{N_c}\frac{\langle \tilde{\jmath}^2\rangle}{\langle \jmath\rangle^2}. \tag{3.2.82}$$

Here $\tilde{\jmath} = \jmath - \langle \jmath \rangle$ is the local intensity fluctuations on the surface of the receiver. According to (3.2.82), an increase of the receiver area S decreases the spatially averaged intensity fluctuations $\overline{\jmath}^S$.

For the conditions of spatial ergodicity we refer the reader to Problem 3.18. These are similar to the conditions of temporal ergodicity and reduce to the requirement that the correlation function of the field under investigation should sufficiently rapidly decrease with distance.

3.2.19 Selfaveraging Quantities

The statistical theory of solids makes a wide use of the new concept of *self-averaging*. A random quantity is self-averaging if their *measured values* are almost devoid of fluctuations. As a rule, these are integral characteristics, i.e, quantities that are themselves integrals of other physical quantities with respect to time or space. Accordingly, one speaks of temporal or spatial self-averaging

The concept of self-averaging can also be useful in statistical optics. It is close to ergodicity, but there is an essential difference. While ergodicity aims at assigning to the experimental data the averages over an *imaginary* statistical ensemble, self-averaging is a *real* process of smoothing.[2]

In the above examples of photocurrent (Section 3.2.16) and the flux of energy P (Section 3.2.18) we dealt with self-averaging in view of integration over time or surface. For example, an increase of the surface S causes the relative flux fluctuations to decrease, i.e., the flux P is a self-averaging quantity. Dividing the measured value of P by the area S we find the surface average intensity $\overline{\jmath}^S = P/S$. By the ergodicity hypothesis we assign it to a statistical ensemble and identify it with $\langle \jmath \rangle$.

Not all quantities, even those subjected to integration over time or space, possess the property of self-averaging, i.e. smoothing, of fluctuations. As an example of a non-self-averaging quantity we consider the intensity of a Gaussian wave field in the focal plane of a large lens. This example is interesting because it contradict a widely accepted opinion that the spatial averaging over the lens surface *always* smoothes fluctuations. This smoothing *may* have place, but by far not in all cases.

Let us revisit the narrow quasi-monochromatic beam wave $u(\mathbf{r},t)$. The complex amplitude $u^F(\mathbf{r}_\perp,t)$ is the focal plane is proportional to the Fourier transform of the amplitude $u(\mathbf{r}_\perp t)$ in the lens plane Σ:

$$u^F(\mathbf{r}_\perp,t) = C\int_\Sigma u(\mathbf{r}'_\perp,t)\exp\left(\frac{ik_0\mathbf{r}_\perp\mathbf{r}'_\perp}{F}\right)d^2r'_\perp, \tag{3.2.83}$$

where F is the focal length of the lens (Fig. 3.11). In agreement with (3.2.83) the

[2] In the solid state physics, one more difference is that self-averaging may occur not only as a result of integration, but also as a result of physical processes such as diffusion, mixing, and scattering.

intensity in the focal plane is

$$\begin{aligned} \jmath^F(\mathbf{r}_\perp,t) &= \left|u^F(\mathbf{r}_\perp,t)\right|^2 \\ &= |C|^2 \int\!\!\int_{\Sigma\Sigma} u(\mathbf{r}'_\perp t) u^*(\mathbf{r}''_\perp,t) \exp\left[\frac{ik_0 \mathbf{r}_\perp(\mathbf{r}'_\perp - \mathbf{r}''_\perp)}{F}\right] d^2 r'_\perp d^2 r''_\perp \\ &= \int\!\!\int u\left(\mathbf{R}'_\perp + \frac{\rho'_\perp}{2},t\right) u^*\left(\mathbf{R}'_\perp - \frac{\rho'_\perp}{2},t\right) \exp\left[\frac{ik_0 \mathbf{r}_\perp \rho'_\perp}{F}\right] d^2\rho'_\perp d^2 R'_\perp . \end{aligned} \tag{3.2.84}$$

In a particular case when the incident wave $u(r,t)$ obeys the normal distribution with zero mean, its relative intensity fluctuations are unity: $\langle \tilde{\jmath}^2\rangle / \langle \jmath \rangle^2 = 1$, where $\tilde{\jmath} = \jmath - \langle \jmath \rangle$ and $\jmath = |u|^2$. This conclusion refers also to the field fluctuations in the focal plane, u^F. In view of the linearity of eqn (3.2.83) these fluctuations are normally distributed with zero mean, and the relative intensity fluctuations of $\jmath^F$ are unity:

$$\left\langle \left(\tilde{\jmath}^{\mathrm{F}}\right)^2 \right\rangle \Big/ \left\langle \jmath^{\mathrm{F}} \right\rangle^2 = 1. \tag{3.2.85}$$

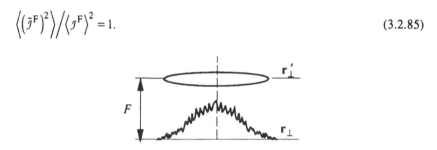

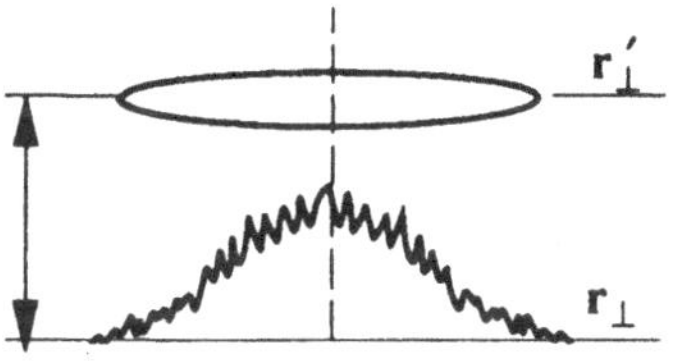

Figure 3.11.

Thus, despite the integration over the lens aperture, no spatial averaging of intensity fluctuations occurs. In other words, the intensity $\jmath^{\mathrm{F}}$ does not exhibit the property of self-averaging when the lens area is increased. Self-averaging is barred by the interference nature of field forming in the focal plane.

The situation changes drastically in the presence of a coherent component in the primary wave when the intensity $\jmath^{\mathrm{F}}$ acquires the property of self-averaging. Let the coherent component be a wave incident normally on the lens so that in the lens plane $u_{\mathrm{coh}} = \langle u(r_\perp)\rangle = \mathrm{const}$. By virtue of (3.2.83), the coherent component in the center of the focal plane $r_\perp = 0$ will be proportional to the lens area $\Sigma = \pi D^2$:

$$u^{\mathrm{F}}_{\mathrm{coh}} = \langle u^{\mathrm{F}}(0)\rangle = C\Sigma \langle u(r_\perp)\rangle.$$

The average intensity in the center of the focal plane $\langle \jmath^{\mathrm{F}}(0)\rangle$ will contain the intensity of the coherent component $\langle \jmath^{\mathrm{F}}(0)\rangle = \left|\langle u^{\mathrm{F}}(0)\rangle\right|^2$ that is proportional to the square of the lens area Σ^2, and the intensity of the incoherent (fluctuating) component, that is proportional only to the first power of Σ. As Σ increases, the contribution of the coherent component becomes dominating, and, in contrast to the considered case of the zero coherent field, the relative fluctuations of intensity will go to zero:

$$\frac{\left\langle\left(\tilde{\jmath}^{\mathrm{F}}\right)^2\right\rangle}{\left\langle\jmath_{\mathrm{F}}\right\rangle^2}=\frac{\left\langle\left[\jmath^{\mathrm{F}}-\left\langle\jmath^{\mathrm{F}}\right\rangle\right]^2\right\rangle}{\left\langle\jmath^{\mathrm{F}}\right\rangle^2}=\frac{\left\langle\left[\jmath^{\mathrm{F}}-\left\langle\jmath^{\mathrm{F}}_{\mathrm{incoh}}\right\rangle-\left\langle\jmath^{\mathrm{F}}_{\mathrm{coh}}\right\rangle\right]^2\right\rangle}{\left(\left\langle\jmath^{\mathrm{F}}_{\mathrm{incoh}}\right\rangle+\left\langle\jmath^{\mathrm{F}}_{\mathrm{coh}}\right\rangle\right)^2}\underset{\Sigma\to\infty}{\to}0. \tag{3.2.86}$$

This implies that, in the presence of a coherent field component, the average intensity in the focal plane will be a self-averaging quantity.

In optical experiments, the coherent component is usually rather small. For example, the average field of a laser in a turbulent atmosphere is attenuated by e times already at a distance of 20 - 50 m (Rytov, Kravtsov, and Tatarskii, 1989b). Therefore, the relative intensity fluctuations (3.2.86) can be asymptotically reduced only if $\left\langle\jmath^{\mathrm{F}}_{\mathrm{coh}}\right\rangle\neq 0$. If the coherent component is absent, then the left hand side of eqn (3.2.86) behaves similar to eqn (3.2.85), that is, the intensity at the center of the focal plane loses the property of self-averaging.

After this analysis it will be an easy matter to establish why the integration over the lens surface does not produce the effect of self-averaging of the intensity $\jmath^{\mathrm{F}}$ in the focal plane at $\langle u\rangle=0$. According to eqn (3.2.83), the lens averages the wave field itself rather than the intensity. Thus, the self-averaging quantity is the field u^{F} that tends to $\langle u^{\mathrm{F}}\rangle$ with the growth of Σ. Accordingly, $\jmath^{\mathrm{F}}$ tends to $\langle\jmath^{\mathrm{F}}_{\mathrm{coh}}\rangle=|\langle u^{\mathrm{F}}\rangle|^2$, and if $\langle\jmath^{\mathrm{F}}_{\mathrm{coh}}\rangle$ vanishes, along with $\langle u^{\mathrm{F}}\rangle$, then the effect of self-averaging disappears.

Estimates of the critical value of Σ, above which the intensity $\jmath^{\mathrm{F}}$ acquires the property of self-averaging, are presented in Problem 3.19.

3.2.20 Measurements of Radiance by a Lens

Suppose that a plane photo detector be placed in the focal plane of a lens, as shown in Fig. 3.11. Speaking of a point detector we mean one with dimensions smaller that the correlation radius of intensity fluctuations in the focal plane l_j. In the order of magnitude, this radius is compared with the radius of an Airy ring, i.e., with the radius of the focal spot $\lambda D/F$. The "coherence area" in the focal plane is then $r_J^2\approx(\lambda D/F)^2$.

According to (3.2.76) the detector with small dimensions measures current proportional to the local intensity $\jmath^{\mathrm{F}}(r_\perp,t)$ averaged over the measurement time T:

$$\overline{i(\mathbf{r}_\perp,t)}^T=\frac{K}{2}\overline{\jmath^F(\mathbf{r}_\perp,t)}^T.$$

Adopting the ergodicity hypothesis we may replace the time average with the ensemble average $\langle\jmath^{\mathrm{F}}\rangle$:

$$\overline{i(\mathbf{r}_\perp,t)}^T \approx \langle i(\mathbf{r}_\perp,t)\rangle = \frac{1}{2}K\langle \mathcal{J}(\mathbf{r}_\perp,t)\rangle$$
$$= \frac{1}{2}K|C|^2 \int d^2\rho'_\perp d^2R'_\perp \Gamma(\rho',\mathbf{R}')\exp\left(\frac{ik_0\mathbf{r}_\perp\rho'_\perp}{F}\right). \tag{3.2.87}$$

In this formula the transition from the first line to the second was made in agreement with formula (3.2.84) that combines the field intensity in the focal plane with the coherence function of the primary field.

Assume that the coherence radius of the primary field l_c is small compared to the lens diameter D. Then the integration limits with respect to the difference variable $\rho'_\perp$ may be easily expanded to infinity. The integral with respect to $R'_\perp$ can be calculated by multiplying the integrand by the lens area and taking the coherence function at the lens center $\mathbf{R}'_c$. Now, the current measured by the detector at a point $\mathbf{r}_\perp$ in the focal plane is proportional to the radiance $I(\mathbf{R}'_c,\mathbf{n}_\perp)$ in the direction $\mathbf{n}_\perp = \mathbf{r}_\perp / F$:

$$\overline{i_T(r_\perp,t)}^T = \frac{K|C|^2}{2} I\left(\mathbf{R}'_c, \frac{\mathbf{r}_\perp}{F}\right). \tag{3.2.88}$$

This formula is the keystone of optical measurements of angular radiance distribution. A similar expression is be obtained for the radiance measured by specular radio antennas.

To close the topic we note that the result of measurements may approach the ensemble average not only because of time averaging but also because of spatial averaging over the aperture of an extended photo detector S. Such an averaging presupposes that S must be large compared to the intensity correlation area $r_J^2 = (\lambda D / F)^2$ in order to accommodate many specles in S. In this case, the measurement is proportional to the radiance integrated over the solid angle $\Delta\Omega \approx S / F$:

$$\overline{i(r_\perp)}^S = \frac{1}{S}\int i(r_\perp)dS \propto \frac{1}{S}\int_S I(\mathbf{R}'_c,\mathbf{r}_\perp / F)d^2r_\perp$$
$$= \frac{F}{S}\int I(\mathbf{R}'_c,\mathbf{n}_\perp)d^2n_\perp . \tag{3.2.89}$$

3.2.21 Spatial Fourier Transform

The radiance $I_\omega(\mathbf{n})$ can be measured indirectly by evaluating the coherence function $\Gamma(\rho)$ and taking its Fourier transform. This type of measurements may be called *spatial Fourier spectroscopy*. This operation is performed in the stellar Michelson interferometer. This interferometer measures the luminosity of the interference pattern formed by the beam waves arriving from two separated points $\mathbf{r}_1$ and $\mathbf{r}_2$, or alternatively, this configuration measures the degree of spatial coherence at these points.Taking the Fourier transform of the coherence function we obtain the spatial (angular) field spectrum, that is, the radiance distribution over the angle (Born and Wolf, 1980).

The inverse procedure — a transition from the radiance (angular spectrum) to the spatial coherence function, is also possible. This procedure is of interest when direct measurements of the spatial coherence function are extremely difficult.

Both the direct and inverse procedures are bounded by the uncertainty relation (Apresyan and Kravtsov, 1984). This relation dictates the smallness of the measuring aperture compared to the scale of field quasi-uniformity: $l_a << L_R$. It thus limits the resolving power $\Delta\theta = \lambda / l_a$ of optical instruments detecting quasi-uniform fields: $\Delta\theta_{min} = \lambda / L_R$. This limitation refers also to the size of the lens if the recording is made with a telescope, or to the length of the base if observations are made with an interferometer.

3.3 ELEMENTS OF MULTIPLE SCATTERING THEORY. THE DYSON AND BETHE–SALPETER EQUATIONS

3.3.1 Operator Form of Field Equations

In order to describe the propagation of a radiation in a scattering medium we have to resort to the equations for the first two statistical moments of the field as we did in handling radiation in a free space. While in the absence of scattering the derivation of such equations was a trivial matter, this is not the case in scattering media. The equations for field moments in scattering media are known as the *Dyson equation* (first moments) and the *Bethe–Salpeter equation* (second moments); they have been widely discussed in the literature, including textbooks (Rytov, Kravtsov, and Tatarskii, 1989). We confine our consideration with the necessary background and quote final results.

Let us recall the general formulation of the problem in the presence of scattering inhomogeneities in the medium. Let the random field $u = u(x)$ satisfy the linear equation

$$Lu \equiv (L_0 - V)u = q. \tag{3.3.1}$$

The wave operator $L = (L_0 - V)$ consists of a deterministic operator L_0 and a random operator V describing inhomogeneities (for brevity we dropped caps in the notation of operators). The function of field sources q may be deterministic or random. It is assumed that eqn (3.3.1) is augmented, when necessary, by the conditions that ensure the uniqueness of the solution.

The separation of L into the undisturbed operator L_0 and the inhomogeneity operator V is to some extent arbitrary and is dictated mainly by physical considerations, and a requirement that for L_0 there exist an inverse operator $G_0 = L_0^{-1}$ such that

$$L_0 G_0 = \hat{1}, \tag{3.3.2}$$

where $\hat{1}$ is the unit operator. The statistical characteristics of the medium, i.e., the characteristics of the operator V are deemed to be specified. In the context of correlation theory, it is required to determine the first and second statistical moments of the field $M = \langle u_1 \rangle = \langle u(x_1) \rangle$ and $\Gamma = \langle u_1 u_2^* \rangle = \langle u(x_1) u^*(x_2) \rangle$ (or M and the correlation function $\Psi = \Gamma - \langle u_1 \rangle \langle u_2^* \rangle$).

As a rule, the complexity of the calculation of statistical moments increases with their order. Therefore, the first moment $\langle u \rangle$ is determined first and the coherence function Γ next, so that when Γ is of interest $\langle u \rangle$ may be deemed known.

The first moments M and Γ are of interest for a wide range of scattering theory applications such as the scattering of scalar, electromagnetic, and elastic waves in a variety of media. To illustrate the general relations we shall consider two example. The first one is the principal problem on the three-dimensional scattering of a monochromatic scalar field $u = u(\mathbf{r})$ at static inhomogeneities of the dielectric permittivity $\varepsilon(\mathbf{r}) = 1 + \tilde{\varepsilon}(\mathbf{r})$, where $\tilde{\varepsilon}$ describes random inhomogeneities. In this example, u satisfies the Helmholtz equation

$$Lu = \left[\Delta + k_0^2 (1 + \tilde{\varepsilon}(\mathbf{r}))\right] u = q, \tag{3.3.3}$$

where $k_0 = \omega / c$ is the wave number, the vector $x = (\mathbf{r}, t)$ is replaced with $x = \mathbf{r}$, and the time factor $\exp(-i\omega t)$ is omitted. For simplicity, we will assume the mean value of the random function $\tilde{\varepsilon}$ is zero, $\langle \tilde{\varepsilon} \rangle = 0$, and the scattering medium is statistically uniform and isotropic — thereby digressing from the effects associated with the boundedness of real scattering media. Clearly, the last assumption may be approximately justified only for points sufficiently distant from the boundaries of the scattering volume.

The second example is connected with the description of polarization of electromagnetic radiation. The point of departure is the equation for the monochromatic component of the electric field strength

$$\begin{aligned} L\mathbf{E} &= \left(-\nabla \times \nabla \times + k_0^2 \varepsilon(\mathbf{r})\right)\mathbf{E} \\ &= -\frac{4\pi}{c} i k_0 \mathbf{j} \equiv \mathbf{q}, \end{aligned} \tag{3.3.4}$$

which evolves from Maxwell's equations. Here, the magnetic permeability is deemed equal to unity, $\varepsilon(\mathbf{r})$ is the dielectric permittivity, and $\mathbf{j}$ is the current density. We shall use eqn (3.3.4) predominantly to illustrate the key points in the description of multicomponent partially polarized radiation.

One more particular case of the general problem (3.3.1) is a quasi-oscillator with fluctuating frequency that will be considered in Problem 3.35. This example does not have a direct bearing on wave scattering, but it is exactly solvable and convenient for illustration of the approximations used.

It is natural to accept as an undisturbed operator for the Helmholtz equation (3.3.3) the operator describing the propagation in a free space:

$$L_0 = \Delta + k_0^2. \tag{3.3.5}$$

The inversion of these operator with allowance for the Zommerfeld radiation conditions gives the operator $G_0 = (\Delta + k_0^2 + i0)^{-1}$ acting by the rule

$$G_0 f(r) = \int G_0(\mathbf{r},\mathbf{r}')f(\mathbf{r}')d^3r', \tag{3.3.6}$$

where

$$G_0(\mathbf{r},\mathbf{r}') = G_0(\mathbf{r}-\mathbf{r}') = -\frac{\exp(ik_0|\mathbf{r}-\mathbf{r}'|)}{4\pi|\mathbf{r}-\mathbf{r}'|} \tag{3.3.7}$$

is the nucleus of G_0 (see Problem 3.23). The action of the operator $V = L_0 - L$ reduces to multiplication by the "scattering potential" $v = -k_0^2\tilde{\varepsilon}$, therefore V has the nucleus

$$V(\mathbf{r},\mathbf{r}') = -k_0^2\tilde{\varepsilon}(\mathbf{r})\delta(\mathbf{r}-\mathbf{r}'). \tag{3.3.8}$$

For the vector problem (3.3.4), the calculation of the unperturbed operator G_0 corresponding to the homogeneous medium leads to an analogous, though more cumbersome, eqn (3.3.7) (Problem 3.24).

Using eqn (3.3.2) we rewrite the general equation (3.3.1) in integral form

$$u = G_0 q + G_0 V u = u^0 + u^{sc}. \tag{3.3.9}$$

Here $u^0 = G_0 q$ and $u^{sc} = G_0 V u$ have the meaning of the incident and scattering waves (Fig. 3.12 depicts the case of a bounded scattering volume and a plane incident wave).

We will seek the solution of eqn (3.3.9) in the form

$$u = Gq, \tag{3.3.10}$$

where $G = L^{-1}$ is the unknown Green's operator of the scattering medium. Substituting eqn (3.3.10) in eqns (3.3.9) and (3.3.1) and dividing both sides of the obtained equations by q, we obtain the operator forms of eqns (3.3.9) and (3.3.1)

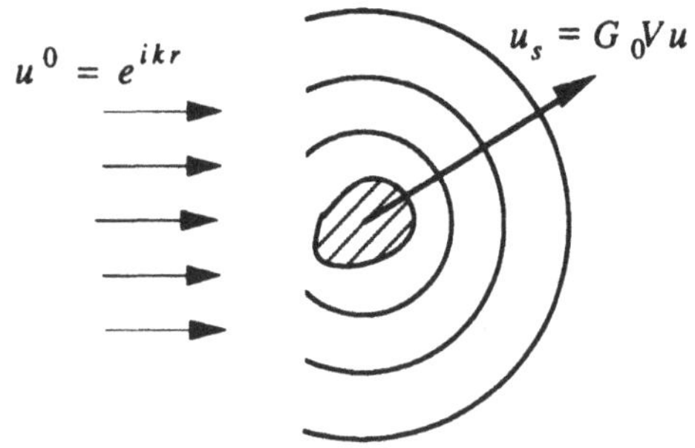

Figure 3.12. Incidence of a plane wave on a bounded scattering volume.

$$G = G_0 + G_0 V G, \tag{3.3.11}$$

$$LG = (L_0 - V)G = \hat{1}. \tag{3.3.12}$$

Let us examine the physical sense of these operator equations for the scattering problem under consideration. The Green's operator G describes the propagation of radiation in inhomogeneous media. If the sources of radiation q are specified, then the field u is expressed in terms of G by the relation (3.3.10). According to (3.3.11), the propagation operator G is constituted by the free propagation operator G_0 that corresponds to the absence of inhomogeneities, and the operator G_0VG describing the scattered wave.

We illustrate eqns (3.3.11) and (3.3.12) in the context of the wave problem (3.3.3). Taking L_0 in the form of (3.3.5) we have

$$G = G_0 + G_0\left(-k^2\tilde{\varepsilon}(r')\right)G, \tag{3.3.13a}$$

$$\left[\Delta + k_0^2(1+\tilde{\varepsilon})\right]G = \hat{1}, \tag{3.3.14a}$$

or in a more expanded form

$$G(\mathbf{r},\mathbf{r}') = \frac{\exp(ik_0|\mathbf{r}-\mathbf{r}'|)}{4\pi|\mathbf{r}-\mathbf{r}'|} + \int \frac{\exp(ik_0|\mathbf{r}-\mathbf{r}'|)}{4\pi|\mathbf{r}-\mathbf{r}'|} k_0^2\tilde{\varepsilon}(\mathbf{r}'')G(\mathbf{r}'',\mathbf{r}')d^3r'', \tag{3.3.13b}$$

$$\left[\Delta + k_0^2(1+\tilde{\varepsilon}(\mathbf{r}))\right]G(\mathbf{r},\mathbf{r}') = \delta(\mathbf{r}-\mathbf{r}'). \tag{3.3.14b}$$

We now introduce for the linear operators A the subscript $i = 1, 2$; and denote by A_1 the operator A acting in the argument x_1, and by A_2^* the complex conjugate of the operator A_2 acting in the argument x_2. Then for any function of two arguments $f_{12} = f(x_1, x_2)$ we have

$$A_1 f_{12} = \int A(x_1, x')f(x', x_2)dx',$$

$$A_2^* f_{12} = \int A^*(x_2, x')f(x_1, x')dx',$$

$$A_1 A_2^* f_{12} = \int A(x_1, x')A^*(x_2, x'')f(x', x'')dx'dx''.$$

We assume that the medium (G) and the sources (q) are statistically independent and we may perform averaging independently for $G_{1,2}$ and $g_{1,2}$. Using the compact operator notation we represent the mean field as

$$\langle u_1 \rangle = \langle G_1 q_1 \rangle = \langle G_1 \rangle \langle q_1 \rangle, \tag{3.3.15}$$

and the coherence function as

$$\Gamma = \langle u_1 u_2^* \rangle = \langle G_1 G_2^* q_1 q_2^* \rangle = \langle G_1 G_2^* \rangle \langle q_1 q_2^* \rangle. \tag{3.3.16}$$

Earlier we considered the case of a deterministic medium and fluctuating sources. In this case, we may drop the angular brackets with G_j in eqns (3.3.15) and (3.3.16). Now we will be interested primarily in the random fluctuations of the *medium*, i.e. G. Therefore, in this section we well assume for simplicity that the sources q do not fluctuate and omit in (3.3.15) and (3.3.16) all symbols of averaging for q_1 and q_2^*.

From eqns (3.3.15) and (3.3.16) we see that to evaluate the field moments $\langle u_1 \rangle$ and $\langle u_1 u_2^* \rangle$ for non-fluctuating sources it suffices to know the analogous moments $\langle G_1 \rangle$ and $\langle G_1 G_2^* \rangle$ of the operator G. This means that we may consider operator equations for G instead of equations and moments for the field u.

3.3.2 Series of Perturbation Theory and the Born Approximation

In the general case, equations (3.3.1) and (3.3.12) cannot be solved exactly. However, a general methods exists that allows one to express the solution of the problem as an infinite series expansion in powers of V. This method assumes that the operator V is small and uses it as a perturbation. The respective series of perturbation theory are also known as Neumann series. They can be obtained by iteration of the integral form of the initial equation. For example, iterating eqn (3.3.11) we have

$$G = G_0 + G_0VG_0 + G_0VG_0VG_0 + G_0VG_0VG_0VG_0 + \ldots = \sum_{n=0}^{\infty} G_0(VG_0)^n. \qquad (3.3.17)$$

We consider the operator V as a small quantity by formally assigning to it a small parameter $\mu \ll 1$, i.e., by replacing V with μV. In final results, μ is set equal to unity. Thus, the expansion (3.8.17) takes on the form of a series in power of μ. The Nth term of such a series formally has the order of $0(\mu^N)$. This may seem to ensure a rapid (geometric) convergence of the series (3.3.17). However, in reality, (3.3.17) is an *operator* series rather than a numerical ones. For operator series, the rigorous conditions of convergence are harder to obtain than for numerical series, because relevant estimates of convergence must rely on the concept of operator norm. Unfortunately, such estimates can be obtained only for a limited number of particular cases. Therefore, for convergence of series like (3.3.17), no complete clarity has been obtained. Moreover, in practice, one often has to use asymptotic series rather than convergent. Even an analysis of the asymptotic behavior with informal physically small parameters may be difficult for operators.

The complexity of sequential terms of the series (3.3.17) rapidly increases, therefore in practical calculations one keeps only a few first terms. The simplest approximation is

$$G \approx G_0 + G_0VG_0. \qquad (3.3.18)$$

It is known as the first Born or Rayleigh–Hans approximation. Formally, it is accurate to terms of order $0(\mu^2)$.

To clarify the physical meaning of the Born approximation (3.3.18) we write the amplitude of the scattered wave as

$$u^s = G^0 Vu, \tag{3.3.19}$$

and compare it with the approximate expression

$$u^s \approx G^0 Vu^0, \tag{3.3.20}$$

following from eqn (3.3.18) (here $u^0 = G^0 q$ is the amplitude of the incident wave). This comparison indicates that the Born approximation corresponds to substituting the incident wave u^0 for the exact value u of the field *inside* the medium's inhomogeneity, i.e., following the action of the operator V.

If we substitute (3.3.18) in (3.3.15) and (3.3.16) we obtain approximate expressions for the moments $\langle u \rangle$ and Γ. However, such an "average Bornean" approximation, as a rule, has a relatively narrow applicability range. Adding a finite number of subsequent terms in the Born series (3.3.17) fails to produce substantial improvements. We estimate the applicability conditions of this approximation by taking into account that the medium's inhomogeneity must not appreciably change the incident field. In the case of the wave equation (3.3.3) we require first of all *optical softness* or that the fluctuations or of the fluctuating medium be small

$$|\delta\varepsilon| << 1. \tag{3.3.21a}$$

For the case of a random medium, it suffices to require that this condition hold in mean square:

$$\sigma_\varepsilon = \langle (\delta\varepsilon)^2 \rangle^{1/2} << 1. \tag{3.3.21b}$$

In the case of large-scale inhomogeneities, the condition that the fluctuating component of the field be small compared with the unperturbed field can be approximately reduced to the requirement of smallness of the phase increment $\delta\varphi = k\int \delta\varepsilon ds$ caused by the medium's inhomogeneities. For a single particle of size $a >> \lambda$, this condition leads to

$$ka|\delta\varepsilon| << 1. \tag{3.3.22a}$$

For a random medium's layer of thickness L containing many inhomogeneities of size l_c we arrive at the inequality

$$ka\sigma_\varepsilon \sqrt{l_c L} << 1. \tag{3.3.22b}$$

When the conditions (3.3.21) and (3.3.22) are satisfied, the amplitude of the scattered wave will be small compared to the amplitude of the incident wave:

$$|u^{sc}| << |u^0|. \tag{3.3.23}$$

Similar inequalities can be obtained also for wave operators with a more complicated structure than the simple d'Alambert operator (3.3.3).

Thus, for the Born approximation be applicable, we need the condition of weak fluctuations (3.3.21) and the condition of small phase increment (3.3.22) that bounds the admissible size of the scatterer. When these conditions are violated the effects of multiple scattering has to be involved into the picture. For statistical problems, there exists two basic approaches: the summation of series like (3.3.17) of perturbation theory, and the solution of equations for moments of the field. We consider the second approach below.

3.3.3 Equations for Field Moments in Scattering Media. Perturbation Theory in the Denominator

The equation for the first moment of the field u, known as the *Dyson equation*, has the form

$$D\langle u\rangle \equiv (L_0 - V^{\text{eff}})\langle u\rangle = q, \tag{3.3.24}$$

or in operator form

$$D\langle G\rangle \equiv (L_0 - V^{\text{eff}})\langle G\rangle = \hat{1}, \tag{3.3.25}$$

where

$$D = L_0 - V^{\text{eff}} \tag{3.3.26}$$

is the *Dyson operator* and V^{eff} is the *mass operator*.

The equation for the second moment Γ, known as the *Bethe–Salpeter equation*, may be written as

$$(D_1 D_2^* - K_{12})\langle u_1 u_2^*\rangle = q_1 q_2^*, \tag{3.3.27}$$

or in operator form

$$(D_1 D_2^* - K_{12})\langle G_1 G_2^*\rangle = \hat{1}. \tag{3.3.28}$$

Here, the operator K_{12} with the nucleus K, known as the *intensity operator*, acts on the functions $f(x_1, x_2)$ in both arguments:

$$K_{12} f(x_1, x_2) = \int K\begin{pmatrix} x_1, x_1' \\ x_2, x_2' \end{pmatrix} f(x_1', x_2')\,dx_1' dx_2'.$$

According to the Dyson equation (3.3.24), $D^{-1} = \langle G\rangle$, therefore the Bethe–Salpeter equation (3.3.28) may be also written in the form

$$\langle G_1 G_2^* \rangle = \langle G_1 \rangle \langle G_2^* \rangle + \langle G_1 \rangle \langle G_2^* \rangle K_{12} \langle G_1 G_2^* \rangle \tag{3.3.29}$$

or

$$\langle \tilde{G}_1 \tilde{G}_2^* \rangle = \langle G_1 \rangle \langle G_2^* \rangle K_{12} (\langle G_1 \rangle \langle G_2^* \rangle + \langle \tilde{G}_1 \tilde{G}_2^* \rangle), \tag{3.3.30}$$

where $\tilde{G} \equiv G - \langle G \rangle$. The operator

$$\langle \tilde{G}_1 \tilde{G}_2^* \rangle = \langle G_1 G_2^* \rangle - \langle G_1 \rangle \langle G_2^* \rangle$$

describes the correlation of Green's operators G_1 and G_2^*.

Formally, the Dyson and Bethe–Salpeter equations (3.3.24) and (3.3.27) are exact equations, however, they contain unknown operators V^{eff} and K_{12} that, in the general case, can be obtained only as series expansions of perturbation theory. Like with the Born series (3.3.17), perturbation theory assumes, in a certain sense, the weakness of scattering fluctuations. The strict requirement of the smallness of the scattered amplitude (3.3.23) is now superfluous.

The efficiency of the moment equations compared to the direct averaging of the Born series (3.3.17) may be explained with a simple example. Let in the initial equation (3.3.11) $G_0 = 1$ and $V = \mu X$, where X is the random parameter, and $\mu \ll 1$ is an auxiliary small parameter. In this case, the operators become ordinary functions, and $G = (1 - \mu X)^{-1}$ is the exact solution of equation (3.3.11). The averaged Born approximation (3.3.18) is associated with the expression

$$\langle G \rangle \approx 1 + \mu \langle X \rangle \tag{3.3.31}$$

linear in μ, whereas the Dyson equation (3.3.22) taken in the same approximation with respect to μ has the form

$$(1 - \mu \langle X \rangle) \langle G \rangle = 1$$

and leads to the expression

$$\langle G \rangle \approx \frac{1}{1 - \mu \langle x \rangle}. \tag{3.3.32}$$

It is intuitively obvious that eqn (3.3.32) must be closer to the exact expression $\langle G \rangle = \left\langle (1 - \mu x)^{-1} \right\rangle$ than the approximation (3.3.31) already because the dependence (3.3.32) on μ is similar to the dependence of individual realizations G on μ. In addition, eqn (3.3.32) crosses over to eqn (3.3.31) as $\mu \to 0$.

One more formal consideration in favor of the Dyson equation is associated with the face that this equation uses perturbation theory to evaluate the effective medium's parameters rather than to evaluate the scattered field that need not be small here.

All these simple considerations concerning the advantages of the *perturbation theory in the denominator* (3.3.24) over the ordinary theory of perturbations (3.3.17) become justified in the case of wave equations far more complicated than the elementary example (3.3.31).

3.3.4 Operators of Effective Inhomogeneities

A comparison of the Dyson equation (3.3.24) with the initial equation (3.3.1) indicates that the first follows from the second on the substitution of the mean value $\langle u \rangle$ for the field u, and V^{eff} for the inhomogeneity operator V. Therefore, the mass operator V^{eff} may be naturally called the *operator of effective medium's inhomogeneity with respect to the mean field.* The meaning of this operator will be best perceived from the expression

$$\langle Vu \rangle \equiv V^{\text{eff}} \langle u \rangle. \tag{3.3.33a}$$

This expression follows from the comparison of the Dyson equation (3.3.24) with the result of averaging of eqn (3.3.12). It may be viewed as a definition of V^{eff}. If we express u in terms of the operator G, then eqn (3.3.33a) can be also represented in operator form

$$\langle VG \rangle \equiv V^{\text{eff}} \langle G \rangle. \tag{3.3.33b}$$

In the general case, the evaluation of V^{eff} is associated with the calculation of the effective parameters of random inhomogeneous media (effective conductivity, viscosity, permittivity, and the like), in particular the effective parameters of mixtures. A problem closely related to this one is the transition from the microscopic to macroscopic description.

The Bethe–Salpeter equation (3.3.28) is obtained form eqn (3.3.12) on a formal substitution of $\langle G_1 G_2^* \rangle$ for G, $D_1 D_2^*$ for L_0, and K_{12} for V^{eff}. The intensity operator K_{12} may be called the operator of effective medium's inhomogeneity with respect to the *correlation* of the field.

The meaning of operator K_{12} can be illustrated by the diagram interpretation of the Bethe–Salpeter equation that dramatically simplifies the analysis of perturbation theory series; expositions of the diagram technique may be found also in Tatarskii (1961), Frisch (1968), and Rytov, Kravtsov, Tatarskii (1989). Let us introduce the diagram notation:

$\langle G_1 \rangle \langle G_2^* \rangle =$  , $K_{12} =$  , (3.3.34)

that will be used in parallel with common algebraic symbols. Now, the Bethe–Salpeter equation (3.3.29) may be written as

$\langle G_1 G_2^* \rangle =$  +  $\langle G_1 G_2^* \rangle$. (3.3.35)

Solving this equation by iterations yields

$\langle G_1 G_2^* \rangle =$

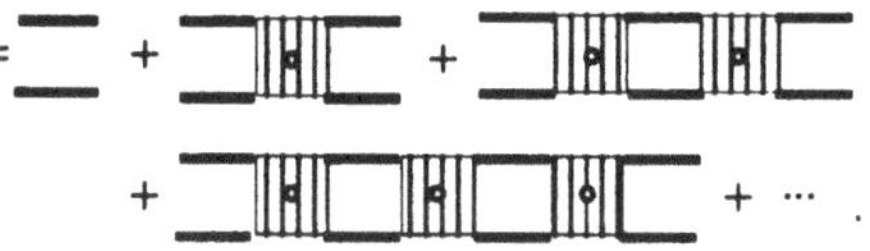

(3.3.36)

The sequential terms of this series may be assigned a simple physical meaning: the parallel solid lines (operators $\langle G_1 \rangle \langle G_2^* \rangle$) imply the propagation of the second moment $\langle u_1 u_2^* \rangle$ in the *effective medium* corresponding to the mean field $\langle u \rangle$, and the rectangles (operators K_{12}) describe scattering at *effective inhomogeneities* which give rise to a correlation.

In the general case, the operators of effective inhomogeneities V^{eff} and K_{12} can be obtained explicitly only in the form of in infinite series expansions in powers of the moments or *cumulants* of V. The simplest method is to consider the Dyson and Bethe–Salpeter equations (3.3.25) and (3.3.28) as the definitions of V^{eff} and K_{12} and calculate them with the use of the known operator relations $(AB)^{-1} = B^{-1}A^{-1}$ and $(1+A)^{-1} = \sum_{n=0}^{\infty}(-A)^n$ (Apresyan, 1974). The first of these relations takes into account that the operators A and B can not commute so that $AB \neq BA$. The second relation is a formal extension of an ordinary Taylor series for the function $(1+z)^{-1}$ to the operator case. In view of (3.3.2) and (3.3.17) we find from eqn (3.3.25)

$$
\begin{aligned}
V^{\text{eff}} &= L_0 - \langle G \rangle^{-1} = L_0 - \left[G_0 \left(1 + \sum_{n=1}^{\infty} \left\langle (VG_0)^n \right\rangle \right) \right]^{-1} \\
&= L_0 - \sum_{m=0}^{\infty} \left(-\sum_{n=1}^{\infty} \left\langle (VG_0)^n \right\rangle \right)^m L_0 = -\sum_{m=1}^{\infty} \left(-\sum_{n=1}^{\infty} \left\langle (VG_0)^n \right\rangle \right)^m L_0 \\
&= \langle V \rangle + \langle \tilde{V} G_0 \tilde{V} \rangle + \ldots, \qquad (3.3.37)
\end{aligned}
$$

where $\tilde{V} = V - \langle V \rangle$ is the random part of V.

Similar algebraic manipulations yield the expansion of the operator K_{12} in powers of moments V when the Bethe–Salpeter equation (3.3.28) is used. The first terms of this expansion has the form

$$
\begin{aligned}
K_{12} &= D_1 D_2^* - \langle G_1 G_2^* \rangle^{-1} \\
&= D_1 D_2^* - L_{01} L_{02}^* \sum_{l} \left(- \sum_{n,m \geq 0; nm \neq 0} \left\langle (G_{01} V_1)^n \left(G_{02}^* V_2^* \right)^m \right\rangle \right)^l \\
&= \langle \tilde{V}_1 \tilde{V}_2^* \rangle + \langle \tilde{V}_1 V_2^* G_{02}^* V_2^* \rangle + \langle V_1 G_{01} V_1 \tilde{V}_2^* \rangle \\
&\quad - \langle V_1 G_{01} + V_2^* G_{02}^* \rangle \langle \tilde{V}_1 \tilde{V}_2^* \rangle - \langle \tilde{V}_1 \tilde{V}_2^* \rangle \langle V_1 G_{01} + V_2^* G_{02}^* \rangle + \ldots, \qquad (3.3.38)
\end{aligned}
$$

where the operators G_{0i} act in the arguments with subscript $i = 1, 2$.

Some general properties of the operators of effective inhomogeneities V^{eff} and K_{12} are considered in Problems 3.25 and 3.26 for a scalar field. Expansions (3.3.37) and (3.3.38) are the result of direct applications of perturbation theory to calculations of effective parameters of the medium. Other methods of calculation of V^{eff} and K_{12} may take into account the particular characteristics of electromagnetic problems under considerations and the structure of particular random media. We revisit such methods associated with discrete media in Section 3.4.

As a reasonable alternative to the "direct" expansion (3.3.37) we now consider the self-consistent scheme that allows one to partially sum the infinite series (3.3.37) (Apresyan, 1974). For this purpose we write the initial equation (3.3.12) as

$$(L_0 - V^{\mathrm{eff}} - V')G = 1,$$

where $V' = V - V^{\mathrm{eff}}$ is the "offset" perturbation operator. The integral form of this equation may be written as

$$G = \langle G\rangle + \langle G\rangle V'G = \langle G\rangle + \eta\langle G\rangle V'G, \tag{3.3.39}$$

where η is the operator of random part selection that acts by the rule $\eta\ a = a - \langle a\rangle$. Here, we took into account that, in agreement with (3.3.33), $\langle V'G\rangle = \langle VG\rangle - V^{\mathrm{eff}}\langle G\rangle = 0$. Iterating eqn (3.3.39) leads to the series

$$G = \left(1 - \eta\langle G\rangle V'\right)^{-1}\langle G\rangle = \sum_{n=0}^{\infty}\left(\eta\langle G\rangle V'\right)^n\langle G\rangle, \tag{3.3.40}$$

in which the free propagation is described by the operator $\langle G\rangle$ rather than by the unperturbed operator G_0 as in the ordinary Born series (3.3.17). Substituting (3.3.40) in (3.3.33) we obtain

$$V^{\mathrm{eff}} = \langle VG\rangle\langle G\rangle^{-1} = \left\langle V(1 - \eta\langle G\rangle V')^{-1}\right\rangle = \sum_{n=0}^{\infty}\left\langle V(\eta\langle G\rangle V')^n\right\rangle. \tag{3.3.41}$$

Letting the mean value of this perturbation equal zero $\langle V\rangle = 0$, in the first nonvanishing approximation we have the expression

$$V^{\mathrm{eff}} \approx \left\langle V\eta\langle G\rangle V'\right\rangle = \left\langle V\langle G\rangle V\right\rangle \tag{3.3.42}$$

known as the Kraichnan approximation (Kraichnan, 1961).

Equation (3.3.41) in which the operator of medium's effective inhomogeneities are expressed in terms of the desired mean $\langle G\rangle$, rather than the free propagation operator G_0, refers to the so-called *self-consistent* schemes yielding nonlinear equations in the unknown operators of effective inhomogeneities. For example, in the Kraichnan approximation (3.3.42), the Dyson equation (3.3.25) takes the form of a nonlinear equation for $\langle G \rangle$. In the general case, eqns (3.3.41) and (3.3.25) form a nonlinear system of two equations for V^{eff} and $\langle G \rangle$.

3.3.5 Cumulant Relations

It is convenient to represent the expansions of the operators of effective inhomogeneities (3.3.37) and (3.3.38) by expressing the moments of the random nucleus V in terms of the respective cumulants. We briefly recall the definition and fundamental properties of cumulants (more detailed treatments of this topic may be found, e.g. in the books of Malakhov, 1978; and Rytov et al., 1987b).

Consider a multicomponent real-valued random quantity $\xi = (\xi_1, \xi_2, \ldots, \xi_N)$ which may be, for example, the value of a random field u at points $x_i : \xi_i = u(x_i)$. For the complete statistical description of ξ it is sufficient to know the *probability distribution density*

$$W(x_1, x_2 \ldots x_N) = \langle \delta(x_1 - \xi_1)\delta(x_2 - \xi_2)\ldots\delta(x_N - \xi_N) \rangle. \tag{3.3.43}$$

In place of this distribution it is often convenient to use the *characteristic function*

$$\psi(\chi) = \left\langle e^{i(\chi,\xi)} \right\rangle \equiv \int e^{i(\chi,\xi)} W(\xi) d\xi_1, d\xi_2 \ldots d\xi_N, \tag{3.3.44}$$

which is the Fourier transform of the distribution density (3.3.43). Here $\chi = (\chi_1, \chi_2, \ldots, \chi_N)$ is auxiliary vector, and $(\chi, \xi) = \sum_{j=1}^{N} \chi_j \xi_j$ is the scalar product. If we write shorthand $\xi_{l_s} \equiv \bar{l}_s$ and specify an infinite series of *moments* $\langle \xi_{l_1}, \xi_{l_2} \ldots \xi_{l_s} \rangle \equiv \langle \bar{l}_1 \bar{l}_2 \ldots \bar{l}_s \rangle$ or a similar series of *cumulants* $\langle \bar{l}_1 \bar{l}_2 \ldots \bar{l}_s \rangle_c$, then we may write the characteristic function (3.3.44) as

$$\begin{aligned} \left\langle e^{i(\chi,\xi)} \right\rangle &= \sum_{s=0}^{\infty} \frac{i^s}{s!} \sum_{l_1, l_2, \ldots, l_s = 1}^{N} \chi_{l_1} \chi_{l_2} \cdots \chi_{l_s} \langle \bar{l}_1 \bar{l}_2 \ldots \bar{l}_s \rangle \\ &= \exp\left\{ \sum_{s=1}^{\infty} \frac{i_s}{s!} \sum_{l_1, l_2 \ldots l_s = 1}^{N} \chi_{l_1} \chi_{l_2} \cdots \chi_{l_s} \langle \bar{l}_1 \bar{l}_2 \ldots \bar{l}_s \rangle_c \right\}. \end{aligned} \tag{3.3.45}$$

The first of these expressions can be obtained by expanding the exponent in a Taylor series and averaging it termwise, and the second may be viewed as the definition of the cumulants $\langle l_1 l_2 \ldots l_s \rangle_c$ as coefficients of the power series for $\ln \psi(\chi)$. We see that the cumulants are expressed as

$$\begin{aligned} \langle \bar{1}\bar{2} \ldots n \rangle_c &= \frac{1}{i^n} \partial_{\chi_1} \partial_{\chi_2} \ldots \partial_{\chi_n} \ln \langle e^{i(\chi,\xi)} \rangle |_{\chi=0} \\ &= \sum_{s=1}^{n} (-1)^{s-1} (s-1)! \sum_{s-\text{part}} \langle \bar{l}_1 \bar{l}_2 \ldots \bar{l}_{p_1} \rangle \langle \bar{l}_{p_1+1} \bar{l}_{p_1+2} \ldots \bar{l}_{p_1+p_2} \rangle \ldots \langle \ldots \bar{l}_n \rangle, \end{aligned} \tag{3.3.46}$$

where the sum is taken over all possible partitions of the product $\bar{1}\bar{2}...\bar{n}$ into s groups consisting of $p_1, p_2, ..., p_s$, terms such that $p_1 + p_2 + ... + p_s = n$. A similar expression of moments in terms of cumulants has the form

$$\langle \bar{1}\bar{2}...n \rangle = \sum_{s=1}^{n} \sum_{s-\text{part}} \langle \bar{l}_1 \bar{l}_2 ... \bar{l}_{p_1} \rangle_c \langle \bar{l}_{p_1+1} \bar{l}_{p_1+2} ... \bar{l}_{p_1+p_2} \rangle_c ... \langle ... l_n \rangle_c . \tag{3.3.47}$$

Using (3.3.46) it is an easy matter to write the first cumulants in terms of moments:

$$\begin{aligned} &\langle \bar{1} \rangle_c = \langle \bar{1} \rangle, \\ &\langle \bar{1}\bar{2} \rangle_c = \langle \bar{1}\bar{2} \rangle - \langle \bar{1} \rangle \langle \bar{2} \rangle \equiv \langle \tilde{\xi}_1 \tilde{\xi}_2 \rangle, \\ &\langle \bar{1}\bar{2}\bar{3} \rangle_c = \langle \bar{1}\bar{2}\bar{3} \rangle - \langle \bar{1} \rangle \langle \bar{2}\bar{3} \rangle - \langle \bar{2} \rangle \langle \bar{1}\bar{3} \rangle - \langle \bar{3} \rangle \langle \bar{1}\bar{2} \rangle + 2\langle \bar{1} \rangle \langle \bar{2} \rangle \langle \bar{3} \rangle \equiv \langle \tilde{\xi}_1 \tilde{\xi}_2 \tilde{\xi}_3 \rangle . \end{aligned} \tag{3.3.48}$$

We see that the second and third cumulants coincide with the respective central moments $\langle \bar{1}\bar{2} \rangle_c \equiv \langle \xi_1 \xi_2 \rangle_c = \langle \tilde{\xi}_1 \tilde{\xi}_2 \rangle$, $\langle \bar{1}\bar{2}\bar{3} \rangle_c \equiv \langle \xi_1 \xi_2 \xi_3 \rangle_c = \langle \tilde{\xi}_1 \tilde{\xi}_2 \tilde{\xi}_3 \rangle$, where $\tilde{\xi} = \xi - \langle \xi \rangle$. Higher order cumulants do not possess this property. For example, if $\langle \xi \rangle = 0$ so that $\xi = \tilde{\xi}$, then

$$\begin{aligned} \langle \bar{1}\bar{2}\bar{3}\bar{4} \rangle_c &= \langle \bar{1}\bar{2}\bar{3}\bar{4} \rangle - \langle \bar{1}\bar{3} \rangle \langle \bar{2}\bar{4} \rangle - \langle \bar{1}\bar{2} \rangle \langle \bar{3}\bar{4} \rangle - \langle \bar{1}\bar{4} \rangle \langle \bar{2}\bar{3} \rangle \\ &= \langle \bar{1}\bar{2}\bar{3}\bar{4} \rangle - \langle \bar{1}\bar{3} \rangle \langle \bar{2}\bar{4} \rangle \neq \langle \tilde{\xi}_1 \tilde{\xi}_2 \tilde{\xi}_3 \tilde{\xi}_4 \rangle . \end{aligned}$$

In contrast to moments $\langle \bar{1}\bar{2}...\bar{n} \rangle$ (or central moments $\langle \tilde{\xi}_1 \tilde{\xi}_2 ... \tilde{\xi}_n \rangle$ for $n > 3$), the cumulant $\langle \bar{1}\bar{2}...\bar{n} \rangle_c$ does not vanish, if the product $\bar{1}\bar{2}...\bar{n}$ within the angular brackets $\langle . \rangle_c$ may be partitioned into two groups of statistically independent factors; this statement can be readily proved with the first equality in (3.3.46). In particular, if one of the $\xi_l = \bar{l}$ does not fluctuate, then the cumulant$\langle \bar{1}\bar{2}...\bar{n} \rangle_c (n \geq 2)$ containing $\bar{l}$ vanishes. Thus, the nth order cumulant $\langle \bar{1}\bar{2}...\bar{n} \rangle_c$ with $n \geq 2$ is the *measure of statistical coupling* in groups of n cofactors $\bar{1}\bar{2}...\bar{n}$ (the first order cumulant coincides with the mean and has no direct bearing on correlations). This property of cumulants is especially important for physical applications. In view of this property cumulants are often referred to as *correlations functions* (Ruelle, 1969) to emphasize their sense as a measure of correlation. The term correlation is understood here in the general sense as a " statistical coupling".

3.3.6 Operator Expansions

The operator of effective inhomogeneities (3.3.37) and (3.3.38) can be written in the diagram notation:

$V = \bullet$, $G_0 =$ —— , $\langle VV...V \rangle_c =$  , $\langle VG_0V \rangle_c =$  , etc., (3.3.49) where $\langle VV \ldots V \rangle_c$ means the operator obtained by replacing nuclei V by their cumulant in the expansion. If in the expansion V^{eff} (3.3.37) we express the moments V in terms of their cumulants, then this expansion takes the form of a sum of all so-called *strongly coupled diagrams* that cannot be separated into two unconnected parts by breaking one "line of free propagation" G_0 (Rytov, Kravtsov, and Tatarskii, 1989b; Tatarskii, 1961; Frisch, 1968):

$V^{\text{eff}} =$ 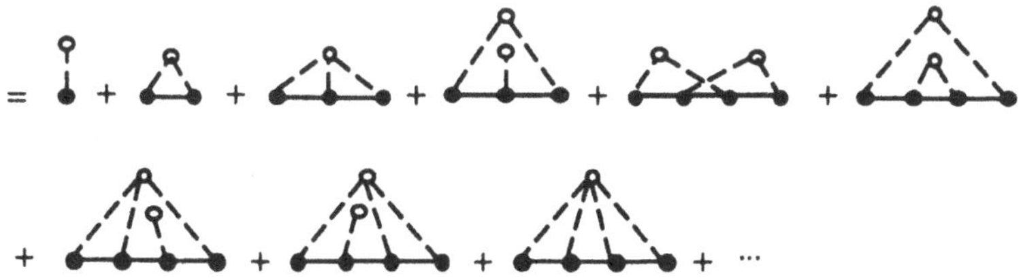

(3.3.50)

Here the dots stand for the subsequent terms of the expansion. If the operator V has a Gaussian kernel with zero mean, then this expansion retains terms with second order cumulants only, so that

$V^{\text{eff}} =$  (3.3.51)

In a similar way, using two-row diagrams we can obtain for the operator K_{12} (3.3.38) the diagram representation

$K_{12} =$ 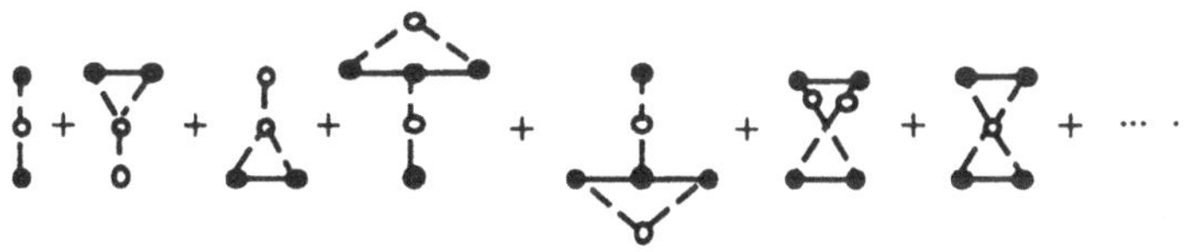

(3.3.52)

For Gaussian fluctuations with $\langle V \rangle = 0$, this series simplifies

$K_{12} =$ 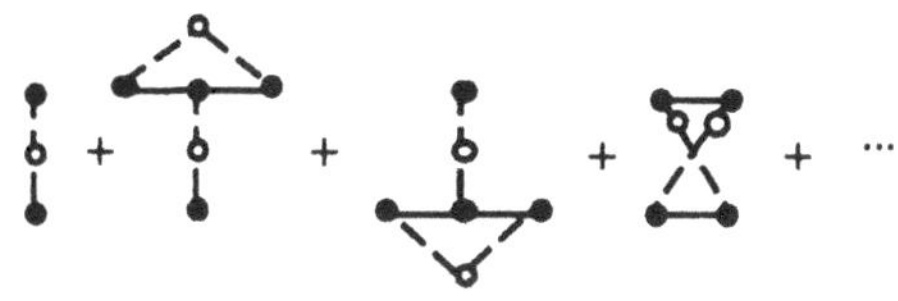 (3.3.53)

The effective inhomogeneity operator V^{eff} is, generally speaking, nonlocal, that is, $V^{\text{eff}}(x,x')$ is nonzero in some domain $x \neq x'$ even when scattering inhomogeneities are local, so that $V(x,x') \propto \delta(x - x')$. This is caused by the fact that the scattering at random inhomogeneities makes the quantity $\langle Vu(x) \rangle$ dependent on values of $\langle u(x) \rangle$ in some spatial domain near point x. The last phenomenon is called the *dispersion of the medium about the mean field* (Tatarskii and Gertsenshtein, 1963; Ryzhov and Tamoikin, 1970).

3.3.7 Energy Conservation Law and the Optical Theorem for Effective Inhomogeneities

If the propagation medium is nondissipative so that the energy of the field is conserved, then this property imposes certain constraints on the form of the operators L_0, V, V^{eff}, and K_{12}. We examine these constraints below.

The conservation of energy can be connected with the condition of the minimum action functional. We will assume that, for the linear equation (3.3.1) and real-valued field $u(x)$, this condition has the form

$$S = \int u(x')L(x',x'')u(x'')dx'dx'' - \int u(x')q(x')dx' = \min, \tag{3.3.54}$$

where $L(x',x'')$ is the real-valued kernel of L. We see that the kernel $L(x,x')$ may be deemed Hermitian:

$$L(x,x') = L^*(x',x). \tag{3.3.55}$$

Indeed, if we write $L(x,x')$ in the form

$$L(x,x') = \frac{1}{2}(L(x,x') + L(x',x)) + \frac{1}{2}(L(x,x') - L(x',x)),$$

then the antisymmetric part of $L(x,x')$ equal to $1/2(L(x,x') - L(x',x))$ disappears from (3.3.54) since the respective term in this equation becomes identically zero.

Equation (3.3.55) may be also written in operator form

$$L = L^+, \tag{3.3.56}$$

where the superscript "+" implies the operator with a Hermite conjugated kernel:

$$(A^+)(x,x') = A^*(x',x).$$

In (3.3.55) we introduced also the symbol of complex conjugation that may be omitted in handling real-valued physical fields and operators. However, in the complex representation of the monochromatic field (time dependence $e^{-i\omega t}$), the kernel L_ω, corresponding to the complex amplitude, will be, generally speaking, complex, and the conjugation symbol in (3.3.55) will then be of informal significance for it.

Thus, in the case of a conserving medium, the operator kernel L may be deemed Hermitian. We will call the operator associated with such kernel Hermitian too.

Assuming that in the absence of fluctuations ($V = 0$) the medium is conserving, we may write the conservation conditions for L_0 and V similar to those in (3.3.56),

$$L_0 = L_0^+ , \quad V = V^+ . \tag{3.3.57}$$

In deriving various operator relations we will widely use the analogy between the operators and noncommuting matrices. However, while finite dimensional matrices may be applied to any vector of the respective dimension, this is, generally speaking, not true for operators which may be treated as a generalization of the concept of matrix to the case of infinite-dimensional space. In infinite dimensional space, certain difficulties occur in connection with the appearance of so-called *unbounded* operators, which cannot be realized in a finite-dimensional space (see, e.g., Taylor, 1972; Richtmyer, 1978). In describing such operators one should exercise certain care, as these may not possess customary properties of finite-dimensional matrices. In particular, using unbounded operators, one has to specify their domain of definition and take care that these functions remain within this domain.

We will not pose here on mathematical niceties whose rigorous solution requires the formalism of functional analysis, instead we adopt a "naive" point of view and consider operators simply as some noncommuting symbols admitting common algebraic operations of addition, inversion, multiplication by one another and by a number. For problem on hand this approach yields correct results, but one should not overlook the possibility of various paradoxes associated with imperfectly correct handling of unbounded operators.

Using the Hermitian property $V = V^+$ and eqn (3.3.11) $G = G_0 + G_0 V G$ we can readily obtain by simple algebra the relation (Problem 3.29)

$$G - G^+ = G^+ \left(\frac{1}{G_0^+} - \frac{1}{G_0} \right) G , \tag{3.3.58}$$

which will be called the *optical theorem for the operator G*.

We establish a correspondence between eqn (3.3.58) and the optical theorem (2.2.53) for a single scatterer considered in Section 2.2.8. To this end we represent (3.3.58) with the scattering operator T defined as

$$T = V + VGV \tag{3.3.59}$$

and expressing G in the form

$$G = G_0 + G_0 T G_0 . \tag{3.3.60}$$

This expression can be readily derived by comparing the Born series (3.3.17) $G = G_0 + G_0 V G_0 + G_0 V G_0 V G_0 + \ldots$ with the similar series for T: $T = V + VGV$ $= V + VG_0V + VG_0VG_0V + \ldots$. Substituting (3.3.60) in (3.3.58) we obtain after some elementary algebra

$$T - T^+ = T^+ (G_0 - G_0^+) T . \tag{3.3.61}$$

According to eqn (3.3.60), now the evaluation of a propagation operator G will be reduced to the evaluation of the operator T describing the scattered wave. For a single scatterer, the kernel of T is directly related to the scattered amplitude. We illustrate this statement for the case of monochromatic radiation.

Expressions (3.3.54) and (3.3.57) hold both for the nonstationary field $u(x) = u(\mathbf{r},t)$ and for monochromatic radiation. In the last case we should arrange that all operators are consistent with the monochromatic dependence of the field u on time, i.e., $u \propto \exp(-i\omega t)$, by replacing the differentiation operator ∂_t with $-i\omega$ and $x = (\mathbf{r},t)$ with $x = \mathbf{r}$. Then, turning to the scalar field model (3.3.3) it is straightforward to demonstrate that the scattering amplitude $f_\omega(\mathbf{n} \leftarrow \mathbf{n}_0)$ that enters in (2.2.53) is related to the scattering operator T by the expression

$$f_\omega(\mathbf{n} \leftarrow \mathbf{n}_0) = \int \frac{d^3r}{4\pi} e^{-i\mathbf{k}\mathbf{r}} T e^{i\mathbf{k}_0\mathbf{r}} \equiv -\frac{T_{\mathbf{k}\mathbf{k}_0}}{4\pi}, \tag{3.3.62}$$

where $k_0 = \omega / c$, and $\mathbf{k} = k_0\mathbf{n}$ and $\mathbf{k}_0 = k_0\mathbf{n}_0$ are the wave vectors of the incident and scattered waves (Problem 3.30). This formula exposes the meaning of T as follows: in the $\mathbf{k}$-space the kernel of this operator $T_{\mathbf{k}\mathbf{k}_0}$ may be viewed, accurate to a constant factor, as an extension of the scattering amplitude f_ω to the case of wave vectors $\mathbf{k}$ and $\mathbf{k}_0$ that lie outside the dispersion surface $k = \omega / c$, i.e., need not satisfy the dispersion relation $k = k_0 = \omega / c$.

By virtue of (3.3.62) it is not hard to demonstrate (see Problem 3.31) that the relation (3.3.61) is a generalized form of the optical theorem (2.2.53) for the case of nondissipative scattering when $\sigma_a = 0$. Thus, eqn (3.3.61) expresses the energy conservation law for the scattering operator T.

We now wish to derive some corollaries of the energy conservation law in the form of the optical theorem (3.3.58) for the statistical characteristics of radiation. Averaging of both sides of eqn (3.3.58) indicates that the energy conservation law relates the first and the second moments of G. This relation can be expressed also in the form of a constraint on the form of the operators of effective inhomogeneities V^{eff} and K_{12} (Problem 3.32):

$$V^{\text{eff}}(x,x_0) - V^{\text{eff}*}(x_0,x) = \int [\langle G\rangle(x'',x') - \langle G\rangle^*(x',x'')] \times K\begin{pmatrix} x', & x_0 \\ x'', & x \end{pmatrix} dx'dx''. \tag{3.3.63}$$

We will call these relations the *optical theorem for the operators of effective inhomogeneities*.

For a statistically uniform and stationary scattering medium, the operators of effective inhomogeneities have difference nuclei that can be written as

$$V^{\text{eff}}(x_1,x_2) = V^{\text{eff}}(\rho),$$
$$K\begin{pmatrix} x_1, & x_1' \\ x_2, & x_2' \end{pmatrix} = K(R - R',\rho,\rho'), \tag{3.3.64}$$

where $R = (x_1 + x_2)/2$, $\rho = x_1 - x_2$, $R' = (x_1' + x_2')/2$, and $\rho' = x_1' - x_2'$. In this case, the formulation of the optical theorem (3.3.63) can be simplified by resorting to the K-representation. This is effected by multiplying both sides of eqn (3.3.63) by $\exp(iKx_0)$ and integrating with respect to x_0:

$$\operatorname{Im} V^{\text{eff}}(K) = |D(K)|^2 \operatorname{Im}\langle G\rangle(K)$$
$$= \int \operatorname{Im}\langle G\rangle(K') K(K' \leftarrow K) d^4K'. \tag{3.3.65}$$

Here

$$A(K) = e^{-iKx} A e^{iKx} \tag{3.3.66}$$

is the kernel of the uniform operator A in the K-representation, and

$$K(K' \leftarrow K) = (2\pi)^{-4} \int K(R', \rho, \rho') e^{iK\rho - iK\rho'} dR' d\rho d\rho'. \tag{3.3.67}$$

We also took into account that $V^{\text{eff}}(K) = L_0(K) - D(K)$, and $D(K) = 1/\langle G\rangle(K)$

For the wave operator $L = \Delta + k^2(1 + \tilde{\varepsilon})$, the optical theorem (3.3.65) may be written in a more general form. We present the respective expression in Problem 3.33.

3.3.8 Field Coherent Function Deep Inside the Medium

For the problem of scattering in a random inhomogeneous medium, the optical theorem (3.3.65) gives, accurate to a constant multiplier, the coherence function of the field deep in the medium (Barabanenkov, 1969a, 1971). In the general case, this function depends upon the boundary and initial conditions, and also upon the form of the sources. While far from the boundaries and sources the medium can be deemed statistically uniform and stationary, in this region, i.e. in the deep propagation mode, in view of multiple scattering the coherence function "forgets" its initial and boundary conditions and its behavior depends now only upon the properties of the scattering medium.

To demonstrate this behavior we take a natural assumption that in the deep propagation mode, the coherence function is uniform and stationary, i.e., $\Gamma_{12} = \Gamma_{12}(x_1 - x_2)$. In the absence of radiation sources this function satisfies the Bethe–Salpeter equation (3.3.27) with $q_1 = q_2 = 0$. It follows that, if we pass over to the K-representation by writing Γ_{12} in the form (3.1.10), then the spectral density J_K will satisfy the equation that coincides with the optical theorem (3.3.65) viewed as an equation for $\operatorname{Im}\langle G\rangle(K)$. We assume that this equation is uniquely solvable and write

$$J_K \equiv (2\pi)^{-4} \int \Gamma_{12}(\rho) e^{-iK\rho} d^4\rho = C \operatorname{Im}\langle G\rangle(K), \tag{3.3.68}$$

where C is a proportionality coefficient. This expression describes the behavior of Γ_{12} as a function of the difference variable ρ when the mean Green's function $\langle G\rangle$ in the scattering medium is specified.

Equation (3.3.68) makes no use of the explicit form of the coefficients in the Bethe–Salpeter equation and is valid for both strong and weak fluctuations of the medium. Actually, eqn (3.3.68) may be viewed as a far-reaching generalization of the condition (3.1.63) of localization of the wave spectrum on the dispersion surface, that has been obtained for propagation in a free space. However, whereas in (3.1.63) the radiation may be anisotropic — in the general case, the spectrum of I_k depends on the direction of the wave vector **k** and in this sense the free propagation retains its memory of the sources, eqn (3.3.68) refers to the deep propagation mode where, in the case of isotropic fluctuations of the medium, the radiation is isotropic and such a memory is lost.

According to eqn (3.3.68), in the case of strong fluctuations of the medium, the spectrum of the deep propagation ceases to be localized, the dispersion surface is washed out and the wave vector is not bounded by the dispersion relation $k = \omega / c$ any longer, but can be nonzero for arbitrary values of $|\mathbf{k}|$.

If we formally pass in eqn (3.3.68) to the case of free propagation, then, in the scalar wave equation, $\mathrm{Im}\langle G \rangle$ becomes

$$\mathrm{Im}\langle G_0 \rangle = \mathrm{Im}\frac{1}{(\omega / c)^2 - k^2 + i0} = -\pi\delta\left[k^2 - (\omega / c)^2\right] \tag{3.3.69}$$

and eqn (3.3.68) passes to the expression for the isotropic fluctuation spectrum that is localized at $k = \omega / c$.

For the case of equilibrium thermal radiation, eqn (3.3.68) directly follows from the fluctuation-dissipation theorem (Problem 3.34). We shall examine this theorem in some detail in Section 3.5.9.

Equation (3.3.68) is the first ever formulation of the statistical wave theory for the field in a scattering medium. This derivation substantially differs from the results of phenomenological transport theory obtained under the assumption that medium's fluctuations are weak and the radiation spectrum is localized on the dispersion surface. The absence of a rigorous relation $k = k(\omega)$ may lead to new physical effects typical of strongly fluctuating media.

By way of example, the theory of nonlinear interactions in homogeneous media intensively used the conditions of phase synchronism that ensure the resonance nonlinear interactions of waves. These conditions relate the wave vectors and frequencies of interacting waves, as in decay interactions: $\mathbf{k}_1 = \mathbf{k}_2 + \mathbf{k}_3$, $\omega_1 = \omega_2 + \omega_3$, where the wave vectors and frequencies satisfy the dispersion relation $k_j = k(\omega_j)$. However, for random inhomogeneous media, in agreement with eqn (3.3.68) strong fluctuations can destroy the dispersion relations, and the effective nonlinear interactions may occur also outside the dispersion surface that ceases to play the dominant role. We note, however, that at present the theory of nonlinear interactions in random inhomogeneous media is insufficiently developed.

The destruction of the dispersion surface becomes significant near the regime of strong (Anderson) localization when owing to strong scattering the extinction length l_{ext} nears the wavelength, so that the fundamental condition of applicability of the classical linear transfer theory $l_{\mathrm{ext}} >> \lambda$ is violated. We shall briefly touch the Anderson localization in Section 5.4.

3.3.9 Group Expansions of Operators of Effective Inhomogeneities

In the general case, the operators V^{eff} and K_{12} are nonlocal, that is, the nuclei of these operators $V^{\mathrm{eff}}(x,x')$ and $K\begin{pmatrix} x_1, & x_1' \\ x_2, & x_2' \end{pmatrix}$ are slowly decaying functions and can appreciably differ from zero at large separations of the arguments (x,x') and (x_1,x_1',x_2,x_2'). This implies, in particular, that the effective inhomogeneities corresponding to V^{eff} and K_{12} depend on field values at all points of the scattering volume and cannot be localized, that is, they cannot be endowed with some characteristic dimension. However, there exist a useful approximation that well responds to physical intuition, in which the effective inhomogeneities are characterized by dimensions in the order of the correlation radii of medium's inhomogeneities. This is the *one-group approximation* set forth by Finkelberg (1967).

This approximation can be obtained by group-summing terms in the cumulant expansions of the effective inhomogeneity operators (3.3.50) and (3.3.52) in a one-group (containing one cumulant) sum, two-group (containing two cumulants) sum, three-group sum, and so on. Denoting the sum of n-group addends in the expansion (3.3.50) by $V^{(n\mathrm{gr})}$, and in (3.3.52) by $K_{12}^{(n\mathrm{gr})}$ we write

$$V^{\mathrm{eff}} = \sum_{n=1}^{\infty} V^{(n\mathrm{gr})}, \tag{3.3.70}$$

$$K_{12} = \sum_{n=1}^{\infty} K_{12}^{(n\mathrm{gr})} \tag{3.3.71}$$

Then in the one-group approximation we have

$V^{\mathrm{eff}} \approx V^{(1\mathrm{gr})} =$  (3.3.72)

and

$K_{12} \approx K_{12}^{(1\mathrm{gr})} =$ 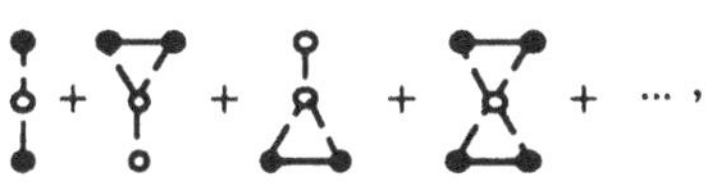 (3.3.73)

This corresponds to discarding all multiple group terms with $n > 1$ in eqns (3.3.70) and (3.3.71).

It is not hard to see that the presence of cumulants in the expansions (3.3.72) and (3.3.73) causes the nuclei of the operators V^{eff} and K_{12} to rapidly decrease (at a rate of about the decay of correlations in the scattering medium) when their arguments are separated. This behavior means that in the one-group approximation, effective inhomogeneities may be assigned with characteristic dimensions in the order of the correlation radii.

On substituting the expansions (3.3.70) and (3.3.71) in the optical theorem (3.3.63) and equating the terms containing identical number of cumulants on the left and right hand sides, we obtain a series of relations the first of which relates $V^{(1gr)}$ and $K_{12}^{(1gr)}$

$$V^{(1gr)}(x,x_0) - V^{(1gr)*}(x_0,x) = \int \left[G_0(x'',x') - G_0(x',x'')\right] \times K_{12}^{(1gr)}\begin{pmatrix} x', & x_0 \\ x'', & x \end{pmatrix} d^4x' d^4x''. \quad (3.3.74)$$

This relation differs from the exact optical theorem (3.3.63) for V^{eff} and K_{12} by G_0 substituted for $\langle G\rangle$. Therefore, the energy conservation law is fulfilled only approximately: for an exact energy matching of $V^{(1gr)}$ and $K_{12}^{(1gr)}$ in (3.3.74) one has to replace G_0 with $\langle G\rangle$ calculated in the one-group approximation, i.e., at $V^{\mathrm{eff}} = V^{(1gr)}$.

In the general case, the one group approximations (3.3.72) and (3.3.73) are infinite sums that are difficult to calculate, but when the kernel V is a Gaussian random function whose cumulants except the first two are zero, then the expansions (3.3.72) and (3.3.73) truncate already after one or two terms and assume extremely simple form

$$V^{\mathrm{eff}} \approx V^{(1gr)} =$$  +  $$= \langle V\rangle + \langle \tilde{V} G_0 \tilde{V}\rangle, \quad (3.3.75)$$

$$K_{12} \approx K_{12}^{(1gr)} =$$  $$= \langle \tilde{V}_1 \tilde{V}_2^{*}\rangle. \quad (3.3.76)$$

The approximation (3.3.75) is known as the *Bourret approximation* and eqn (3.3.76) as the *ladder approximation* (Rytov et al., 1989b). As in the general case of one-group approximations, these too obey the relation (3.3.74), i.e., they are not exactly energy matched. The ladder approximation is exactly energy matched the Kraichnan approximation (3.3.42) that along with (3.3.76) satisfies the exact optical theorem (3.3.63).

3.3.10 Applicability of the One-Group Approximations

In the simple case of Gaussian fluctuations of V, the one group approximations (3.3.75) and (3.3.76) can be formally derived by retaining only quadratic terms in the expansions of V^{eff} and K_{12} in powers of the cumulant of the inhomogeneity operator V. Clearly, to be applicable, the one-group approximations of the fluctuation of V must be small in a certain sense.

Discussing the applicability conditions of the one-group approximations we have to use non-rigorous estimates even if some natural simple constraints are imposed. For example, from the physical point of view, it is natural to require that the last retained term of the expansion be large compared to the first discarded term. For Gaussian fluctuations, when the expansion of V^{eff} has the form (3.3.51), this condition may be written as

(3.3.77)

where $\|\ldots\|$ denotes the norm, i.e. a number estimating this quantity. Unfortunately, a verification of this seemingly simple condition is a rather hard problem. Therefore, we confine our consideration to simple and physically transparent heuristic reasoning yielding rough estimations of the necessary applicability conditions of the one-group approximations for V^{eff} and K_{12}. For definiteness, we consider the scattering of waves at large (compared to the size of inhomogeneities) scattering volumes when boundary effects associated with the boundedness of the medium may be disregarded.

We suppose that the absorption is absent and the radiation in the medium is characterized by the free path length L_{f}, i.e., by the mean distance between wave collisions with medium's inhomogeneities. If one differs collisions leading to the variations in the amplitude and the phase, then the free path lengths with respect to variation of the amplitude, L_A, and with respect to variations of the phase, L_{ph}, may be introduced. The free path length L_{f} will be on the order of the least of these two quantities, $L_{\text{f}} \sim \min\{L_A, L_{\text{ph}}\}$.

Successive terms of the group expansions (3.3.70) and (3.3.71) may be assigned the sense of operators that take into account the effects of scattering at groups of n correlated inhomogeneities. In order that evaluation of successive scattering events make sense, these scattering events must occur at uncorrelated inhomogeneities, i.e. the free path length L_{f} must be large compared to the characteristic dimension of inhomogeneities l^*. Indeed, in the opposite case the wave undergoes scattering earlier than it leaves the inhomogeneity, the effective inhomogeneities begin to overlap and the group expansions (3.3.70) and (3.3.71) become physically meaningless. Thus, it is natural to require that

$$L_{\text{f}} \gg l^*. \tag{3.3.78}$$

Consider the elementary estimate of the free path length L_{f} in the context of the one-group approximation for the mean field. We recall that in this approximation, the size of effective inhomogeneities, i.e., the radii of nonlocality $V^{(1\text{gr})}$ and $K_{12}^{(1\text{gr})}$ have the order of magnitude of the correlation radii of medium's fluctuations: $l^* \sim l_c$. Let us examine the Dyson equation (3.3.24) in the one-group approximation. Assume that the medium is statistically uniform and stationary, the sources are absent ($q = 0$) and the initial field is monochromatic (we suppress the factor $e^{-i\omega t}$). Substituting the plane wave solution $\langle u \rangle = e^{i\mathbf{kr}}$ in the Dyson equation (3.3.24) yields the dispersion equation for the wave vector $\mathbf{k}$

$$D(\mathbf{k}) = 0, \tag{3.3.79}$$

where

$$D(\mathbf{k}) = e^{-i\mathbf{kr}} D e^{i\mathbf{kr}} = e^{-i\mathbf{kr}}(L_0 - V^{\text{eff}})e^{i\mathbf{kr}} = L_0(\mathbf{k}) - V^{\text{eff}}(\mathbf{k}). \tag{3.3.80}$$

Suppose that $k^{(1)}$ is a solution to (3.3.79) which in the absence of scattering ($V \equiv 0$) coincides with the wave vector of the considered wave k_0. Then $k^{(1)}$ may be treated as the wave vector k_0 due to scattering at random inhomogeneities. (In the general case, the effect of scattering on the mean field does not reduce to the substitution of $k^{(1)}$ for k_0, but for rough estimation, this reasoning is adequate.) The free path length may be defined as

$$L_{\text{f}} = \left|k_0 - k^{(1)}\right|^{-1}, \tag{3.3.81}$$

because this quantity has the order of magnitude of the characteristic length over which the amplitude or phase of the wave changes appreciably due to scattering (the scattering event is actually such a collision).

Thus, the heuristic necessary condition of applicability of the one group approximation (3.3.78) may be given as

$$l_c \Delta k \equiv l_c \left|k^{(1)} - k_0\right| << 1. \tag{3.3.82}$$

It should be obvious that, in the case of the first and second moments, the strict applicability conditions of one-group approximations will differ. However, from physical considerations we expect that the condition (3.3.82) obtained for the first moment must satisfy for the second moment as well. Indeed, in both cases we use one the same pattern of scattering at localized inhomogeneities, or, formally, used similar expansions in the number of correlation groups. This simple guess about the consistency of one-group approximations for the first and second moments in scattering problems has been proved by a more rigorous analyses (Barabanenkov, 1974; Ishimaru, 1984).

We illustrate the condition (3.3.82) with reference to the scalar field (3.3.3) (see also Problem 3.35). Let $\tilde{\varepsilon}$ be a Gaussian random quantity with the correlation function of fluctuations

$$B_\varepsilon(\rho)^2 \equiv \left\langle \tilde{\varepsilon}(\rho)^2 \tilde{\varepsilon}(0)\right\rangle = \sigma^2 \exp\left(-|\rho| / l_c\right). \tag{3.3.83}$$

In this case, $L_0(k) = -k^2 + k_0^2$, and the effective inhomogeneity operator in the k-representation reduces to multiplication by

$$V^{\text{eff}}(k) \equiv e^{-i\mathbf{kr}} V^{\text{eff}} e^{i\mathbf{kr}} = k_0^4 \int d^3r' \langle \tilde{\varepsilon}(\mathbf{r}) G_0(\mathbf{r} - \mathbf{r}')\tilde{\varepsilon}(\mathbf{r}')\rangle e^{-i\mathbf{k}(\mathbf{r}-\mathbf{r}')}$$

$$= k_0^4 \int G_0(\rho) B_\varepsilon(\rho) e^{i\mathbf{k}\rho} d^3\rho = -\left[(1 - ik_0 l_c)^2 + (kl_c)^2\right]^{-1} k_0^4 \sigma^2 l_c^2 ,$$

so that the dispersion equation (3.3.79) can be written in the form

$$k^2 = k_0^2 + \left[(1 - ik_0l_c)^2 + (kl_c)^2\right]^{-1} k_0^4\sigma^2 l_c^2.$$

It has the solution

$$k = k^{(1)} = k_0\left[1 + \frac{2(\sigma k_0 l_c)^2}{\sqrt{4k_0^4 l_c^4\sigma^2 + (1 - i2k_0l_c)^2} + 1 - 2ik_0l_c}\right]^{1/2},$$

that becomes equal to the undisturbed solution $k^{(1)} = k_0$ as $\sigma \to 0$. For small-scale fluctuations $k_0l_c << 1$, the condition (3.3.82) holds for

$$\left(\sigma k_0 l_c\right)^2 << \left(k_0 l_c\right)^{-2}, \tag{3.3.84}$$

whereas, for large-scale inhomogeneities $k_0l_c >> 1$, from eqn (3.3.82) we have

$$(\sigma k_0 l_c)^2 << 1. \tag{3.3.85}$$

We emphasize that the condition (3.3.82) is a fairly general and may be applied to any arbitrary stochastic equation of the form $Lu = 0$ for which, in the absence of fluctuations, the presence of natural waves is characteristic. However, this condition does not take into account certain aspects in the problem formulation, such as initial or boundary conditions, and is, in fact, equivalent to elementary estimates derivable from dimensionality considerations. Accordingly, the condition (3.3.82) can be used only for a rough estimation of applicability criteria of the one-group approximation and, generally speaking, is necessary but not sufficient. More rigorous estimates could hardly be obtained in the general form without no reference to the specific problem formulation. From physical reasoning we can expect that when the condition (3.3.82) is violated, the one-group approximation will certainly be inapplicable. A simple statistical problem. for which the Bourret approximation becomes exact independently of whether or not the condition (3.3.82) is satisfied, may be found in Frisch et al., 1973.

This reasoning can be illustrated by taking the above Helmholtz equation with the correlation function of fluctuations (3.3.83). In this case, for small-scale fluctuations $k_0l_c << 1$, the condition (3.3.84) allows considerable deviations of the real part of $k^{(1)}$ from k_0, that is, does not require that $\Delta k \equiv |k^{(1)} - k_0| << k_0$. Should the condition (3.3.82) be not only necessary, but also sufficient for the validity of Bourret's approximation, then, for $kl_c << 1$, one could obtain large corrections to the effective permittivity of the medium. Frisch (1968) derived from dimensional considerations a condition coincident with (3.3.84). Tatarskii (1961) obtained a stronger condition $(\sigma k_0 l_c)^2 << (k_0 l_c)^{-1}$ that follows from inequalities like (3.3.77) and admits appreciable deviations of $k^{(1)}$ from k_0. Other authors (Ryzhov and Tamoikin, 1970; Finkelberg, 1967) require a stronger constraint $(\sigma k_0 l_c)^2 << 1$ that does not allow large values of

$\Delta k / k_0$ any longer. These estimates differ because no sufficient applicability criteria are available for the one-group approximation.

For Gaussian fluctuations and the scalar wave equation, the more elaborate analysis of the applicability conditions for the one-group approximation of the Dyson equation has been given in the book by Rytov, Kravtsov and Tatarskii (1989b), where the sufficient condition for large and small inhomogeneities is eqn (3.3.85) leading to small corrections to effective medium's parameters. Following this treaties we will call eqn (3.3.85) the sufficient condition of the Bourret approximation.

3.3.11 One-Group Approximation for the Dyson Equation. The Effective Wave Number

Consider the mean Green's function in a scattering medium for a scalar wave field satisfying the Helmholtz equation (3.3.3). This function is expressed in terms of the effective inhomogeneity operator V^{eff} as

$$\langle G\rangle(r,r') = \left(L_0 - V^{\text{eff}}\right)^{-1}(r,r')$$
$$= (2\pi)^{-3}\int\left[-k^2 + k_0^2 - V^{\text{eff}}(\mathbf{k}) + i0\right]^{-1} e^{i\mathbf{k}(\mathbf{r}-\mathbf{r}')}d^3k\,. \tag{3.3.86}$$

For V^{eff}, we use the one group-approximation which, for Gaussian fluctuations, reduces to the Bourret approximation (3.3.75). The integral can be taken as the sum of residues if we assume that the integrand has only simple poles, alternatively if the dispersion equation for the mean field

$$k^2 = k_0^2 - V^{(1\text{gr})}(k) \tag{3.3.87}$$

has only simple roots. In the general case, this equation is transcendental and its explicit form depends upon the statistical properties of scattering inhomogeneities.

A general analysis of such an equation would lead us far astray, therefore we limit this discussion by a simple assumption that eqn (3.3.87) can be solved by iterations letting in the zero approximation $k = k_0$. Then, the first iteration yields

$$k^2 = k_0^2 - V^{(1\text{gr})}(k_0^2), \tag{3.3.88}$$

and the second

$$k^2 = k_0^2 - V^{(1\text{gr})}\left(k_0^2 - V^{(1\text{gr})}(k_0^2)\right)$$
$$= k_0^2 - V^{(1\text{gr})}(k_0^2)\left(1 - \frac{dV^{(1\text{gr})}(k_0^2)}{dk_0^2}\right) + \ldots, \tag{3.3.89}$$

where dots stand for terms with higher powers of $V^{(1\text{gr})}$, and $V^{(1\text{gr})}(\mathbf{k}^2) = V^{(1\text{gr})}(k^2)$ in view of the assumption that the medium is isotropic.

We see that when

$$\left|\frac{dV^{(1gr)}(k_0^2)}{dk_0^2}\right| << 1 \tag{3.3.90}$$

we may confine ourselves with the approximate expression (3.3.88). If in addition, $|V^{(1gr)}(k_0)| << k_0^2$, then

$$k = k^{\text{eff}} \approx \sqrt{k_0^2 - V^{(1gr)}(k_0^2)} \approx k_0 - \frac{V^{(1gr)}(k_0^2)}{2k_0}. \tag{3.3.91}$$

A more detailed analysis conducted by Tatarskii (1961) indicates that the dispersion equation (3.3.87) can have several roots $k = k^{(i)}$, but, if the condition (3.3.90) is satisfied, then the main root is the approximate result (3.3.91). The other roots describe terms that rapidly decay when the observation points moves away from the sources.

For Gaussian fluctuations $\tilde{\varepsilon}$ and the simplest exponential correlation function (3.3.83), the condition (3.3.90) may be written as

$$\left|\frac{dV^{(1gr)}(k)}{d(k^2)}\right|_{k=k_0} = \frac{k_0^4\sigma^2 l_c^4}{1+4k_0^2 l_c^2} << 1. \tag{3.3.92}$$

In is not hard to see that this inequality follows from the applicability conditions of the one-group approximation for small-scale fluctuations, eqn (3.3.84), and large scale inhomogeneities, eqn (3.3.85). Under the above assumptions, for the case of eqn (3.3.83), the transition to the one-group approximation in the Dyson equation boils down mainly to the substitution of the effective wave number k^{eff} (3.3.91) for the wave number k_0. This operation is equivalent to the replacement of the nonlocal operator of effective inhomogeneities V^{eff} in the Dyson equation by the local operator V_l^{eff} whose kernel, for small scale fluctuations $k_0 l_c << 1$, is

$$V_l^{\text{eff}}(\mathbf{r},\mathbf{r}') = \delta(\mathbf{r}-\mathbf{r}')\int V^{\text{eff}}(\rho)d^3\rho. \tag{3.3.93}$$

This substitution is equivalent to the neglect of the spatial dispersion of the mean field in the random inhomogeneous medium (Barabanenkov, 1975; Rytov, Kravtsov and Tatarskii, 1989b).

It can be demonstrated that the real part of k^{eff} exceeds k_0, the wave number in a free space. This corresponds to an increase of the mean optical path due to random inhomogeneities. The imaginary part of the effective wave number k^{eff} owes its existence to the attenuation of the mean field because of scattering at random inhomogeneities.

In the general case, the transition to the one-group approximation is not limited by the change to the effective wave number k^{eff}. However, far from the boundaries and sources in a scattering medium, the use of k^{eff} allows one to correctly describe the effects covered by the one-group approximation.

3.4 DISCRETE SCATTERING MEDIA IN ELECTRODYNAMIC PROBLEMS

3.4.1 Media with Discrete Scatterers

The equations for moments (3.3.25) and (3.3.29) retain their form in media containing discrete scattering particles (inclusions). These particles may be macroscopic formations on which the medium parameters change jumpwise (for example, aerosol particles, rain drops, snowflakes, grain media like polycrystallites or mixtures of fine powders) and microscopic objects — atoms and molecules. Thus, for discrete media, the calculation of the effective inhomogeneity operators V^{eff} and K_{12}, or alternatively, the calculation of effective parameters of the random inhomogeneous media is extremely diverse. It involves a variety of problems from the calculation of the effective parameters of mixtures to the fundamental problems associated with the transition from the microscopic to macroscopic model. This diversity explains the existence of very large number of approaches to this problem in various fields of physics, including electrodynamics, acoustics, elasticity theory, the theory of solids, liquids, gases and dielectrics, and in the theory of scattering in random inhomogeneous media. These approaches have much in common and differ in some details. This similarity is especially pronounced when the problem is formulated in an abstract operator form.

Even a plain review of all available derivations of the equations of moments in discrete scattering media would require too large a space in this volume and would lead us far astray from the radiative transfer theory. Therefore, we briefly outline only some approximations directly related with the electrodynamic problem. For more substantial expositions we refer the reader to the reviews of Frisch (1968), Barabanenkov et al. (1971), Barabanenkov (1975a) to multiauthor volumes edited by Garland and Tanner (1978), to the paper of and Burridge at al. (1982) and to the literature appended at the end of the next section [we note the early papers of Foldy (1945) and Lax (1951, 52)].

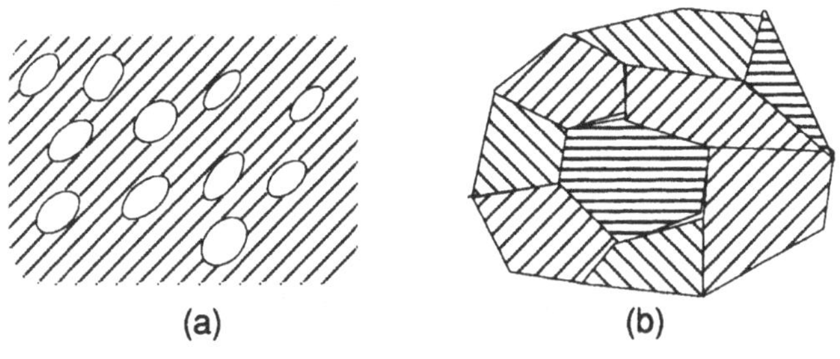

Figure 3.13. Structure of scattering media with discrete imbeddings: (a) cermet topology, (b) aggregate topology.

Let us consider a simple classification of discrete scattering media.Two qualitatively different classes of discrete media are usually distinguished, the *media with cermet topology* and the *media with aggregate topology*. The former contain discrete impregnations separated from each other by the matrix material of the host medium (Fig. 3.13a). These media are asymmetric about the replacement of matrix parameters with the parameters of inclusions. Typical examples of such media may be particulate media, such as aerosols, rain, and cermet materials that lend their name to these media.

Media with aggregate topology are symmetric. They include media consisting of cells with different parameters, where no one cell is preferable to others (Fig. 3.13b). Powder mixtures are examples of such media.

As a rule, speaking of discrete media we imply media with cermet topology. In the general case, such media can contain micro or macro particles of different size, form, and composition. We assume, for simplicity, that these particles differ not very far so that they may be endowed with a characteristic size a, average distance between the particles l_{av}, and correlation radius l_c. In the general case, for correlated solid particles, $l_c \geq a$.

We digress for a moment from correlation effects and assume that $l_c \sim a$ (thus we exclude from our consideration media with far order for which $l_c >> a$, specifically, media with periodic arrangement of particles). Then the structure of the medium will be characterized by two main length parameters a and l_{av}. Depending on the ratio of these parameters to one another and to the characteristic wavelength λ in the host material, we introduce a classification convenient for description of scattering as follows: we call media with $l_{av} >> a$ *sparse*, media with $l_{av} \sim a$ *densely packed*, media with $l_{av} >> \lambda$ *strongly discrete*, and media with $l_{av} << \lambda$ *quasi-continuous*. (The two last terms intuitively correspond to the condition that the concept of a continuous medium in the electrodynamic problem should refer to the wavelength.)

According to the above definition, media with cermet topology may be categorized into four main types, as shown in Fig 3.14. The boundaries separating one domain from another are, of course, conditional and somewhat blurred. The intermediate case of $l_{av} \sim \lambda$ may be called the region of interparticle resonances, but we exclude resonances of this kind from our rough categorization.

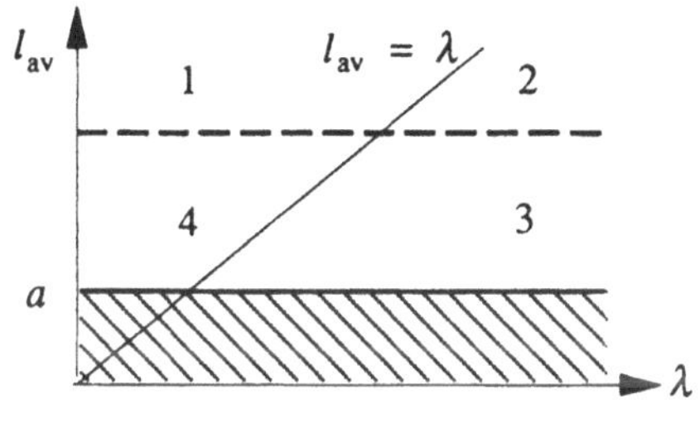

Figure 3.14. Classification of media with cermet topology: (1) sparse strongly discrete, $l_{av} > a, \lambda$; (2) sparse quasi-continuous, $\lambda > l_{av} > a$; (3) densely packed quasi-continuous, $\lambda > l_{av} \sim a$; (4) densely packed, strongly discrete, $l_{av} \sim a > \lambda$. The hatched area $l_{av} < a$ is non-physical.

With reference to Fig. 3.14 we briefly characterize each medium type. Region *1* $l_{av} > a, \lambda$ belongs to sparse strongly discrete media. This case covers the large variety of natural media on Earth (clouds, aerosols, snow, various suspensions — to be studied in the optical range) and in space (interplanetary and interstellar dust — to be studied in the optical and radio wave ranges). This case is simplest from the electrodynamic standpoint. For sparse or strongly rarefied media with $l_{av} >> a$, the extinction length l_{ext} is usually large compared to the wavelength, so that the of effect of mutual influence of particles is small and the parameters of the effective propagation medium differ from the interparticle medium's parameters insignificantly.

The case of small scattering volumes, V_{sc}, when the characteristic dimension of the scattering medium, $L_{sc} \sim V_{sc}^{1/3}$, is small compared to the extinction length, may be modeled by the single scattering approximation which represents the field as a sum of independent contributions from individual particles (see Section 3.4.6). In the opposite case of extended medium with $L_{sc} < l_{ext}$, multiple scattering may be described by a radiative transfer equation with scattering cross-section taken as a sum of similar cross-sections for individual particles [see eqn (2.2.50)]. The effective wave number k^{eff} may be represented by the "weak dispersion formula" (2.2.59) that provides small deviations of k^{eff} from the unperturbed value $2\pi / \lambda$.

Sparse quasi-continuous media of region *2*, where $\lambda > l_{av} > a$, may be the same clouds, rains, aerosols, and snow, but already in the radio wave range. Here, one wavelength covers a large number of particles, so that the neglect of the interference of the fields scattered by different particles may be justified only conditionally. In strongly rarefied media, the extinction length also usually considerably exceeds the wavelength, $l_{ext} > \lambda$, but the differences of the effective medium parameters from the unperturbed values need not necessarily be small. In the general case, the correlations of particle positions may be significant in such media. Accordingly, in the transfer equation, like in the equation for the effective medium parameters, we have to introduce corrections for the correlations.

Regions *3* and *4* represent, respectively, densely packed ($l_{av} \sim a$) quasi-continuous media ($\lambda >> l_{av}$), and strongly discrete media ($\lambda << l_{av}$). In these regions, the effective parameters of random inhomogeneous media, as a rule, strongly differ from their no-scatterers counterparts. In the last case, the Hulst "weak" dispersion formula (2.2.59) will be already inapplicable.

The quasi-continuous region *3* includes, in particular, the *quasi-static scattering* that can be realized also in the case of sparse media from region *2*. In this scattering limit, the spatial scales of the medium are small compared not only with the wavelength λ in the host-medium, but also with the characteristic wavelengths inside scattering particles λ_i

$$\lambda_i, \lambda >> l_{av}, a. \tag{3.4.1}$$

A host of special literature has been devoted to the calculation of the effective medium parameters in connection with the evaluation of the mean field $\langle \mathbf{E} \rangle$ in the quasi-static mode. To determine the effective dielectric permittivity, as an initial approximation one may use the classic Lorenz–Lorentz formula and its modifications that will be considered in some more detail in the ensuing section. As to the higher moments of the field, specifically field correlations, for densely packed media, these have been poorly reported.

Region *4* ($l_{av} \sim a > \lambda$) seems to be the most complicated for analysis. In this region, the densely packed medium contains large (in the wavelength scale) scatterers, so that the spatial dispersion with respect to statistical field moments has to be taken into account. As a result, the Lorenz–Lorentz type formulas are no longer applicable here, and the results will be strongly dependent on the medium model chosen.

At present, no general theory of scattering exists for densely-packed random inhomogeneous media. Such a theory should take into account the mutual influence of, specifically, the correlations of positions of scatterers. However, even simple evaluations of such correlations in real media have been faced with unsurmountable difficulties. For example, even in a relatively simple case of a liquid in thermal equilibrium, the second (lowest) correlations of scatterer positions are known only for the model of solid spheres (the solution due to Wertheim, 1963; and Thiele, 1963).

The above categorization of discrete random inhomogeneous media does not pretend to be exhaustive. First, it fails to observe the electrodynamic properties of scatterers and the possibility of various kinds of resonances. For instance, the case of optically soft scatterers admits a relatively full examination with the Born series (3.3.17) or with various modifications of the familiar parabolic equation. Second, this classification refers, strongly speaking, to the limiting cases of $l_{av} >> a, l_{av} \sim a$ and $l_{av} >> \lambda$ or $l_{av} << \lambda$. Nonetheless, it can be useful as an initial orientation in the subject.

3.4.2 Effective Medium Parameters in the Quasi-Static Limit

The quasi-static limit corresponds to the case given by eqn (3.4.1) when all spatial scales of the medium are small compared with the characteristic wavelengths. This limit can be realized in media with cermet topology and in media with aggregate topology. In this limit, the effective parameters are not uniquely defined by the composition of the medium, but rather may appreciably depend on the spatial distribution of its components. This can be seen already in a simple example of a capacitor filled with a layered dielectric, as shown in Fig. 3.15. The effective permeability of such a system depends on layers' orientation with respect to the electric field. If **E** is parallel to the layers of the dielectric, the effective capacity of the system in Fig. 3.15a is the sum of sectional capacities $C = \sum C_i$, and the effective permittivity is

$$\varepsilon^{\text{eff}} = \sum \varepsilon_i p_i \equiv \langle \varepsilon \rangle_V ,$$

where p_i is the volume fraction of the i th dielectric. Conversely, if **E** is perpendicular to the layers of dielectric, as in Fig. 3.15b, then the inverse capacities are summed $C^{-1} = \sum C_i^{-1}$ and, respectively,

$$\frac{1}{\varepsilon^{\text{eff}}} = \sum \frac{p_i}{\varepsilon_i} \equiv \left\langle \frac{1}{\varepsilon} \right\rangle_V .$$

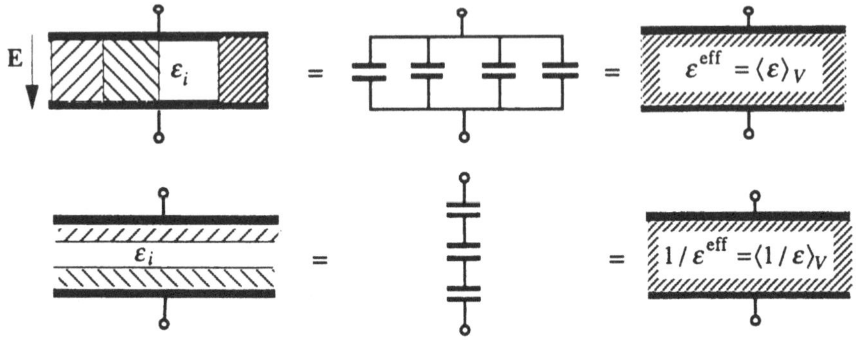

Figure 3.15. Calculating the effective dielectric permittivity ε^* of a capacitor with layered dielectric for **E** parallel to the dielectric layers (top line), and **E** perpendicular to these layers (bottom line).

Let us consider the main approximations known for the quasi-static mode from the aspect of the effective medium permittivity, for definiteness. Similar equations may be put down for other effective parameters, for example, for electrical conductivity, medium's magnetic permeability, and such.

For a microscopically inhomogeneous medium (composite material) of permittivity $\varepsilon(\mathbf{r})$, the effective permittivity ε^* in the quasi-static mode may be determined with the aid of relations similar to eqn (3.3.33), viz.,

$$\langle \varepsilon(\mathbf{r})E(\mathbf{r})\rangle = \int \varepsilon^{\text{eff}}(\mathbf{r},\mathbf{r}')\langle E(\mathbf{r}')\rangle d^3r' \approx \int \varepsilon^{\text{eff}}(\mathbf{r},\mathbf{r}')d^3r'\langle E(\mathbf{r})\rangle \equiv \varepsilon^* \langle E(\mathbf{r})\rangle, \quad (3.4.2)$$

where ε^* is expressed as the integral of the kernel $\varepsilon^{\text{eff}}(\mathbf{r},\mathbf{r}')$ of the effective permittivity operator $\hat{\varepsilon}^{\text{eff}}$:

$$\varepsilon^* = \int \varepsilon^{\text{eff}}(\mathbf{r},\mathbf{r}')d^3r'. \quad (3.4.3)$$

For a statistically homogeneous medium, this limit is independent of **r**.

Assuming that the spatial parameters of the medium are small compared to the similar mean field parameters we factor $\langle \mathbf{E}(\mathbf{r})\rangle$ outside the integral sign in (3.4.2). Thus, we tacitly assume that the integral (3.4.3) converges, which is true not for all models of effective parameters (see Ramshow, 1984, where other pitfalls of the theory are discussed).

We will refer to the replacement of the action of the operator ε^{eff} by the operation by ε^* as the local limit of ε^{eff}. A rigorous transition to the local limit is possible in the case of a homogeneous range-independent mean field [we considered this limit above in connection with the effective wave number concept, eqn (3.3.93)].

The definition (3.4.2) of ε^* remains valid also for microscopic medium models of the type of point dipole models popular in molecular dielectric theory. In the last case, the fluctuating local quantity $\varepsilon(\mathbf{r})$ becomes senseless and, in eqn (3.4.2), we need to consider the mean displacement vector $\langle \mathbf{D} \rangle$ in place of the mean $\langle \varepsilon \mathbf{E} \rangle$.

We have already noted that in the quasi-static mode, the *Lorenz–Lorentz formula* and allied generalizations are of great significance for permittivity calculations. The traditional derivation of this formula is described in many textbooks on electrodynamics (see, e.g., Jackson, 1975) and is based on the difference of the effective field acting on the scatterer $\mathbf{E}^{\text{eff}}$ from the mean field $\langle \mathbf{E} \rangle$. In the simplest case, $\mathbf{E}^{\text{eff}} = \langle \mathbf{E} \rangle + 4\pi / 3 \langle \mathbf{P} \rangle$, where $\mathbf{P}$ is the medium's polarization. In the model of point dipoles with polarizability α, the Lorenz–Lorentz formula has the form

$$\frac{\varepsilon^* - \varepsilon_{\text{m}}}{\varepsilon^* + 2\varepsilon_{\text{m}}} = \frac{4\pi}{3} \frac{\nu \alpha}{\varepsilon_{\text{m}}}, \tag{3.4.4}$$

where ε_{m} is the dielectric permittivity of the matrix or host material, and ν is the number of scatterers per unit volume. For a static permittivity, his expression is also known as the Clausius–Mosotti formula (a historical outline of the Lorenz–Lorentz formula may be found in Landauer, 1978).

Equation (3.4.4) is widely used to calculate the permittivity in the theory of liquids and solids, where $\varepsilon_{\text{m}} = 1$, and the modifier "effective" is usually dropped because the calculated permittivity is related to the transition from a "discrete" microscopic description to a "continuous" macroscopic description (see, e.g., Born and Wolf, 1980).

In contrast to the Hulst "weak" dispersion formula (2.2.59), the Lorenz–Lorentz formula describes strong deviations of the effective dielectric permittivity ε^* from the matrix permittivity ε_{m}. These formulas differ quantitatively and qualitatively; while in the case of (3.4.4) the difference of $\text{Im}\,\varepsilon^*$ from zero may be associated only with the presence of true absorption in the medium (i.e., with the nonzero imaginary parts of ε_{m} or of α), then in the case of (2.2.59), $\text{Im}\,\varepsilon^{\text{eff}}$ describes not only absorption, but also the losses of the coherent field due to scattering into the incoherent component. Notwithstanding to these differences, formulas (3.4.4) and (2.2.59) agree with one another in the limiting case of strongly rarefied and weakly scattering media (Problem 3.36).

If in place of point dipoles we consider small spheres with permittivity ε_1, then substituting in eqn (3.4.4) the familiar expression for the sphere polarizability

$$\alpha = a^3 \, \frac{\varepsilon_1 - \varepsilon_{\text{m}}}{\varepsilon_1 + 2\varepsilon_{\text{m}}}, \tag{3.4.5}$$

where a is the sphere radius, we arrive at the Maxwell Garnett formula (1904):

$$\frac{\varepsilon^* - \varepsilon_{\text{m}}}{\varepsilon^* + 2\varepsilon_{\text{m}}} = f_1 \frac{\varepsilon_1 - \varepsilon_{\text{m}}}{\varepsilon_1 + 2\varepsilon_{\text{m}}} \equiv \left\langle \frac{\varepsilon_1 - \varepsilon_{\text{m}}}{\varepsilon_1 + 2\varepsilon_{\text{m}}} \right\rangle_\nu, \tag{3.4.6}$$

which is widely used for engineering estimations of the permittivity of mixtures. In this formula, f_1 is the volume fraction of scattering spheres so that the angular brackets correspond to volume averaging. This form may be viewed as an extension of the Maxwell Garnett formula to the case of an arbitrary distribution of $\varepsilon(\mathbf{r})$ over the volume if one treats the volume averaging as the calculation of the integral

$$\langle f(\mathbf{r})\rangle_v = \int\limits_v f(\mathbf{r})\frac{d^3r}{V}. \tag{3.4.7}$$

However, for nonspherical scatterers, this generalization has no sufficiently convincing justifications.

Better substantiated generalizations of the Lorenz–Lorentz formula are associated with considering the polarizability α of complex scatterers, for example, spheres in an envelope which are used for the description of scattering at hail (Sihvola, 1989).

The Maxwell Garnett formula is perhaps the most popular representative of the wide class of available formulas for computing the effective dielectric permittivity of macroscopic random inhomogeneous media in the quasi-static mode. This kind of formula can be deduced by different methods and under different assumptions about the medium's properties; normally they have the form

$$f(\varepsilon^*) = \langle f(\varepsilon(\mathbf{r}))\rangle_v, \tag{3.4.8}$$

where $f(x)$ is some function [such as,e.q.,$x^{1/2}$, $x^{1/3}$, $\log x$, and $(x+1/x)/2$], the, ε^* is the effective permittivity, $\varepsilon(\mathbf{r})$ is the local permittivity, and $\langle\cdot\rangle_v$ stands for volume averaging.

One more popular formula to define ε^* is due to Bruggeman (1935)

$$\left\langle \frac{\varepsilon(\mathbf{r}) - \varepsilon^*}{\varepsilon(\mathbf{r}) + 2\varepsilon^*} \right\rangle = 0, \tag{3.4.9}$$

who was the first to come up with effective medium theory.[3] Unlike the asymmetric Maxwell Garnet formula (3.4.6), eqn (3.4.9) is symmetric, which fact is consistent with the cermet topology, indeed, it does not contain a selected host material. As a consequence, they describe different physical effects. For instance, for the case of a medium containing small metallic spheres, the Maxwell Garnet formula describes the *optical resonance* also called the *conductivity resonance* (Marton and Lemon, 1971). This resonance is exhibited by a single particle and is associated with a growth of sphere polarizability near the Fröhlich surface mode where the denominator in eqn (3.4.6) becomes close to zero: $\varepsilon_1 + 2\varepsilon_m \approx 0$ (Bohren and Huffman, 1983). Formula (3.4.6) may be said to describe a modification of the Fröhlich resonance associated with the mutual influence of particles (Problem 3.37).

The Bruggeman formula (3.4.9) does not give an optical resonance, but instead it describes the *threshold of conductance*, which means the occurrence of a static conductance in a medium consisting of dielectric with metal particles when the metal component attains some critical value (see, e.g., Kirkpatrick, 1973).

[3] In the Russian literature on the subject, eqn (3.4.9) is often referred to as the Maxwell–Odelevskii formula (see Odelevskii, 1951). In applications, a popular formula is that of Polder–Van Santen (1946), which, as Bohren and Battan (1980) have noted, is equvalent to eqn (3.4.9).

Both eqn (3.4.6) and eqn (3.4.9) can be obtained from different physical models. Specifically, Straud (1975) has demonstrated that both formulas can be obtained as different approximations of one and the same integral equation (see Section 3.4.10). Sheng (1980) has suggested a heuristic theory of effective dielectric permittivity of granulated composite materials that describes both the conductance threshold and the occurrence of an optical anomaly.

Thus, for media with densely packed inhomogeneities, two key initial approximations derived for the quasi-static mode are used: these are the Lorenz–Lorentz, Clausius–Mosotti, and Maxwell–Garnet formulas on the one hand, and the Bruggeman–Odelevskii formula on the other.

Both the Lorenz effective field theory and the Bruggeman effective medium method have gained multiple followers that refine and develop these approaches. In the molecular theory of dielectrics, the first results in this direction were obtained in the 30's. Kirkwood (1936) and Yvon (1937) generalized the Lorenz–Lorentz formula as

$$\frac{\varepsilon^*-1}{\varepsilon^*+2}=\frac{4\pi}{3}\nu\alpha\begin{cases}1+S_K & \text{(Kirkwood)}\\ \dfrac{1}{1-S_Y} & \text{(Yvon)}\end{cases}, \qquad (3.4.10)$$

where S_K and S_Y are expressed as expansions in formal series in powers of α. Later, many authors have examined similar expansions in powers of some small parameters with allowance for different complicating factors such as multipole interactions, quantum effects, violations of the quasistatic approximation, and building more detailed models of inhomogeneous media. The extension of the Kirkwood expansion onto the case of non-point correlated scatterers beyond the framework of the quasistatic approximation will be examined in Section 3.4.10.

Convergence of expansions similar to (3.4.10), unfortunately, is almost not amenable to analysis, and the sequential terms are expressed in the form of multiple integrals whose complexity rapidly increases with the order of the term. We list only a few approaches to the calculation of the effective parameters of discrete media along with the relevant literature:

- coherent potential approximation (Landauer, 1978),
- effective medium approximation (Roth, 1974),
- quasi-crystalline approximation (Lax, 1952),
- iterated dilute approximation (Sen *et al.*, 1981)
- average field approximation (Polder and van Santen, 1946),
- a variety of cumulant expansions (Finkelberg, 1964; Hori, 1977; Felderhof *et al.*, 1983; Ramshaw, 1984),
- Twersky's muiltiple scattering theory (Twersky, 1962a, b; Tsolakis et al., 1985),
- Bergman's analytical representation for the effective permittivity of two-phase composites (Bergman, 1982),
- the theory of strong fluctuations in the electromagnetic problem (Tatarskii and Gertsenshtein, 1963; Ryzhov and Tamolikin, 1970).

We may refer here also the different theories of feasible boundaries for effective medium parameters (Hashin and Shtrikman, 1962; Bergman, 1978; Milton, 1981; Golden and Papanicolaou, 1983; and Kohler and Papanicolaou, 1982).

3.4.3 Exact Solution of the Scattering at a Cloud of Point Dipoles

In order to clarify the general relations we describe an important case of scattering of monochromatic radiation at a cloud of stationary point dipoles for which an exact solution can be given. We consider the distribution of polarizability in the form $\alpha(r)=\sum_l \alpha_l \delta(\mathbf{r}-\mathbf{r}_l)$, where $\mathbf{r}_l$ and α_l is the radius vector and polarizability of dipole l. We assume the permittivity of the medium be equal to unity. Then for the matrix Green's operator G we may write the equation similar to eqn (3.3.12):

$$LG=\left(L_0-\sum_l V^{(1)}\right)G=\left[-\nabla\times\nabla\times+k_0^2\left(1+4\pi\sum_l \alpha_l\delta(\mathbf{r}-\mathbf{r}_l)\right)\right]G=\hat{1}. \tag{3.4.11}$$

The integral form (3.4.11) of this equation has the form

$$G=G_0+G_0\left(-4\pi k_0^2\sum_l \alpha_l\delta(\mathbf{r}-\mathbf{r}_l)\right)G, \tag{3.4.12}$$

where $G_0=(-\nabla\times\nabla\times+k_2^2+i0)^{-1}$ is the Green's operator of free propagation. The action of the inhomogeneity operator for the lth scatterer $V^{(l)}$ reduces to the multiplication by $-4\pi k_0^2\alpha_l\delta(\mathbf{r}-\mathbf{r}_l)$.

However, this description of point scatterers is insufficiently correct. Indeed, real scatterers occupy a finite volume. The concept of a point dipole is feasible only with respect to the region outside the scatterer and becomes senseless inside it. Therefore, eqns (3.4.11) and (3.4.12) are applicable only for $\mathbf{r}\neq\mathbf{r}_l$, and at $\mathbf{r}=\mathbf{r}_l$ we have to consider the field incident on the lth scatterer from outside. This means that if we would solve eqn (3.4.12) by iterations, then the quantity $G_0(\mathbf{r}_l,\mathbf{r}_l)$ formally equal to infinity should be replaced with zero. This replacement corresponds to the elimination of the intrinsic field of the dipole. If the problem involves only one particle, then the first iteration of (3.4.12) yields

$$\begin{aligned}G(\mathbf{r},\mathbf{r}_0)&=G_0(\mathbf{r},\mathbf{r}_0)+G_0(\mathbf{r},\mathbf{r}_l)(-4\pi k_0^2\alpha_l)G_0(\mathbf{r}_l,\mathbf{r}_0)\\&\quad+(4\pi k_0^2\alpha_l)^2G_0(\mathbf{r},\mathbf{r}_l)G_0(\mathbf{r}_l,\mathbf{r}_l)G_0(\mathbf{r}_l,\mathbf{r}_0)\\&=G_0(\mathbf{r},\mathbf{r}_l)-G_0(\mathbf{r},\mathbf{r}_l)4\pi k_0^2\alpha_lG_0(\mathbf{r}_l,\mathbf{r}_0).\end{aligned} \tag{3.4.13}$$

Comparing this expression with $G=G_0+G_0t^{(1)}G_0$ we see that the polarizability of a point dipole $\alpha_l\delta(\mathbf{r}-\mathbf{r}_l)$ formally introduced in eqn (3.4.11) corresponds to the operator $-t^{(l)}/(4\pi k_0^2)$. Hence, the action of $t^{(l)}$ in the point dipole model α_l reduces to multiplication by $-4\pi k_0^2\alpha_l\delta(\mathbf{r}-\mathbf{r}_l)$, viz.,

$$t^{(l)}=-4\pi k_0^2\alpha_l\delta(\mathbf{r}-\mathbf{r}_l)\hat{1}, \tag{3.4.14}$$

so that the kernel $t^{(l)}$ has the form

$$t^{(l)}(\mathbf{r},\mathbf{r}_0) = -4\pi k_0^2 \alpha_l \delta(\mathbf{r}-\mathbf{r}_l)\delta(\mathbf{r}-\mathbf{r}_0).$$

Substituting (3.4.14) in the multiple scattering system (3.4.32), to be derived in Section 3.4.5 below for a cloud of N scattering dipoles, we obtain

$$T^{(N,l)}(\mathbf{r},\mathbf{r}_0) = -4\pi k_0^2 \alpha_l \left[\delta(\mathbf{r}_l - \mathbf{r}_0) + \int G_0(\mathbf{r}_l - \mathbf{r}') \sum_{m \neq 1} T^{(N,m)}(\mathbf{r}',\mathbf{r}_0) d^3 r' \right]$$

$$\equiv -4\pi k^2 \alpha_l \delta(\mathbf{r}-\mathbf{r}_l) C_l\,, \tag{3.4.15}$$

Now, for the auxiliary parameter C_l characterizing the magnitude of dipole moments induced in the lth particle, we obtain the system of linear algebraic equations

$$C_l \equiv \delta(\mathbf{r}_l - \mathbf{r}_0) + \int G_0(\mathbf{r}_l - \mathbf{r}') \sum_{n \neq 1} T^{(N,n)}(\mathbf{r}',\mathbf{r}_0) d^3 r'$$

$$= \delta(\mathbf{r}_l - \mathbf{r}_0) - 4\pi k_0^2 \sum_{n \neq 1} G_0(\mathbf{r}_l - \mathbf{r}_n)\alpha_n C_n\,, \tag{3.4.16}$$

which is equivalent to the system of equations due to Watson (1960).

The solution of this system may be written as the inversion of a block matrix:

$$C_l = \sum_{p=1}^{N} (\delta_{nm} + 4\pi k_0^2 G_{nm}\alpha_m)_{lp}^{-1} \delta(\mathbf{r}_p - \mathbf{r}_0), \tag{3.4.17}$$

where $G_{lm} = G_0(\mathbf{r}_l - \mathbf{r}_m)$ for $l \neq m$ and $G_{ll} = 0$. Substituting this expression in eqn (3.4.15) and using eqn (3.4.29) we finally obtain for the kernel T

$$T(\mathbf{r},\mathbf{r}_0) = -\sum_{l,p} 4\pi k_0^2 \alpha_l \delta(\mathbf{r}-\mathbf{r}_l)(\delta_{nm} + 4\pi k_0^2 G_{nm}\alpha_m)_{lp}^{-1} \delta(\mathbf{r}_p - \mathbf{r}_0)\,. \tag{3.4.18}$$

Thus, to calculate the T-operator for the case of point dipoles it is sufficient to invert one $3N \times 3N$ matrix. The complexity of this problem rapidly increases with the number of particles N. This property makes the solution (3.4.18) prohibitively time consuming for $N >> 1$, i.e., for the description of scattering at systems of many particles, especially if the position of particles may vary at random. To attack such problems one have to resort to statistical methods.

3.4.4 Distribution Functions of Discrete Particles and the Thermodynamic N/V Limit

In Section 3.3.5 we considered a sequence of cumulants describing the statistical properties of arbitrary random quantities ξ_l. These sequences may be used in the case of discrete scatterers. However, instead of the cumulants of random positions of scatterers $\mathbf{R}_l$ it is convenient to consider multiparticle distribution functions.

For simplicity we confine our consideration to the case when all scatterers are identical and each scatterer is characterized only by one random parameter — the radius vector of its "center" $\mathbf{R}_l$ (in the general case $\mathbf{R}_l$ should be augmented by random internal parameters, describing the scatterer's shape, orientation, composition, and the like). Then the summary statistical description of a system of N particles reduces to specifying the N-particle distribution function

$$P_{12\ldots N} = P(\mathbf{r}_1, \mathbf{r}_2, \ldots \mathbf{r}_N) = \langle \delta(\mathbf{r}_1 - \mathbf{R}_1)\delta(\mathbf{r}_2 - \mathbf{R}_2)\ldots\delta(\mathbf{r}_N - \mathbf{R}_N)\rangle. \tag{3.4.19}$$

However, for practical applications it suffices to consider lower order distribution functions $P_{12\ldots s}$, $s = 1,2,\ldots,N$, each of which is expressed in terms of $P_{1,2,\ldots,N}$ by integration with respect to the superfluous argument

$$\begin{aligned} P_1 &= \langle \delta(\mathbf{r}_1 - \mathbf{R}_1)\rangle = \int P_{12\ldots N} d\mathbf{r}_2 d\mathbf{r}_3 \ldots d\mathbf{r}_N \,, \\ P_{12} &= \langle \delta(\mathbf{r}_1 - \mathbf{R}_1)\delta(\mathbf{r}_2 - \mathbf{R}_2)\rangle = \int P_{12\ldots N} d\mathbf{r}_3 \ldots d\mathbf{r}_N \,, \\ P_{12\ldots s} &= \langle \delta(\mathbf{r}_1 - \mathbf{R}_1)\delta(\mathbf{r}_2 - \mathbf{R}_2)\ldots\delta(\mathbf{r}_s - \mathbf{R}_s)\rangle . \end{aligned} \tag{3.4.20}$$

All the distribution functions are normalized to unity

$$\int P_{12\ldots s} d\mathbf{r}_1 d\mathbf{r}_2 \ldots d\mathbf{r}_s = 1 \tag{3.4.21}$$

and are of dimension L^{-3s}.

In practical applications the number of scatterers N is very large, and the concentration of scatterers $\nu = N / V_{sc}$ is bounded. A specific number N become immaterial for the description of such systems, and so one may pass over to the so-called *thermodynamic* or N / V- *limit*: $N \to \infty$, $V_{sc} \to \infty$, $\nu = N / V_{sc} = \text{const}$. The first name owes its existence to the statistical substantiation of thermodynamics.

In order to describe the limiting transition as $N \to \infty$ and $V_{sc} \to \infty$ it is convenient to consider in place of $P_{12\ldots s}$ the dimensionless distribution functions

$$F_{12\ldots s} = V^s P_{12\ldots s}, \tag{3.4.22}$$

for which

$$F_{12\ldots s} = \int F_{12\ldots s+1} \frac{d\mathbf{r}_{s+1}}{V^s}, \tag{3.4.23}$$

and

$$\int F_{12\ldots s} \frac{d\mathbf{r}_1 \ldots d\mathbf{r}_s}{V^s} = 1. \tag{3.4.24}$$

Treating the delta function $\delta(\mathbf{r}_i - \mathbf{R}_i)$ as a random quantity ξ_i and using the results of Section 3.3.5 we may introduce the *correlation distribution functions* $q_{12...s}$ by expressing the moment $P_{12...s} = \langle \xi_1 \xi_2 \cdots \xi_s \rangle$ in terms of the respective cumulant $\langle \xi_1 \xi_2 ... \xi_s \rangle_c$. Specifically, in agreement with (3.3.47), the low order correlation functions are expressed as

$$q_1 = \langle \delta(\mathbf{r}_1 - \mathbf{R}_1) \rangle_c = P_1 ,$$
$$q_{12} = \langle \delta(\mathbf{r}_1 - \mathbf{R}_1)\delta(\mathbf{r}_2 - \mathbf{R}_2) \rangle_c = P_{12} - P_1 P_2 , \qquad (3.4.25)$$
$$\cdots$$

From the general properties of cumulants it follows that the correlation function $q_{12...s}$ is a measure of statistical relations in the group of s particles, and

$$\int q_{12...s} d\mathbf{r}_s = 0. \qquad (3.4.26)$$

To prove this relation, it is sufficient to take into account that in integration with respect to $\mathbf{r}_s$, the delta function $\delta(\mathbf{r}_s - \mathbf{R}_s)$ becomes unity, i.e., gives a nonfluctuating quantity.

In the simplest approximation of independent particles, all correlation functions $q_{12...s}$ are assumed to be zero for $s \geq 2$. To this approximation,

$$P_{12...s} = P_1 P_2 ... P_s , \qquad (3.4.27)$$

so that the specification of one particle distribution function P_1 specifies all the statistical properties of the particle system.

The approximation (3.4.27) becomes inapplicable for particles of finite size and small separations of the arguments $\mathbf{r}_i - \mathbf{r}_j$. Indeed, in the case of solid particles with which one has to do in common applications, the interparticle spacing $|\mathbf{r}_i - \mathbf{r}_j|$ cannot be smaller than particle's size. The allowance for this condition is known as the *hole correction* (Problem 3.38).

3.4.5 The Series of Multiple Scattering at an Ensemble of Particles

For a medium constituted by discrete scatterers, it is natural to study the problem of scattering at an ensemble of particles in steps: first solve the scattering at one particle, then use this solution to attack the scattering at an ensemble of particles as the interaction of the fields scattered by individual particles.

To this end it is convenient to use instead of the the Green operator G the above scattering T-operator (3.3.59) that is widely used in quantum mechanical problems (see, e.g.,Taylor, 1972). Substituting the definition of this operator (3.3.59) in eqn (3.3.11) yields the equation for T

$$T = V + V G_0 T. \qquad (3.4.28)$$

We associate with single particle l a scattering inhomogeneity $V^{(l)}$, and with the entire medium the inhomogeneity operator $V = \sum_{l=1}^{N} V^{(l)}$. Then eqn (3.4.28) takes the form

$$T = \sum_l V^{(l)}(1 + G_0 T) \equiv \sum_l T^{(N,l)}, \tag{3.4.29}$$

where the operator

$$T^{(N,l)} = V^{(l)}(1 + G_0 T) \tag{3.4.30}$$

corresponds to the scattering at particle l with allowance for mutual irradiation of all N scattering particles. We introduce the operator $t^{(l)} \equiv T^{(1,l)}$ that describes the scattering at single particle l and satisfies the equation

$$t^{(l)} = V^{(l)}\left(1 + G_0 t^{(l)}\right). \tag{3.4.31}$$

Assuming that this operator is known, from eqn (3.4.30) we obtain the system

$$T^{(N,l)} = t^{(l)} + \sum_{m \neq l} t^{(l)} G_0 T^{(N,m)}, \tag{3.4.32}$$

where the operator $t^{(l)}$ the plays role of the medium's inhomogeneity operator V. In the case of discrete particles, this system is taken to be our starting system in place of eqn (3.4.11). The operator $t^{(l)}$ explicitly takes into account the solution of eqn (3.4.31) of scattering at a single particle.

Solving the system (3.4.32) by iterations, we obtain a series in the scattering multiplicity

$$T^{(N,l)} = t^{(l)} + \sum_{m \neq l} t^{(l)} G_0 t^{(m)} + \sum_{l \neq m \neq n} t^{(l)} G_0 t^{(m)} G_0 t^{(n)} + \dots. \tag{3.4.33}$$

In contrast to the Born series (3.3.17), here every scattering event corresponds to a specific single particle, rather than a formal inhomogeneity operator V.

3.4.6 Cross Section of Scattering at an Ensemble of Particles. The Single Scattering Approximation

If the scattering volume is sufficiently far from the observation point, then, as in the case of single scatterers (3.3.62), it is convenient to change from the T-operator to the respective scattering amplitude. In agreement with eqn (3.4.29), the total scattering amplitude is represented as a sum of terms

$$f_\omega(\mathbf{n} \leftarrow \mathbf{n}_0) = \sum f_l, \tag{3.4.34}$$

where the amplitudes f_l are related to $T^{(N,l)}$ by an equation of the type of (3.3.62).

It is convenient to split the mean scattering cross section $\sigma_\omega = \langle |f_\omega|^2 \rangle$ corresponding to (3.4.34) into the coherent part $\sigma_{\rm coh} = |\langle f_a \rangle^2|$ that describes the scattering of the coherent field $\langle u \rangle$, and into the incoherent part $\sigma_{\rm incoh} = \langle |\tilde{f}_\omega|^2 \rangle = \langle |f_\omega|^2 \rangle - |\langle f_\omega \rangle|^2$, $\tilde{f}_\omega = f_\omega - \langle f_\omega \rangle$, corresponding to the scattering of the fluctuating component $\tilde{u} = u - \langle u \rangle$.

If the medium is constituted by identical particles, and their contributions in the amplitude f_ω is about equal, then the coherent scattering cross section may be written as

$$\sigma_{\rm coh} = \left|\left\langle \sum f_l \right\rangle\right|^2 \approx N^2 \left|\langle f_1 \rangle\right|^2, \tag{3.4.35}$$

where N is the total number of particles. The incoherent component $\sigma_{\rm incoh}$ may in turn be represented as the sum

$$\sigma_{\rm incoh} = \sigma_{\rm indep} + \sigma_{\rm coll} , \tag{3.4.36}$$

where the summand $\sigma_{\rm indep}$ corresponds to the scattering at independent particles:

$$\sigma_{\rm indep} = \sum \left\langle \left|\tilde{f}_l\right|^2 \right\rangle = N \left\langle \left|\tilde{f}_1\right|^2 \right\rangle \tag{3.4.37}$$

and the term $\sigma_{\rm coll}$ corresponds to the collective effects due to the correlations of particle positions

$$\sigma_{\rm coll} = \left\langle \sum_{l \neq m} \tilde{f}_l \tilde{f}_m^* \right\rangle = \left(N^2 - N\right) {\rm Re} \left\langle \tilde{f}_1 \tilde{f}_2^* \right\rangle. \tag{3.4.38}$$

Thus, the cross section of scattering at the system of particles is represented in the form

$$\begin{aligned} \sigma_\omega(\mathbf{n} \leftarrow \mathbf{n}_0) &= \sigma_{\rm coh} + \sigma_{\rm indep} + \sigma_{\rm coll} \\ &= N^2 |\langle f_1 \rangle|^2 + N \langle |\tilde{f}_1|^2 \rangle + (N^2 - N) {\rm Re} \langle \tilde{f}_1 \tilde{f}_2^* \rangle . \end{aligned} \tag{3.4.39}$$

In the general case, the separation of the incoherent scattering cross section (3.4.36) into the independent and collective components is conditional because, owing to multiple scattering, each summand in (3.4.39) depends on the positions of all scattering particles. This separation becomes physically meaningful only in the single scattering approximation which corresponds to the case of strongly rarefied medium. In agreement with the classification of Section 3.4.1, such a medium may be both strongly discrete, when the interparticle distance $l_{\rm av}$ if large compared to the wavelength, $l_{\rm av} >> \lambda$, and quasi- continuous, when $l_{\rm av} << \lambda$ (the interval in between is occupied by the region of interparticle resonances, $l_{\rm av} \sim \lambda$).

The single scattering approximation retains in (3.4.33) only the first term so that

$$T^{(N,l)} \approx t^{(l)}. \tag{3.4.40}$$

In this approximation, each particle scatters independently of the others. This model corresponds to a total neglect of the mutual irradiation of particles. Nonetheless, the mutual influence of the fields scattered by different particles is retained because of the interference terms σ_{coll} (3.4.38) caused by the correlation of scatterer positions is nonzero.

In the single scattering approximation (3.4.40), the single particle amplitude $t^{(l)}$ is expressed in terms of f_l by the relation (3.3.62), whence it follows

$$f_l(\mathbf{n} \leftarrow \mathbf{n}_0) = f_0(\mathbf{n} \leftarrow \mathbf{n}_0)\exp(-i\mathbf{q}\mathbf{r}_l). \tag{3.4.41}$$

Here $\mathbf{q} = k(\mathbf{n} - \mathbf{n}_0)$ is the scattering vector, and f_0 is the amplitude of scattering at a particle placed at the origin (Problem 3.39). Substituting (3.4.41) in (3.4.39) yields

$$\sigma_\omega(\mathbf{n} \leftarrow \mathbf{n}_0) = \sigma_{0\omega}(\mathbf{n} \leftarrow \mathbf{n}_0)\{\ldots\}, \tag{3.4.42}$$

where

$$\begin{aligned}\{\ldots\} &= N^2|\langle\xi_1\rangle|^2 + N\langle|\tilde{\xi}_1|\rangle^2 + (N^2 - N)\text{Re}\langle\tilde{\xi}_1\tilde{\xi}_2^*\rangle \\ &= N^2|\langle\xi_1\rangle|^2 + N\left(1 - |\langle\xi_1\rangle|^2\right) + (N^2 - N)\left(\text{Re}\langle\tilde{\xi}_1\tilde{\xi}_2^*\rangle - |\langle\xi_1\rangle|^2\right).\end{aligned} \tag{3.4.43}$$

Here $\sigma_{0\omega} = |f_0|^2$ is the scattering cross section for a particle at the origin, $\xi_l = \exp(i\mathbf{q}\mathbf{r}_l)$, and $\tilde{\xi}_l = \xi_l - \langle\xi_l\rangle$ is the fluctuation of ξ_l .

For the forward scattering, $\mathbf{q} = 0$, $\xi_l = 1$, and $\tilde{\xi}_l = 0$, so that in eqn (3.4.43), the second and third summands, describing the incoherent scattering, vanish. We see that, in the single scattering approximation, $\sigma_\omega(\mathbf{n}_0 \leftarrow \mathbf{n}_0) = \sigma_{\text{coh}} = N^2\sigma_{0\omega}(\mathbf{n}_0 \leftarrow \mathbf{n}_0)$.

The physical meaning of all the terms in (3.4.43) is rather transparent. The first term in (3.4.43) describes the coherent scattering and corresponds to the diffraction of the incident wave at the whole scattering volume. The directional pattern for this summand is concentrated near the forward scattering direction, exactly in the region where the incoherent summand σ_{incoh} vanishes. The first term has an angular width about λ / L_{sc}, where L_{sc} is the characteristic dimension of the scattering volume.

The second term corresponds to the contribution of independent particles and is almost isotropic - the anisotropy of scattering at a single particle is taken into account in (3.4.42) by the factor $\sigma_{0\omega}(\mathbf{n} \leftarrow \mathbf{n}_0)$. The third, "collective" term describes the interference effects due to the correlation of particle positions and is other than zero in the range of angles around λ / l_{cor}, where l_{cor} is the correlation radius of particles. For the forward scattering direction, the second and third summands vanish in the region with the angular width around λ / L_{sc} (see also Problem 3.40).

We note that in contrast to separation of the field into the coherent and incoherent components, the representation of the incoherent scattering cross section in the form (3.4.36) plays a secondary role, because the quantity measured here is the whole sum (3.4.36) rather than each term individually. Thus, the collective term σ_{coll} may be less than zero, whereas the cross section σ_{incoh} is always nonnegative.

3.4.7 Dynamic Group Expansions and Mutual Influence Operators

When particles get closer to one another, the single scattering approximation (3.4.40) becomes inadequate and one has to take into account the mutual radiation of scatterers. In going beyond the single scattering formalism, it is natural, as a first step, to attempt to take into account the effects associated with the scattering at groups of 2, 3, etc. particles. This has been done by Finkelberg (1967) who have constructed the dynamic group expansions. The term *dynamic* is used here to emphasize that these expansions are not related to statistical averaging.

To construct this expansion we introduce the T-operator of mutual influence $T^{(\text{gr}12\ldots\text{N})}$ by the relations

$$T^{(1)} = T^{(\text{gr}1)},$$

$$T^{(2)} = T^{(\text{gr}1)} + T^{(\text{gr}2)} + T^{(\text{gr}12)},$$

$$\ldots$$

$$T^{(12\ldots N)} = \sum_{i} T^{(\text{gr}1)} + \sum_{(ij)} T^{(\text{gr}ij)} + \sum_{(ijk)} T^{(\text{gr}ijk)} + \ldots + T^{(\text{gr}12\ldots\text{N})}, \tag{3.4.44}$$

where the symbol (ijk...l) should be understood as $1 \le i < j < k \ldots < l \le N$.

Solving this system sequentially for $T^{(\text{gr}1)}$, $T^{(\text{gr}12)}$, ..., $T^{(\text{gr}12\ldots N)}$, it is not hard to obtain the explicit expressions for the T-operators of mutual influence:

$$T^{(\text{gr}12\ldots N)} = T^{(12\ldots N)} - \sum_{(i_1 i_2 \ldots i_{N-1})} T^{(i_1 i_2 \ldots i_{N-1})} + \sum_{(i_1 i_2 \ldots i_{N-2})} T^{(i_1 i_2 \ldots i_{N-2})} + \ldots + (-)^N \sum T^{(i)}. \tag{3.4.45}$$

Thus, to find $T^{(\text{gr}12\ldots N)}$ we need to know the solutions for the scattering at one, two, three, etc. up to N particles.

From the construction of eqn (3.4.44) it is not hard to see that the T-operator of mutual influence $T^{(\text{gr}12\ldots N)}$ must tend to zero, if the system of scatterers 1,2,..., N may be divided into two sufficiently far subsystems whose mutual irradiation can be neglected. Consequently, the operator $T^{(\text{gr}12\ldots N)}$ takes into account the mutual influence in the group of N scatterers. One may similarly construct dynamic group expansions for other operators similar with T, that depend upon the coordinates of scattering particles.

The expansion (3.4.44) may be viewed as a natural generalization of the single scattering approximation (3.4.40) in which we sequentially took into account the effects of mutual irradiation in groups of 2, 3, etc. particles. For instance, the simplest approximation allowing for the mutual irradiation for pairs, rather than say for triples, of particles results if we discard in (3.4.45) $T^{(\mathrm{gr}12\ldots s)}$ for $s \geq 3$:

$$T^{(12\ldots N)} \approx \sum T^{(i)} + \sum_{(ij)} T^{(\mathrm{gr}ij)} = \sum_{(ij)} T^{(ij)} - (N-2)\sum_{i} T^{(i)}. \tag{3.4.46}$$

The use of group expansions (3.4.44) in the general scattering theory seems very attractive, since, for a system of a large number of particles, this approach allows the solutions with a far smaller number of particles. It is true, however, that the evaluation of the two-particle operator $T^{(12)}$ for the scattering at non-point particles is a hard task that has ben solved only for some particular cases. This aspect severely limits the practical value of the group expansions (3.4.46).

3.4.8 Effective Inhomogeneity Operators for Sparse Discrete Media

In sparse discrete media, the effect of mutual irradiation is sufficiently small and the operators of effective inhomogeneities V^{eff} and K_{12} may be obtained with different forms of perturbation theory. As a zero approximation one usually takes the independent particle approximation adjacent to the single scattering approximation, and the effects of correlation and mutual irradiation are introduced as small corrections. These expansions can be obtained with the aid of the diagram technique adapted to the case of discrete media (Frisch, 1968). This technique provides some geometric insight and gives a better understanding of the physics of multiple scattering processes. However, if from the very beginning, one aims at obtaining the equation of moments in the Dyson form (3.3.25) or in the Bethe–Salpeter form (3.3.28), then the formal algebraic approach may be used that enables one to readily write down the desired expansions for V^{eff} and K_{12} (Problem 3.41).

Real scattering media usually contain a large number of scattering particles: $N >> 1$ This enables some simplification of the problem by passing to the *N/V* limit: $N \to \infty$, $V_{\mathrm{sc}} \to \infty$, $v_0 = N / V_{\mathrm{sc}} = \mathrm{const}$, where V_{sc} is the scattering volume, and v_0 is the number of particles per unit volume. In this limit, the first terms of expansion of the effective inhomogeneity operators in powers of $t^{(l)}$ have the form

$$V^{\mathrm{eff}} = N\left\langle t^{(1)}\right\rangle + N^2\left\langle \tilde{t}^{(1)} G_0 \tilde{t}^{(2)}\right\rangle + \ldots, \tag{3.4.47}$$

$$K_{12} = N\left\langle t_1^{(1)} t_2^{(1)*}\right\rangle + N^2\left\langle \tilde{t}_1^{(1)} \tilde{t}_2^{(2)*}\right\rangle + \ldots, \tag{3.4.48}$$

where, as before, $\tilde{t} = t - \langle t \rangle$, the superscript of $t_j^{(l)}$ stands for the random coordinate of particle *l*, and the subscript denotes the argument in which $t_j^{(l)}$ acts.

If we express in the expansions (3.4.47) and (3.4.48) the moments of operators $t_j^{(l)}$ in terms of the respective cumulants, then, as in the case of continuous media, one can separate the summands constituting V^{eff} and K_{12} into *one-group*, i.e. containing one correlation function each, and *multiple group*, where the number of correlation functions is more than unity (Finkelberg, 1967). In contrast to the case of a continuous medium when these expansions contain the cumulant functions of the scattering potential, now they will contain the cumulant functions of scatterers positions. The one-group Finkelberg approximation retains in V^{eff} and K_{12} only terms containing one correlation function each. The properties of this approximation are analogous to that of the approximation of the same name for continuous media, considered in Section 3.3.9. Specifically, one-group approximation $V^{(1\text{gr})}$ and $K^{(1\text{gr})}$ turn out to be approximately consistent in energy, in the sense that they obey the approximate optical theorem (3.3.74).

In the general case, $V^{(1\text{gr})}$ and $K^{(1\text{gr})}$ have the form of infinite series. However, in the case of uncorrelated scatterers, these expressions drastically simplify and have the form

$$V^{(1\text{gr})} = N\left\langle t^{(1)} \right\rangle, \tag{3.4.49}$$

and

$$K^{(1\text{gr})} = N^2\left\langle t_1^{(1)} t_2^{(1)*} \right\rangle, \tag{3.4.50}$$

which means that only the leading terms will be retained in (3.4.47) and (3.4.48). The approximation(3.4.49) is known as the approximation due to Foldy (1945).

A more exact model of correlated scatterers takes into assount weak mutual irradiation. This approximation keeps in $V^{(1\text{gr})}$ and $K^{(1\text{gr})}$) only terms containing two-particle correlations (Barabanenkov, 1975):

$$V^{(1\text{gr})} \approx N\left\langle t^{(1)} \right\rangle + N^2\left\langle \tilde{t}^{(1)} G_0 \tilde{t}^{(2)} \right\rangle, \tag{3.4.51}$$

$$K^{(1\text{gr})} \approx N\left\langle t_1^{(1)} t_2^{(1)*} \right\rangle + N^2\left\langle \tilde{t}_1^{(1)} \tilde{t}_2^{(2)*} + t_1^{(1)} G_{01} t_1^{(1)} \tilde{t}_2^{(2)*} + \tilde{t}_1^{(1)} t_2^{(2)*} G_{02}^{*} t_2^{(2)*} \right\rangle. \tag{3.4.52}$$

where $K^{(1\text{gr})}$ retains only terms of the third order of smallness so that $V^{(1\text{gr})}$ and $K^{(1\text{gr})}$ exactly satisfy the optical theory (3.3.74) with the propagation operator G_0. The first two terms on the right-hand side of (3.4.52) similar to the expressions (3.4.37) and (3.4.38) for cross sections σ_{indep} and σ_{coll} derived in the single scattering approximation. Below we demonstrate that, keeping these terms in the transfer equation yields exactly the cross section in the single scattering approximation. Subsequent terms in eqn (3.4.52) give a more accurate scattering cross section.

3.4.9 One-Group Approximation and Weak Dispersion Formula for Correlated Particles

The expression (3.4.51) for $V^{(1\text{gr})}$ enables us to generalize the weak dispersion formula (2.2.59) on the case of correlated particles. Indeed, in the effective wave number approximation, the base model (3.3.3) of the scalar field yields the mean-field dispersion equation (3.3.91) which in this case takes the form

$$\begin{aligned} k^{\text{eff}} &\approx k_0\left(1-\frac{V^{(1\text{gr})}(k_0^2)}{2k_0^2}\right) \\ &= k_0\left(1+\frac{2\pi v_0}{k_0^2}f_\omega(\mathbf{n}_0 \leftarrow \mathbf{n}_0)-\frac{v_0^2}{16\pi k_0^2}\int t_{\mathbf{k}\mathbf{k}'}G^0_{\mathbf{k}'}t_{\mathbf{k}'\mathbf{k}_0}\tilde{q}(\mathbf{k}_0-\mathbf{k}')d^3k'\right). \end{aligned} \tag{3.4.53}$$

Here $G_k^0 = (k^2 - k_0^2 + i0)^{-1}$, $t_{\mathbf{k}\mathbf{k}_0}$ is the kernel (written in the K-representation) of the T-operator of scattering at a single particle placed at the origin (Problem 3.39), viz.,

$$\begin{aligned} t_{\mathbf{k}\mathbf{k}_0} &= \int d^3r e^{-i\mathbf{k}\mathbf{r}} t^{(1)} e^{i\mathbf{k}_0\mathbf{r}} \\ &= \int d^3r d^3r_0 t^{(1)}(\mathbf{r},\mathbf{r}_0)\exp(-i\mathbf{k}\mathbf{r}+i\mathbf{k}_0\mathbf{r}_0), \end{aligned} \tag{3.4.54}$$

and

$$\tilde{q}(\mathbf{k}) = V_{\text{sc}}^2\int q_{12}(\rho)e^{-i\mathbf{k}\rho}d^3\rho \tag{3.4.55}$$

is the Fourier transform of the two-particle correlation function $V_{\text{sc}}^2 q_{12}$ with $q_{12} = P_{12} - P_1P_2 \approx q_{12}(\mathbf{r}_1 - \mathbf{r}_2)$. In deriving eqn (3.4.53) we assumed that the particles are uniformly distributed over a sufficiently large volume V_{sc} containing a large number of particles $N >> 1$.

The second summand on the right-hand side of (3.4.53) corresponds to the formula (2.2.60) and the third is a correction to the effective wave number associated with the correlations of scatterers positions. Although in eqn (3.4.53) this correction is assumed to be small, its inclusion allows us to describe qualitatively new effects, for example, anisotropic properties of the medium due to the anisotropy of the scatterer-position correlation function [in the last case, in place of (3.3.91) one should better consider the dispersion equation (3.3.87)].

3.4.10 Operators of Effective Inhomogeneities for Media with Strong Fluctuations

The approximations considered in the preceding sections refer to the case of sparse, or strongly rarefied, media (regions *1* and *2* according to the classification of Fig. 3.13). If the medium is not sparse, then both the effective parameters and the explicit form of the effective inhomogeneity operators especially strongly depend on the scattering medium structure.

In Section 3.4.2 we described two main approximations for the effective dielectric permittivity in the quasi-static mode, these were the Lorenz–Lorentz type formulas and Bruggeman's formula related to media with different topologies and symmetries. We demonstrate how these approximation can be obtained by regular procedures. Formally, these procedures are exact, but, as in the case of sparse media above, they lead to infinite sums. The latter cannot be summed, but can be used to compute small corrections. An estimate of these corrections may be used as a yardstick of the applicability of the main approximations. In this section we will develop the approach of Ryzhov, Tamoikin and Tatarskii (1965) by including some new unpublished results due to Apresyan.

Consider eqn (3.3.4) for the EM field vector in a medium with fluctuating permittivity $\varepsilon(\mathbf{r})$ which will be written in the form

$$\left[-\nabla\times\nabla\times+k_0^2\varepsilon_0+k_0^2\left(\varepsilon(\mathbf{r})-\varepsilon_0\right)\right]\mathbf{E}=\mathbf{q}, \tag{3.4.56}$$

where ε_0 is an auxiliary parameter having the meaning of a "seed" permittivity value to be defined below. As an unperturbed operator describing some reference medium we choose $L_0=-\nabla\times\nabla\times+k_0^2\varepsilon_0$, so that the perturbation is $V=-k_0^2\left(\varepsilon(r)-\varepsilon_0\right)$.

From eqn (3.4.56) it is not hard to change in a usual manner to the integral operator equation

$$G=G_0+G_0VG. \tag{3.4.57}$$

In contrast to the scalar wave field, the Green's function G_0 of the electromagnetic problem has a strong singularity. This singularity can be regularized by introducing the respective limit transition operation. The procedure of such a regularization is not one-one and depends on the form of the infinitesimally small "regularizing volume" (Problem 3.24). We confine our consideration to the case of a statistically isotropic medium. Then, it is natural to take the a regularizing volume in the form of an infinitesimally small sphere to write the action of G_0 as

$$G_0=\mathcal{P}G_0+\frac{\hat{1}}{3k_0^2\varepsilon_0}\equiv G_0'+\frac{\hat{1}}{3k_0^2\varepsilon_0}, \tag{3.4.58}$$

where $\mathcal{P}$ stands for the integration in the sense of the principal value, $G_0'=\mathcal{P}G_0$ is the regularized operator G_0, and $\hat{1}$ is the unit operator (Problem 3.24).

Substituting eqn (3.4.58) in (3.4.57) we change from eqn (3.4.57) to the equation

$$F=G_0+G_0'\kappa F \tag{3.4.59}$$

for the operator

$$F\equiv\left(1-V/3k_0^2\varepsilon_0\right)G. \tag{3.4.60}$$

Here

$$\kappa=V\frac{1}{1-V/3k_0^2\varepsilon_0}=-3k_0^2\varepsilon_0\frac{\varepsilon(r)-\varepsilon_0}{\varepsilon(r)+2\varepsilon_0}. \tag{3.4.61}$$

We call this procedure the $\varepsilon\rightarrow\kappa$ *transition.*

In place of the singular operator G_0V appearing in eqn (3.4.57), eqn (3.4.59) contains the regularized operator $G_0'\kappa$. Accordingly, we may expect that eqn (3.4.60) will give better results than the original equation (3.4.57). Starting with eqn (3.4.59) it is straightforward to write the respective Dyson equation for $\langle F\rangle$

$$\langle F\rangle = G_0 + G_0'\kappa^{\text{eff}}\langle F\rangle, \tag{3.4.62}$$

where the operator κ^{eff} is given by a condition similar to eqn (3.3.33a):

$$\langle \kappa F\rangle = \kappa^{\text{eff}}\langle F\rangle. \tag{3.4.63}$$

The Dyson equation (3.4.62) for $\langle F\rangle$ turns out to be directly connected with the ordinary Dyson equation (3.3.25) for $\langle G\rangle$. Indeed, averaging the obvious relation $\kappa F = VG$ and using eqn (3.4.63) and (3.3.33b) we have

$$\kappa^{\text{eff}}\langle F\rangle = V^{\text{eff}}\langle G\rangle. \tag{3.4.64}$$

Similarly, averaging of eqn (3.4.60) yields

$$\langle F\rangle = \left(1 - \frac{V^{\text{eff}}}{3k_0^2\varepsilon_0}\right)\langle G\rangle. \tag{3.4.65}$$

Whence it follows that the effective inhomogeneity operators V^{eff} and κ^{eff} are connected by the same relation as V and κ, viz.,

$$\begin{aligned} \kappa^{\text{eff}} &= V^{\text{eff}}\left(1 - \frac{V^{\text{eff}}}{3k_0^2\varepsilon_0}\right)^{-1}, \\ V^{\text{eff}} &= \left(1 + \frac{\kappa^{\text{eff}}}{3k_0^2\varepsilon_0}\right)^{-1}\kappa^{\text{eff}}. \end{aligned} \tag{3.4.66}$$

Once some approximate expression for κ^{eff} has been obtained, it is not hard to obtain the respective approximate expression for V^{eff}. The expression for κ^{eff} may be obtained with the ordinary perturbation theory in the denominator (from Section 3.3.3) considering now eqn (3.4.59) for F as the initial expression and treating the operator $G_0'\kappa$ as a perturbation.

Owing to the presence of the free parameter ε_0, this scheme proves to be sufficiently flexible and allows generalizations of the Lorenz–Lorentz type formulas and Bruggeman approximation. We demonstrate this fact by considering first the case with macroscopic fluctuating permittivity $\varepsilon(\mathbf{r})$.

3.4.11 Effective Parameters — Non-selfconsistent Approximation

As already noted the choice of an effective parameter calculation scheme depends on medium's topology. Considering the case of cermets it is natural to take as the permittivity of a reference medium ε_0 the permittivity of the matrix ε_m.

Letting $\varepsilon_0 = \varepsilon_m$ we put down κ^{eff} in the Bourret approximation (3.3.75)

$$\kappa^{\text{eff}} = \langle \kappa \rangle + \langle \tilde{\kappa} G_0' \tilde{\kappa} \rangle. \tag{3.4.67}$$

Here, the first term $\langle \kappa \rangle$ is *local*, i.e., proportional to the unit operator, and the second is related to the correlations of fluctuations and is described by an integral operator allowing for the spatial dispersion about the mean field. If we neglect these correlations, i.e., discard the second term on the right-hand side of (3.4.67),and express κ and κ^{eff} in terms of ε and ε^{eff}, we arrive at the formula

$$\frac{\varepsilon^{\text{eff}} - \varepsilon_0}{\varepsilon^{\text{eff}} + 2\varepsilon_0} = \left\langle \frac{\varepsilon(r) - \varepsilon_0}{\varepsilon(r) + 2\varepsilon_0} \right\rangle, \tag{3.4.68}$$

coinciding with the generalized Maxwell Garnett formula (3.4.6) if we assume that the statistical averaging is equivalent to the volume averaging. Thus, eqn (3.4.68) may be viewed as an extension of the formula (3.4.6) onto the case when correlations are taken into account.

3.4.12 Effective Parameters — Selfconsistent Approximation

In the case of a symmetric medium with aggregate topology (Section 3.4.1) no one medium's component can be preferred to the others and taken as a reference. In order to choose ε_0 it is natural then to impose an additional condition

$$\langle \kappa \rangle \equiv -\left\langle 3k_0^2 \varepsilon_0 \frac{\varepsilon(r) - \varepsilon_0}{\varepsilon(r) + 2\varepsilon_0} \right\rangle = 0. \tag{3.4.69}$$

This equation for ε_0 as can be easily seen, coincides with the Bruggeman equation (3.4.9) for the effective permittivity ε^{eff}.

In this case, in the current approximation (3.4.67) for κ^{eff}, the first term vanishes, and we have only the second term related to the correlations of fluctuations. We now express κ^{eff} in terms of ε^{eff}, using eqn (3.4.66) and the formula $V^{\text{eff}} = -k_0^2(\varepsilon^{\text{eff}} - \varepsilon_0)$, and apply the Bourret approximation (3.4.67) for κ^{eff} to get

$$\varepsilon^{\text{eff}} = \varepsilon_0 \frac{1 - \dfrac{2}{3} \dfrac{\langle \tilde{\kappa} G_0' \tilde{\kappa} \rangle}{k_0^2 \varepsilon_0}}{1 + \dfrac{1}{3} \dfrac{\langle \tilde{\kappa} G_0' \tilde{\kappa} \rangle}{k_0^2 \varepsilon_0}}. \tag{3.4.70}$$

This formula refines the Bruggeman approximation allowing for the paired correlations of medium fluctuations. Higher correlations can be incorporated by extending the expansion of κ^{eff} (3.4.67).

3.4.13 Generalized Lorenz–Lorentz Formula

We now turn to the case of a medium constituted by discrete scatterers distributed in a free space. We assume that the properties of these scatterers are well known, i.e., that the T-operator of scattering at a single particle $t^{(j)}$ is specified. Here, it is again natural to let ε_0 equal to the medium's permittivity; in this case, $\varepsilon_0 = 1$.

We take eqn (3.4.59) as an initial equation for F and repeat the reasoning of Section 3.4.10 to express κ^{eff} in terms of discrete particle characteristics (Problem 3.42). As a result, for $N >> 1$, we arrive at the expression (Apresyan, 1992)

$$\frac{\varepsilon^{\text{eff}} - \varepsilon_m}{\varepsilon^{\text{eff}} + 2\varepsilon_m} = -\frac{N}{3k_0^2\varepsilon_m}\langle t^{(1)}\rangle - \frac{N^2}{3k_0^2\varepsilon_m}\langle \tilde{t}^{(1)} G_0' \tilde{t}^{(2)}\rangle. \tag{3.4.71}$$

This expression generalizes the classic Lorenz–Lorentz formula in three aspects. First, the role of the point dipole dipole moment α is played here by the mean value of the T-operator of scattering at a non-point particle $t^{(1)}$. Second, an additional two-particle term appears that describes the correlations of particles positions. Finally, eqn (3.4.71) does not require passing to the quasi-static limit, so that, in the general case, ε^{eff} is not a number but rather an operator taking into account the spatial dispersion of the medium about the mean field.

Assuming that the scattering medium is statistically uniform and passing to the K-space we rewrite eqn (3.4.71) in a form similar to eqn (3.4.53), viz.,

$$\frac{\varepsilon_{\mathbf{k}}^{\text{eff}} - \varepsilon_m}{\varepsilon_{\mathbf{k}}^{\text{eff}} + 2\varepsilon_m} = -\frac{v_0}{3k_0^2\varepsilon_m} \, t_{kk} - \frac{v_0^2}{3k_0^2\varepsilon_m}\int t_{\mathbf{kk}'} G_{\mathbf{k}'}^{0\prime} t_{\mathbf{k}'\mathbf{k}} \tilde{q}(\mathbf{k}-\mathbf{k}') \frac{d^3k'}{(2\pi)^3}. \tag{3.4.72}$$

Here, $t_{kk'}$ and $\tilde{q}(\mathbf{k})$ are defined by the relations (3.4.54) and (3.4.55), whereas the quantity

$$\varepsilon_{\mathbf{k}}^{\text{eff}} = e^{-i\mathbf{kr}}\varepsilon^{\text{eff}} e^{i\mathbf{kr}} \tag{3.4.73}$$

is, in the general case, a tensor depending on the wave vector $\mathbf{k}$.

In the particular case of point dipoles with polarizability α, for traveling waves with $k \approx k_0$ and $t_{k_0k_0} = -4\pi k_0^2\alpha$, the formula (3.4.72) becomes

$$\frac{\varepsilon_{\mathbf{k}}^{\text{eff}} - \varepsilon_m}{\varepsilon_{\mathbf{k}}^{\text{eff}} + 2\varepsilon_m} = \frac{4\pi v_0}{3\varepsilon_m}\alpha_l - \frac{2v_0^2\alpha_l^2}{3\pi\varepsilon_m}\int G_{\mathbf{k}'}^{0\prime}\tilde{q}(\mathbf{k}-\mathbf{k}')d^3k'. \tag{3.4.74}$$

It is not hard to see that eqn (3.4.72) may be viewed as an extension of the weak dispersion formula (2.2.59) to the case of dense media. The role of the scattering amplitude, entering eqn (2.2.59) and describing the radiation due to distant particles, is played by the T-operator of scattering corresponding to arbitrary distances to particles. Thus, eqn (3.4.72) bridges the gap between the weak dispersion formula (2.2.59) that describes small deviations of the effective medium's permittivity from the unperturbed value $\varepsilon = \varepsilon_m$, and the classic Lorenz–Lorentz formula (3.4.4) that remains valid also for densely packed media. It should be noted that, in the quasi-static limit as $k \to 0$, the integral summand in (3.4.74) vanishes, and an incorporation of subsequent terms in powers of k leads to a Kirkwood type expansion.

3.4.14 Applicability of Expansions of Effective Inhomogeneity Operators to Discrete Scatterers

For discrete scattering media, the applicability conditions of the expansions of effective inhomogeneity operators turn out to be a more involved problem than in the case of a continuous scattering medium. Indeed, changing to V^{eff} and K_{12} implies a transition from a discrete dynamic description to a continuous statistical description in the N/V limit. Whereas in a continuous medium, the disappearance of all correlation of the scattering potential implies a crossover to a nonfluctuating medium, in the discrete case, even for independent scatterers with uncorrelated positions, the scattering medium remain random all the same.

The applicability conditions of the considered approximations are yet to be developed. Such conditions may be expected to follow from the estimates of the discarded terms that have been derived with the theory of perturbations in the denominator. Since no rigorous conditions for the N/V limit validity and the group expansions of V^{eff} and K_{12} in the scattering theory are available (see Ruelle, 1969) we confine ourselves here to giving simple qualitative estimates of the conditions weakness of mutual radiation in the model of dipole scatterers with polarizabilities α_l. If the distribution of particles over a volume V_{sc} is uniform, then the average interparticle distance is

$$l_{\text{av}} \sim v_0^{-1/3},$$

where $v_0 = N/V_{sc}$. Recalling that the near field of an electromagnetic dipole varies as $u_l = \alpha_l u^0 / r^3$, for $r \sim l_{\text{av}} \sim v_0^{-1/3}$, we obtain the estimate

$$u_1 \sim \alpha_l v_0 u^0.$$

For a plane incident wave $u^0 = \exp(i\mathbf{kr})$, the condition of smallness of the effects of mutual radiation of scatterers $|u_l| << |u^0|$ will be

$$\alpha v_0 << 1. \tag{3.4.75}$$

This inequality serves as a necessary applicability criterion of the expansions (3.4.47) and (3.4.48). It is not sufficient because it fails to take into account the correlations between scatterers. In the case of densely packed media, the inequality (3.4.75) breaks down and the effects of mutual radiation have to be taken into account. At simplest idealizations, this account may lead to the Lorenz–Lorentz formula for the effective dielectric permittivity.

3.5 DERIVATION OF THE RADIATIVE TRANSFER EQUATION FOR A SCATTERING MEDIUM

3.5.1 Historical Background

Wave derivations of the radiative transfer equation have been made under a variety of initial assumptions by a great number of authors including Bugnolo (1960); Watson (1960,1969); Peacher and Watson (1970); Law and Watson (1970); Gnedin and Dolginov (1963); Dolginov et al. (1970); Barabanenkov (1967, 1969, 1975); Barabanenkov and Finkelberg (1967, 1968); Sazonov and Tsytovich (1968); Zhekeznyakov (1968); Stott (1968); Galinas and Ott (1970); Apresyan (1973,1981); Howe (1973); Ovchinnikov (1973); Erukhomov and Kirsh (1973); Furutsu (1975); Acquista and Anderson (1977); and Diener (1982) whose results are partially presented in the monographs by Ishimaru (1978) and Rytov, Kravtsov, and Tatarskii (1989b). We will follow the work of Apresyan (1973) who proposed a consistent procedure of asymptotic expansions that explicitly uses the condition of quasi-homogeneity.

3.5.2 Starting Equations

Consider the coherence function of a field in a scattering medium. This function obeys the Bethe–Salpeter equation (3.3.27) that generally takes into account both diffraction and scattering of waves. The strict general solution to this equation is unavailable at present. Moreover, it is unavailable even for a single scatterer when the diffraction is free from multiple scattering effects. Meanwhile, under the condition of "weak disorder", when the extinction length in the medium L_{ext} is much greater than the characteristic radiation wavelength λ, $L_{\text{ext}} >> \lambda$, a considerably simpler approximate which, in essence, contains for the quasi-uniform part of the radiation spectrum follows from the Bethe–Salpeter equation. This equation is equivalent to the equation of radiative transfer.

We start with the linear general form of wave equation (3.3.1) $Lu = q$. To make general relations clear, we use the scalar wave equation

$$Lu \equiv (\Delta - \partial_{ct}^2 \varepsilon)u = (4\pi / c^2)\partial_t j \equiv q. \tag{3.5.1}$$

The mean value of mediums permittivity $\varepsilon = \varepsilon_0 + \tilde{\varepsilon}$ is assumed to be other from unity, i.e. $\langle \varepsilon \rangle = \varepsilon_0 \neq 1$. The field u and the "current" j may be approximately (in a homogeneous medium exactly) treated as components of the electric field strength $\boldsymbol{E}$ and current $\boldsymbol{j}$, say $u = E_x$ and $j = j_x$. We assume that the radiation sources q (current j) fluctuate remaining statistically independent of medium fluctuations. Simultaneously, we abstract ourselves from boundary effects assuming the mediums to be homogeneous and isotropic. All these assumptions are not necessary so that the majority of our results hold also for more complicated cases including the electromagnetic problem.

First, let the mean values for both the source function q and the field u be equal to zero, $\langle q \rangle = 0$, $\langle u \rangle = 0$. Then, the coherence function Γ_{12} coincides with the correlation function: $\Gamma_{12} = \langle u_1 u_2^* \rangle = \langle \tilde{u}_1 \tilde{u}_2^* \rangle = \Psi_{12}$, in which case, after averaging over fluctuations of sources q the Bethe–Salpeter equation (3.3.27) may be written in the form

$$D_1 D_2^* \Psi_{12} = K_{12}\Psi_{12} + \langle q_1 q_2^* \rangle, \tag{3.5.2}$$

where

$$D = L_0 - V^{\text{eff}}. \tag{3.5.3}$$

We assume $L_0 = \langle L \rangle - \Delta - \partial_{ct}^2 \varepsilon_0$, $V = \partial_{ct}^2 \tilde{\varepsilon}$ for the wave equation (3.5.1). Operators of effective inhomogeneities are taken in the one-group approximation, i.e., $V^{\text{eff}} \approx V^{(1\text{gr})}$; $K_{12} \approx K_{12}^{(1\text{gr})}$. The applicability conditions for these approximations are assumed to be satisfied. For the sake of simplicity, we focus on the case of free radiation setting the sources q equal to zero in (3.5.1) and (3.5.2). The dependence on q will be restored in final results.

For further analysis, it is important that, in the single-group approximation, V^{eff} and K_{12} be weakly nonlocal (or compact; Barabanenkov and Finkelberg, 1967) operators with nonlocality radii on the order of medium correlation radii. This implies that due to the cumulant functions in expansions (3.3.72) and (3.3.73), the kernels $V^{(1\text{gr})}$ and $K_{12}^{(1\text{gr})}$ tend to zero when separations between their arguments exceed the correlation radii of medium's fluctuations, the rate of their decay being determined by the rate of correlation decay. Therefore, the quantity $V^{(1\text{gr})}$ depends upon values of $u(x)$ in a neighborhood of point x whose size is about the correlation radius l_c and the correlation time τ_c, and $K_{12}^{(1\text{gr})}\Psi_{12}$ depends upon values of $\Psi_{12} = \langle \tilde{u}(x_1)\tilde{u}^*(x_2)\rangle$ in a similar neighborhood of points x_1 and x_2 and, falls off rapidly when the separation between x_1 and x_2 exceeds the correlation radii. It is these features of the one-group approximation that enable us to eventually introduce the scattering cross section of a unit volume as a *local* property of a scattering medium.

3.5.3 Necessary Conditions for the Quasi-Uniformity of the Wave Field in a Scattering Medium

In view of the compactness of $K_1^{(1\text{gr})}$, for small separations, $|\rho| = |\mathbf{r}_1 - \mathbf{r}_2| \leq l_c$, $|\tau| = |t_1 - t_2| \leq \tau_c$, it is sufficient to take the second moment Ψ_{12} in describing the scattering events in the one-group approximation. Let fluctuations of the scattered field be quasi-uniform for such $|\rho|$ and τ, so that inequalities of the form (3.1) be satisfied:

$$|\partial_R \Psi_{12}| << |\partial_\rho \Psi_{12}|. \tag{3.5.4}$$

This implies that, for not too large $\rho = (\rho, \tau)$, the function Ψ_{12} varies rapidly along the difference coordinate ρ as compared to its slow variations along the center of gravity coordinate $R = (x_1 + x_2)/2$.

In Section 3.2 we have shown that, in a homogeneous medium, the conditions of the form (3.5.4) reduce to inequalities (3.2.8c) expressing the quasi-uniformity of the field in the scale of the characteristic wavelength $\bar{\lambda}$ or its quasi-stationarity in the scale of the characteristic radiation period $\bar{\tau}$. To give specific expression for condition (3.5.4), consider first a monochromatic radiation [we omit factors $\exp(-i\omega t)$] in a static (i.e., time-independent) medium.

In a scattering medium, the propagation of the second moment Ψ_{12} in the interval between scattering events may be deemed to occur in an *effective medium* which is described by the Dyson operator. In the one-group approximation, the transition to the effective medium reduces mainly to replacing the free-space wave number k_0 by the effective wave number k^{eff}. Taking this into account, we assume for estimations the random wave field to have the form of the quasi-plane wave

$$u = u(\mathbf{r}) = A(\mathbf{r})e^{i\mathbf{k}^{\text{eff}}\mathbf{r}}, \tag{3.5.5}$$

where $\mathbf{k}^{\text{eff}} = k^{\text{eff}}\mathbf{n}$, $\mathbf{n}$ is the unit vector the a wave propagation direction, and $A(\mathbf{r})$ is the random amplitude ($\langle A\rangle = 0$) varying slightly over a wavelength; the characteristic scale of its variations $L_A \sim |\nabla \ln A|^{-1} << |k^{\text{eff}}|^{-1}$. For small $|\rho| << L_A$, we have

$$\begin{aligned}\Psi_{12}(\mathbf{R},\rho) &= \langle \tilde{u}(\mathbf{R}+\rho/2)\tilde{u}^*(\mathbf{R}-\rho/2)\rangle \\ &\approx \langle |A(\mathbf{R})|^2\rangle \exp\{-2\mathbf{R}\,\mathrm{Im}\,\mathbf{k}^{\text{eff}} + i\rho\,\mathrm{Re}\,\mathbf{k}^{\text{eff}}\},\end{aligned} \tag{3.5.6}$$

and condition (3.5.4) may be written as

$$|\partial_R \Psi_{12}| \sim \left(2\left|\mathrm{Im}\,k^{\text{eff}}\right| + 2/L_A\right)|\Psi_{12}| << |\partial_\rho \Psi_{12}| \sim \left|\Psi_{12}\,\mathrm{Re}\,k^{\text{eff}}\right|. \tag{3.5.7}$$

Here the magnitude $|2\,\mathrm{Im}\,k^{\text{eff}}|^{-1}$ has the meaning of the extinction length L_{ext}.

Let us introduce the characteristic scale of Ψ_{12} varying in the argument $\mathbf{R}$ as

$$L_R \sim \min(L_A, L_{\text{ext}}). \tag{3.5.8}$$

Condition (3.5.7) then may be written in the form

$$L_R >> 2\pi/|\mathrm{Re}\,k^{\text{eff}}| \equiv \lambda^{\text{eff}} \tag{3.5.9a}$$

or more roughly

$$\begin{aligned}L_{\text{ext}} &>> \lambda^{\text{eff}}, \\ L_A &>> \lambda^{\text{eff}},\end{aligned} \tag{3.5.9b}$$

where λ^{eff} is the wavelength in the effective medium.

The first inequality represents the requirement of weak disorder. When this condition is violated, i.e. $L_{\text{ext}} \sim \lambda^{\text{eff}}$ (this is the so-called Ioffe–Regel condition), we arrive in the range of strong localization where the conventional theory of radiative transfer is inapplicable (see Section 5.4). The second condition (3.5.9) does not allow for strongly nonuniform radiation, that is for abrupt changes in the field statistics within the effective wavelength.

Similarly to (3.5.9), for a nonmonochromatic field, we can write

$$T_R >> \overline{\tau}, \tag{3.5.10}$$

where $\overline{\tau}$ is the characteristic radiation period, and T_R is the characteristic time of nonstationarity, i.e., of the variation of Ψ_{12} in the coordinate $T = (t_1 + t_2)/2$.

Thus, in accordance with (3.5.9) and (3.5.10), condition (3.5.4) implies the quasi-uniformity on the scales of the characteristic wavelength λ^{eff} and the radiation period $\overline{\tau}$ and has a simple physical meaning: *on the average* the scattered field must have the form of traveling waves with amplitudes changing insignificantly within the wavelength λ^{eff} over time $\overline{\tau}$.

3.5.4 Geometrical Optics Asymptotics for the Second Moments of Radiation: Expansion of the Bethe–Salpeter Equation in Small Parameters

The quasi-uniformity conditions (3.5.9) and (3.5.10) along with the necessary applicability condition (3.3.78) for the one-group approximation (i.e. the inequality $L_f >> l_c$ where $L_f \sim L_R$) almost coincide with the applicability conditions of geometrical optics in describing the mean value (the first moment) of the field $\langle u \rangle$ (see Appendix A). Therefore, it is not unnatural to use the geometrical-optics asymptotic approach for finding the second moment Ψ_{12}. The applicability of the geometrical optics for describing *mean* quantities does not require, obviously, its applicability to describing random field realizations u. Small parameters to be used in expanding into a series are the dimensionless quantities $\lambda^{\text{eff}} / L_R$ and l_c / L_R, i.e. the ratios of the characteristic wavelength λ^{eff} and the correlation radius to the characteristic inhomogeneity scale L_R, as well as quantities $\overline{\tau} / T_R$ and τ_c / T_R equal to the ratios of the radiation period τ and the correlation time τ_c to the characteristic time of problem nonstationarity, respectively.

Let us write the starting equation (3.5.2) in the form

$$D_1 \Psi_{12} = (D_2^*)^{-1} K_{12}^{(1gr)} \Psi_{12}. \tag{3.5.11}$$

In view of the assumption on statistical uniformity of the medium, the kernels of the operators D and $K_{12}^{(1\text{gr})}$ are of difference type and can be written as in (3.3.64)

$$D(x,x') = D(x - x'), \quad K_{12}^{(1\text{gr})} \begin{pmatrix} x_1, & x_1' \\ x_2, & x_2' \end{pmatrix} = K_{12}^{(1\text{gr})}(R - R', \rho, \rho'),$$

where $R = (x_1 + x_2)/2$, $R' = (x_1' + x_2')/2$ and ,$\rho = x_1 - x_2$, $\rho' = x_1' - x_2'$, are the sum and difference coordinates, respectively.

Let us introduce an auxiliary dimensionless small parameter μ to describe all the assumptions on smallness. The quasi-homogeneity condition (3.5.4) can then be written as $|\partial_R \Psi_{12}| \sim \mu |\partial_\rho \Psi_{12}|$, or, alternatively, as $\Psi_{12} = \Psi_{12}(\mu R, \rho)$ (μR is the slow variable). Substituting this expression into (3.5.11), we expand the two sides of the resultant equation in a power series of the small gradient $\partial_R \sim \mu$. For the left-hand side of (3.5.11), such an expansion gives

$$\begin{aligned} D_1\Psi_{12} &= \int D(x')\Psi_{12}(R - x'/2, \rho - x')d^4x' \\ &= \int D(x')(1 - (x'/2)\partial_R + \ldots)e^{iK'(\rho - x')}W(K', R)d^4K'd^4x' \\ &= \int \left[D(K') + (2i)^{-1}\left(\partial_{K'}D(K')\right)\partial_R + \ldots\right]W(K', R)e^{iK'\rho'}d^4K'. \end{aligned} \quad (3.5.12)$$

Here, $D(K)$ is the Fourier transform of $D(x)$

$$\begin{aligned} D(K) &= \int D(x')e^{-iKx'}d^4x' = e^{-iKx}De^{iKx} \\ &= L_0(K) - V^{\text{eff}}(K) = k_0^2\varepsilon_0 - K^2 - V^{\text{eff}}(K), \end{aligned} \quad (3.5.13)$$

and

$$W(K, R) \equiv (2\pi)^{-4}\int \Psi_{12}(R, \rho')e^{-iK\rho'}d^4\rho' \quad (3.5.14)$$

is the spectrum of scattered field fluctuations defined by Wigner's function.

Expanding similarly the right-hand side of (3.5.11), we multiply the two sides of (3.5.11) by $(2\pi)^{-4}\exp(-iK\rho)$ and integrate over ρ. The region of large ρ, where the quasi-uniformity condition can be violated, is assumed to contribute to Wigner's function (3.5.14) insignificantly. As a result, we obtain

$$\begin{aligned} &\left(D(K) + (2i)^{-1}(\partial_K D(K))\partial_R + \ldots\right)W(K, R) \\ &= (D^*(K))^{-1}\{K^{(1\text{gr})}W(K, R)\} + \ldots \\ &\equiv (D^*(K))^{-1}\int K^{(1\text{gr})}(K \leftarrow K')W(K', R)d^4K' + \ldots, \end{aligned} \quad (3.5.15)$$

Here the kernel $K^{(1\text{gr})}(K \leftarrow K')$ is determined by the relation of the form (3.3.67):

$$K^{(1\text{gr})}(K \leftarrow K') = \int (2\pi)^{-4}d^4x_1 \exp(-iK(x_1 - x_2))K_{12}^{(1\text{gr})}\exp(iK'(x_1 - x_2)), \quad (3.5.16)$$

and the dots denote higher order terms due to the statistical nonuniformity of Ψ_{12}.

Finally, we take into account that the Dyson operator describes the propagation of the mean field $\langle u \rangle$, its imaginary part $\operatorname{Im} D(K) \equiv D^a(K)$ being related to the extinction, that is the attenuation due to scattering and absorption. In the one-group approximation, this quantity may naturally be regarded small in comparison with the real part $D^h = \operatorname{Re} D(K)$:

$$|D^a| << |D^h|. \tag{3.5.17}$$

More exactly, D^a must be small compared to $k_0^2\varepsilon$ or k^2 appearing in D^h because the case of $D^h = 0$ is treated below. For the scalar model (3.4.1), this condition may be written approximately as

$$|D^a| = |k_0^2\varepsilon_0^a - \operatorname{Im} V^{\text{eff}}(K)| << |D^h| \sim |k_0^2\varepsilon_0^h|,$$

where ε_0^h and ε_0^a are the real and imaginary parts of permittivity ε_0, $\varepsilon_0 = \varepsilon_0^h + i\varepsilon_0^a$. When (3.5.17) is satisfied, D^a may be referred to the first-order terms in μ so that $D(K) = D^h + i\mu D^a$. Similarly, the operator $K_{12}^{(1\text{gr})}$ associated with scattering is deemed to be small, $K_{12}^{(1\text{gr})} \sim \mu$.

3.5.5 Equations of Successive Approximations

Having determined the orders of magnitude for terms appearing in (3.5.15), we will seek a solution to equation (3.5.15) in the form of an expansion similar to that used in the conventional geometrical optics approach

$$W(K,\mu R) = \sum_{n=0}^{\infty} \mu^n W^{(n)}(K,\mu R). \tag{3.5.18}$$

Substituting this expansion in (3.5.15), we equate all the terms of the same order in μ to zero, and then put $\mu = 1$. As a result, we come to the chain of equations

$$D^h(K)W^{(0)}(K,R) = 0. \tag{3.5.19}$$

$$D^h(K)W^{(1)}(K,R) = -\left(iD^a(K) + (2i)^{-1}(\partial_K D^h(K)\right)\partial_R)W^{(0)}(K,R)$$
$$+\left(D_1^*(K)\right)^{-1}\{K^{(1\text{gr})}W^{(0)}(K,R)\}. \tag{3.5.20}$$

Formally, this procedure is equivalent to passing to the limit $\lambda \to 0$ like in geometrical optics, but for the second statistical moment $W(K,R)$ rather than for individual realizations.

According to the zeroth order equation (3.5.19), the principal term of expansion (3.5.18) must be localized on the dispersion surface $D^h(K) = 0$, that is

$$W^0(K,R) = I_k'(R)\delta\left(D^h(K)\right) = I_{\omega n}(R)k^{-2}\delta(k - k_1(\omega)), \tag{3.5.21}$$

where $I_k'(R)$ is the spectral density, $I_{\omega n} = I_k' k^2 / |\partial_k D^h|$ is the angular-frequency spectrum, and $k_1(\omega)$ is the solution of the dispersion equation $D^h(K) = 0$. Relation (3.5.21) is quite similar to eqn (3.1.63) for a homogeneous medium. The approximate dispersion equation $D^h = 0$ corresponds to a *nonabsorbing effective medium*, and the attenuation of the average field due to scattering is taken into account by the term with iD^a in the first approximation equation. Formally, this corresponds to carrying over the term iD^a appearing in the exact mean-field dispersion equation $D^h + iD^a = 0$ into the first approximation equation (3.5.20).

It can be easily shown that the solution $k_1(\omega)$ of the approximate dispersion equation $D^h(K) = 0$ in the zeroth approximation in μ to relate to the solution k^{eff} of the exact dispersion equation $D(K) = D^h + iD^a = 0$ by $k_1(\omega) \approx \operatorname{Re} k^{\text{eff}}$. Therefore, according to eqns (3.5.21) and (3.5.14), in the zeroth approximation, the correlation function of the scattered field can be written as

$$\Psi_{12}(R,\rho) \equiv \langle \tilde{u}(R+\rho/2)\tilde{u}^*(R-\rho/2)\rangle = \int J_K^{(0)}(R)\exp(iK\rho)d^4K$$
$$= \int I_{\omega\mathbf{n}}(R)\exp[i(\mathbf{n}\rho \operatorname{Re} k^{\text{eff}} - \omega\tau)]d\omega d\Omega_{\mathbf{n}}. \qquad (3.5.22)$$

3.5.6 Transition to the Radiative Transfer Equation

In this section we show that the angular-frequency spectrum $I_{\omega\mathbf{n}}$ satisfies the radiative transfer equation. For this purpose, we substitute (3.5.21) into (3.5.20) and take into account that both sides of eqn (3.5.20) must vanish on the dispersion surface $D^h(K) = 0$. We evaluate $(D^*(K))^{-1}$ using the familiar Sokhotski formula:

$$(D^*(K))^{-1} = (D^h - iD^a\mu)^{-1} \approx \mathcal{P}(D^h)^{-1} + i\pi\delta(D^h),$$

which is satisfied in the limit $\mu \to 0$ for $D^a > 0$. The last inequality corresponds to the attenuation of $\langle u\rangle$. As a result, canceling out the common factor $\delta(D^n)$, we obtain the equation

$$[(\partial_K D^h)\partial_R - 2D^a]I_{\omega\mathbf{n}} = -|\partial_K D^h|^{-1} 2\pi k^2 \{K^{(1\text{gr})} I_{\omega\mathbf{n}} k^{-2}\delta(k-k_1)\}, \qquad (3.5.23)$$

both sides of which are taken on the dispersion surface $D^h = 0$.

Let us introduce ray equations corresponding to the dispersion equation $D^h = 0$ for waves in the effective medium. Choosing the arc length s of spatial ray projection as a parameter, we may write these equations in the form (see Appendix A):

$$\dot{R} \equiv d_s R = \frac{-\partial_K D^h}{|\partial_k D^h|}, \quad \dot{K} \equiv d_s K = \frac{\partial_R D^h}{|\partial_k D^h|} = 0, \qquad (3.5.24)$$

where $\dot{K}$ is equal to zero by virtue of the assumption on statistical homogeneity of the medium.

So far, we have treated free radiation in the absence of sources ($q = 0$). This restriction can be lifted if, in the presence of sources, the structure of the coherence function remains almost identical to that in case of free propagation. Formally, this implies that the second moment $\langle q_1 q_2^* \rangle$ in eqn (3.5.2) should be regarded as a small quantity of the order of μ and included into the first approximation equation (3.5.20). As additional term proportional to $\langle q_1 q_2^* \rangle$ will then appear on the right-hand side of (3.5.23). Further, we can perform integration with respect to k' on the right-hand side of (3.5.23) using delta function properties, so that the triple integral over frequency and directions remains. As a result, in view of (3.5.24) eqn (3.5.23) can be written in the final form of the transfer equation for the angular-frequency spectrum of:

$$(d_s + \alpha_t) I_{\omega\mathbf{n}} = \int \sigma'(\omega, \mathbf{n} \leftarrow \omega', \mathbf{n}) I_{\omega'\mathbf{n}'}(R) d\omega' d\Omega_{\mathbf{n}'} + \varepsilon'_q$$
$$\equiv \hat{\sigma} I_{\omega\mathbf{n}} + \varepsilon'_q, \tag{3.5.25}$$

where d_s is the operator of differentiation along the ray:

$$d_s = -\frac{\partial_K D^h}{|\partial_k D^h|}\partial_R = \frac{(\partial\omega D^h)\partial_T - (\partial_K D^h)\partial_R}{|\partial_k D^h|} = n\nabla + v_g^{-1}\partial_T . \tag{3.5.26}$$

The extinction coefficient

$$\alpha^{\text{ext}} = \alpha^{abs} + \alpha^{\text{sc}} = \frac{2D^a}{|\partial_k D^h|} = 2\,\text{Im}\frac{\left(L_0(K) - V^{(1\text{gr})}(K)\right)}{|\partial_k D^h|} \tag{3.5.27}$$

is the sum of the absorption coefficient α^{abs} and the scattering coefficient α^{sc}:

$$\alpha^{\text{abs}} = \frac{2\,\text{Im}\,L_0(K)}{|\partial_k D^h|}, \quad \alpha^{\text{sc}} = -\frac{2\,\text{Im}\,V^{(1\text{gr})}(K)}{|\partial_k D^h|}. \tag{3.5.28}$$

With allowance for frequency change is the scattering cross section

$$\sigma'(\omega, \mathbf{n} \leftarrow \omega', \mathbf{n}') = c_1 K^{(1\text{gr})}(K \leftarrow K'), \tag{3.5.29}$$

where $c_1 = 2\pi k^2 / |\partial_k D^h|^2$. Finally, the source function

$$\varepsilon'_q = c_1 \Gamma_q(R, K) \tag{3.5.30}$$

is proportional to the coherence function of wave sources Γ_q:

$$\Gamma_q(R, K) = \int \left\langle q\left(R + \frac{\rho}{2}\right) q^*\left(R - \frac{\rho}{2}\right)\right\rangle \exp(-iK\rho)\frac{d^4\rho}{(2\pi)^4}, \tag{3.5.31a}$$

which can be written also way as

$$\Gamma_q(R,K)$$
$$= \int \left\langle q\left(\mathbf{R}+\frac{\rho}{2},\omega+\frac{\omega'}{2}\right)q^*\left(\mathbf{R}-\frac{\rho}{2},\omega-\frac{\omega'}{2}\right)\right\rangle \exp(-i\omega T - i\mathbf{k}\rho)\frac{d^3\rho d\omega'}{(2\pi)3} \quad (3.5.31b)$$

in terms of the frequency spectrum of the sources

$$q(\mathbf{r},\omega) = \int q(x)e^{i\omega t}(2\pi)^{-1}dt.$$

The vectors $K = (\omega, k\mathbf{n})$ and $K' = (\omega', k'\mathbf{n}')$ appearing in (3.5.26)-(3.5.31) satisfy the dispersion equation $D^h = D^h(K) = 0$ (so that $k = k' \approx \mathrm{Re}\, k^{\mathrm{eff}}$), and $\mathbf{v}_g = -\partial_k D^h / \partial_\omega D^h = d_{\mathbf{k}}\omega$ is the group velocity corresponding to this equation.

As we have seen in Section 3.3, the dispersion equation can be solved by iterations in the one-group approximation. In the first approximation, we have $D^h \approx k_0^2 \varepsilon_0^h - k^2 - \mathrm{Re}\,\mathbf{V}^{(1\mathrm{gr})}(k_0^2)$, so that $|\partial_k D^n| \approx 2k$, $c_1 = \pi/2$, and eqns (3.5.27), (3.5.29) and (3.5.30) take the form

$$\alpha^{\mathrm{ext}} = \alpha^{\mathrm{abs}} + \alpha^{\mathrm{sc}} = k_0^2 k^{-1} \varepsilon_0^a - k^{-1}\,\mathrm{Im}\,\mathbf{V}^{(1\mathrm{gr})}(k_s^2),$$
$$\sigma'(\omega,\mathbf{n} \leftarrow \omega',\mathbf{n}') = \frac{\pi}{2}K^{(1\mathrm{gr})}(K \leftarrow K'), \quad (3.5.32)$$
$$\varepsilon_q' = \frac{\pi}{2}\Gamma_q(R,K).$$

The frequency-angular spectrum of $I_{\omega\mathbf{n}}$ appears as the radiance in the transfer equation (3.5.25). This spectrum enables us to approximately express the mean radiance I in a scattering medium if we ignore the correlation between medium and field fluctuations in the expression for energy density by setting $\langle \varepsilon |E|^2 \rangle \approx \langle \varepsilon \rangle \langle |E|^2 \rangle$. In this approximation, the radiance is expressed in terms of the frequency-angular spectrum $I_{\omega\mathbf{n}}$ in the same manner as for a homogeneous medium with permittivity $\langle \varepsilon \rangle$, that is, $I \approx aI_{\omega\mathbf{n}}$, where $a = a(\omega)$ is a factor dependent on the nature of the field [see (3.1.86); for example, in the electromagnetic case we have $a = (\varepsilon_0^e / 8\pi)v_g$ according to (3.1.99)]. Substituting $I_{\omega\mathbf{n}} = I/a$ into eqn (3.5.25), we obtain the transfer equation for radiance I:

$$(d_s + \alpha^{\mathrm{ext}})I = a\hat{\sigma}'\left(\frac{I}{a}\right) + a\varepsilon_q' \equiv \hat{\sigma}I + \varepsilon_q. \quad (3.5.33)$$

It follows that in the transfer equation for radiance I, the source function $\varepsilon_q = a\varepsilon_q'$ and the cross section σ differ from similar quantities ε_q' and σ' appearing in the equation for $I_{\omega\mathbf{n}}$ by some factors. For scattering without change in frequency, the difference between the cross sections is easily seen to disappear, so that $\sigma = \sigma'$. For simplicity, sometimes we will drop primes of ε' and q'.

Scattering without change in frequency corresponds to a stationary medium for which the kernel $K_{12}^{(\mathrm{gr})}$ depends upon difference $\tau - \tau'$:

$$K_{12}^{(1\mathrm{gr})}(R-R',\rho,\rho') = K_{12}^{(1\mathrm{gr})}(R-R',\mathrm{p},\mathrm{p}',\tau-\tau').$$

According to (3.5.16), we have

$$K^{(1\mathrm{gr})}(K \leftarrow K') = \delta(\omega-\omega')\int \frac{d^3\rho}{(2\pi)^3} e^{-iK\rho} K_{12}^{(1\mathrm{gr})} e^{iK'\rho}$$
$$\equiv \delta(\omega-\omega')K_{\omega}^{(1\mathrm{gr})}(\mathbf{n} \leftarrow \mathbf{n}'). \tag{3.5.34}$$

and the scattering cross section from (3.5.29) may then be written as

$$\sigma(\omega,\mathbf{n} \leftarrow \omega',\mathbf{n}') = \left(\frac{\pi}{2}\right) K_{\omega}^{(1\mathrm{gr})}(\mathbf{n} \leftarrow \mathbf{n}')\delta(\omega-\omega'). \tag{3.5.35}$$

The delta function $\delta(\omega-\omega')$ appearing here does correspond to the absence of frequency variations in scattering.

The assumption of a homogeneous and stationary medium is not necessary. In order to go to the transfer equation, it is sufficient to assume that the mean medium properties vary smoothly within the characteristic radiation wavelengths, periods, and of correlation radii of medium fluctuations. This implies that not only the field but also the medium must be statistically quasi-homogeneous so that the statistical inhomogeneity scales of the medium must obey inequalities similar to those for the statistical nonuniformity scales of the field. As will be recalled that the *statistical nonuniformity scales* should not be confused with sizes of medium's *inhomogeneities*. The former characterize statistical nonuniformity and are equal to infinity for a statistically homogeneous medium whereas the latter describe the correlation properties of medium fluctuations. For a smoothly inhomogeneous medium, the parameters α and σ of the transfer equation will depend smoothly upon R, and the operator of differentiation along a rectilinear ray in (3.5.25) should be replaced with the operator of the type (C.7), given in Appendix C, that takes into account the refractive curvature of rays.

3.5.7 Remark on the Optical Theorem

By taking into account the Sokhotski formula and going over to the limit of weak extinction $D^a \to 0$, eqn (3.3.65) can readily be transformed into the form

$$\operatorname{Im} V^{\mathrm{eff}}(K) = -\pi \int \mathrm{K}(K' \leftarrow K)\delta\big(D^h(K')\big)d^4K'$$
$$= -\pi \int \mathrm{K}(K' \leftarrow K)|\partial_{k'} D^h|^{-1} k'^2 d\omega' d\Omega_{\mathbf{n}'}, \tag{3.5.36}$$

where $K(K' \leftarrow K)$ is expressed in terms of K_{12} according to (3.5.16), and

$K(K \leftarrow K')$ is replaced by $K(K' \leftarrow K)$ by virtue of section 3.3.7.

Comparing (3.5.36) with (3.5.27) and (3.5.28), we can see that if the exact values of V^{eff} and K_{12} rather than their one-group approximations are used in (3.5.27) and (3.5.28), then in scattering with small changes in frequency, $|\omega - \omega'| << \omega$, such that $k^2 / |\partial_k D^h|$ may be factored outside the integral sign, the scattering coefficient α^{sc} may be expressed as the total (over all directions and frequencies) scattering cross section per unit volume:

$$\alpha^{\text{sc}} = \int \sigma(\omega', \mathbf{n}' \leftarrow \omega, \mathbf{n}) d\omega' d\Omega_{\mathbf{n}'}. \tag{3.5.37}$$

This relationship also remains approximately valid in the one-group approximation for $V^{\text{eff}} \approx V^{(\text{1gr})}$ and $K_{12} \approx K_{12}^{(\text{1gr})}$ when the optical theorem (3.3.65) becomes eqn (3.3.74), and the relation similar to (3.5.36) involves the delta function $\delta(L_0)$ instead of $\delta(D^h)$. This can be easily shown to be satisfied when the quantity $\text{K}(K' \leftarrow K)$ is a sufficiently smooth function of K'.

Relation (3.5.37) coincides with the optical theorem (2.2.33) derived above in the framework of the phenomenological theory thus serving as the statistical basis for eqn (2.2.33)

3.5.8 Transfer Equation and the Coherent (Average) Field

So far, the mean field $\langle u \rangle$ has been assumed zero. This assumption is not necessary for deriving a transfer equation which enables us to describe not only the correlation Ψ_{12} but also the coherent part $\langle u_1 \rangle \langle u_2^* \rangle$ of Γ_{12}. Indeed, the quasi-homogeneity of fluctuation (Ψ_{12}) was the key assumption in deriving the transfer equation. However, for the mean field $\langle u \rangle$ described in the framework of geometrical optics, the coherent part $\langle u_1 \rangle \langle u_2^* \rangle$ has been shown (see section 3.1.9) to be locally quasi-uniform. Hence, the coherence function, $\Gamma_{12} = \Psi_{12} + \langle u_1 \rangle \langle u_2^* \rangle$, will also be locally quasi-uniform. Naturally, the transfer equation will then describe the total coherence function Γ_{12} rather than its fluctuation part Ψ_{12} alone.

In the simplest case of one-ray propagation (when only one ray arrives at each point located far from the source) under ignoring the interference rays (Section 3.3), the coherent part $\langle u_1 \rangle \langle u_2^* \rangle$ differs from the fluctuation part $\langle \tilde{u}_1 \tilde{u}_2^* \rangle$ in that at each point R its frequency-angular spectrum $I_{\omega\mathbf{n}}^{\text{coh}}$ (coherent radiance) proves to be localized in the direction $\mathbf{n}_R$, that is, involves the delta function $\delta_2(\mathbf{n}\mathbf{n}_R)$:

$$I_{\omega\mathbf{n}}^{\text{coh}} = I^0 \delta_2(\mathbf{n}\mathbf{n}_R). \tag{3.5.38}$$

This means that the coherent radiation propagates at each point R along a single ray arriving from the source. In the absence of sources, the coherent radiance $I_{\omega\mathbf{n}}^{\text{coh}}$ obeys the homogeneous transfer equation

$$(d_s + \alpha^{\text{ext}}) I_{\omega\mathbf{n}}^{\text{coh}} = 0 \tag{3.5.39}$$

and varies along a ray as $\exp(-\int \alpha^{\text{ext}} ds)$ that corresponds to the mean field attenuation due to scattering and absorption.

In accordance with the division of Γ_{12} into coherent and incoherent parts

$$\Gamma_{12} = \langle u_1 \rangle \langle u_2^* \rangle + \Psi_{12},$$

a similar division for the spectral radiance $I_{\omega n}$ and for source function ε_q' seems to be expedient in the presence of sources:

$$I_{\omega n} = I_{\omega n}^{\text{coh}} + I_{\omega n}^{\text{incoh}}, \quad \varepsilon_q' = \varepsilon'^{\text{coh}} + \varepsilon'^{\text{ihcoh}}. \tag{3.5.40}$$

If the total radiance $I_{\omega n}$ satisfies the transfer equation (3.5.25)

$$(d_s + \alpha^{\text{ext}}) I_{\omega n} = \hat{\sigma}' I_{\omega n} + \varepsilon_q', \tag{3.5.41}$$

the coherent part of the radiance obeys the transfer equation

$$(d_s + \alpha^{\text{ext}}) I_{\omega n}^{\text{coh}} = \varepsilon'^{\text{coh}}, \tag{3.5.42}$$

with the coherent source

$$\varepsilon^{\text{coh}} = c_1 \int \left\langle q\left(R + \frac{\rho}{2}\right)\right\rangle \left\langle q^*\left(R - \frac{\rho}{2}\right)\right\rangle e^{iK\rho} \frac{d^4\rho}{(2\pi)^4}, \tag{3.5.43}$$

that is proportional to the product $\langle q_1 \rangle \langle q_2^* \rangle$ of source mean values.

The incoherent part of radiance is governed by the equation

$$(d_s + \alpha^{\text{ext}}) I_{\omega n}^{\text{incoh}} = \hat{\sigma}'(I_{\omega n}^{\text{incoh}} + I_{\omega n}^{\text{coh}}) + \varepsilon'^{\text{incoh}} \tag{3.5.44}$$

with the source

$$\varepsilon^{\text{incoh}} = \varepsilon_q' - \varepsilon'^{\text{coh}} = c_1 \int \left\langle \tilde{q}\left(R + \frac{\rho}{2}\right) \tilde{q}^*\left(R - \frac{\rho}{2}\right)\right\rangle e^{iK\rho} \frac{d^4\rho}{(2\pi)^4}, \tag{3.5.45}$$

which is determined by source fluctuations $\tilde{q} = q - \langle q \rangle$.

It is customary to solve the transfer equation without dividing the radiance into coherent and incoherent parts. The arbitrariness in choosing a solution to the homogeneous transfer equation is removed by imposing some or other boundary conditions. More often than not, boundary conditions correspond to a coherent wave incident on the scattering volume from the outside. In this case the solution of the transfer equation describes automatically not only the incoherent part of radiance but also the geometrical optics structure of the coherent part propagating along the rays in the effective medium. Even a single scattering transforms the coherent part into the incoherent part which acquires the meaning of the radiance of scattered radiation. These topics have been discussed by Ishimaru (1978).

The transfer equation (3.4.39) involves the geometrical optics description of the coherent (mean) part and does not take into account diffraction phenomena. At the same time equation (3.4.44) involves scattering, thus describing the diffraction, that is, changes in the direction of propagation due to scattering by random inhomogeneities.

In principle, transfer theory enables one to take into account the diffraction of the average field. First, the nonclassical radiometry is implied where, at the expense of abandoning source radiance nonnegativity, one can choose for the transfer equation a boundary condition such that provides the correct description for the coherent function in the far zone as noted of Section 3.2.5. Second, in order to find the average field, one can use the Dyson equation which, in principle, describes diffraction effects for $\langle u \rangle$. By substituting the corresponding expression for the of coherent part radiance I^{coh} into the "incoherent" transfer equation (3.5.44), one can determine the incoherent radiance I^{incoh}. As far as we know, such a scheme has not yet been reported yet in the literature.

Remark on terminology. We have called the solution of the homogeneous transfer equation (3.5.39) the coherent radiance proceeding from the fact that, in the geometrical optics limit, an identical equation governs the quasi-uniform part of the radiance spectrum associated with the coherent field $\langle u \rangle$. In the general case, the solution of the homogeneous equation (3.5.39) is specified by the properties of a primary (incident) wave. *Incoherent radiation* with a broad angular spectrum may also appear as such a wave. In this case, it is more appropriate to use the terminology accepted by Ishimaru (1978) who calls the solution to the homogeneous transfer equation the *attenuated incident intensity*, and referred to the radiance as the of scattered radiation *diffuse intensity*. In the context of problems treated here, our terminology seems to be justified for we consider a coherent incident wave with a narrow directional pattern.

3.5.9 Sources of Thermal Radiation. The Fluctuation-Dissipation Theorem

In the case of equilibrium thermal radiation, the phenomenological radiative transfer theory represents the source function ε_q in terms of absorption coefficient and the radiance of equilibrium thermal radiation by eqn (2.2.66). We now demonstrate that this relation is a corollary of the general fluctuation-dissipation theorem.

For our purpose, it is convenient to state the fluctuation-dissipation theorem in the following general form. Consider the spectral component of an equilibrium thermal field $u = u(\mathbf{r}, \omega)$ as a resulting of the action of distributed microscopic thermal sources $q(\mathbf{r}, \omega)$. In the linear approximation the field $u(\mathbf{r}, \omega)$ is related to $q = q(\mathbf{r}, \omega)$ by the general equation $L_\omega u = q$ where the operator L_ω, unlike the operator L from eqn (3.5.1), corresponds to a monochromatic component of u, $L_\omega = e^{i\omega t} L e^{-i\omega t}$. In the t-representation, the average energy dissipation in a unit volume per unit time may be written in the form

$$Q = -\int \langle u \partial_t B^{-1} q \rangle d^3 r = \int \langle (\partial_t u) B^{-1} q \rangle d^3 r, \tag{3.5.46}$$

where B^{-1} is the deterministic stationary linear operator. In the specific case of $B^{-1}=1$, the field q will be the Lagrangian conjugate relative to u, and case we come to the fluctuation-dissipation theorem in its conventional form (a detailed description may be found in the book by Rytov, Kravtsov, and Tatarskii 1989).

According to the fluctuation-dissipation theorem the correlation of thermal sources q is expressed through the kernel of the anti-Hermitian part $A_\omega^a(\mathbf{r},r')=(A_\omega-A_\omega^*)/2i$ of the operator $A_\omega=L_\omega B_\omega^+$ by the relation

$$\langle q(\mathbf{r},\omega)q^*(\mathbf{r},\omega)\rangle_T=-\frac{\theta(\omega,T)}{\pi\omega}A_\omega^a(\mathbf{r},\mathbf{r}')\delta(\omega-\omega'), \tag{3.5.47}$$

where $Q(\omega,T)=(\hbar\omega/2)\coth(\hbar\omega/2kT)$ is the mean energy of the quantum oscillator, and the angular brackets $\langle\ldots\rangle_T$ imply averaging over equilibrium thermal fluctuations. This averaging may be treated as the conditional averaging at fixed macroscopic parameters of the medium. Both sides of eqn (3.5.47) may remain random in the macroscopic sense as in the case of macroscopic random media.

Equation (3.5.47) shows the properties of thermal sources to be related to the absorbing properties of the medium or, more exactly, to the form of the operator A^a. So far, however, we deemed the properties of sources to be statistically independent of medium's properties. For the field $u=Gq$ the coherence function may be represented as the sum $\Gamma_{12}=\Gamma_{12}^{(1)}+\Gamma_{12}^{(2)}$. The function $\Gamma_{12}^{(1)}=\langle G_1G_2^*\rangle\langle q_1q_2^*\rangle$ corresponds to the approximation of the uncorrelated sources q and the medium G and is governed by a transfer equation that can be derived. Calculation of the function $\Gamma_{12}^{(2)}=\Gamma_{12}-\Gamma_{12}^{(1)}$ with allowance for correlation calls for a special treatment.

Let us eliminate the function $\Gamma_{12}^{(2)}$, thus ignoring the correlation (weak by our assumption) of fluctuations of dissipative and nondissipative medium's parameters, and average additionally both sides of eqn (3.5.47) over macroscopic fluctuations on the assumption of a statistically uniform medium (extension onto the case of a quasi-homogeneous medium is trivial). As a result we have

$$\begin{aligned}\langle q(\mathbf{r},\omega)q^*(\mathbf{r}',\omega')\rangle &= -\frac{\theta(\omega,T)}{\pi\omega}\left\langle A_\omega^a(\mathbf{r}_1,\mathbf{r}_2)\right\rangle\delta(\omega-\omega') \\ &\equiv -\frac{\theta(\omega,T)}{\pi\omega}\overline{A_\omega^a}(\mathbf{r}_1-\mathbf{r}_2)\delta(\omega-\omega').\end{aligned} \tag{3.5.48}$$

Using this relationship, we get from (3.5.30) and (3.5.31a) we obtain

$$\varepsilon_q'=-c_1(2\pi)^{-3}\left(\frac{\theta(\omega,T)}{\pi\omega}\right)\overline{A^a}(K), \tag{3.5.49}$$

where

$$\overline{A^a}(K)=e^{i\mathbf{kr}}\,\overline{A_\omega^a}e^{i\mathbf{kr}}=e^{-iKx}Ae^{iKx}.$$

For the *scalar model* (3.5.1), where the field u is treated as a component of the electric strength vector $\mathbf{E}$, the radiation source is $q = (4\pi / c^2)\partial_t j$ and the mean energy dissipation rate may be expressed through currents as

$$Q = -\int \langle jE \rangle d^3r = -\int \langle ju \rangle d^3r. \tag{3.5.50}$$

A comparison of this expression eqn with (3.5.46) shows that $j = \partial_t B^{-1} q = (4\pi / c^2)\partial_t B^{-1}\partial_t j$ or, for monochromatic components, $j = -4\pi k_0^2 B_\omega^{-1} j$ so that the operator B_ω reduces to the factor $-4\pi k_0^2$. Expressing $\overline{L^a}(K) = \operatorname{Im} L_0(K)$ through the absorption coefficient α^{abs}, we obtain with allowance for (3.5.28)

$$\begin{aligned} \overline{A^a}(K) &= (\overline{L_\omega B_\omega^+})^a(K) = -4\pi k_0^2 \overline{L^a}(K) \\ &= -4\pi k_0^2 k \alpha^{\text{abs}}. \end{aligned}$$

In addition we note that $c_1 = \pi / 2$ for eqn (3.5.1) and use eqn (3.5.49) to write the source function in the transfer equation for energy radiance (3.5.33) in the form

$$\begin{aligned} \varepsilon_q &= a\varepsilon_q' = \frac{\varepsilon_0^{\text{e}} v_g}{8\pi}\varepsilon_q' = \frac{1}{(2\pi)^3}\frac{\varepsilon_0^{\text{e}} k_0^2}{4}\theta(\omega,T)\alpha_a \\ &= I_\omega \frac{\alpha_a}{2}. \end{aligned} \tag{3.5.51}$$

where $I_\omega = \varepsilon_0^{\text{e}} k_0^2 \theta(\omega,T) / 2(2\pi)^3$ is the mean energy of a quantum oscillator for frequency ranging in the interval $-\infty < \omega < \infty$.

This expression differs from above relations (2.2.66) and (2.1.29), first, in the allowance for null oscillations (by replacing θ' with θ) and, second, in the multiplier 1/2 which is the tribute to the fact that the scalar model being not completely appropriate to Maxwell's equations because it describes only one of the two independent components the transverse electromagnetic field. These discrepancies can be removed by going from the scalar model (3.3.1) to Maxwell's equations. In the "wave" derivation of the radiative transfer equation on the basis of Maxwell's equations, expression (2.2.66) for the equilibrium thermal source function ε_q can be shown to be a direct corollary of the fluctuation-dissipation theorem.

3.5.10 Scattering Cross Section in the One Group Approximation

Substituting expressions for $K_{12}^{(1\text{gr})}$ into eqn (3.5.29), we can explicitly write down an expansion for the scattering cross section in terms of cumulants of continuous fluctuations or coordinates of scattering particles. In the case of continuous fluctuations, using the expansion of $K_{12}^{(1\text{gr})}$ (3.3.73) with allowance for (3.5.16), we have from (3.5.29)

$$\sigma(\omega,\mathbf{n} \leftarrow \omega',\mathbf{n}) \approx \frac{\pi}{2}\int \frac{d^4x}{(2\pi)^4}\exp(-iK(x_1-x_2))$$

$$\times\Big\{$$

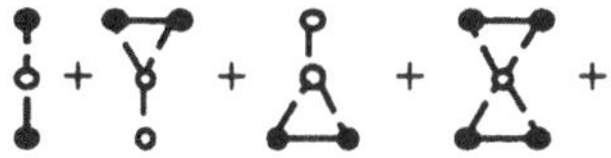

$$+\ \cdots\Big\}\exp\left(iK_1'(x_1-x_2)\right)$$

$$= \frac{\pi}{2}\int \frac{dx_1^4}{(2\pi)^4}\exp(-iK(x_1-x_2))\{\langle\tilde{V}_1\tilde{V}_2^*\rangle + \langle\tilde{V}_1 G_{01}\tilde{V}_1\tilde{V}_2^*\rangle$$

$$+ \langle\tilde{V}_1\tilde{V}_2^* G_{02}^*\tilde{V}_2^*\rangle + \langle\tilde{V}_1 G_{01}\tilde{V}_1\tilde{V}_2^* G_{02}^*\tilde{V}_2^*\rangle_c + \ldots\}\exp\left(iK_1'(x_1-x_2)\right). \quad (3.5.52)$$

Here the wave vectors K and K' are taken on the dispersion surface $D^{\mathrm{h}}(K)=0$ and the coincidence of the second and third cumulants with the respective central moments is taken into account [see eqn (3.3.47)].

This expression is different from the scattering cross section in the Born approximation in two respects. First, it takes account of not only the second cumulant but also higher cumulants of scattering fluctuations $\tilde{V}$, i.e. enables us to estimate small corrections to the cross section due to non-Gaussian $\tilde{V}$ (in the case of Gaussian fluctuations, these corrections vanish). Second, the wave number in free space, k_0, is replaced with the efficient wave number k^{eff}, that is due to the use of the one-group approximation for the mean field. The last difference can be expected to be not significant. Indeed, the heuristic condition for applicability of one-group approximations (Section 3.3.10) restricts $\Delta k = |k^{\mathrm{eff}} - k_0|$ to inequality (3.3.82): $\Delta k << 1/l_c$. However, the characteristic scale of changes of the right-hand side of eqn (3.5.52) in the argument k is around $1/l_c$. Therefore, a difference of k^{eff} from k_0 in eqn (3.5.52) is actually small. Thus, the major distinction between expression (3.5.52) for the cross section and the conventional Born approximation is the allowance for higher cumulants of scattering fluctuations $\tilde{V}$.

Similarly, for a medium with a sparse distribution of discrete scatterers, using the one-group approximation $K_{12}^{(1\mathrm{gr})}$ [see eqn (3.4.52)] we have from eqn (3.5.29)

$$\sigma(\omega,\mathbf{n} \leftarrow \omega',\mathbf{n}') = \frac{\pi}{2}\int \frac{d^4x_1}{(2\pi)^4}\exp(-iK(x_1-x_2))\{N\langle t_1^{(1)}t_2^{(1)*}\rangle$$

$$+ N^2\langle\tilde{t}_1^{(1)}\tilde{t}_2^{(2)*}\rangle + \ldots\}\exp\left(iK(x_1-x_2)\right). \quad (3.5.53)$$

(here we took into consideration that in the N/V limit $\langle\tilde{t}_1^{(1)}t_2^{(1)*}\rangle \approx \langle t_1^{(1)}t_2^{(1)*}\rangle$).
In the case of static scatterers, when frequency does not change in scattering, we can integrate with respect to the time argument in (3.5.52) and (3.5.53) so that these take the form

$$\sigma(\omega,\mathbf{n} \leftarrow \omega',\mathbf{n}')$$

$$= \delta(\omega-\omega')\frac{\pi}{2}\int \frac{d^3r_1}{(2\pi)^3}\exp(-ik(r_1-r_2))\{\ldots\}\exp\left(i\mathbf{k}(\mathbf{r}_1-\mathbf{r}_2)\right). \quad (3.5.54)$$

Here the expression in braces coincides formally with the similar expressions in eqn (3.5.52) or in eqn (3.5.53) whereas the operators V and G need to be treated as corresponding to the monochromatic radiation with time dependence $\exp(-i\omega t)$. The delta function in this expression corresponds to the frequency-independent scattering.

3.5.11 Scattering Cross Section and Sources of EM Radiation

The above approaches to the derivation of the transfer equation can be applied without any modifications to the case of polarized electromagnetic radiation. For a statistically isotropic medium the expression for tensor scattering cross sections differs from those for the scalar model [eqns (3.5.52) and (3.5.53)] by the projection tensors

$$\hat{\rho} = \hat{1} - \mathbf{n} \otimes \mathbf{n}$$

which allows for the transverse nature of electromagnetic waves (in the phenomenological theory, it was discussed in item Section 2.3.8):

$$\sigma(\omega, \mathbf{n} \leftarrow \omega', \mathbf{n}') = \rho \otimes \rho \sigma(\omega, \mathbf{n} \leftarrow \omega', \mathbf{n}') \rho' \otimes \rho'. \qquad (3.5.55)$$

Here $\hat{\rho}' = \hat{1} - \mathbf{n}' \otimes \mathbf{n}'$, and $\sigma(\omega, \mathbf{n} \leftarrow \omega', \mathbf{n}')$ is given by the tensor expression (3.5.52) in the case of continuous scattering inhomogeneities and by the tensor expression (3.5.53) in the case of discrete inhomogeneities.

Quite similarly, in the case of quasi-stationary electromagnetic radiation, the source function ε_q' (3.5.30) converts into the matrix

$$\varepsilon_q' = \frac{k_0^2}{2\pi c^2} \hat{\rho} \int \left\langle \mathbf{j}\left(R + \frac{\rho'}{2}\right) \otimes \mathbf{j}^*\left(R - \frac{\rho}{2}\right) \right\rangle e^{-iK\rho} d^4\rho \hat{\rho}, \qquad (3.5.56)$$

which involves the tensor product of vector currents $\mathbf{j}$.

3.5.12 Scattering Cross Section for Discrete Scatterers

We now consider the scattering cross section for discrete scatterers (3.5.53) in more detail. It may be written as the sum

$$\sigma = \sigma_{\text{indep}} + \sigma_{\text{coll}}, \qquad (3.5.57)$$

where σ_{indep} corresponds to the first term on the right-hand side of eqn (3.5.53) and describes scattering by independent particles, and σ_{coll} relates to the contribution from the other terms and describes the collective effects due to the second and higher correlations. If we eliminate all the terms that are not written down in (3.5.53), then σ_{coll} will describe two-particle correlations. In this approximation, eqn (3.5.53) differs from the single-scattering cross section (3.4.36) (Section 3.4.6) by k_0 replaced with k^{eff}

3.5.13 Scattering from Moving Inhomogeneities

Expressions (3.5.52) and (3.5.53) for scattering cross sections are applicable to scattering from static inhomogeneities and to scattering with a frequency change. We consider the latter separately for scalar and electromagnetic radiation in more detail.

In the scalar model (3.5.1), the perturbation operator is

$$V = \tilde{V} = c^{-2}\partial_t^2 \tilde{\varepsilon} = \frac{\tilde{\varepsilon}}{c^2}\partial_t^2, \tag{3.5.58}$$

where we used the quasi-static approximation, i.e. neglected the derivatives of $\tilde{\varepsilon}(x)$ with respect to time assuming $\tilde{\varepsilon}$ to be slowly varying within a time interval on the order of the characteristic period of radiation ($\sim \omega^{-1}$). If $\tilde{\varepsilon}$ is a Gaussian random quantity, only the first term will remain in braces of eqn (3.5.52) so that the scattering cross section takes the form

$$\begin{aligned}\sigma(\omega,\mathbf{n} \leftarrow \omega',\mathbf{n}) &= \frac{\pi}{2}\int \frac{d^4x_1}{(2\pi)^4}\exp(-iK(x_1-x_2))\langle \tilde{V}_1 \tilde{V}_2^* \rangle \exp(iK'(x_1-x_2)) \\ &= \frac{\pi}{2}k_0^4 \int \frac{d^4\rho}{(2\pi)^4}\langle \tilde{\varepsilon}(\rho)\tilde{\varepsilon}(0)\rangle \exp(i(K-K')\rho),\end{aligned} \tag{3.5.59}$$

or in more detail

$$\begin{aligned}\sigma(\omega,\mathbf{n} \leftarrow \omega',\mathbf{n}') = \frac{\pi}{2}k_0^4 \int \frac{d^3\rho d\tau}{(2\pi)^4}\langle \tilde{\varepsilon}(\rho,\tau)\tilde{\varepsilon}(0,0)\rangle \\ \times \exp[i(\mathbf{k}-\mathbf{k}')\rho - i(\omega-\omega')\tau].\end{aligned} \tag{3.5.60}$$

For static fluctuations, the correlation function $\langle \tilde{\varepsilon}(\rho,\tau)\tilde{\varepsilon}(0,0)\rangle$ does not depend upon τ, so that (3.5.60) becomes

$$\begin{aligned}\sigma(\omega,\mathbf{n} \leftarrow \omega',\mathbf{n}') = \delta(\omega-\omega')\frac{\pi}{2}k_0^4 \int \frac{d^3\rho}{(2\pi)h3}\langle \tilde{\varepsilon}(\rho)\tilde{\varepsilon}(0)\rangle \\ \times \exp[i\,\mathrm{Re}\,k^{\mathrm{eff}}(\mathbf{n}-\mathbf{n}')\rho],\end{aligned} \tag{3.5.61}$$

which is only distinguished from the conventional Born approximation by k_0 replaced with k^{eff} in the exponent (Rytov *et al.*, 1989).

Consider now an example of electromagnetic scattering from moving dipoles. As shown in Section 3.3, for a stationary dipole, the T-operator has the form (3.4.14). If a dipole velocity is small compared to the velocity of light, so that the quasi-static approximation is valid, it is sufficient to replace $\mathbf{r}_l$ by $\mathbf{r}_l(t)$ and $k_0 = \omega / c$ by $-i\partial_t / c$ in (3.4.14) to obtain

$$t^{(l)} = \frac{4\pi\alpha_l}{c^2}\delta(\mathbf{r} - \mathbf{r}_l(t))\partial_t^2. \tag{3.5.62}$$

For such an operator $t^{(l)}$, the quantity σ, appearing in the expression for cross section (3.5.55) takes the form

$$\sigma(\omega,\mathbf{n} \leftarrow \omega',\mathbf{n}') = \frac{\pi}{2}k_0^4(4\pi\alpha_l)^2 \int \frac{d^4x_1}{(2\pi)^4}\exp(iQ(x_1 - x_2))$$
$$\times\left\{N\langle\delta(\mathbf{r}_1 - \mathbf{r}_1(t_1))\delta(\mathbf{r}_2 - \mathbf{r}_1(t_2))\rangle + N^2\langle\tilde{\delta}(\mathbf{r}_1 - \mathbf{r}_1(t_1))\tilde{\delta}(\mathbf{r}_2 - \mathbf{r}_1(t_2))\rangle + \ldots\right\}, \tag{3.5.63}$$

where

$$Q = K - K' = (\omega - \omega', k\mathbf{n} - k'\mathbf{n}) \equiv (\Delta\omega, \mathbf{q}) \tag{3.5.64}$$

is the 4-vector of scattering. Here the first term in braces is due to scattering from independent particles, the second term allows for pair correlations, $\tilde{a} \equiv a - \langle a \rangle$, and dots denote terms due to higher correlations.

The expression (3.5.63) differs from the single-scattering approximation by k_0 replaced with $\mathrm{Re}\,k^{\mathrm{eff}}$ in the expression for the scattering vector. This difference is not substantial as in the case of the Born cross section (3.5.61). Also, it allows for a frequency change due to movement of particles. Finally, the expansion (3.5.63) enables higher correlations of particle positions (terms represented by dots) to be taken into account, which is impossible within the framework of the conventional single-scattering approximation.

3.6 CORRELATION FUNCTIONS OF THE SCATTERED FIELD. DIAGRAM INTERPRETATION OF THE TRANSFER EQUATION

3.6.1 Correlation Scattering Cross Sections. An Isolated Scatterer in a Free Space

In discussing the physical meaning of the transition from the Bethe–Salpeter equation to the equation of radiative transfer, it is useful to somewhat extend the conventional problem statement of scattering theory and to introduce the concept of the correlation scattering cross section.

Suppose that two plane waves with wave vectors $\mathbf{k}_1$ and $\mathbf{k}_2$, $\mathbf{k}_{1,2} = k_0\mathbf{n}_{1,2}$, where $k_0 = \omega / c$ is the free-space wave number,

$$u^0(\mathbf{r}) = A_1 e^{i\mathbf{k}_1\mathbf{r}} + A_2 e^{i\mathbf{k}_2\mathbf{r}} \tag{3.6.1}$$

are incident upon a bounded scattering volume V_s (Fig.3.16). In this section we restrict ourselves to the simplest case of a monochromatic field which satisfies the scalar wave equation (the factor $\exp(-i\omega t)$ is omitted).

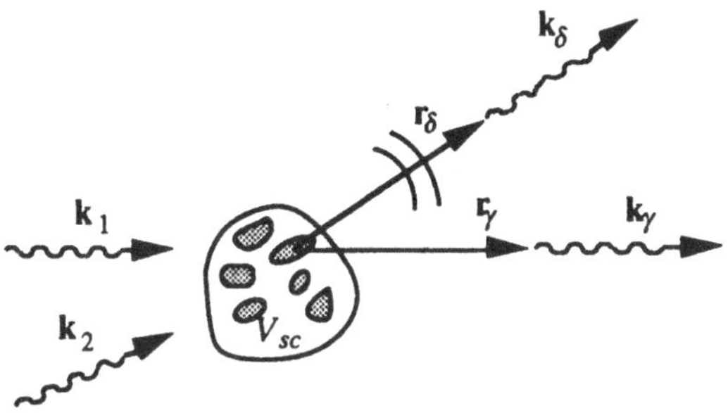

Figure 3.16. Incidence of two plane waves on a bounded scattering volume.

In view of the linearity of the problem, at a point $\mathbf{r}_\delta$ far from V_{sc}, the scattered field u_{sc} may be expressed in the form of two diverging spherical waves:

$$u_{sc}(\mathbf{r}_\delta) = f(\mathbf{k}_\delta \leftarrow \mathbf{k}_1)A_1 r_\delta^{-1}\exp(ikr_\delta) + f(\mathbf{k}_\delta \leftarrow \mathbf{k}_2)A_2 r_\delta^{-1}\exp(ikr_\delta), \tag{3.6.2}$$

where $f(\mathbf{k}_\delta \leftarrow \mathbf{k}_j)$ are the random scattering amplitudes and $\mathbf{k}_\delta = k_0\mathbf{r}_\delta / r_\delta$. It is natural to treat this expression as a single spherical wave with the amplitude $f(\mathbf{k}_\delta \leftarrow \mathbf{k}_1)A_1 + f(\mathbf{k}_\delta \leftarrow \mathbf{k}_2)A_2$. Later, however, it will be more convenient to deal with *two* scattered waves corresponding to *two* incident waves.

Let us consider two points r_δ and r_γ far from V_{sc}. From (3.6.2) we have for the correlation function of the scattered field at these points

$$\left\langle \tilde{u}_{sc}(\mathbf{r}_\delta)u_{sc}^*(\mathbf{r}_\gamma)\right\rangle = \frac{\exp\left[ik_0(r_\delta - r_\gamma)\right]}{r_\delta r_\gamma} \sum_{\alpha,\beta=1,2} \sigma(k_\delta, k_\gamma \leftarrow k_\alpha, k_\beta)A_\alpha A_\beta^*. \tag{3.6.3}$$

where $\tilde{u} = u - \langle u \rangle$ is the random part of u. It is natural to call the quantities

$$\sigma\left\langle \mathbf{k}_\delta, \mathbf{k}_\gamma \leftarrow \mathbf{k}_\alpha, \mathbf{k}_\beta \right\rangle = \left\langle \tilde{f}(\mathbf{k}_\delta \leftarrow \mathbf{k}_\alpha)\tilde{f}^*(\mathbf{k}_\gamma \leftarrow \mathbf{k}_\beta)\right\rangle \tag{3.6.4}$$

appearing in (3.6.3) *the correlation scattering cross sections.*

The physical meaning of these parameters is evident: they describe the correlation of scattered plane waves with wave vectors $\mathbf{k}_\delta$ and $\mathbf{k}_\gamma$ in the far zone (associated with the incidence of two plane waves with wave vectors $\mathbf{k}_\alpha$ and $\mathbf{k}_\beta$), or alternatively, the correlation of scattering processes $\mathbf{k}_\alpha \to \mathbf{k}_\delta$ and $\mathbf{k}_\beta \to \mathbf{k}_\gamma$. It follows that for determining the correlation scattering cross section (3.6.4) in the case of a stationary medium, the simultaneous incidence of two plane waves is unnecessary [case (3.6.1)]. Instead, we may successively treat two scattering processes which are statistically correlated with one another owing to the scattering at the same fluctuations of the medium.

If the observation points are assumed to coincide ($\delta = \gamma$) and there is only one incident wave ($A_2 = 0$), the quantity $\sigma(\mathbf{k}, \mathbf{k} \leftarrow \mathbf{k}', \mathbf{k}')$ is readily seen to be the usual cross section of incoherent scattering of a plane wave with wave vector $\mathbf{k}'$ into a wave with wave vector $\mathbf{k}$:

$$\sigma(\mathbf{k},\mathbf{k} \leftarrow \mathbf{k}',\mathbf{k}') \equiv \sigma(\mathbf{k} \leftarrow \mathbf{k}') = \sigma(\omega,\mathbf{n} \leftarrow \omega',\mathbf{n}'). \tag{3.6.5}$$

3.6.2 Weak Isolated Scatterer

Let us consider the relationship between the correlation scattering cross section (3.6.4) and the intensity operator K_{12} assuming as before that K_{12} may be taken in the one-group approximation (for simplicity the superscript is henceforth omitted).

Suppose that the scattering volume V_{sc} is sufficiently small, and inhomogeneities are rather weak in the sense that the average field within V_{sc} may be assumed equal to the field of the incident wave. To justify this assumption, it is necessary that the size of the scattering volume $V_{sc}^{1/3}$ be small compared to the extinction length L_{ext}, $V_{sc}^{1/3} << L_{ext}$. In the Bethe–Salpeter equation (3.3.35), the average Green's function $\langle G\rangle$ may then be replaced with the Green's function of free-space propagation G_0. Omitting all terms beginning from the third term in the iterative series for the Bethe–Salpeter equation (3.3.36) and acting by this expression onto a source function $q_1 q_2^*$, we obtain for the scattered-field correlation function

$$\begin{aligned}\left\langle \tilde{u}_{sc}(\mathbf{r}_1)\tilde{u}^*_{sc}(\mathbf{r}_2)\right\rangle &\approx G_{01}G^*_{02}K_{12}u_1^0 u_2^{0*}\\ &= \int G_0(\mathbf{r}_1-\mathbf{r}_1')G_0^*(\mathbf{r}_2-\mathbf{r}_2')K_{12}\begin{pmatrix}\mathbf{r}_1', & \mathbf{r}_1''\\ \mathbf{r}_2', & \mathbf{r}_2''\end{pmatrix}\\ &\times u^0(\mathbf{r}_1'')u^{0*}(\mathbf{r}_2'')d\mathbf{r}_1'd\mathbf{r}_2'd\mathbf{r}_1''d\mathbf{r}_2''\\ &= G_0(r_1)G_0^*(r_2)\int \exp[-i\mathbf{k}_1\mathbf{r}_1' + i\mathbf{k}_2\mathbf{r}_2']K_{12}\begin{pmatrix}\mathbf{r}_1', & \mathbf{r}_1''\\ \mathbf{r}_2', & \mathbf{r}_2''\end{pmatrix}\\ &\times u^0(\mathbf{r}_1'')u^{0*}(\mathbf{r}_2'')d\mathbf{r}_1'd\mathbf{r}_2'd\mathbf{r}_1''d\mathbf{r}_2''\\ &= \frac{\exp[ik_0(r_1-r_2)]}{r_1 r_2}\sum_{\alpha,\beta=1,2}(4\pi)^{-2}\tilde{K}_{12}\begin{pmatrix}\mathbf{k}_1, & \mathbf{k}_\alpha\\ \mathbf{k}_2, & \mathbf{k}_\beta\end{pmatrix}A_\alpha A_\beta^*.\end{aligned} \tag{3.6.6}$$

Here the last expression relates to the case where the observation points $\mathbf{r}_1$ and $\mathbf{r}_2$ are in the far zone with respect to scattering inhomogeneities, so that the Fraunhofer approximation could be used in evaluating the Green function

$$\begin{aligned}G_0(\mathbf{r}_\alpha - \mathbf{r}_\alpha') &= -4\pi\left|\mathbf{r}_\alpha - \mathbf{r}_\alpha'\right|^{-1}\exp(ik_0|\mathbf{r}_\alpha - \mathbf{r}_\alpha'|)\\ &\approx G_0(r_\alpha)\exp(i\mathbf{k}_\alpha\mathbf{r}_\alpha'),\end{aligned} \tag{3.6.7}$$

where

$$|\mathbf{r}_\alpha| >> |\mathbf{r}_\alpha'|,\ \mathbf{k}_\alpha = k_0\mathbf{r}_\alpha / \mathbf{r}_\alpha,\ \alpha = 1,2.$$

The quantity

$$\tilde{K}_{12}\begin{pmatrix} \mathbf{k}_1, & \mathbf{k}_3 \\ \mathbf{k}_2, & \mathbf{k}_4 \end{pmatrix} = \int \exp(-i\mathbf{k}_1\mathbf{r}_1' + i\mathbf{k}_2\mathbf{r}_2')K_{12}\begin{pmatrix} \mathbf{r}_1', & \mathbf{r}_1'' \\ \mathbf{r}_2', & \mathbf{r}_2'' \end{pmatrix}\exp(i\mathbf{k}_3\mathbf{r}_1'' - i\mathbf{k}_4\mathbf{r}_2'')d\mathbf{r}_1'd\mathbf{r}_2'd\mathbf{r}_1''d\mathbf{r}_2''$$
$$= \int d\mathbf{r}_1 d\mathbf{r}_2 \exp(-i\mathbf{k}_1\mathbf{r}_1 + i\mathbf{k}_2\mathbf{r}_2)K_{12}\exp(i\mathbf{k}_3\mathbf{r}_1 - i\mathbf{k}_4\mathbf{r}_2) \tag{3.6.8}$$

has the meaning of the kernel of the intensity operator K_{12} in the k-representation.

A comparison of eqns (3.6.3) and (3.6.6) shows that, in this approximation, the correlation scattering cross section is identical, accurate to the factor $(4\pi)^{-2}$, with the kernel of the intensity operator taken on the "energy surface" $k_1 = k_2 = k_3 = k_4 = k_0$ in the k-space :

$$\sigma(\mathbf{k}_1,\mathbf{k}_2 \leftarrow \mathbf{k}_3,\mathbf{k}_4) = (4\pi)^{-2}\tilde{K}_{12}\begin{pmatrix} \mathbf{k}_1, & \mathbf{k}_3 \\ \mathbf{k}_2, & \mathbf{k}_4 \end{pmatrix}. \tag{3.6.9}$$

3.6.3 Scattering at Continuous Gaussian Fluctuations

The above approximation is directly contiguous with the single-scattering approximation. For non-Gaussian fluctuations, however, this is likely to be somewhat more exact because for the intensity operator K_{12},the one-group approximation takes into account not only the second cumulants but also cumulants of higher orders. For Gaussian fluctuations,eqn (3.6.9) becomes to the single-scattering approximation. We consider the last case in more detail.

In the simple case (3.3.8) where the action of a perturbation operator V reduces to the multiplication by the Gaussian random quantity $k_0^2\tilde{\varepsilon}$, we have

$$V(\mathbf{r},\mathbf{r}') = -k_0^2\tilde{\varepsilon}(\mathbf{r})\delta(\mathbf{r}-\mathbf{r}'),$$
$$K_{12} = \langle V_1 V_2^* \rangle, \tag{3.6.10}$$
$$K_{12}\begin{pmatrix} \mathbf{r}_1, & \mathbf{r}_1' \\ \mathbf{r}_2 & \mathbf{r}_2' \end{pmatrix} = k_0^4\langle\tilde{\varepsilon}(\mathbf{r}_1)\tilde{\varepsilon}(\mathbf{r}_2)\rangle\delta(\mathbf{r}_1-\mathbf{r}_1')\delta(\mathbf{r}_2-\mathbf{r}_2').$$

The kernel K_{12} is then expressed in terms of the Fourier transform of the fluctuation correlation function

$$\tilde{B}(\mathbf{q}_R,\mathbf{q}_\rho) = \int B(\mathbf{R},\rho)\exp(i\mathbf{q}_R\mathbf{R} + i\mathbf{q}_\rho\rho)d^3R d^3\rho, \tag{3.6.11}$$

$$B(\mathbf{R},\rho) = \left\langle \tilde{\varepsilon}\left(\mathbf{R}+\frac{\rho}{2}\right)\tilde{\varepsilon}\left(\mathbf{R}-\frac{\rho}{2}\right)\right\rangle, \tag{3.6.12}$$

as follows

$$\tilde{K}_{12}\begin{pmatrix} k_1, & k_3 \\ k_2, & k_4 \end{pmatrix}$$
$$= k_0^4 \int d^3r_1 d^3r_2 \exp(-i\mathbf{k}_1\mathbf{r}_1 + i\mathbf{k}_2\mathbf{r}_2)\langle\tilde{\varepsilon}(\mathbf{r}_1)\tilde{\varepsilon}(\mathbf{r}_2)\rangle\exp(i\mathbf{k}_3\mathbf{r}_1 - i\mathbf{k}_4\mathbf{r}_2)$$
$$= k_0^4\tilde{B}\left(-\mathbf{k}_1 + \mathbf{k}_2 + \mathbf{k}_3 - \mathbf{k}_4, \frac{1}{2}(-\mathbf{k}_1 - \mathbf{k}_2 + \mathbf{k}_3 + \mathbf{k}_4)\right). \quad (3.6.13)$$

The respective scattering cross section determined by eqns (3.6.13) and (3.6.5) proves to be identical to that in the Born approximation.

3.6.4 Quasi-Homogeneous Weak Scatterer. Analogy with the Van Cittert–Zernike Theorem

Consider a quasi-homogeneous scattering volume V_{sc} which contains a large number of random inhomogeneities with characteristic sizes $l_c << V_{sc}^{1/3}$ and at the same time is rather small compared to the extinction length, $V_{sc}^{1/3} << L_{ext}$, so that the effect of scattering remains weak.

For the case of *continuous Gaussian fluctuations* (3.6.10) - (3.6.13), the assumption of the quasi-uniformity implies that the correlation function of fluctuations $B(\mathbf{R},\rho)$ changes slowly with $\mathbf{R}$ compared to its fast decay with an increase in $|\rho|$. For the sake of simplicity suppose that within the scattering volume V_{sc} the fluctuations are statistically uniform, that is, the correlation function $B(\mathbf{R},\rho)$ is factored:

$$B(\mathbf{R},\rho) = B_\varepsilon(\rho)\theta_V(\mathbf{R}), \quad (3.6.14)$$

where $B_\varepsilon(\rho)$ is the correlation function of fluctuations, and $\theta_V(R)$ is the characteristic function of the scattering volume which is equal to unity in V_{sc} and is zero outside V_{sc} Then eqn (3.6.11) yields

$$\tilde{B}(\mathbf{q}_R,\mathbf{q}_\rho) = \tilde{B}_\varepsilon(\mathbf{q}_\rho)(2\pi)^3\delta_V(\mathbf{q}_R), \quad (3.6.15)$$

where

$$\tilde{B}_\varepsilon(\mathbf{q}) = \int B_\varepsilon(\rho)e^{i\mathbf{q}\rho}d^3\rho,\ \delta_V(\mathbf{q}) = \int e^{i\mathbf{q}\mathbf{R}}\theta_V(\mathbf{R})(2\pi)^{-3}d^3R. \quad (3.6.16)$$

In passing to a statistically homogeneous unbounded medium ($V_{sc} \to \infty$), the function $\delta_V(\mathbf{q})$ becomes to the conventional delta function in accordance with (3.6.16):

$$\delta_V(\mathbf{q}) \xrightarrow[V_{sc}\to\infty]{} \delta(\mathbf{q}). \quad (3.6.17)$$

Thus, for large V_{sc}, the function $\delta_V(q_R)$ may be deemed different from zero in the region of small q_R only : $q_R << V_{sc}^{-1/3}$. On the other hand, the characteristic variation scale for $\tilde{B}_\varepsilon(q_\rho)$ is on the order of $q_\rho^* \sim l_c^{-1} >> V_{sc}^{-1/3}$. This enables us to set $q_R = 0$ in the argument of $\tilde{B}_\varepsilon(\mathbf{q}_\rho)$ in the quasi-homogeneous limit when substituting eqn (3.6.15) into eqn (3.6.13). Then

$$\tilde{K}_{12}\begin{pmatrix}\mathbf{k}_1, & \mathbf{k}_3 \\ \mathbf{k}_2 & \mathbf{k}_4\end{pmatrix} \approx k_0^4 \tilde{B}_\varepsilon\left(\frac{1}{2}(-\mathbf{k}_1 - \mathbf{k}_2 + \mathbf{k}_3 + \mathbf{k}_4)\right)$$
$$\times(2\pi)^3 \delta_V(-\mathbf{k}_1 + \mathbf{k}_2 + \mathbf{k}_3 - \mathbf{k}_4)$$
$$= k_0^4 \tilde{B}_\varepsilon(\mathbf{k}_3 - \mathbf{k}_1)(2\pi)^3 \delta_V(-\mathbf{k}_1 + \mathbf{k}_2 + \mathbf{k}_3 - \mathbf{k}_4), \qquad (3.6.18)$$

so that in accordance with eqn (3.6.9)

$$\sigma(\mathbf{k}_1, \mathbf{k}_2 \leftarrow \mathbf{k}_3, \mathbf{k}_4)$$
$$\approx k_0^4 (4\pi)^{-2} \tilde{B}_\varepsilon(\mathbf{k}_3 - \mathbf{k}_1)(2\pi)^3 \delta_V(-\mathbf{k}_1 + \mathbf{k}_2 + \mathbf{k}_3 - \mathbf{k}_4)$$
$$= \sigma(\mathbf{k}_1 \leftarrow \mathbf{k}_3)(2\pi)^3 \delta_V(-\mathbf{k}_1 + \mathbf{k}_2 + \mathbf{k}_3 - \mathbf{k}_4), \qquad (3.6.19)$$

Here $\sigma(\mathbf{k}_1 \leftarrow \mathbf{k}_3)$ is identical to the usual Born scattering cross section per unit volume.

For one incident wave with a wave vector $\mathbf{k}_1$, i.e. at $A_2 = 0$, in accordance with eqns (3.6.3) and (3.6.19) we may write (for $A_1 = 1$)

$$\left\langle \tilde{u}_s(\mathbf{r}_\delta)\tilde{u}_s^*(\mathbf{r}_\gamma)\right\rangle = \frac{\exp\left[ik_0(\mathbf{r}_\delta - \mathbf{r}_\gamma)\right]}{\mathbf{r}_\delta \mathbf{r}_\gamma} V_{sc}\sigma(\mathbf{k}_\delta \leftarrow \mathbf{k}_1)\left\{(2\pi)^3 V_{sc}^{-1} \delta_V(\mathbf{k}_\gamma - \mathbf{k}_\delta)\right\}. \quad (3.6.20)$$

When the observation points coincide, i.e. $\mathbf{r}_\gamma = \mathbf{r}_\delta$, the factor in braces brackets is easily seen to reduce to unity and this relationship then yields the usual expression for the mean of scattered field intensity $\langle |\tilde{u}_s^0(r_\delta)|^2 \rangle$ in the Born approximation. An angular distribution of this intensity is determined by a directional pattern for scattering by an effective inhomogeneity (i.e., with the cross section σ) and has a width $\theta' \sim \lambda / l_c$, as shown in Fig 3.17.

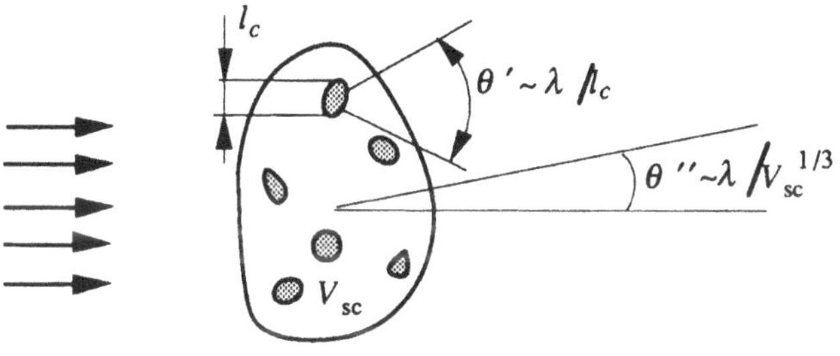

Figure 3.17. Scattering of a plane wave: θ' is the characteristic angle of scattering and θ'' is the characteristic angle of correlation of scattered waves.

The correlation properties of scattered radiation are described (abstracting from the trivial factor $\exp[ik_o(r_\delta - r_\gamma)]/r_\delta r_\gamma$) by a delta-like function of $\mathbf{k}_\gamma - \mathbf{k}_\delta$ and are determined by the diffraction at the *entire* volume V_{sc}, the correlation of scattered waves vanishing upon separation of $\mathbf{k}_\gamma$ and $\mathbf{k}_\delta$ by an angle $\theta'' \sim \lambda / V_{sc}^{1/3}$.

The formula (3.6.20) which describes the field correlation properties is an analog of the Van Cittert-Zernike theorem relating the coherence functions for radiation of short correlated sources to the solution of the wave diffraction on a region occupied by the sources (see Section 3.2). The distinction consists in that in eqn (3.6.20) sources are secondary (induced by an incident wave) and *volume* whereas the Van Cittert-Zernike theorem is usually stated for primary and surface sources.

3.6.5 General Case of Non-Gaussian Fluctuations

We us now write relations similar to (3.6.18) without being attached to any specific model for a quasi-homogeneous scattering medium.

If a scattering medium is statistically homogeneous (and therefore unbounded), the kernel of the intensity operator remains unchanged under simultaneous and uniform transtation of all its arguments,

$$K_{12}\begin{pmatrix} \mathbf{r}_1 + \mathbf{R}, & \mathbf{r}_3 + \mathbf{R} \\ \mathbf{r}_2 + \mathbf{R}, & \mathbf{r}_4 + \mathbf{R} \end{pmatrix} = K_{12}\begin{pmatrix} \mathbf{r}_1, & \mathbf{r}_3 \\ \mathbf{r}_2, & \mathbf{r}_4 \end{pmatrix},$$

that is, it depends only upon the differences $\mathbf{r}_i - \mathbf{r}_j$. As a consequence, using eqn (3.6.8) one can readily show that in the k-representation this kernel involves a delta function:

$$\tilde{K}_{12}\begin{pmatrix} \mathbf{k}_1, & \mathbf{k}_3 \\ \mathbf{k}_2, & \mathbf{k}_4 \end{pmatrix} \propto \delta(-\mathbf{k}_1 + \mathbf{k}_2 + \mathbf{k}_3 - \mathbf{k}_4).$$

In a more general case of a bounded but large (compared to the size of inhomogeneities) scattering volume ($V_{sc} >> l_c^3$) which is statistically homogeneous far from its boundaries, the delta function in $\tilde{K}_{12}$ transforms to the function $\delta_V(\mathbf{q}_R)$(3.6.16), so that the kernel of the intensity operator may be written in the form

$$\tilde{K}_{12}\begin{pmatrix} \mathbf{k}_1, & \mathbf{k}_3 \\ \mathbf{k}_2, & \mathbf{k}_4 \end{pmatrix} = V_{sc}^{-1}\tilde{K}_{12}\begin{pmatrix} \mathbf{k}_1, & \mathbf{k}_3 \\ \mathbf{k}_2, & \mathbf{k}_4 \end{pmatrix}(2\pi)^3 \delta_V(-\mathbf{k}_1 + \mathbf{k}_2 + \mathbf{k}_3 - \mathbf{k}_4). \tag{3.6.21}$$

This expression extends eqn (3.6.18) to the case of non-Gaussian fluctuations of the medium.

3.6.6 Diagram Interpretation of the Radiative Transfer Equation

The transition from the Bethe–Salpeter equation to the radiative transfer equation by asymptotic expansions, discussed in Section 3.5.4, represents a regular procedure that is suitable in the general case, rather a formal one. Here we consider the diagram interpretation of this transition suggested by Barabanenkov and Finkelberg (1967). Although the diagram approach does not enable one to directly estimate corrections to the transfer equation, it is very transparent and, in addition, leads to some new consequences which are rather difficult to derive by using the technique of asymptotic expansions. Analytical estimates of the corrections to the transfer equation derived by Barabanenkov and Finkelberg (1967) with the diagram approach may be found in the work of Barabanenkov (1973).

For simplicity we restrict ourselves to the monochromatic case where $L_0 = \Delta + k_0^2$ and $V = -k_0^2\tilde{\varepsilon}$ in the scalar problem (3.3.3). The iterative series for the Bethe–Salpeter equation (3.3.30) taken in the one-group approximation may be written as

$$\langle \tilde{G}_1 G_2^* \rangle = \langle G_1 \rangle \langle G_2^* \rangle K_{12}^{(1gr)} \langle G_1 \rangle \langle G_2^* \rangle + \langle G_1 \rangle \langle G_2^* \rangle K_{12}^{(1gr)} \langle G_1 \rangle \langle G_2^* \rangle K_{12}^{(1gr)} \langle G_1 \rangle \langle G_2^* \rangle + \ldots =$$

= 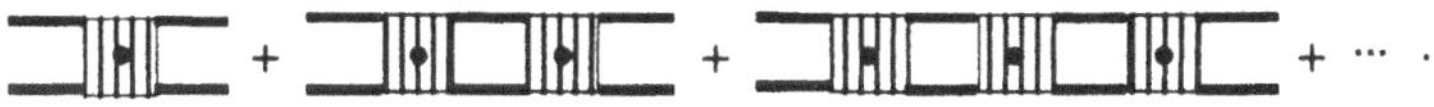 (3.6.22)

Here the operators $K_{12}^{(1gr)}$ describe the effective medium's inhomogeneities with sizes about l_c they appear as

$K_{12}^{(1gr)}$ =  , (3.6.23)

and parallel heavy lines correspond to operators $\langle G_1 \rangle \langle G_2^* \rangle$.

According to (3.3.76), in the simple case of a continuous medium with Gaussian fluctuations, in the one-group approximation, the intensity operator $K_{12}^{(1gr)}$ degenerates into $\langle \tilde{V}_1 \tilde{V}_2^* \rangle$ for which the previous graphical representation is used:

$\langle \tilde{V}_1 \tilde{V}_2^* \rangle$ =  . (3.6.24)

Now, the series (3.6.22) becomes the sum

$\langle \tilde{G}_1 \tilde{G}_2^* \rangle$ = 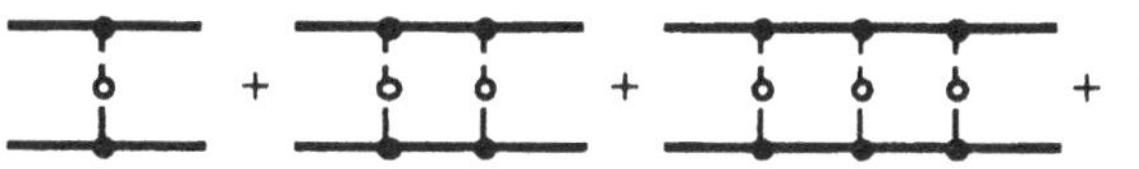 (3.6.25)

which is usually called the ladder approximation (Rytov et al., 1989b).

Barabanenkov and Finkelberg (1968) observed that the transition from the Bethe–Salpeter equation to the radiative transfer equation corresponds to the assumption that the major contribution into a scattered field is given by distant inhomogeneities which are in the Fraunhofer zone with respect to each other. This enables us to replace such inhomogeneities with effective point scatterers, that is to switch from the description of *partially* coherent act of scattering at effective inhomogeneities, $K_{12}^{(\mathrm{lgr})}$, with a size about l_c to the description of *completely* incoherent acts of scattering on the distant point inhomogeneities. Symbolically, such a replacement may be represented as a contraction of the effective inhomogeneity symbols in (3.6.22) into a point :

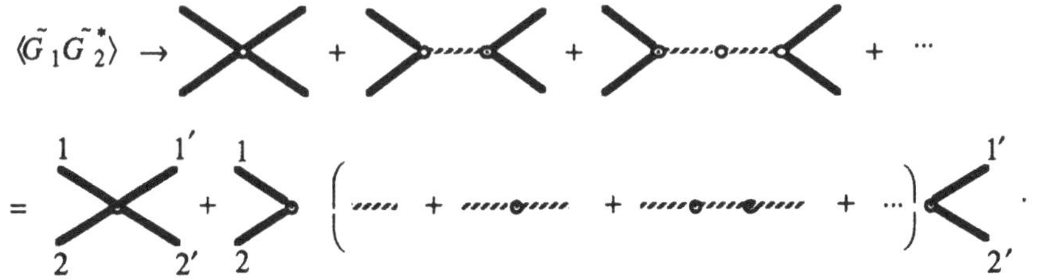

(3.6.26)

Here the circle with a dot represents a point scatterer, and the hatched strip describes the product of Green's functions $\langle G_1\rangle\langle G_2^*\rangle$ with coincident ends $x_1 = x_2$ and $x_1' = x_2'$ (subscripts 1' and 2' label the sources and subscripts 1 and 2 label observation points).

The sum entering eqn (3.5.26)

(3.6.27)

satsfies the equations

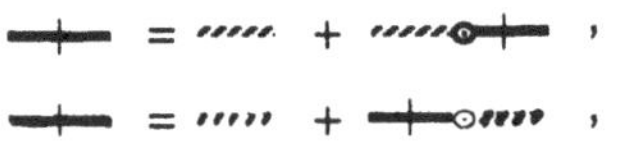

(3.6.28)

because their iterations lead to the series (3.6.27).

A more rigorous transition to the Fraunhofer approximation involves also a transition from the space of functions of two arguments (x_1, x_2) corresponding to the second moments $\langle u_1 u_2^*\rangle$, to the space of functions of x and $\mathbf{n}$ corresponding to the radiance $I = I(x, \mathbf{n})$. A detailed description of this transition is given in Section 3.5 where relations (3.6.28) are shown to be a diagram representation of the integral formulation of the radiative transfer equation.

From the diagram interpretation of the transition to the transfer equation we can draw an important conclusion. As seen from eqn (3.6.26) and (3.6.27), if one of equations (3.6.28) is solved, that is, the sum (3.6.27) is found, the second moment of Green's function can be expressed in terms of this sum as

$$\langle G_1 G_2^*\rangle = \langle G_1\rangle\langle G_2^*\rangle + \langle\tilde{G}_1\tilde{G}_2^*\rangle =$$

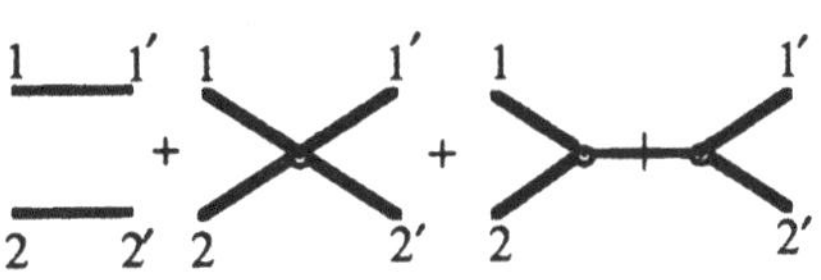

. (3.6.29)

This relationship enables us to write the coherence function Γ_{12} in terms of the solution to the transfer equation (3.6.28).

The first term on the right-hand side of eqn (3.6.29) is due to the free propagation of the coherent wave in the effective medium, the second describes the singly scattered field, and the third, in accordance with (3.6.27), takes into account an infinite number of completely incoherent scattering events and satisfies equation (3.6.28).

We have shown that the solution to the transfer equation, which was derived in Section 3.5 using asymptotic expansions, yields the coherence function of the scattered field for moderate separations $\rho = x_1 - x_2$ between observation points such that permits this function to be considered quasi-uniform. The relationship (3.6.29) extends this result because it implies that the solution to the transfer equation expresses the coherence function of scattered field for *arbitrary* separations ρ and *arbitrary* sources q (i.e., arbitrary incident wave). The requirement of complete quasi-uniformity of the scattered field, which was imposed in Section 3.6.3, is in fact unnecessarily strict. Actually, it is sufficient to demand that the field be quasi-uniform in a scattering event, that is, for separations ρ smaller than the inhomogeneity size l_c.

In order to follow the relationship between (3.6.29) and the transfer equation derived in Section 3.5.6 using the asymptotic expansion technique, we assume that the source points 1′ and 2′ as well as the observation points 1 and 2 are *sufficiently close* and we show this symbolically by bringing these points together in the diagrams (3.6.29). Then, the second moment $\langle G_1 \rangle \langle G_2^* \rangle$ will be directly expressed as a solution to the transfer equation:

$\langle G_1 G_2^* \rangle_{x_1 = x_2,\, x_1' = x_2'} =$ 

(3.6.30)

A more detailed discussion of the diagram relationship (3.6.29) will be given below in Section 3.6.9.

The diagram interpretation enables us also to qualitatively obtain the necessary condition for an allowable transition to the radiative transfer equation. Indeed, since successive scattering events are assumed to occur in the Fraunhofer zone with respect to each other, the separation between them (i.e., the length of the free path L_f) must satisfy the far zone condition:

$$L_f >> k l_c^2, \tag{3.6.31}$$

which implies that every subsequent scattering event must occur only after forming the directional pattern of the scattering at the effective inhomogeneity. For small-scale inhomogeneities ($kl_c << 1$) this condition sets up automatically by virtue of the assumption $L_f >> l_c$ which is necessary for the applicability of the one-group approximation (see Section 3.3.10) whereas for large-scale inhomogeneities ($kl_c >> 1$) the condition (3.6.31) is stronger than the inequality $L_f >> l_c$. Note that more detailed estimations of Barabanenkov (1975b) show the condition (3.6.31) to be replaced by a weaker restriction $L_f >> kl_c^2 (l_c / L_f)$ in the case of the coherence function. This is due to the fact that in the transition to the transfer equation, in phase expansions, the terms quadratic in l_c cancel out because of the symmetry of the coherence function.

3.6.7 Radiative Transfer Equation and the Correlation Cross Section in Scattering Media

Here we proceed to describe radiation in a scattering medium. Under certain conditions this radiation is described by the transfer equation derived above. Consider algebraic expressions corresponding to the diagram interpretation of the transfer equation from the preceding section.

For this purpose we somewhat modify the approach of Barabanenkov and Finkelberg (1967) and assume that the intensity operator K_{12} is expressed in the form of an integral of a certain distribution density over the scattering volume (in other words, the specific intensity operator) $K_{12}^0 \equiv K_{12}^0(\mathbf{R})$:

$$K_{12} = \int K_{12}^0(\mathbf{R}) d^3R. \tag{3.6.32}$$

The auxiliary density K_{12}^0 serves only to maintain the symmetry and otherwise is not necessary. For our consideration, it will be sufficient to deem the medium quasi-homogeneous so that the kernel of the intensity operator

$$K_{12}\begin{pmatrix} \mathbf{r}_1, & \mathbf{r}_3 \\ \mathbf{r}_2, & \mathbf{r}_4 \end{pmatrix}$$

depends significantly upon three argument differences $\mathbf{r}_i - \mathbf{r}_j$ only and almost does not change with a variation of the fourth independent argument.

The kernel

$$K_{12}^0\begin{pmatrix} \mathbf{r}_1, & \mathbf{r}_3 & \mathbf{R} \\ \mathbf{r}_2, & \mathbf{r}_4 & \end{pmatrix}$$

of the operator K_{12}^0 vanishes if one of the arguments $\mathbf{r}_j$ is far from $\mathbf{R}$ at a distance greater than the characteristic correlation scale of the medium. It is this property together with the locality of K_{12} in the one-group approximation that enables us to assign to $K_{12}^0(\mathbf{R})$ the meaning of the density of K_{12} in the neighborhood of $\mathbf{R}$. The choice of a specific intensity operator K_{12}^0 with these properties is ambiguous. However, this ambiguity does not affect the final results within the accuracy of the theory under consideration.

For definiteness we take this operator equal to

$$K_{12}^0(\mathbf{R}) = \delta\left(\mathbf{R} - \frac{\mathbf{r}_1 + \mathbf{r}_2}{2}\right) K_{12}, \tag{3.6.33}$$

assuming the density K_{12}^0 to be localized exactly midway between of points $\mathbf{r}_1$ and $\mathbf{r}_2$. In particular, for the case (3.6.10), this yields

$$\tilde{K}_{12}^0\begin{pmatrix}\mathbf{k}_1, & \mathbf{k}_3 \\ & & \mathbf{R} \\ \mathbf{k}_2, & \mathbf{k}_4\end{pmatrix} \equiv \int d^3r_1 d^3r_2 \exp(-ik_1\mathbf{r}_1 + ik_2\mathbf{r}_2) K_{12}^0(\mathbf{R}) \exp(ik_3\mathbf{r}_1 - ik_4\mathbf{r}_2)$$

$$= \exp\left[-i(\mathbf{k}_1 - \mathbf{k}_2 - \mathbf{k}_3 + \mathbf{k}_4)\mathbf{R}\right] k_0^4 \int B(\mathbf{R}, \rho)$$

$$\times \exp\left[-i(\mathbf{k}_1 + \mathbf{k}_2 - \mathbf{k}_3 - \mathbf{k}_4)\rho\right] d^3\rho . \tag{3.6.34}$$

It follows from (3.6.32) and (3.6.9) that, for a weak isolated scatterer, the correlation cross section can be written as an integral over the scattering volume:

$$\sigma(\mathbf{k}_1, \mathbf{k}_2 \leftarrow \mathbf{k}_3, \mathbf{k}_4) = \int \sigma_R(\mathbf{k}_1, \mathbf{k}_2 \leftarrow \mathbf{k}_3, \mathbf{k}_4) d^3R, \tag{3.6.35}$$

where the quantity

$$\sigma(\mathbf{k}_1, \mathbf{k}_2 \leftarrow \mathbf{k}_3, \mathbf{k}_4) = \tilde{K}_{12}^0\begin{pmatrix}\mathbf{k}_1, & \mathbf{k}_3 \\ & & \mathbf{R} \\ \mathbf{k}_2, & \mathbf{k}_4\end{pmatrix}(4\pi)^{-2} \tag{3.6.36}$$

has the meaning of the correlation scattering cross section per unit volume.

Description the field within a scattering medium, one should take into account that between scattering events the radiation propagates not in a free space but rather in some effective medium corresponding to the average field. In the one-group approximation, the transition to the effective medium reduces basically to replacing the wave number in free space k_0 with the effective wave number k^{eff}. Accordingly, we take the Green's function $\langle G\rangle(\mathbf{r}_1 - \mathbf{r}_2) \equiv \overline{G}_{12}$ describing the propagation of the mean field in a statistically homogeneous medium to be approximately equal to

$$\overline{G}_{12} \approx -\left(4\pi|\mathbf{r}_1 - \mathbf{r}_2|\right)^{-1} \exp\left(ik^{\text{eff}}|\mathbf{r}_1 - \mathbf{r}_2|\right). \tag{3.6.37}$$

Now, without further elucidation we write the algebraic relations that correspond to the Fraunhofer approximation using the contraction-to-a-point operator of effective inhomogeneities (3.6.22). With the aid of the representation (3.6.32) we obtain

$$= \overline{G}_{11'}\overline{G}_{22'}^* K_{12}^0\begin{pmatrix}\mathbf{r}_1' & \mathbf{r}_1'' \\ & & \mathbf{r}_0' \\ \mathbf{r}_2' & \mathbf{r}_2''\end{pmatrix}\overline{G}_{1''\tilde{1}}\overline{G}_{2''\tilde{2}}^* \rightarrow$$

$$= \overline{G}_{10'}\overline{G}_{20'}^* \exp\left[i\mathbf{k}_{20'}(\mathbf{r}_0' - \mathbf{r}_1') - i\mathbf{k}_{20'}^*(\mathbf{r}_0' - \mathbf{r}_2')\right] K_{12}^0\begin{pmatrix}\mathbf{r}_1' & \mathbf{r}_1'' \\ & & \mathbf{r}_0' \\ \mathbf{r}_2' & \mathbf{r}_2''\end{pmatrix}\overline{G}_{0'\tilde{1}}\overline{G}_{0'\tilde{2}}^*$$

$$\times \exp\left[i\mathbf{k}_{0'\tilde{1}}(\mathbf{r}_0' - \mathbf{r}_2'') - i\mathbf{k}_{0'\tilde{2}}(\mathbf{r}_0' - \mathbf{r}_2'')\right]$$

$$= \overline{G}_{10'}\overline{G}^*_{20'} \exp\left[i\left(\mathbf{k}_{10'} - \mathbf{k}^*_{20'} - \mathbf{k}_{0'\bar{1}} + \mathbf{k}^*_{0'\bar{2}}\right)\mathbf{r}_{0'}\right]$$

$$\times K^0_{12}\begin{pmatrix} \mathbf{k}_{10'}, & \mathbf{k}_{0'\bar{1}} & \\ & & \mathbf{r}'_0 \\ \mathbf{k}^*_{20'}, & \mathbf{k}_{0'\bar{2}} & \end{pmatrix}\overline{G}_{0'\bar{1}}\overline{G}^*_{0'\bar{2}}$$

$$= \overline{G}_{10'}\overline{G}^*_{20'}(4\pi)^2\sigma_{\mathbf{r}'_0}(\mathbf{k}_{10'},\mathbf{k}_{20'} \leftarrow \mathbf{k}_{0'\bar{1}},\mathbf{k}_{0'\bar{2}})\overline{G}_{0'\bar{1}}\overline{G}^*_{0'\bar{2}} \,. \tag{3.6.38}$$

Here $\overline{G}_{11'} = \langle G\rangle(\mathbf{r}_1,\mathbf{r}'_1)\,, \overline{G}_{22'} = \langle G\rangle(\mathbf{r}_2,\mathbf{r}'_2)$ and so on,

$$\mathbf{k}_{\alpha\beta} = k^{\text{eff}}(\mathbf{r}_\alpha - \mathbf{r}_\beta)|\mathbf{r}_\alpha - \mathbf{r}_\beta|^{-1},$$

the integration is performed with respect to the primed arguments, and the quantity $\sigma_R(\mathbf{k}_1,\mathbf{k}_2 \leftarrow \mathbf{k}_3,\mathbf{k}_4)$ related to K^0_{12} by

$$\sigma_R(\mathbf{k}_1,\mathbf{k}_2 \leftarrow \mathbf{k}_3,\mathbf{k}_4) = \exp\left[i(\mathbf{k}_1 - \mathbf{k}^*_2 - \mathbf{k}_3 + \mathbf{k}^*_4)\mathbf{R}\right]\tilde{K}^0_{12}\begin{pmatrix} \mathbf{k}_1, & \mathbf{k}_3 & \\ & & \mathbf{R} \\ \mathbf{k}^*_2, & \mathbf{k}^*_4 & \end{pmatrix}, \tag{3.6.39}$$

has the meaning of the correlation scattering cross section per unit volume. Substituting here eqn (3.6.34), we find for the model of Gaussian fluctuations

$$\sigma_R(\mathbf{k}_1,\mathbf{k}_2 \leftarrow \mathbf{k}_3,\mathbf{k}_4) = k_0^4\int B(\mathbf{R},\rho')\exp\left(i\frac{-\mathbf{k}_1 - \mathbf{k}_2 + \mathbf{k}_3 + \mathbf{k}_4}{2}\rho'\right)d^3\rho'. \tag{3.6.40}$$

The relation (3.6.39) differs in two respects from the previously considered relation (3.6.36) which is valid for a weak scatterer in free space. First, eqn (3.6.39) has an exponential factor due to the fact that the point $\mathbf{R}$ does not coincide with the origin of coordinates in the general case. Second, in (3.6.39), the wave number for the effective medium $k^{\text{eff}} = k_1^{\text{eff}} + ik_2^{\text{eff}}$ arises instead of the free-space wave number k_0. We recall that the imaginary part k_2^{eff} describes the attenuation of the beam wave because of scattering. The one-group approximation, which was used in deriving the radiative transfer equation, requires that of this attenuation be small over the length of a characteristic inhomogeneity; that is, $k_2^{\text{eff}} l_c << 1$. Therefore substituting k^{eff} into the expression for the correlation scattering cross section we may neglect the imaginary part of k^{eff} and let $k^{\text{eff}} \approx k_1^{\text{eff}}$. Quite similarly, we may write the expressions for other terms of the expansion (3.6.22) in the Fraunhofer approximation.

From this derivation it follows that the point R, at which the correlation cross section is taken, is defined to an accuracy of distances on the order of many correlation lengths l_c. The cross section (3.6.39) therefore has an informal meaning when applied to large scattering volumes $V_{sc} >> l_c^3$ containing many inhomogeneities.

3.6.8 Algebraic and Diagram Forms of the Transfer Equation

Let us introduce the following notation for the operators and the kernels corresponding to the foregoing diagrams:

$$\cdots\cdots = F_0 \leftrightarrow F_0(\mathbf{r},\mathbf{n},\mathbf{r}',\mathbf{n}') = |\langle G\rangle(r,r')|^2\,\delta_2(\mathbf{n}\mathbf{n}_{rr'})\delta_2(\mathbf{n}\mathbf{n}'),$$

$$-\!\!+\!\!- = F \leftrightarrow F(\mathbf{r},\mathbf{n},\mathbf{r}',\mathbf{n}'),$$

$$\circ \;= \tau \leftrightarrow \tau(\mathbf{r},\mathbf{n};\mathbf{r}',\mathbf{n}') = \delta(\mathbf{r}-\mathbf{r}')\;\tilde{K}_{12}^{0}\begin{pmatrix} k^{\mathrm{eff}}\mathbf{n}, & k^{\mathrm{eff}}\mathbf{n}' & \\ & & \mathbf{r} \\ k^{\mathrm{eff}*}\mathbf{n}, & k^{\mathrm{eff}*}\mathbf{n}' & \end{pmatrix}$$

$$= \delta(\mathbf{r}-\mathbf{r}')(4\pi)^2\,\sigma(k^{\mathrm{eff}}\mathbf{n},k^{\mathrm{eff}}\mathbf{n} \leftarrow k^{\mathrm{eff}}\mathbf{n}',k^{\mathrm{eff}}\mathbf{n}')$$

$$= \delta(\mathbf{r}-\mathbf{r}')(4\pi)^2\,\sigma(k^{\mathrm{eff}}\mathbf{n} \leftarrow k^{\mathrm{eff}}\mathbf{n}). \tag{3.6.41}$$

Here $|\mathbf{n}|=|\mathbf{n}'|=1$, $\mathbf{n}_{rr'} = (\mathbf{r}-\mathbf{r}')/|\mathbf{r}-\mathbf{r}'|$, all the operators act in the space of functions depending upon $\mathbf{r}$ and the unit vector$\mathbf{n}$, and $\delta_2(\mathbf{n}\mathbf{n}')$ is the delta function on a unit sphere which may be defined by

$$\delta_2(r\mathbf{n} - r_1'\mathbf{n}_1') = \delta_2(\mathbf{n}\mathbf{n}')\delta(r-r')/r^2, \tag{3.6.42}$$

so that

$$\delta_2(\mathbf{n}\mathbf{n}') = 0, \quad \mathbf{n} \neq \mathbf{n}',$$

$$\int f(\mathbf{n}')\delta_2(\mathbf{n}\mathbf{n}')d\Omega_{\mathbf{n}'} = f(\mathbf{n}). \tag{3.6.43}$$

In these designations the transfer equation (3.6.28) can be written in operator form as

$$F = F_0 + F_0\tau F \tag{3.6.44}$$

or in the form of equation for kernels

$$F(\mathbf{r},\mathbf{n};\mathbf{r}_0,\mathbf{n}_0) = F_0(\mathbf{r},\mathbf{n};\mathbf{r}_0,\mathbf{n}_0) + \int F(\mathbf{r},\mathbf{n};\mathbf{r}',\mathbf{n}')\tau(\mathbf{r}',\mathbf{n}';\mathbf{r}'',\mathbf{n}'')$$
$$\times F(\mathbf{r}'',\mathbf{n}'';\mathbf{r}_0,\mathbf{n}_0)d^3r'd^3r''d\Omega_{\mathbf{n}'}d\Omega_{\mathbf{n}''}. \tag{3.6.45}$$

It is easy to verify that the kernel F_0 satisfies the equation

$$n\nabla F_0(\mathbf{r},\mathbf{n};\mathbf{r}_0,\mathbf{n}_0) = -2\,\mathrm{Im}\,k^{\mathrm{eff}}F_0(\mathbf{r},\mathbf{n};\mathbf{r}_0,\mathbf{n}_0) + (4\pi)^{-2}\delta(\mathbf{r}-\mathbf{r})\delta_2(\mathbf{n}\mathbf{n}_0) \tag{3.6.46}$$

or in operator form

$$n\nabla F_0 = -2\,\mathrm{Im}\,k^{\mathrm{eff}}F_0 + (4\pi)^{-2}\hat{1}. \tag{3.6.47}$$

Taking this relationship into account, we act by the operator $\mathbf{n}\nabla$ on both sides of equation (3.6.44) to obtain

$$n\nabla F = -2\,\mathrm{Im}\,k^{\mathrm{eff}} F + (4\pi)^{-2}\,\tau F + (4\pi)^{-2}\hat{1} \tag{3.6.48}$$

or in the form of an equation for the kernel F

$$\begin{aligned} n\nabla F(\mathbf{r},\mathbf{n};\mathbf{r}_0,\mathbf{n}_0) = {} & -2\,\mathrm{Im}\,k^{\mathrm{eff}} F(\mathbf{r},\mathbf{n};\mathbf{r}_0,\mathbf{n}_0) \\ & + \int \sigma_r(k^{\mathrm{eff}}\mathbf{n} \leftarrow k^{\mathrm{eff}}\mathbf{n}')F(\mathbf{r},\mathbf{n}';\mathbf{r}_0,\mathbf{n}_0)\,d\Omega_{\mathbf{n}} \\ & + (4\pi)^{-2}\,\delta(\mathbf{r}-\mathbf{r}_0)\,\delta_2(\mathbf{n}\mathbf{n}_0). \end{aligned} \tag{3.6.49}$$

Here

$$\sigma_r(k^{\mathrm{eff}}\mathbf{n} \leftarrow k^{\mathrm{eff}}\mathbf{n}') = (4\pi)^{-2}\;\tilde{K}^0_{12}\begin{pmatrix} k^{\mathrm{eff}}\mathbf{n}, & k^{\mathrm{eff}}\mathbf{n}' \\ k^{\mathrm{eff}*}\mathbf{n}, & k^{\mathrm{eff}}\mathbf{n}' \end{pmatrix}\mathbf{r} \tag{3.6.50}$$

is the scattering cross section per unit volume.

The relationship (3.6.49) represents the conventional transfer equation with the source $(4\pi)^{-2}\delta(\mathbf{r}-\mathbf{r}_0)\delta_2(\mathbf{n}\mathbf{n}_0)$, hence the quantity $(4\pi)^2 F(\mathbf{r},\mathbf{n};\mathbf{r}_0,\mathbf{n}_0)$ is the Green function for the transfer equation corresponding to the unit source $\delta(\mathbf{r}-\mathbf{r}_0)\delta_2(\mathbf{n}\mathbf{n}_0)$. We now write the explicit relationship between this Green function and the second moment of the Green function for the wave equation $\langle G_1 G_2^*\rangle$ which was previously obtained in diagram designations. For this purpose we represent (3.6.29) in the form

$$\langle \tilde{G}_1\tilde{G}_2^*\rangle \equiv \langle \tilde{G}_1\tilde{G}_2^*\rangle - \langle G_1\rangle\langle G_2^*\rangle =$$ 

$+$ 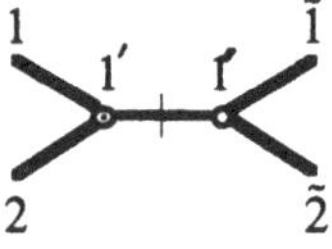

. (3.6.51)

The first term on the right-hand side of this relationship describes a singly scattered field represented in detail in (3.6.38). For the second term, in full analogy with (3.6.38), we have

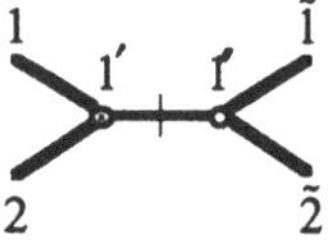

$$\begin{aligned} & = \overline{G}_{11'}\overline{G}^*_{21'}\exp\left[i(\mathbf{k}_{11'} - \mathbf{k}^*_{21'} - k^{\mathrm{eff}}\mathbf{n}' + k^{\mathrm{eff}*}\mathbf{n}')\mathbf{r}'_1\right] \\ & \times \tilde{K}^0_{12}\begin{pmatrix} \mathbf{k}_{11'}, & k^{\mathrm{eff}}\mathbf{n}' \\ \mathbf{k}^*_{21'}, & k^{\mathrm{eff}*}\mathbf{n}' \end{pmatrix}\mathbf{r}'_1 \\ & \times F(\mathbf{r}'_1,\mathbf{n}';\mathbf{r}''_1,\mathbf{n}'')\exp\left[i(k^{\mathrm{eff}}\mathbf{n}'' - k^{\mathrm{eff}*}\mathbf{n}'' - \mathbf{k}_{1''\tilde{1}} + \mathbf{k}^*_{1''\tilde{2}})\mathbf{r}''_1\right] \end{aligned}$$

$$\times \tilde{K}_{12}^{0}\begin{pmatrix} k^{\text{eff}}\mathbf{n}'', & \mathbf{k}_{1''\tilde{1}} \\ k^{\text{eff}}\mathbf{n}'', & \mathbf{k}^{*}_{1''\tilde{2}} \end{pmatrix} \mathbf{r}_1'' \Big) \overline{G}_{1''\tilde{1}} \overline{G}^{*}_{1''\tilde{2}} \,. \qquad (3.6.52)$$

As before, here the integration with respect to the primed arguments is implied.

3.6.9 Radiance and Correlation Functions of the Scattered Field

In the preceding section we noted that the solution to the radiative transfer equation enables one to express the correlation function of the scattered field not only for sufficiently small separations $\rho = \mathbf{r}_1 - \mathbf{r}_2$, when we may directly use the relationship (2.1.25), but also for arbitrary separations between observation points and for an arbitrary (rather than quasi-uniform only) incident wave. This is clearly seen from the diagram interpretation for the transfer equation which expresses the second moment of the field Green function through the Green function of the radiative transfer equation using (3.6.29). Consider next the corresponding algebraic relations for the second moments of the field.

For the sake of simplicity we restrict ourselves to the case of an incident coherent wave described in the geometrical optics approximation. As shown in Section 3.1.9, the second moment of such a wave is quasi-uniform and hence may be treated within the framework of transfer theory. In the diagram interpretation, the transition to transfer theory corresponds to bringing together the end points of respective diagrams, therefore, in order to obtain the correlation function, it is sufficient to bring source points on the diagrams (3.6.29) together. In view of the transfer equation (3.6.28) this yields for the correlation of Green's function

$\langle \tilde{G}_1 \tilde{G}_2^{*} \rangle =$  $+$  $=$  , (3.6.53)

We now act by two sides of (3.6.53) on the source function $q_1 q_2^{*}$ corresponding to the second moment of an incident wave and obtain the following relation for the correlation function of the scattered field:

$\langle \tilde{u}_1 \tilde{u}_2^{*} \rangle = \langle \tilde{G}_1 \tilde{G}_2^{*} \rangle q_1 q_2^{*} =$  $q_1 q_2^{*}$

$$= \int \overline{G}_{12} \overline{G}^{*}_{20} (4\pi)^2 \sigma_{r_0} (\mathbf{k}_{10}, \mathbf{k}_{20} \leftarrow k^{\text{eff}}\mathbf{n}', k^{\text{eff}}\mathbf{n}') I(\mathbf{r}_0, \mathbf{n}') d^3 r_0 d\Omega_{\mathbf{n}'} \,. \qquad (3.6.54)$$

We take into account that, in this case, the quantity $F q_1 q_2^{*}$ may be assumed equal to the radiance:

$F q_1 q_2^{*} =$ $q_1 q_2^{*} = I(\mathbf{r}, \mathbf{n})$. (3.6.55)

The physical meaning of eqn (3.6.54) is quite clear: it describes the propagation of the field second moment in the effective medium (operator $\langle G_1\rangle\langle G_2\rangle$) following the last scattering at an effective inhomogeneity for the field that has already undergone an infinite number of single scattering events (solution of the transfer equation I). In the general case, as seen from (3.6.54), in order to express the correlation $\langle \tilde{u}_1\tilde{u}_2^*\rangle$ it is necessary to use the correlation scattering cross section $\sigma_r(\mathbf{k}_1,\mathbf{k}_2 \leftarrow \mathbf{k}_3,\mathbf{k}_4)$ whereas it is not sufficient to know the conventional scattering cross section $\sigma_r(\mathbf{k} \leftarrow \mathbf{k}')$ which is expressed in terms of the correlation cross section at $\mathbf{k}_1 = \mathbf{k}_2$ and $\mathbf{k}_3 = \mathbf{k}_4$. It is only when scattering inhomogeneities are basically in the Fraunhofer zone relative to $\rho = \mathbf{r}_1 - \mathbf{r}_2$, including the observation points $\mathbf{r}_1$ and $\mathbf{r}_2$ (that is, the distance from the observation points to the main part of scattering inhomogeneities is large, $R \geq \max(k_0\rho^2,\rho)$, and the contribution from closer inhomogeneities is negligible), we may assume that one and the same plane wave ($k_1 \approx k_2$) arrives at both observation points and the second moment can then be expressed through the solution of the transfer equation similarly to eqn (3.5.22). The respective expression corresponds to an incomplete contraction of the ends on the diagram in eqn (3.6.54).

An algebraic relationship corresponding to eqn (3.5.22) can be obtained from eqn (3.6.54) by introducing the sum and difference coordinates for the observation points, $\mathbf{r}_{1,2} = \mathbf{R} \pm \rho/2$, and using the Fraunhofer approximation in the Green's functions $\overline{G}_{10}$ and $\overline{G}_{20}$:

$$\begin{aligned}\langle u_1u_2^*\rangle &= \int \overline{G}(\mathbf{R}+\rho/2-\mathbf{r}_0)\overline{G}^*(\mathbf{R}-\rho/2-\mathbf{r}_0)(4\pi)^2\\ &\times \sigma_{r_0}(\mathbf{k}_{10},\mathbf{k}_{20} \leftarrow k^{\text{eff}}\mathbf{n}',k^{\text{eff}}\mathbf{n}')I(\mathbf{r}_0,\mathbf{n}')d^3r_0d\Omega_{\mathbf{n}'}\\ &\approx \int |\overline{G}(\mathbf{R}-\mathbf{r}_0)|^2 \exp(i\operatorname{Re}\mathbf{k}_{R_0}\rho)(4\pi)^2\sigma_{r_0}(\mathbf{k}_{R_0},\mathbf{k}_{R_0} \leftarrow k^{\text{eff}}\mathbf{n}',k^{\text{eff}}\mathbf{n}')\\ &\times I(\mathbf{r}_0,\mathbf{n}')d^3r_0d\Omega_{\mathbf{n}'} = \int \exp(i\operatorname{Re}k^{\text{eff}}\mathbf{n}'\rho)I(\mathbf{R},\mathbf{n}')d\Omega_{\mathbf{n}'}. \qquad (3.6.56)\end{aligned}$$

Here, we let

$$\mathbf{k}_{10} \approx \mathbf{k}_{20} \approx \mathbf{k}_{R_0} = k^{\text{eff}}(\mathbf{R}-\mathbf{r}_0)/|\mathbf{R}-\mathbf{r}_0|,$$

observe that, according to eqns (3.6.55) and (3.6.44), the radiance satisfies the integral transfer equation

$$\begin{aligned}I &= Fq_1q_2^* = (F_0\tau F - F_0)q_1q_2^* = F_0\tau I - F_0q_1q_2^*\\ &= \int |\overline{G}(\mathbf{r}-\mathbf{r}_0)|^2(4\pi)^2\sigma_{r_0}(\mathbf{k}_{rr_0},\mathbf{k}_{rr_0} \leftarrow k^{\text{eff}}\mathbf{n}',k^{\text{eff}}\mathbf{n}')\\ &\times I(\mathbf{r}_0,\mathbf{n}')d^3r_0d\Omega_{\mathbf{n}'} - F_0q_1q_2^*, \qquad (3.6.57)\end{aligned}$$

and neglect the term $-F_0q_1q_2^*$, thus ignoring the coherent component of the field.

PROBLEMS

Problem 3.1. The coherence function $\Gamma = \langle u(x_1)u^*(x_2)\rangle$ is positive definite

$$\int \Gamma(x_1,x_2)v(x_1)v^*(x_2)dx_1dx_2 = \left\langle \left| \int u(x_1)v(x_1)dx_1 \right|^2 \right\rangle \geq 0, \tag{1}$$

where $v(x)$ is an arbitrary function. This property enables $\Gamma(x_1,x_2)$ to be treated as the kernel of a positive definite Hermitian operator $\hat{\Gamma}$. Show that this property implies the possibility of expanding the coherence function $\Gamma(x,x_2)$ into a sum of mutually incoherent components (the *Karhunen–Loeve expansion*).

Solution. For Hermitian kernels, the following expansion is valid (see e.g. Korn and Korn, 1961):

$$\Gamma(x_1,x_2) = \sum_n \varphi_n(x_1)\varphi_n^*(x_2)\lambda_n, \tag{2}$$

where λ_n are the eigenvalues and $\varphi_n(x)$ are the eigenfunctions of the operator $\hat{\Gamma}$ that satisfy the orthonormality condition

$$\int \varphi_n(x)\varphi_m^*(x)dx = \delta_{nm}. \tag{3}$$

In statistical theory, relation (2) is called the Karhunen–Loeve expansion. It corresponds to the expansion of the random field into a sum of many mutually incoherent (orthogonal) components with intensities λ_n (see Problem 3.2) similarly to the decomposition of a vector field into two orthogonal polarizations. In the general case, the summation in (2) is completed by the integration over the continuous spectrum of eigenvalues.

The Wiener–Khinchin theorem (3.1.10) written in the form

$$\Gamma = \int J_k \exp(ikx_1)\exp(-ikx_2)dk, \tag{4}$$

is obviously a special case of the Karhunen–Loeve expansion of a statistically uniform field where the eigenfunctions $\varphi_n(x)$ have the form of plane waves $\exp(ikx)$, and the spectrum J_k is simply the intensity of such waves.

The Karhunen–Loeve expansion (2) is valid in the general case of statistically nonuniform fluctuations but this expansion is rather difficult to be used in solving practical problems. While, in the uniform case, the eigenfunctions φ_n are plane waves, in the general case, the functions $\varphi_n(x)$ are nearly unknown. Therefore even the knowledge of the spectrum λ_n does not guarantee the restoration of the coherence function Γ.

Problem 3.2. Using the expansion (2) of Problem 3.1, prove the *Karhunen–Loeve theorem*: in the expansion of the random field $u(x)$

$$u(x) = \sum_n u_n \varphi_n(x) \tag{1}$$

in terms of the basis functions $\varphi_n(x)$ the random coefficients u_n are uncorrelated with each other (the mean value of u is assumed to be zero).

Solution. In view of the orthonormality condition for the eigenfunctions φ_n, we have

$$u_n = \int \varphi_n^*(x) u(x) dx. \tag{2}$$

It follows that $\langle u_n \rangle = 0$, and by virtue of eqn (2) from Problem 3.1 we obtain

$$\langle u_l u_m^* \rangle = \int \varphi_l^*(x_1) \varphi_m(x_2) \Gamma(x_1, x_2) dx_1 dx_2 = \delta_{lm} \lambda_l. \tag{3}$$

Thus, the coefficients u_l and u_m are uncorrelated for $l \neq m$, and the quantities $\lambda_l = \langle |u_l|^2 \rangle$ represent the average intensities of coefficients u_l.

Problem 3.3. The function Q serves as a kernel of the integral operator (3.1.15) in the definition of the local spectrum. What kind of constraint is imposed on the function Q by the real-valuedness of the spectrum $J_K(R)$?

Solution. Note that $\Gamma^*(R,\rho) = \Gamma(R,-\rho)$, and the real-valuedness condition of the spectrum has the form $J_K(R) = J_K^*(R)$. Applying these conditions to (3.1.15), we obtain the desired relationship

$$Q(R,K;R',\rho') = Q^*(R,K;R',-\rho')$$

Problem 3.4. The Wigner function (3.1.17) is one of the possible Fourier transforms for the coherence function $\Gamma(R,\rho)$, namely, the Fourier transform with respect to the difference argument ρ. Find the relationship between the Wigner function and other Fourier transforms (with respect to the first argument R and with respect to both arguments R and ρ).

Solution. Denote taking the Fourier transform with respect to x by a symbolic replacement of the coordinate x with its conjugate argument K_x. The Wigner function is then written symbolically as $\Gamma(R,K_\rho)$. The Fourier transforms with respect to the first, second, and both arguments then have the form (with allowance of the definition for the coherence function):

(*i*) $\Gamma(R,\rho)=\left\langle u\left(R+\frac{\rho}{2}\right)u*\left(R-\frac{\rho}{2}\right)\right\rangle$, coherence function

(*ii*) $\Gamma(R,K_\rho)=\hat{F}_{K_\rho\leftarrow\rho}\Gamma(R,\rho)$, Wigner function

(*iii*) $\Gamma(K_R,\rho)=\hat{F}_{K_R\leftarrow R}\Gamma(R,\rho)$, ambiguity function

(*iv*) $\Gamma(K_R,K_\rho)=\hat{F}_{K_R\leftarrow R}\Gamma(R,\rho)$, full Fourier transform.

Here, the variables R and ρ are related to the original arguments $x_{1,2}$ by $R=(x_1+x_2)/2$, $\rho=x_1-x_2$. If K_1 and K_2 are the conjugates of x_1 and x_2, then $K_R=K_1-K_2$ and $K_\rho=(K_1+K_2)/2$.

Similarly to the Wigner function (*ii*), the ambiguity function (*iii*) obeys equation (3.1.25) which is known as the constancy condition for the *body* of the ambiguity function in this case (Skolnic, 1970). [Radar theory, where this terminology is used, treats *signals* $\xi(t)$ without spatial argument in place of the fields $u(\mathbf{r},t)$].

Problem 3.5. Prove the uncertainty relation (3.1.32).

Solution. In order to prove this relation, it is sufficient to consider the nonnegativity condition for the quadratic form:

$$\langle\int|\alpha r_z u(x)+d_{r_z}u(x)|^2\,d^4x\rangle=\alpha^2(\Delta z)^2-\alpha+(\Delta k_z)^2\geq 0.$$

Problem 3.6. The velocity potential φ of *an acoustic* field in a homogeneous and stationary medium obeys the wave equation (3.1.52)

$$L_\varphi=(\Delta-c^{-2}\partial_\tau^2)\varphi=q,$$

the energy density w and its flux $\mathbf{S}$ are expressed as follows:

$$w=\left(\frac{\rho_0}{2}\right)\left[(\nabla\varphi)^2+c^{-2}(\partial_t\varphi)^2\right],$$

$$\mathbf{S}=-\rho_0(\nabla\varphi)(\partial_t\varphi),$$

where c is the sound velocity and ρ_0 the average density of the medium. Using the radiometric relationships (2.1.16) and (2.1.17), find the relation between the radiance I and the spectral density $I_{\omega n}$ from eqn (3.1.78).

Solution. Let us represent the real-valued velocity potential φ as the real part of the complex field, $\varphi=\mathrm{Re}\,u$, and introduce the *second coherence function* $\tilde{\Gamma}=\langle u_1u_2\rangle$ along with the coherence function $\Gamma=\langle u_1u_2^*\rangle$. The second moment of φ will then be

$$\langle\varphi_1\varphi_2\rangle \equiv \langle\varphi(x_1)\varphi(x_2)\rangle = 1/2\,\mathrm{Re}(\Gamma+\tilde{\Gamma}), \tag{1}$$

and the ensemble average of the energy density may be written in the form

$$\begin{aligned}\langle w\rangle &= \frac{1}{2}\rho_0(\nabla_1\nabla_2 + c^{-2}\partial_{t_1}\partial_{t_2})\langle\varphi_1\varphi_2\rangle|_{x_1=x_2=x}\\ &= \frac{1}{2}\rho_0\left(\frac{1}{4}\nabla_R^2 - \nabla_\rho^2 + \frac{1}{4}c^{-2}\partial_T^2 - c^{-2}\partial_\tau^2\right)\frac{1}{2}\mathrm{Re}(\Gamma+\tilde{\Gamma})|_{\rho=0}\,.\end{aligned} \tag{2}$$

Similarly, for the average flux density we have

$$\langle \mathbf{S}\rangle = \frac{1}{2}\rho_0\left(\frac{1}{2}\nabla_R + \nabla_\rho\right)\left(\frac{1}{2}\partial_T - \partial_\tau\right)\mathrm{Re}(\Gamma+\tilde{\Gamma})|_{\rho=0}, \tag{3}$$

where $R = (x_1 + x_2)/2 = (\mathbf{R},T)$, $\rho = x_1 - x_2 = (\rho,\tau)$.

Consider the case of a *homogeneous and stationary* field u. In practice, its energy characteristics, averaged over time interval Δt, are usually of interest. The averaging interval should exceed the typical period of the signal $T_0 \approx 2\pi/\omega_0$ where ω_0 is the typical frequency of the signal; it can correspond, for example, to the peak of the spectral density. The average quantities are denoted by overbars

$$\overline{\langle w\rangle} = \frac{1}{\Delta t}\int_0^{\Delta t}\langle w\rangle dt, \quad \overline{\langle \mathbf{S}\rangle} = \frac{1}{\Delta t}\int_0^{\Delta t}\langle \mathbf{S}\rangle dt. \tag{4}$$

$\tilde{\Gamma}$ in (1) and (2) oscillates in time with a frequency on the order of $2\omega_0$, that is, $\tilde{\Gamma} \sim \cos 2\omega_0 t$. Having averaged $\langle w\rangle$ and $\langle \mathbf{S}\rangle$ over time in accordance with (4), we may neglect the contribution from $\tilde{\Gamma}$ in comparison with Γ. Further, by virtue of the assumption on the uniformity of the field, the coherence function Γ is independent of R. Taking this into account and substituting Γ from eqn (3.1.78) into eqns (2) and (3), we have

$$\langle w\rangle = c^{-1}\int\left(\frac{\omega^2\rho_0}{2c}\right)I_{\omega\mathbf{n}}\,d\omega d\Omega_{\mathbf{n}},$$

$$\langle \mathbf{S}\rangle = \int\left(\frac{\omega^2\rho_0}{2c}\right)\mathbf{n}I_{\omega\mathbf{n}}\,d\omega d\Omega_{\mathbf{n}}.$$

By way of comparison of these expressions with the classical radiometric relationships (2.1.16) and (2.1.17), we find that the radiance I is expressed through $I_{\omega\mathbf{n}}$ by

$$I \equiv I_\omega = \left(\frac{\omega^2 \rho_0}{2c}\right) I_{\omega\mathbf{n}}$$

which is a specific case of eqn (3.1.86).

Problem 3.7. Prove the asymptotic relation (3.2.12).

Solution. Far from the source, the field forms a divergent spherical wave which may be treated locally as a plane wave. The average energy flux of a scalar field is given by $\mathbf{S} = \langle u^* \nabla u - u \nabla u^* \rangle / 2ik_0$. The amplitude is normalized so that the unit flux $|\mathbf{S}| = 1$ corresponds to a plane wave $\exp(i\mathbf{kr})$, and the flux (equal to the average intensity $\langle |u|^2 \rangle$) corresponds to a locally plane wave with amplitude u. Then the energy flux per unit solid angle $J(\mathbf{n})$ is expressed as

$$J(\mathbf{n}) = \lim_{r\to\infty} r^2 |\mathbf{S}| = \lim_{r\to\infty} r^2 \langle |u|^2 \rangle \tag{1}$$

which proves the relationship (3.2.12). Strictly speaking, this proof is only valid within the framework of the phenomenological ray approach. Indeed, eqn (3.2.11) relates the radiant intensity to the integral of the radiance over the plane of sources $z = 0$, whereas the asymptotic relationship (1) expresses it in terms of an amplitude distribution *far* from this plane. The equivalence of these relationships is quite obvious only if the geometrical-optics concept of energy flowing along the rays is assumed. A more convincing derivation of (1) can be given in the framework of the statistical approach.

Problem 3.8. Show that the use of the approximation (3.2.18) in (3.2.15) leads to (3.2.17).

Solution. Substitute (3.2.18) into (3.2.15) and write $\Gamma^0(\mathbf{R}'_\perp, \boldsymbol{\rho}'_\perp)$ in terms of the Wigner function $W^0(\mathbf{k}_{\rho_\perp}, \mathbf{R}'_\perp)$. Then

$$\Gamma = \int |G(\mathbf{R} - \mathbf{R}'_\perp)|^2 \exp(ik_0 \mathbf{n}\boldsymbol{\rho})(2\pi)^2 \Gamma^0(\mathbf{R}'_\perp, k\mathbf{n}) d^2 R'_\perp . \tag{1}$$

In this formula we go to the integration over solid angle $d\Omega_\mathbf{n}$ around the direction $\mathbf{n} = (\mathbf{R} - \mathbf{R}')/|\mathbf{R} - \mathbf{R}'|$ using the relations

$$d^2\mathbf{n}_\perp = n_z d\Omega_\mathbf{n}, \quad n_z = \sqrt{1 - \mathbf{n}_\perp^2}, \quad \mathbf{R}'_\perp = \mathbf{R}_\perp - \frac{\mathbf{n}_\perp R_z}{n_z}, \quad \left|\frac{\partial \mathbf{n}_\perp}{\partial \mathbf{R}'_\perp}\right| = \frac{R_z^2}{|\mathbf{R} - \mathbf{R}'|^4} . \tag{2}$$

The latter expresses the Jacobian of the transition from $R'_\perp$ to $d\Omega_\mathbf{n}$. As a result, we come to (3.2.17) where the radiance I^0 is given by (3.2.19).

Problem 3.9. Derive eqn (3.2.19) for the radiance I^0 of plane sources with the aid of the Rayleigh expansion in terms of plane waves.

Solution. The Rayleigh solution to the problem of diffraction in a half-space is

$$u(\mathbf{r}) = \int_{-\infty}^{\infty} \exp\left(i\mathbf{k}_\perp \mathbf{r}_\perp + iz\sqrt{k_0^2 - \mathbf{k}_\perp^2}\right) u^0(\mathbf{k}_\perp) d^2 k_\perp, \tag{1}$$

where

$$u^0(\mathbf{k}_\perp) = \int_{-\infty}^{\infty} \exp(-i\mathbf{k}_\perp \mathbf{r}_\perp) u^0(\mathbf{r}_\perp) \frac{d^2 r_\perp}{(2\pi)^2}, \tag{2}$$

represents the Fourier transform of a monochromatic field $u^0(\mathbf{r}_\perp)$ on the plane $z = 0$.

The formula (1) describes the propagation of noninteracting plane waves

$$\exp(i\mathbf{kr}) = \exp\left(i\mathbf{k}_\perp \mathbf{r}_\perp + iz\sqrt{k_0^2 - \mathbf{k}_\perp^2}\right).$$

Here,

$$|k_\perp| < k_0, \; k_z = \sqrt{k_0^2 - \mathbf{k}_\perp^2} > 0$$

for *traveling* waves, and

$$|\mathbf{k}_\perp| > k_0, \; k_z = i\sqrt{k_\perp^2 - \mathbf{k}_0^2}$$

for *nonuniform* waves. For observation points at a distance of a few wavelengths from the plane $z = 0$, the contribution from exponentially attenuated nonuniform waves can be neglected. This enables the integration limits in (1) to be restricted by the condition $k_\perp^2 < k_0^2$. Formally, it can be done by introducing into the integral (1) the factor $\theta(k_0^2 - \mathbf{k}_\perp^2)$ where $\theta(x)$ is the Heaviside step function $\theta(x) = 1$ for $x \geq 0$ and $\theta(x)$ for $x \leq 0$.

Somewhat cumbersome calculations yield for the coherence function in the half-space $z > 0$ (Apresyan, 1975):

$$\Gamma(\mathbf{R}, \rho) = \int J\left(\mathbf{n}, \mathbf{R}_\perp - \mathbf{n}_\perp \frac{R_z}{n_z}, \mathbf{k}_{R_\perp}\right) \exp\left(ik_\rho(\mathbf{n}_\perp, \mathbf{k}_{R_\perp})\mathbf{n}\rho\right) d^2 k_{R_\perp} d\Omega_{\mathbf{n}}, \tag{3}$$

where

$$J(\mathbf{n},\mathbf{R}_\perp,\mathbf{k}_{R_\perp}) = \frac{\exp(i\mathbf{k}_{R_\perp}\mathbf{R}_\perp)k_\rho^2\theta(k_1^2)\theta(k_2^2)}{\left|n_z - \partial_{k_\rho}\frac{k_1+k_2}{2}\right|}$$

$$\times\int \Gamma^0(\mathbf{R}_\perp,\mathbf{\rho}_\perp)\exp(-i\mathbf{k}_{R_\perp}\mathbf{R}_\perp - ik_\rho\mathbf{n}\mathbf{\rho})\frac{d^2R_\perp d^2\rho_\perp}{(2\pi)^4}, \tag{4}$$

$$k_{1,2} = \sqrt{k_0^2 - \left(k_\delta\mathbf{n}_\perp \pm \frac{\mathbf{k}_{R_\perp}}{2}\right)^2},$$

$$k_\rho = k_\rho(\mathbf{n},\mathbf{k}_{R_\perp}) = \sqrt{k_0^2 - \left(\frac{\mathbf{k}_{R_\perp}}{2}\right)^2 - \left(\frac{\mathbf{k}_{R_\perp}n_\perp}{2n_z}\right)}.$$

The longitudinal and transverse (relative to the z axis) components of $\mathbf{a}$ are denoted by a_z and $a_\perp$, respectively.

The meaning of eqn (3) is as follows. Evaluating the coherence function brings about correlations between the amplitudes $\langle u(\mathbf{k}_1)u^*(\mathbf{k}_2)\rangle$ of traveling waves with wave vectors $\mathbf{k}_1$ and $\mathbf{k}_2$. The coordinates ρ and $\mathbf{R}$ are Fourier conjugated with the variables $\mathbf{k}_\rho = k_\rho\mathbf{n} = (\mathbf{k}_1 + \mathbf{k}_2)/2$ and $\mathbf{k}_R = \mathbf{k}_1 - \mathbf{k}_2$ (Fig.3.18). In the general case, the magnitude of $\mathbf{k}_\rho$ is not constant and varies with the direction of $\mathbf{n}$ and the magnitude of $\mathbf{k}_{R_\perp}$. Therefore, $\mathbf{k}_\rho$ does not satisfy the dispersion equation for waves propagating in free space:$|\mathbf{k}_\rho| \neq k_0$. For a statistically uniform field, however, $\langle u(\mathbf{k}_1)u^*(\mathbf{k}_2)\rangle = I(\mathbf{k}_1)\delta(\mathbf{k}_1 - \mathbf{k}_2)$ so that the correlation $\langle u(\mathbf{k}_1)u^*(\mathbf{k}_2)\rangle$ vanishes if $\mathbf{k}_R = \mathbf{k}_1 - \mathbf{k}_2 \neq 0$, and $k_\rho = k_\rho(\mathbf{n},\mathbf{k}_{R_\perp})$ changes to $k_\rho(\mathbf{n},0)$ belonging to the dispersion

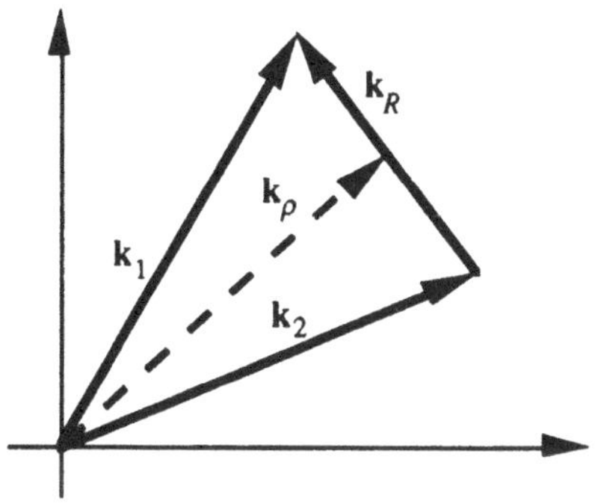

Figure 3.18. Wave vectors of traveling waves in a halfspace.

surface $k_\rho(\mathbf{n},0) = k_0$. This result remains approximately valid for the quasi-uniform field thus enabling a transition from the rigorous wave theory to the theory of radiative transfer.

The expression (3) takes the form (3.2.17) with I^0 taken from (3.2.19) if $\mathbf{k}_\rho(\mathbf{n},\mathbf{k}_{R_\perp})$ is replaced with $\mathbf{k}_\rho(\mathbf{n},0) = \mathbf{k}_0$ in its exponent. At $\rho = 0$, this can always be done, therefore the radiative transfer equation and the wave equation lead to an identical average intensity $\Gamma|_{\rho=0} = \langle |u|^2 \rangle$. To a certain extent, this coincidence is casual and associated with the choice of Wigner's function as a fluctuation spectrum and with a special form of the wave equation (3.2.2) for which the transfer equation $n\nabla I = 0$ is exact.

If $\rho \neq 0$, it is sufficient that the quasi-uniformity condition $|k_{R_\perp}| << k_0$ be satisfied in the region essential for integration in eqn (3).

Problem 3.10. Show that, in the case of diffraction of coherent radiation at an aperture, the quasi-uniformity conditions (3.2.21) and (3.2.22) do not require a passage to the Fraunhofer zone.

Solution. For this problem the vector ρ in eqn (3.2.21) specifies the separation between observation points, the vector $\rho'_\perp$ specifies the separation of points on the aperture ($|\rho'_\perp| \leq a$), and the unit vector $\mathbf{n}$ characterizes the direction of diffraction ($n_z \sim \cos\theta$, where θ is the characteristic diffraction angle). In order to satisfy eqn (3.1.21), it is sufficient to simultaneously satisfy the following inequalities: $R_z >> \rho\cos\theta \sim \rho$, i.e. the separation between observation points must be small compared to their distance from the source, and $R_z >> a\cos\theta$, requiring that the source size $a\cos\theta$ visible in the direction θ, must be small in comparison with R_z (see Fig.3.5). For coincident observation points ($\rho = 0$), the condition (3.2.22) yields $R_z >> a\cos\theta\sqrt{k_0 a \sin\theta}$. This inequality is far weaker (roughly by $\sqrt{k_0 a}$ times) than the condition for the Fraunhofer zone $R_z >> k_0 a^2$, therefore the applicability of the generalized radiance does not require going away to the Fraunhofer zone of the aperture.

Problem 3.11. Find the coherence function for a uniform and isotropic radiance.

Solution. Evaluating the integral that expresses the coherence function of a uniform wave field through the isotropic radiance $I = I^0 = \text{const}$, we obtain the function

$$\Gamma = \oint I \exp(ik_0\mathbf{n}\rho)d\Omega_\mathbf{n} = I^0 \int_0^{2\pi} d\varphi \int_{-1}^{1} d\mu \exp(ik_0\rho\mu) = 4\pi I^0 \frac{\sin(k_0\rho)}{k_0\rho} \tag{1}$$

with the characteristic dependence on the wavelength $\sin(k_0\rho)/k_0\rho$ and the correlation radius on the order of the wavelength $\lambda = 2\pi/k_0$.

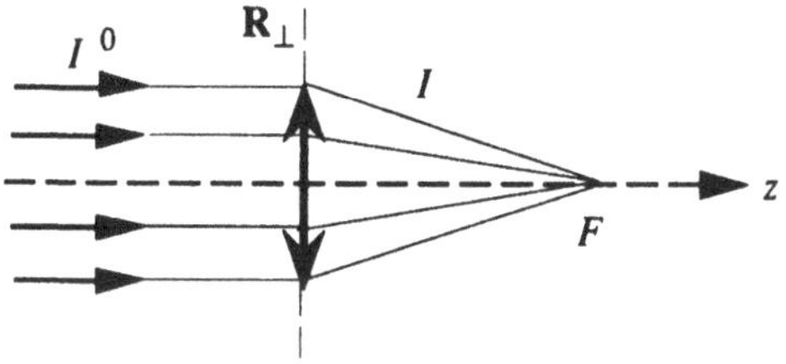

Figure 3.19. Transmission of generalized radiance through a thin lens.

Problem 3.12. In the geometrical-optics approximation, a paraxial beam passing through a thin lens can be described by multiplying its complex amplitude by the phase factor $\exp(-ikr_\perp^2/2F)$, where F is the focal length of the lens. Consider the geometrical meaning of this operation as applied to the generalized radiance of the beam.

Solution. The introduction of the phase factor corresponds to the refraction of the generalized radiance passing through a lens. Indeed, if we relate the radiance behind the lens to the radiance before the lens by eqn (3.2.46), the ray spread function entering eqn (3.2.46) will be

$$\mathcal{P}(\mathbf{n}_\perp,\mathbf{R}_\perp,\mathbf{n}'_\perp,\mathbf{R}'_\perp)=\delta(\mathbf{R}_\perp-\mathbf{R}'_\perp)\delta\left(\mathbf{n}_\perp-\mathbf{n}'_\perp-\frac{\mathbf{R}}{F}\right). \tag{1}$$

We can give a simple geometrical interpretation of this expression which describes a change in the direction of the generalized radiance passing through a thin lens. Let us assume that $\mathbf{R}_\perp$ and $\mathbf{n}_\perp$ describe the coordinate and the slope of the generalized ray, respectively. The first factor in eqn (1) may then be treated as the conservation of the coordinate $\mathbf{R}_\perp$, i.e., the continuity condition for the generalized ray passing through the lens, and the second factor describes changes in the direction $\mathbf{n}_\perp$ that cause the convergence of rays to the lens focus (Fig.3.19).

Problem 3.13. Estimate the longitudinal and transverse (relative to the direction of propagation) correlation distances for the field in the small-angle approximation.

Solution. Let us consider the quasi-homogeneous radiation propagating in a small angular sector near the z axis (narrow beam wave) and replace approximately the integration over $d\Omega_\mathbf{n}$ by the integration over $d^2n_\perp$ in the general expression (3.1.84). Assuming in addition that $n_z=\sqrt{1-\mathbf{n}_\perp^2}\approx 1-\mathbf{n}_\perp^2/2$, we have

$$\Gamma\approx\int I\exp\left[ik_0\left(\mathbf{n}_\perp\rho_\perp+\left(1-\frac{n_\perp^2}{2}\right)\rho_z\right)\right]d^2n_\perp. \tag{1}$$

In view of the Fourier transform properties, the transverse correlation radius $l_\perp$ can be estimated from $l_\perp k_0|\mathbf{n}_\perp|\sim l_\perp k_0\delta\vartheta\geq 1$, whence, omitting numerical multipliers, we have $l_\perp \sim \lambda / \delta\vartheta$, where $\delta\vartheta \sim |\mathbf{n}_\perp|$ has the meaning of the angular beam width. Similarly, the longitudinal correlation radius $l_\parallel$ is estimated from $k_0 n_\perp^2 l_\parallel / 2 \sim k_0 (\delta\vartheta)^2 l_\parallel / 2 \sim 1$ as $l_\parallel \sim \lambda / (\delta\vartheta)^2$.

Problem 3.14. About 150 years ago Talbot (1836) described the effect of self-imaging of a diffraction grating under coherent illumination at distances that are multiples of $2d^2/\lambda$ where d is the grating constant. About fifty years later, Lord Rayleigh (1881) observed this effect in contact photographs of diffraction gratings. This effect was harnessed in the so-called Talbot interferometers (the general class of self-imaging objects was described by Montgomery, 1967).

By way of example of a simple one-dimensional diffraction grating with the transmittance $1 + \cos 2\pi x / d$ show that treating the Talbot effect with the formalism of the generalized radiance brings about interference rays located in the between the geometrical-optics rays (Ojeda–Castaneda and Sicre, 1985).

Solution. Directly behind the grating, the field has the amplitude $u^0(x) = 1 + \cos k_d x$, $k_d = 2\pi / d$. It is convenient to consider the corresponding generalized radiance in the x–z plane expressing this as a one-dimensional analog of the two-dimensional integral (3.2.43)

$$
\begin{aligned}
I^0(n_\perp, R_\perp) &= \frac{k_0}{2\pi} \int u^0\left(R_\perp + \frac{\rho_\perp}{2}\right) u^{0*}\left(R_\perp - \frac{\rho_\perp}{2}\right) \exp(-ikn_\perp \rho_\perp) d\rho_\perp \\
&= \delta(n_\perp)[1 + \cos(2k_d R_\perp)] + \left[\delta\left(n_\perp - \frac{k_d}{2k}\right) + \delta\left(n_\perp + \frac{k_d}{2k}\right)\right] \\
&\times \cos(k_d R_\perp) + \frac{1}{4}\left[\delta\left(n - \frac{k_d}{k}\right) + \delta\left(n + \frac{k_d}{k}\right)\right].
\end{aligned} \tag{1}
$$

In accordance with the small-angle transfer equation $(\partial_z + n_\perp \nabla) I = 0$, the radiance $I(n_\perp, R_\perp, R_z)$ may be written as $I^0(n_\perp, R_\perp - R_z n_\perp)$ for $R_z > 0$, which immediately yields the Talbot effect, that is, the periodic self-imaging of the grating at $R_z = 2(d^2/\lambda)m$, $m = 1, 2, \ldots$.

The function I^0 is concentrated in five rays with the directions $n_\perp = 0, \pm k_d / k$, and $\pm k_d / 2k$. The first three rays coincide with the conventional geometrical-optics rays (diffraction spectra), and the last two rays represent the interference rays due to the nonlocality of the argument R (see discussion in Sections 3.1.9 and 3.2.12).

Problem 3.15. To allow for polarization in eqn (3.2.51), the products of the scalar amplitudes and f should be replaced by the scalar product of the electric strength and the antenna current, $uf^* \to \mathbf{E}\mathbf{f}^*$. Show that when the polarizations of $\mathbf{E}$ and $\mathbf{f}$ are constant over the aperture, the allowance for polarization in the expression (3.2.61) for the received power reduces to multiplying the integral (3.2.61) by the factor $1 + \mathbf{P}_E \cdot \mathbf{P}_f$, where $\mathbf{P}_f$ and $\mathbf{P}_E$ are the Stokes polarization vectors for $\mathbf{f}$ and for the projection of $\mathbf{E}$ onto the aperture plane, respectively.

Solution. The expression for P acquires an additional factor $|\mathbf{E}\cdot\mathbf{f}^*|^2|E|^{-2}|f|^{-2}$ that is equal to $1+\mathbf{P}_E\cdot\mathbf{P}_f$ (see Problem 2.22). In particular, it follows that when the polarization vectors $\mathbf{P}_E$ and $\mathbf{P}_f$ are opposite, $\mathbf{P}_E\mathbf{P}_f=-1$, i.e. the polarizations of the antenna and the radiation are orthogonal to each other, then the antenna response vanishes. In other words, the antenna receives only one of the two orthogonal polarizations.

Problem 3.16. On the assumption that the antenna is sufficiently large in comparison to the wavelength, prove the relationship (3.2.66): $D=4\pi S_e/\lambda^2$.

Solution. Substituting eqns (3.2.62) and (3.2.56) into eqn (3.2.65), we estimate the resultant integral in the limit of a vanishingly small wavelength, $\lambda\to 0$. As a result, we have

$$\Omega_A=\left(\frac{2\pi}{k_0}\right)^2\frac{\int|f(r_\perp)|^2d^2r_\perp}{\left|\int f(r'_\perp)d^2r'_\perp\right|^2}. \tag{1}$$

The ratio of the integrals in eqn (1) may be estimated as $1/S_e$ so that for $D=4\pi/\Omega_A$ we obtain $D=4\pi S_e/\lambda^2$.

Problem 3.17. Determine the dependence of the effect of smoothing of light-flux fluctuation on the diameter of the aperture S.

Solution. In the small-angle approximation, the energy flux density is proportional to the wave intensity $J=|u|^2$ so that the energy flux through the aperture S is determined by the integral (3.2.80) of the intensity J, and its fluctuation $\tilde{P}$ is determined by the integral of the intensity fluctuation $\tilde{J}=J-\langle J\rangle$:

$$\tilde{P}=\int_S \tilde{J}d^2r.$$

As a result, the average square $\langle\tilde{P}^2\rangle$ can be expressed through the intensity covariance $\psi_J(\mathbf{r}_1,\mathbf{r}_2)=\langle\tilde{J}(\mathbf{r}_1)\tilde{J}(\mathbf{r}_2)\rangle$:

$$\langle\tilde{P}^2\rangle=\iint_S \Psi_J(r_1,r_2)d^2r_1d^2r_1. \tag{1}$$

If the aperture S is large compared to the squared intensity correlation distance l_J^2, then the integration can be performed with respect to the difference coordinate $\rho=\mathbf{r}_1-\mathbf{r}_2$ using the relationship

$$\int\Psi_J(\rho)d^2\rho=\langle\tilde{J}^2\rangle l_J^2, \tag{2}$$

which serves essentially as a definition for l_J^2 similarly to the definition (3.2.73) for the correlation time. The subsequent integration over $R = (r_1 + r_2)/2$ yields the aperture area S so that $\langle \tilde{P}^2 \rangle = \langle \tilde{J}^2 \rangle l_J^2 S$.

Taking into account that $\langle P \rangle = \langle J \rangle S$, we find that the relative fluctuation of the flux

$$\frac{\langle \tilde{P}^2 \rangle}{\langle P \rangle^2} = \frac{\langle \tilde{J}^2 \rangle}{\langle J \rangle^2} \frac{l_J^2}{S} << 1 \tag{3}$$

is by S / l_J^2 times smaller than the relative fluctuation of intensity and abates as $1/S$.

In the opposite case of a small aperture $S << l_J^2$, the energy flux is proportional to the intensity, $P = JS$, so that

$$\frac{\langle \tilde{P}^2 \rangle}{\langle P \rangle^2} = \frac{\langle \tilde{J}^2 \rangle}{\langle J \rangle^2}; \tag{4}$$

that is, for a small aperture, the spatial smoothing effect is negligible. The transition from (3) to (4) occurs at $S \sim l_J^2$ when the aperture area becomes comparable with the area of a single speckle.

Problem 3.18. Write down the condition for spatial ergodicity of a random field.

Solution. For definiteness we write the ergodicity condition on a plane. Using the same reasoning as in the case of time correlation, we can conclude that the closeness between the spatial average $\overline{\eta}^S$ and the statistical average $\langle \eta \rangle$ is ensured by the smallness of the correlation area l_c^2 for η in comparison with the averaging area S. This property was used in Problem 3.17 for the energy flux fluctuation P. In view of the definition for the correlation area [formula (2) in the preceding problem], the condition for $\overline{\eta}^S$ and $\langle \eta \rangle$ to be close may be written as

$$\frac{l_c^2}{S} = \frac{1}{S} \frac{\int \Psi_\eta(\rho) d^2\rho}{\langle \tilde{\eta}^2 \rangle} \xrightarrow[S \to \infty]{} 0,$$

where $\psi_\eta = \psi_\eta(\mathbf{r}_1 - \mathbf{r}_2)$ is the correlation function of η. This is a necessary and sufficient condition.

Problem 3.19. Estimate the smallest lens aperture Σ_{cr} at which the smoothing of intensity fluctuations in the lens focus begins in the presence of a coherent field.

Solution. In the focal plane, the intensity of the coherent field is $J_{\text{coh}}^{\text{F}} = |C|^2 \Sigma^2 |\langle u \rangle|^2$. The average incoherent intensity may be estimated by

$$\langle J_{\text{incoh}}^{\text{F}} \rangle = |C|^2 \left\langle |\tilde{u}|^2 \right\rangle l_c^2 \Sigma,$$

where l_c is the coherence distance of the field in the lens plane. The mean square deviation $\Delta J^F = J^F - \langle J^F \rangle$ is approximately (for Gaussian fluctuations, exactly) equal to this value. Therefore,

$$\beta^2 = \frac{\langle (\Delta J^F)^2 \rangle}{\langle J^F \rangle^2} \approx \frac{\left[\langle |\tilde{u}|^2 \rangle l_c^2 \Sigma\right]^2}{\left[\langle |\tilde{u}|^2 \rangle l_c^2 \Sigma + |\langle u \rangle|^2 \Sigma^2\right]^2}.$$

From the condition $\beta^2 \approx 1/2$ we obtain an estimate for the critical area

$$\Sigma_{\rm cr} \approx l_c^2 \langle |\tilde{u}|^2 \rangle / |\langle u \rangle|^2 ,$$

beginning from which the scintillation index decreases in proportion with Σ. For a weak coherent field, $\Sigma_{\rm cr}$ may significantly exceed the correlation area l_c^2.

Problem 3.20. Extend the formula (3.2.19) for radiance to the case of a nonmonochromatic field.

Solution. Let us revert to the time-dependent field by replacing the spatial amplitude $u(\mathbf{r})$ in (3.2.19) with the space-time amplitude $u(\mathbf{r},t) = u(x)$. Then eqn (3.2.19) transforms to

$$\begin{aligned} I^0 &= I^0(\mathbf{R}_\perp, t; \omega, \mathbf{n}) \\ &= (\omega / c)^2 \cos\theta \int \langle u^0(\mathbf{R}_\perp + \rho_\perp / 2, t + \tau / 2) u^{0*}(\mathbf{R}_\perp - \rho_\perp / 2, t - \tau / 2) \rangle \\ &\quad \times \exp[-i(\omega / c)\mathbf{n}_\perp \rho_\perp + i\omega\tau](2\pi)^{-3} d^2\rho_\perp d\tau . \end{aligned} \tag{1}$$

Here I^0 is the radiance of the sources in the plane $z = 0$, and $u^0(\mathbf{R}_\perp, t)$ is the field in this plane. In the case of the monochromatic radiation, $u^0 = u^0(\mathbf{R}_\perp)\exp(-i\omega_0 t)$, and

$$\begin{aligned} I^o &= \delta(\omega - \omega_0)(\omega_0 / c)^2 \cos\theta \int \left\langle u^0(\mathbf{R}_\perp + \rho_\perp / 2) u^{0*}(\mathbf{R}_\perp - \rho_\perp / 2)) \right\rangle \\ &\quad \times \exp[-i(\omega_0 / c)\mathbf{n}\rho_\perp](2\pi)^{-2} d^2\rho_\perp . \end{aligned}$$

This formula coincides with the above expression (3.2.19), where the delta function $\delta(\omega - \omega_0)$ was omitted for brevity.

It is natural that for the applicability of (1), in addition to the requirement of the spatial quasi-uniformity of the field, we need to satisfy the quasi-stationarity condition $T >> \tau_c$, where T is the characteristic time scale of the correlator $\langle u^0(\mathbf{R}_{1\perp}, t + \tau / 2) u^0 * (\mathbf{R}_{2\perp}, t - \tau / 2) \rangle$, and τ_c is the correlation time which is on the order of the reciprocal radiation bandwidth $(\Delta\omega)^{-1}$.

Problem 3.21. Determine the conditions for the rate of photon counting in the Fabry–Perot interferometer to be proportional to the spectral radiance I_ω (Eberly and Wodkiewicz, 1977).

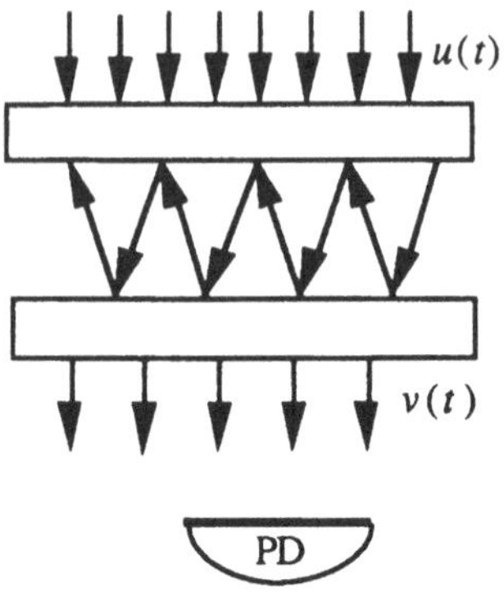

Figure 3.20. Frequency spectrum measurement with a Fabry–Perot interferometer.

Solution. Let the measuring system consist of a photon counter and a Fabry–Perot interferometer (Fig.3.20) hit by a normally incident plane wave

$$u(x) = u(t)\exp\left(i(\overline{\omega} / c)\mathbf{n}_0\mathbf{r}\right).$$

This field is quasi-monochromatic with a characteristic frequency $\overline{\omega}$ and with a bandwidth $\Delta\omega << \overline{\omega}$. Such a wave is spatially coherent and may be described in terms of the radiance localized in the direction $\mathbf{n}_0$

$$I = I_{\omega}(t)\delta_2(\mathbf{n}\mathbf{n}_0).$$

Here, the frequency spectrum $I_{\omega}(t)$ is related with the amplitude $u(t)$ by

$$I_{\omega}(t) = \int \langle u(t+\tau/2)u^*(t-\tau/2)\rangle \exp(i\omega\pi)d\tau / 2\pi. \tag{1}$$

The radiation emerging from the interferometer and incident on the detector is structurally close to a plane wave. Therefore, the Fabry–Perot interferometer may be thought of as a frequency filter which transforms the incident amplitude $u(t)$ into the emerging amplitude $v(t)$:

$$v(t) = \int_{-\infty}^{t} H(t-t')u(t')dt', \tag{2}$$

where the response $H(t)$ is the Fourier transform of the spectral function $H(\omega)$:

$$H(\omega) = \int H(t)\exp(i\omega t)dt = \frac{(1-r^2)\exp[i(\omega/c)d]}{1-r^2\exp[2i(\omega/c)d]}.$$

In this relationship, known as the Airy formula, d is the distance between interferometer plates and r the amplitude reflection coefficient which may be assumed close to unity, $r \sim 1$.

The function $H(\omega)$ markedly differs from zero near resonance frequencies only, i.e. at $\omega \approx \omega_m = m\pi c / d$, where $m = 1,2,\ldots$. If the mean frequency $\overline{\omega}$ is close to one of these frequencies, $\overline{\omega} \approx \omega_m$, the function $H(\omega)$ may be approximated by

$$H(\omega) \approx \frac{\gamma \exp[i(\omega / c)d]}{\gamma - i(\omega - \omega_m)},$$

where $\gamma = c(1 - r^2) / 2d$ has the meaning of the effective attenuation of the resonator and simultaneously characterizes the transmission bandwidth of the interferometer. The corresponding response function is

$$H(t) = \int H(\omega) e^{-i\omega t} (2\pi)^{-1} d\omega \cong \theta(t) \gamma e^{-(\gamma + i\omega_m)t}. \tag{3}$$

The Heaviside step-function [$\theta(t) = 1$ for $t \geq 0$, and $\theta(t) = 0$ for $t < 0$] stands here to reflect the causality principle: the response $v(t)$ in (2) depends only on values of $u(t')$ in the preceding time $t' < t$, and the time origin is chosen at $t = d / c$.

The rate of photon counting $R(t)$ is proportional to the squared amplitude of the field incident on the photon counter. Therefore, accurate to a constant factor, we can write

$$R(t) = \langle |v(t)|^2 \rangle.$$

Substituting eqn (2) and (3) into this relation yields

$$\begin{aligned} R(t) &= \int H(t - t') H^*(t - t'') \langle u(t') u^*(t'') \rangle dt' dt'' \\ &= \gamma^2 \int \theta\left(1 - T - \frac{|\tau|}{2}\right) \exp[-i\omega_m \tau - 2\gamma(t - T)] \\ &\quad \times \langle u(T + \tau / 2) u^*(T - \tau / 2) \rangle dT d\tau. \end{aligned} \tag{4}$$

Let an incident radiation be turned on at the instant $t = 0$ so that

$$u(t) = \theta(t) u^0(t). \tag{5}$$

As we move away from $t = 0$, the local spectrum of $u(t)$ approaches the frequency spectrum of the stationary process $u^0(t)$:

$$\begin{aligned} I_\omega(t) &= \int_{|\tau| < 2t} \langle u^0(t + \tau / 2) u^{0*}(t - \tau / 2) \rangle e^{i\omega\tau} (2\pi)^{-1} d\tau \\ &\xrightarrow[t \to \infty]{} \int_{-\infty}^{\infty} \Gamma^0(\tau) e^{i\omega\tau} (2\pi)^{-1} d\tau \equiv I_\omega^0, \end{aligned}$$

where $\Gamma^0(\tau) = \langle u^0(t+\tau/2)u^{0*}(t-\tau/2)\rangle$.

Substituting (5) into the expression for the rate of counting (4) we find

$$R(t) = \gamma^2 \int \theta(t-T-|\tau|/2)\exp[-i\omega_m\tau - 2\gamma(t-T)]$$

$$\times\theta(T-|\tau|/2)\Gamma^0(\tau)dTd\tau$$

$$= (\gamma/2)\int_{-t}^{t} \exp(i\omega_m\tau)\{\exp(-\gamma|\tau|) - \exp[-\gamma(2t-|\tau|]\}\Gamma^0(\tau)d\tau.$$

This expression approaches the spectrum $I^0(\omega)$ accurate to the constant factor $\gamma/2$ when

$$t >> \gamma^{-1} >> \tau_c, \tag{6}$$

where $\tau_c \sim (\Delta\omega)^{-1}$ is the characteristic correlation time of $u^0(t)$. The first of these inequalities implies that the observation time t must be large compared to the relaxation time γ^{-1} of the interferometer, otherwise the memory about the time zero of observation $t = 0$ does not have enough time to decay. The second inequality requires that the correlation of the field must decay before the attenuation of the resonator becomes significant. This inequality can be stated in a more instructive frequency form

$$\tau_c^{-1} \sim \Delta\omega >> \gamma,$$

which implies that the interferometer transmission bandwidth γ should be small compared to the spectrum bandwidth $\Delta\omega$ of u^0. It is obvious that only in this case we may treat the interferometer as a narrow-band filter and neglect the variations of the spectrum of u^0 within the transmission bandwidth.

Thus, when inequalities (6) are satisfied, the photon counting rate $R(t)$ is independent of time t and is equal (accurate to a constant factor) to the frequency spectrum of the incident radiation I_ω^0 at the frequency ω_m. If inequalities (6) are not satisfied, $R(t)$ may be regarded only as the result of linear filtering of the spectrum I_ω^0.

Problem 3.22. Wolf (1986) demonstrated that the spectrum of light from a partially coherent source changes with the distance from the source (the Wolf effect). This spectrum varies even in free-space propagation. Write down the general relationships connecting the spectrum of the field at some distance from the source with the spectrum of the source. Consider special cases of the far-field spectrum and the spectrum of coherent sources.

Solution. Following Wolf and Fienup (1991), we consider the transmission of monochromatic wave from the plane $z = z_0$ [input plane with the field $u(\rho',\omega)$] to the plane $z = z_1$ [output plane with the field $v(\rho,\omega)$, Fig.3.21]. Here, ρ and ρ' are vectors in these planes.

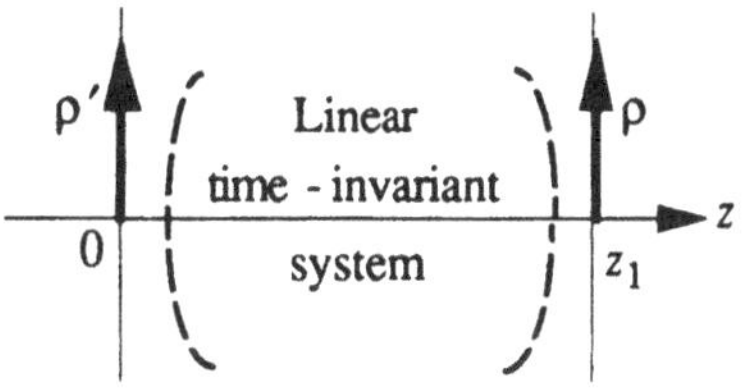

Figure 3.21. Schematic representation of an optical system transforming the input field $u(\rho',\omega)$ into the output field $v(\rho,\omega)$.

The fields $u(\rho',\omega)$ and $v(\rho,\omega)$ are connected by the linear relation (3.2.45):

$$v(\rho,\omega)=\int P(\rho,\rho',\omega)u(\rho',\omega)d^2\rho', \tag{1}$$

where we indicated explicitly the dependence on frequency ω entering eqn (1) as a parameter. The optical system need not be an imaging system. It may be a prism, a diffraction grating, or even a free space. The coherence functions in the two planes may be written as

$$\Gamma^0(\rho_1',\rho_2',\omega)=\langle u(\rho_1,\omega)u^*(\rho_2,\omega)\rangle,$$

$$\Gamma(\rho_1,\rho_2,\omega)=\langle v(\rho_1,\omega)v^*(\rho_2,\omega)\rangle.$$

By virtue of eqn (1), these functions are related by

$$\Gamma(\rho_1,\rho_2,\omega)=\iint P(\rho_1,\rho_1',\omega)P^*(\rho_2,\rho_2',\omega)\Gamma^0(\rho_1',\rho_2',\omega)d^2\rho_1'd^2\rho_2'..$$

Assume for simplicity that at all points of the input plane the field spectrum is identical and equal to $S^0(\omega)\equiv\Gamma^0(\rho',\rho',\omega)$. For the output spectrum $S^0(\rho,\omega)\equiv\Gamma(\rho,\rho,\omega)$, we then have

$$S^0(\rho,\omega)\equiv S^0(\omega)\iint\gamma^0(\rho_1',\rho_2',\omega)$$
$$\times P(\rho,\rho_1',\omega)P^*(\rho,\rho_2',\omega)d^2\rho_1'd^2\rho_2', \tag{2}$$

where

$$\gamma^0(\rho_1',\rho_2',\omega)=\frac{\Gamma^0(\rho_1',\rho_2',\omega)}{S^0(\omega)}.$$

is the degree of spatial coherence for the input radiation. According to eqn (2), in the output plane, the spectrum differs generally from the input spectrum by two reasons:

(i) because of the transmission properties of the system described by the function $P(\rho,\rho',\omega)$,

(ii) because of the effect of the spatial coherence of the input radiation.
In the case of the *totally coherent source* ($\gamma^0 = 1$), the output spectrum

$$S(\rho,\omega) = S^0(\omega)\left|\int P(\rho,\rho',\omega)d^2\rho'\right|^2$$

differs from the input spectrum solely because of the transmission properties of the optical system which performs filtering of the input optical signal.

In another particular case of free-space propagation, when no optical system exists and the free space plays the role of the filter, the effect of spatial coherence becomes crucial. In this case, eqn (2) may be written in the form (3.2.15). In the far zone, $|z_1 - z_0| >> ka^2$, where a is the size of the region A occupied by the sources in the input plane, from eqns (3.2.14), (3.2.15), and (3.2.18) we obtain for the initial field $[\gamma^0 = \gamma^0(\rho_1' - \rho_2', \omega)]$ uniform in the region A :

$$S(\rho,\omega) = S^0(\omega)\left(\frac{\omega\cos\theta}{cr}\right)^2 A\tilde{\gamma}^0(k\mathbf{n}_\perp,\omega), \tag{3}$$

where

$$\tilde{\gamma}^0(\mathbf{q},\omega) = \int \gamma^0(\rho',\omega)e^{-i\mathbf{q}\rho'}\frac{d^2\rho'}{(2\pi)^2}$$

is the two-dimensional Fourier transform of the degree of spatial coherence, and A is the area occupied by sources.

Equation (3) contains two frequency-dependent factors: ω^2 due to the impulse response function, and $\tilde{\gamma}^0$ depending on the degree of spatial coherence of the initial field. Controlling the latter factor, for example, with a variable aperture, one can achieve a desirable change in the frequency spectrum. In particular, the center of radiation line can be shifted toward the low-frequency end (red shift) or toward the high-frequency end (blue shift) (Wolf, 1987). Available experiments to control the profile of the spectrum using the correlation of the sources have been described by Wolf (1991).

Problem 3.23. The wave operator $L_0 = \Delta + k_0^2$ is degenerate; i.e., the equation $L_0 u = 0$ has nontrivial physical (bounded at infinity) solutions $u = \exp(i\mathbf{kr})$, $k^2 = k_0^2$. Certain conditions must be satisfied to invert such an operator, that is, to find Green's function $G_0 = L_0^{-1}$. As a rule, the Sommerfeld radiation condition is invoked or the principle of limited absorption is used. In the latter case, the free space is treated as a limiting case of an absorbing medium with a vanishingly small absorption, so that k_0 may be replaced with $k_+ = k_0 + i0$, and k_0^2 replaced with $k_+^2 = k_0^2 + i0$. Using this replacement, show that the operator $\hat{G}_0 = (\Delta + k_0^2 + i0)^{-1}$ has the kernel (3.3.7).

Solution. The kernel G_0 is homogeneous (i.e. $G_0(\mathbf{r},\mathbf{r}_0) = G_0(R)$ where

$R = \mathbf{r} - \mathbf{r}_0$) and expressed as the action of $\hat{G}_0$ on the delta function $\delta(R)$:

$$G_0(R) = \hat{G}_0\delta(R) = \frac{1}{\Delta + k_0^2 + i0}\int e^{i\kappa\mathbf{R}}\frac{d^3\kappa}{(2\pi)^3}$$

$$= \int_0^\infty\int_0^\pi\int_0^{2\pi} \frac{e^{i\kappa R\cos\theta}\kappa^2 d\kappa \sin\theta d\theta d\varphi}{-\kappa^2 + k_0^2 + i0} = \frac{1}{4\pi^2 Ri}\int_{-\infty}^{\infty}\frac{e^{i\kappa R}\kappa d\kappa}{-\kappa^2 + k_0^2 + i0}. \tag{1}$$

This integral is readily calculated if we assume κ complex and close the integration contour by an infinite semi-circle in the upper κ half-plane. The integral can then be found by evaluating the residue in the region $\operatorname{Im}\kappa > 0$. This evaluation yields the expression (3.3.7).

Problem 3.24. Using the operator technique, i.e., treating the operator ∇ as a conventional vector, find the kernel of the unperturbed operator $G_0 = [-\nabla\times\nabla\times + k^2]^{-1}$ for the electromagnetic problem (3.3.4).

Solution. In the operator approach, calculations of G_0 reduce to inverting the matrix $[-\nabla\times\nabla\times + k^2]$. In accordance with the formula $\mathbf{a}\times\mathbf{b}\times\mathbf{c} = \mathbf{b}(\mathbf{a}\cdot\mathbf{c}) - \mathbf{c}(\mathbf{a}\cdot\mathbf{b})$, we can write

$$-\nabla\times\nabla\times = -\nabla\otimes\nabla + \Delta\hat{1}, \tag{1}$$

where $\hat{1}$ is the unit matrix. Let us seek G_0 in a form similar to eqn (1) using the method of undetermined coefficients:

$$G_0 = \nabla\otimes\nabla A + B\hat{1}, \tag{2}$$

where A and B are scalar operators. Substituting (1) and (2) into the equation for G_0

$$\left[-\nabla\times\nabla\times + \hat{1}k^2\right]G_0 = \hat{1} \tag{3}$$

and equating the coefficients of $\nabla\otimes\nabla$ and $\hat{1}$ we obtain

$$G_0 = \left(1 + \frac{\nabla\otimes\nabla}{k^2}\right)\frac{1}{\Delta + k^2}, \tag{4}$$

where the operator $1/(\Delta + k^2)$ has the kernel $-[\exp(ikR)/4\pi R]$ with $R = |\mathbf{r} - \mathbf{r}_0|$ obtained in the preceding problem.

The use of eqn (4) for the field in the source region leads to certain difficulties associated with the strong singularity of G_0, namely, $G_0(R) \sim 1/R^3$, at $R = |r - r_0| \to 0$. This question has been thoroughly discussed in the literature (Yaghjian, 1980; Collin, 1986). The simplest way to overcome this difficulty is to interpret eqn (4) in the sense of generalized functions. It is not hard to demonstrate that, for generalized functions,

$$\nabla\otimes\nabla\frac{1}{R}=\eta_v(R)\left\{\nabla\otimes\nabla\frac{1}{R}\right\}-\hat{L}_S\delta(R). \tag{5}$$

Here $\{\nabla\otimes\nabla(1/R)\}$ means the differentiation in the common sense, the function $\eta_v(R)$ is unity outside and zero inside an infinitesimal volume V centered at $R=0$, and

$$\hat{L}_S=\oint_S\frac{\mathbf{R}\otimes d\mathbf{S}}{R^3}=\oint\frac{\mathbf{R}\otimes\mathbf{n}}{R}d\Omega_{\mathbf{n}} \tag{6}$$

is the matrix of the dyadic solid angle, where the integration is performed over the surface S of the volume V and $\mathbf{n}$ is the outward normal to S (Fig.3.22).

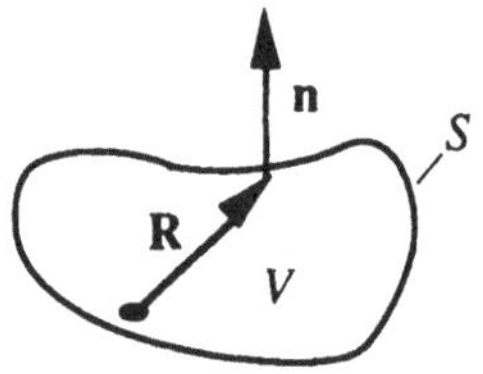

Figure 3.22. To eliminate the singularity of Green's dyadic function.

The relationship (5) helps us to exclude the singularity near the sources when eqn (4) is used. It is important that the left-hand side of eqn (5) is independent of the shape of the volume V used to exclude the singularity, whereas the terms on the right-hand side of eqn (5) do depend upon the shape of V. More often than not V is a sphere. In view of symmetry, the integral in eqn (6) is easily evaluated: $\hat{L}_S=(4\pi/3)\hat{1}$. Indeed, in this case, $\hat{L}_S=4\pi\langle\mathbf{n}\otimes\mathbf{n}\rangle_n$ where $\langle\cdot\rangle_n$ means averaging over the directions of the isotropic vector $\mathbf{n}$. The expression for G, similar to eqn (5), then takes the form

$$G=\mathcal{P}\left\{\left(1+\frac{\nabla\otimes\nabla}{k^2}\right)\frac{e^{ikR}}{(-4\pi R)}\right\}+\frac{\hat{1}}{3k^2}\delta(\mathbf{R}). \tag{7}$$

Here the symbol $\mathcal{P}$ implies that the singularity is taken in the sense of the principal value. In the general case, in place of eqn (7), we have

$$G=\eta_V(R)\left\{\left(1+\frac{\nabla\otimes\nabla}{k^2}\right)\frac{e^{ikR}}{(-4\pi R)}\right\}+\frac{\hat{L}_S}{4\pi k^2}\delta(\mathbf{R}). \tag{8}$$

Here, similarly to eqn (5), some terms depend on the shape of the infinitesimal volume used to exclude the singularities.

Problem 3.25. Express the scattered field $u_s = G_0 V u$ appearing in eqn (3.3.9) in terms of the incident wave field $u^0 = G_0 q$ and Green's operator G of the scattering medium.

Solution. From the representation of G as the series (3.3.17) it follows that G obeys the equation

$$G = G_0 + GVG_0. \tag{1}$$

Note that this equation is different from eqn (3.3.18) by the order of the operators G and G_0. By virtue of eqn (1) we have

$$u_s = u - u^0 = Gq - G_0 q = GVG_0 q = GVu^0. \tag{2}$$

Thus, the scattered field u_s is expressed through GV and u^0 in a manner quite similar to that the total field u [eqn (3.3.10)] is expressed through G and q. One more expression for u_s in terms of u_0 may be obtained using the T-operator of scattering.

Problem 3.26. The propagation of a random incident wave in a random medium is governed by eqn (3.3.10):

$$u = Gq, \tag{1}$$

where G is the random Green's operator of the scattering medium, and q is the random function of sources describing the fluctuation of the incident wave. Given that the fluctuations of the medium (G) and incident wave (q) are statistically independent, express the coherence function associated with the incoherent component $\tilde{u} = u - \langle u \rangle$ in terms of the mean values and fluctuations of G and q.

Solution. In this case, the full statistical averaging includes the averaging over fluctuations of both G and q. Taking into account that these fluctuations are uncorrelated, so that $\langle \tilde{G}\tilde{q} \rangle = \langle \tilde{G} \rangle \langle \tilde{q} \rangle = 0$, we have

$$\langle u \rangle = \langle G \rangle \langle q \rangle, \tag{2}$$

that is, the field component $\langle u \rangle$, coherent with respect to the full statistical ensemble, is expressed in terms of the means $\langle G \rangle$ and $\langle q \rangle$.

The total fluctuation $\tilde{u}$ is represented as the sum of three terms:

$$\tilde{u} = \langle G \rangle \tilde{q} + \tilde{G} \langle q \rangle + \tilde{G}\tilde{q}. \tag{3}$$

It is easily seen that although not all of these components are mutually statistically independent, they are uncorrelated, and so the respective coherence function $\Gamma^{\text{inc}} = \langle \tilde{u}_1 \tilde{u}_2^* \rangle$ splits into three terms

$$\Gamma^{\text{inc}} = \Gamma^{(q)} + \Gamma^{(G)} + \Gamma^{(Gq)}, \tag{4}$$

where

$$\Gamma^{(q)} = \langle G_1\rangle\langle G_2^*\rangle\langle\tilde{q}_1\tilde{q}_2^*\rangle \tag{5}$$

corresponds to the fluctuation of the incident wave,

$$\Gamma^{(G)} = \langle\tilde{G}_1\tilde{G}_2^*\rangle\langle q_1\rangle\langle q_2^*\rangle \tag{6}$$

corresponds to the fluctuation of the medium, and

$$\Gamma^{(Gq)} = \langle\tilde{G}_1\tilde{G}_2^*\rangle\langle\tilde{q}_1\tilde{q}_2^*\rangle \tag{7}$$

corresponds to the fluctuations of both the incident wave and medium. This example may be related to the description of the partial coherence or, in the case of electromagnetic radiation, the partial polarization of a wave scattered from a random medium. For this purpose, it is necessary to use instead of (1) the expression for u_s obtained in terms GV and u_0 in the preceding problem.

Problem 3.27. The kernel of the operator $G_0 = L_0^{-1}$, where $L_0 = \Delta - \partial_{ct}^2$ is d'Alembertian, satisfies the reciprocity relationship $G_0(\mathbf{r},t;\mathbf{r}_0,t_0) = G_0(\mathbf{r}_0,-t_0;\mathbf{r},-t)$, that is, the arguments $x = (\mathbf{r},t)$ and $x_0 = (\mathbf{r}_0,t_0)$ of the kernel $G_0(x,x_0)$ may be transposed changing simultaneously the sign of time t. For the time-harmonic dependence $\exp(-i\omega t)$, this relationship simplifies:

$$G_{0\omega}(\mathbf{r},\mathbf{r}_0) \equiv \left(\Delta + (\omega/c)^2\right)^{-1}(\mathbf{r},\mathbf{r}_0) = G_{0\omega}(\mathbf{r}_0,\mathbf{r}).$$

Formulate reciprocity relationships for the effective inhomogeneity operators V^{eff} and K_{12}.

Solution. Using the perturbation series (3.3.17), it is easy to show that, for the operator G, and hence for the mean $\langle G\rangle$, the reciprocity relationships similar to those for G_0 are satisfied: $\langle G\rangle(x,x_0) = \langle G\rangle(\tilde{x}_0,\tilde{x})$, where $\tilde{x} = (\mathbf{r},-t)$, $\tilde{x}_0 = (\mathbf{r}_0,-t_0)$, and $\langle G_\omega\rangle(\mathbf{r},\mathbf{r}_0) = \langle G_\omega\rangle(\mathbf{r}_0,\mathbf{r})$. Taking this into account and using the definition (3.3.33) of the operator V^{eff}, we can easily show that similar reciprocity relationships are also satisfied for the kernel V^{eff}.

For the kernel of the operator K_{12}, these relationships give the following symmetry properties :

$$K\begin{pmatrix} x_1, & x_1' \\ x_2, & x_2' \end{pmatrix} = K\begin{pmatrix} \tilde{x}_1', & \tilde{x}_1 \\ x_2, & x_2' \end{pmatrix} = K\begin{pmatrix} x_1 & x_1' \\ \tilde{x}_2', & \tilde{x}_2 \end{pmatrix} = K\begin{pmatrix} \tilde{x}_1', & \tilde{x}_1 \\ \tilde{x}_2', & \tilde{x}_2 \end{pmatrix}, \tag{1}$$

where tildes denote sign reversals of the time argument, and, in the monochromatic case,

$$K\begin{pmatrix} \mathbf{r}_1, & \mathbf{r}_1' \\ \mathbf{r}_2, & \mathbf{r}_2' \end{pmatrix} = K\begin{pmatrix} \mathbf{r}_1', & \mathbf{r}_1 \\ \mathbf{r}_2, & \mathbf{r}_2' \end{pmatrix} = K\begin{pmatrix} \mathbf{r}_1, & \mathbf{r}_1' \\ \mathbf{r}_2', & \mathbf{r}_2 \end{pmatrix} = K\begin{pmatrix} \mathbf{r}_1', & \mathbf{r}_1 \\ \mathbf{r}_2', & \mathbf{r}_2 \end{pmatrix}. \quad (2)$$

For a homogeneous and stationary medium, eqn (1) suggests in particular that, in the $R-\rho$ variables,

$$K(R,\rho,\rho') = K(-\tilde{R},\tilde{\rho}',\tilde{\rho}). \quad (3)$$

One more property of K is due to the symmetry of Γ with respect to the transposition of u_1 and u_2 simultaneously with complex conjugation:

$$K\begin{pmatrix} x_1, & x_1' \\ x_2, & x_2' \end{pmatrix} = K^*\begin{pmatrix} x_2, & x_2' \\ x_1, & x_1' \end{pmatrix}, \quad (4)$$

or, for the kernel (3),

$$K(R,\rho,\rho') = K^*(R,-\rho,-\rho'). \quad (5)$$

Problem 3.28. Show that for the effective inhomogeneity operators V^{eff} and K_{12} the following commutation relationships are valid:

$$\begin{gathered} G_0 V^{\text{eff}} \langle G \rangle = \langle G \rangle V^{\text{eff}} G_0 \\ \langle G_1 \rangle \langle G_2^* \rangle K_{12} \langle G_1 G_2^* \rangle = \langle G_1 G_2^* \rangle K_{12} \langle G_1 \rangle \langle G_2^* \rangle . \end{gathered} \quad (1)$$

Solution. Write the Dyson equation (3.3.25) in the integral formulation as

$$\langle G \rangle = G_0 + G_0 V^{\text{eff}} \langle G \rangle. \quad (2)$$

Iterating this equation, we find the series

$$\langle G \rangle = G_0 + G_0 V^{\text{eff}} G_0 + G_0 V^{\text{eff}} G_0 V^{\text{eff}} G_0 + \ldots. \quad (3)$$

Substituting this series for $\langle G \rangle$ in eqn (2) yields the first relations in eqn (1). Similarly, by iterating the Bethe–Salpeter equation (3.3.29) to obtain $\langle G_1 G_2^* \rangle$, one can verify the second relation in eqn (1). (Strictly speaking, such a proof is not complete, because the convergence of the formal series (3) are generally unknown.)

Problem 3.29. Prove the optical theorem (3.3.58).

Solution. Assuming that for operators as well as for conventional matrices the operations of inversion and Hermitian conjugation are permutable, that is, $(1/A)^+ = 1/A^+$, we represent V as $V = 1/G - 1/G_0$ in the condition of the nondissipative perturbation $V = V^+$. We have

$$\frac{1}{G^+} - \frac{1}{G} = \frac{1}{G_0^+} - \frac{1}{G_0}.$$

Premultiplying this expression by G and postmultiplying by G^+, we obtain the optical theorem (3.3.58).

Problem 3.30. For the scalar wave problem (3.3.3), prove that the scattering amplitude is related to the T-operator by eqn (3.3.62). Extend this relation to the electromagnetic case.

Solution. The scattering amplitude determines the asymptotics of the scattered wave far from a single scatterer in agreement with eqn (2.2.51)

$$u^s \underset{R\to\infty}{\approx} f_\omega(n \leftarrow n_0)\frac{e^{ikR}}{R}. \tag{1}$$

On the other hand, from eqns (3.3.10) and (3.3.60) we have

$$u^s = G_0 T u^0, \tag{2}$$

where $u^0 = G_0 q = \exp(ik_0\mathbf{n}_0\mathbf{r})$ is the incident wave field. Switching from the operators to their kernels and taking into account that the kernel $G_0(r,r')$ may be replaced with $-[\exp(ik_o R - ik_0\mathbf{nr}')/4\pi R]$ far from the scatterer, we compare eqns (1) and (2) and obtain eqn (3.3.62).

In the electromagnetic problem, the scalar amplitude u^s becomes the vector amplitude $\mathbf{E}^s$, and the operator G_0 becomes the tensor operator G_0 considered in Problem 3.24. Thus, instead of eqn (3.3.62) we have

$$\mathbf{f}_\omega(n \leftarrow n_0) = -(1 - \mathbf{n}\otimes\mathbf{n})\frac{T_{kk_0}}{4\pi}\mathbf{E}_0, \tag{3}$$

where $\mathbf{E}_0$ is the polarization vector of the incident wave.

Problem 3.31. Show that, for a single scatterer, the optical theorem (2.2.53) follows directly from the condition of energy conservation for the T-operator (3.3.61).

Solution. In the absence of dissipation, using (2.2.52), the optical theorem (2.2.53) may be written as

$$\sigma_s \equiv \int |f_\omega(\mathbf{n} \leftarrow \mathbf{n}_0)|^2 d\Omega_{\mathbf{n}} = \frac{4\pi}{k_0} \mathrm{Im}\, f_\omega(\mathbf{n}_0 \leftarrow \mathbf{n}_0). \tag{1}$$

Transforming the left-hand side of eqn (3.3.61) with allowance for eqn (3.3.62), we have

$$\int d^3r \exp(-i\mathbf{k}_0\mathbf{r})(T - T^+)\exp(i\mathbf{k}_0\mathbf{r}) = -4\pi\left[f_\omega(\mathbf{n}_0 \leftarrow \mathbf{n}_0) - f_\omega^*(\mathbf{n}_0 \leftarrow \mathbf{n}_0)\right]. \tag{2}$$

A similar transformation on the right-hand side of eqn (3.3.61) yields

$$\int d^3r \exp(-i\mathbf{k}_0\mathbf{r})T^+(G_0 - G_0^+)T \exp(i\mathbf{k}_0\mathbf{r}) = \int \frac{d^3k}{4\pi^3} |T_{\mathbf{k}\mathbf{k}_0}|^2 \,\mathrm{Im}\, G_{\mathbf{k}}^0, \tag{3}$$

where $G_{\mathbf{k}}^0 = [(\omega / c)^2 - k^2 + i0]^{-1}$. Using the familiar relation of the theory of generalized functions $\mathrm{Im}(a + i0)^{-1} = -\pi\delta(a)$, we integrate eqn (3) with respect to the modulus of k and express $T_{\mathbf{k}\mathbf{k}_0}$ in terms of $f_\omega(\mathbf{n} \leftarrow \mathbf{n}_0)$. The result is the optical theorem (1).

Problem 3.32. Show that the optical theorem for effective inhomogeneities (3.3.63) follows from the optical theorem for the operator (3.3.58) (Barabanenkov and Finkelberg, 1968).

Solution. Introduce an operator S which acts on a function of two arguments $f(x_1, x_2)$ by the rule

$$Sf(x_1, x_2) = \int f(x, x) dx. \tag{1}$$

It is easily seen that eqn (3.3.58) is equivalent to the two-index relation

$$S(G_1 - G_2^*) = S[(G_{02}^*)^{-1} - (G_{01})^{-1}]G_1 G_2^*. \tag{2}$$

Indeed, acting by both sides of eqn (2) on the function $\delta(x_1 - x_1')\delta(x_2 - x_2')$, we get (3.3.58) written as an equation for kernels of the respective operators.

Average both sides of eqn (2) and postmultiply the result by the operator $\langle G_1 G_2^* \rangle^{-1} = D_1 D_2 - K_{12}$ (see 3.3.28):

$$\begin{aligned} S\big(\langle G_1\rangle - \langle G_2^*\rangle\big)\big(D_1 D_2 - K_{12}\big) &= S\big[D_2^* - D_1 - (\langle G_1\rangle - \langle G_2^*\rangle)K_{12}\big] \\ &= S\big[(G_{02}^*)^{-1} - (G_{01})^{-1}\big]. \end{aligned} \tag{3}$$

Expressing here D_j as $G_{0j}^{-1} - V_j^{\text{eff}}$ in accordance with the Dyson equation (3.3.25), we obtain the expression

$$S(V_1^{\text{eff}} - V_2^{\text{eff}}) = S\langle G_1 - G_2^* \rangle K_{12}, \qquad (4)$$

that is equivalent to the optical theorem (3.3.63).

Problem 3.33. For the wave equation (3.3.3) $L = \Delta + k^2(1+\varepsilon)$, with the local perturbation operator $V = -k^2\varepsilon$, the optical theorem from the preceding problem can be strengthened: the operator S may be replaced with the operator

$$S' f(x_1, x_2) = f(x_1, x_1), \qquad (1)$$

that does not involve integration. Write the respective expression for the enhanced optical theorem in the K-space.

Solution. Repeating the reasoning of the preceding problem, we arrive at the optical theorem in the form (4) with operator S' in place of S. Switching to the K-representation, we obtain

$$V^{\text{eff}}\left(K_\rho + \frac{K_R}{2}\right) - V*\left(K_\rho - \frac{K_R}{2}\right)$$
$$= \int \left(\left\langle G\left(K'_\rho + \frac{K_R}{2}\right)\right\rangle - \left\langle G*\left(K'_\rho - \frac{K_R}{2}\right)\right\rangle \right) K(K_R, K'_\rho, K_\rho) d^4 K'_\rho . \qquad (2)$$

$$K(K_R, K'_\rho, K_\rho) = \int K(R', \rho', \rho) \exp\left(-i(K_R R' + K'_\rho \rho' - K_0 \rho)\right) \frac{d^4 R' d^4 \rho' d^4 \rho}{(2\pi)^4}. \qquad (3)$$

The meaning of wave vectors K_R and K_0 is given in Problem 3.4. (Here $K(\cdot)$ as a function should not be confused with a 4-D wave vector K.)

The optical theorem (3.3.65) is a special case of eqn (2) taken at $K_R = 0$, and eqn (3.3.67) is then expressed as

$$K(K' \leftarrow K) = K(K_R = 0,\ K', K). \qquad (4)$$

We recognize that, in the R-space, the substitution $K_R = 0$ is equivalent to the integration with respect to R. Therefore, the optical theorem (3.3.65) may be called integral in contrast to the local optical theorem (2) where such integration is not performed. Note that in quantum mechanics a relation similar to eqn (2) is known as the Ward identity.

Problem 3.34. Show that, in the case of equilibrium thermal radiation, the relation (3.3.68) expressing the spectrum of wave field fluctuations in the depth regime follows directly from the fluctuation-dissipation theorem (FDT).

Solution. The equilibrium thermal radiation field u and its fluctuation sources q are energy conjugates in the sense that the small force q results in the mean dissipation of energy per unit time

$$Q = \int \langle q \partial_t u \rangle_T d^3 r. \tag{1}$$

The angular brackets denote here the averaging over an ensemble of microscopic thermal fluctuations. In the monochromatic case the spectral component u_ω is expressed in terms of q_ω as

$$u_\omega = G_\omega q_\omega \equiv \int G_\omega(\mathbf{r}, \mathbf{r}') q_\omega(\mathbf{r}') d^3 r', \tag{2}$$

where $G_\omega(\mathbf{r}, \mathbf{r}')$ is the Green function. In accordance with the FDT, the spectral density of correlation of u_ω at two points $\mathbf{r}_1$ and $\mathbf{r}_2$ is expressed in terms of the kernel of the anti-Hermitian part of $G_\omega^a = (G_\omega - G_\omega^+)/2i$ as

$$\langle u_\omega(\mathbf{r}) u_\omega^*(\mathbf{r}_0) \rangle_T = \frac{\theta(\omega, T)}{\pi\omega} G_\omega^a(\mathbf{r}, \mathbf{r}_0), \tag{3}$$

where $\theta(\omega, T) = (\hbar\omega/2)\coth(\hbar\omega/2\kappa T)$ is the mean energy of the quantum oscillator. A more general formulation of the FDT for the correlation of sources q_ω was discussed in Section 3.5.9.

In the general case, the kernel G_ω^a entering eqn (3) may depend upon macroscopic medium's parameters fluctuating in space. (For the wave operator (3.3.3), such a parameter is the permittivity ε.) The correlation of thermal fluctuations u_ω entering eqn (3) is then also a nonuniform fluctuating quantity. We may then additionally average eqn (3) over macroscopic fluctuations to obtain

$$\left\langle \langle u_\omega(\mathbf{r}) u_\omega^*(\mathbf{r}_0) \rangle_T \right\rangle = \frac{\theta(\omega, T)}{\pi\omega} \left\langle G_\omega^a(\mathbf{r}, \mathbf{r}_0) \right\rangle. \tag{4}$$

This relationship represents an extension of eqn (3.3.68) to the case of equilibrium thermal fluctuations. The correlation (4) generally remains nonuniform even after statistical averaging over macroscopic fluctuations because of the nonuniformity of the operator G_ω^a that is especially pronounced near interfaces. If the medium may be deemed statistically homogeneous far from interfaces, so that $\langle G_\omega^a(\mathbf{r}, \mathbf{r}_0) \rangle$ depends only on the difference $\mathbf{r} - \mathbf{r}_0$ in this region, the field correlation on the left-hand side of eqn

(4) will also be uniform. The Fourier transformation of eqn (4) yields then the spectrum of field fluctuations (3.3.68) with $c = \theta(\omega,T)/\pi\omega$.

Problem 3.35. To illustrate the general relations involved in the derivation of moment equations, it is useful to employ the model of a quasi-oscillator with a fluctuating frequency $\omega(t) \equiv \omega_t$ for which the initial equation (3.6.1) has the form

$$Lu = (i^{-1}d_t - \omega_t)u(t) = q(t) \tag{1}$$

completed with the initial condition

$$u|_{t=0} = u_0. \tag{2}$$

(Recall that a real oscillator is governed by a second order differential equation). For the problem (1) and (2), an exact solution is available

$$u(t) = \exp\left(i\int_0^t \omega_{t'}dt'\right)u_0 + i\int_0^t \exp\left(i\int_{t'}^t \omega_{t''}dt''\right)q(t')dt', \tag{3}$$

which enables one to easily check the correctness of the heuristic considerations on the validity of various approximations.

Consider the case of Gaussian fluctuations of frequency ω_t with the correlation function

$$b(t) = \langle\tilde{\omega}_t\tilde{\omega}_0\rangle = \sigma^2\exp(-|t/\tau_{\mathbf{k}}|). \tag{4}$$

Write the Bourret approximation for this problem and estimate the validity conditions for this equation.

Solution. Let L_0 be the mean of the operator L:

$$L_0 = \langle L\rangle = \left\langle i^{-1}d_t - \omega_t\right\rangle = i^{-1}d_t - \overline{\omega},$$

where $\overline{\omega} = \langle\omega_t\rangle = \text{const}$. Inverting this operator for the problem with the initial condition (2) yields the operator G_0 whose kernel may be chosen in the form

$$G_0(t,t') = (i^{-1}d_t - \overline{\omega})^{-1}(t,t') = i\theta(t-t')e^{i\overline{\omega}(t-t')}. \tag{5}$$

Here, $\theta(t)$ is the Heaviside step function, and the action of G_0 on an arbitrary function $f(t)$ is given by

$$G_0 f(t) = \int G_0(t,t')f(t')dt' = \int_{-\infty}^{t} \exp(i\omega(t-t'))f(t')dt'. \tag{6}$$

In addition, the action of the perturbation operator reduces to the multiplication by the random part of the frequency ω_t, i.e.

$$V = L_0 - L = \omega_t - \overline{\omega} \equiv \tilde{\omega}(t),$$

so that the kernel of V is equal to $V(t,t') = \tilde{\omega}(t)\delta(t-t')$.

Equations (3.3.11) and (3.3.12) take the form

$$G \equiv (i^{-1}d_t - \omega_t)^{-1} = (i^{-1}d_t - \overline{\omega})^{-1} + (i^{-1}d_t - \overline{\omega})^{-1}\tilde{\omega}_t G, \qquad (i^{-1}d_t - \omega_t)G = \hat{1},$$

or, for kernels of the corresponding operators,

$$G(t,t') = i\theta(t-t')\exp(i\overline{\omega}(t-t') + \int_{t'}^{t} i\exp(i\overline{\omega}'(t-t'))\tilde{\omega}_{t''}G(t'',t')dt'',$$

$$(i^{-1}d_t - \omega_t)G(t,t') = \delta(t-t').$$

Assuming $u_0 = 0$ for simplicity, from a comparison of eqn (3.3.10) with eqn (3), we find for the kernel of G:

$$G(t,t') = i\theta(t-t')\exp\left(i\int_{t'}^{t} \omega_{t''}dt''\right).$$

The calculation of the moments of a random field u reduces then to the calculation of the moments of this kernel.

In this case, the frequency ω acts as of the wave vector $\mathbf{k}$, and the correlation time acts as of the correlation radius l_c. By the same token, here we deal with a random process $u(t)$ with a specified initial value u_0 rather than with a traveling wave. Nevertheless, considerations leading to the condition (3.3.82) remain basically valid and need only an insignificant modification: a dispersion equation of the type (3.3.79) may be used to describe the solution for large times t, which is an analog of the free-space propagation far from interfaces. In the Bourret approximation, we have the equation for $\langle u \rangle$:

$$(L_0 - \langle \tilde{V}G_0\tilde{V}\rangle)\langle u \rangle = (i^{-1}d_t - \overline{\omega})\langle u \rangle - i\left\langle \tilde{\omega}_t \int_0^t \exp(i\overline{\omega}(t-t'))\tilde{\omega}_{t'} \right\rangle \langle u(t')\rangle dt' = q.$$

The substitution of $q=0$ and $u=\exp(i\omega t)$ yields the dispersion equation

$$\omega-\overline{\omega}-i\left\langle \tilde{\omega}_t\int_0^t \exp(i(\omega-\overline{\omega})(t-t'))\tilde{\omega}_{t'}\right\rangle dt'=\omega-\overline{\omega}-i\int_0^t \exp(i(\omega-\overline{\omega})t')b(t')dt'=0$$

For large $t>>\tau_c$, we may put $t=\infty$ in the upper limit of integration, so that this expression becomes time-independent. Substituting $b(t)$ from eqn (4), we have

$$\omega-\overline{\omega}-i\int_0^\infty \exp(i(\omega-\overline{\omega})t')b(t')dt'=\omega-\overline{\omega}-\frac{i\sigma^2\tau_c}{1-i(\omega-\overline{\omega})\tau_c}=0.$$

This equation has two roots:

$$\omega^{(1,2)}=\overline{\omega}+\left(\frac{i}{2\tau_c}\right)\left(\pm\sqrt{1+(2\sigma\tau_c)^2}-1\right)$$

the root $\omega^{(1)}$ with plus sign becomes the unperturbed solution $\omega=\overline{\omega}$ when fluctuations disappear ($\sigma=0$). Therefore, replacing k with ω and l_c with τ_c in eqn (3.3.82), we obtain the necessary applicability condition for the Bourret approximation:

$$\tau_c|\omega^{(1)}-\overline{\omega}|=\frac{1}{2}\left|\sqrt{1+(2\sigma\tau_c)^2}-1\right|<<1$$

or more roughly

$$(\sigma\tau_c)^2<<1. \tag{7}$$

Under this condition, $\omega^{(1)}\approx\overline{\omega}+i\sigma^2\tau_c$, and, in the Bourret approximation, for large $t>>\tau_c$, the mean value $\langle u\rangle$ varies as $\exp(i\omega^{(1)}t)$. In this case, the exact value of $\langle u\rangle$ can be directly evaluated as

$$\langle u\rangle=\left\langle \exp(i\int_0^t \omega_{t'}dt')\right\rangle u_0=u_0\exp\left(i\overline{\omega}t-\frac{1}{2}\int_0^t\int_0^t b(t'-t'')dt'dt''\right)$$

$$=u_0\exp\left\{i\overline{\omega}t-\sigma^2\left[(t-\tau_c)\tau_c\left(1-\exp(-t/\tau_c)\right)+t\tau_c\exp(-t/\tau_c)\right]\right\}$$

$$\underset{t>>\tau_c}{\approx} u_0\exp\left\{i\overline{\omega}t-\sigma^2(t-\tau_c)\tau_c\right\}=u_0\exp\left(\sigma^2\tau_c^2+i\omega^{(1)}t\right).$$

Here we observed that the fluctuations are assumed to be Gaussian, and u is averaged by eqn (3.3.45). Under the condition (7) the Bourret approximation describes $\langle u \rangle$ correctly for large $t >> \tau_c$, whereas when eqn (7) is violated ($\sigma\tau_c \geq 1$), this approximation becomes invalid for large t. Thus, for this example, the condition (7) is necessary for the applicability of the Bourret approximation.

Problem 3.36. The Lorenz-Lorentz formula is mathematically derived on the assumption that the mean distance between particles is small compared to the wavelength, thus allowing for the interaction of scatterers due to the near field. However, this formula must remain also valid in the case of sparse media where such an interaction is absent. Given that, for small dipoles, the forward scattering amplitude is expressed in terms of the polarizability as

$$f_\omega(\mathbf{n}_0 \leftarrow \mathbf{n}_0) \approx k_0^2 \alpha, \tag{1}$$

show that in the latter case the Lorenz-Lorentz formula transforms into the dispersion formula (2.2.60) (van de Hulst, 1957).

Solution. In sparse media, $\nu\alpha << 1$, so that at $\varepsilon_m = 1$ the Lorenz-Lorentz formula (3.4.4) yields the expression

$$\varepsilon^* = \frac{1 + \frac{8}{3}\pi\nu\alpha}{1 - \frac{4}{3}\pi\nu\alpha} \approx 1 + 4\pi\nu\alpha, \tag{2}$$

which coincides with a similar relation following from (2.2.59) in view of eqn (1).

This coincidence is due to the fact that the approximation (1), in which $f_\omega(\mathbf{n}_0 \leftarrow \mathbf{n}_0)$ turns out to be real-valued, does not allow for the attenuation of radiation due to scattering and, as a consequence, violates the optical theorem. In the next order in $k_0^3\alpha$, f_ω acquires an imaginary part on the order of $k_0^5\alpha^2$. As a result, the optical theorem becomes valid, and an effective absorption, absent in the Lorenz–Lorentz formula, occurs.

Problem 3.37. A perfect material for absorbers of solar energy must efficiently absorb radiation in the visible and ultraviolet ranges which cover the larger portion of solar energy (the temperature of the sum ~ 5000 K corresponds to the wavelength of the radiation maximum ~ 100 Å); it must also weakly radiate (hence must weakly absorb in accordance with Kirchhoff's law) in the infrared range that corresponds to temperatures of the earth (Seraphin, 1979). For cermets consisting of small (≤ 100 Å) metal particles in a dielectric host matrix, these conditions can be satisfied by shifting the optical resonance to the optical range.

Given the free electron model for metal particles to be $\varepsilon_1 = 1 - (\omega_p / \omega)^2$, where

ω_p is the plasma frequency, and $\varepsilon_m \approx 1$ for a matrix material, find the dependence of the optical-resonance frequency on the volume fraction of particles P.

Solution. The conductivity proportional to the imaginary part of the effective medium permittivity strongly increases at the optical-resonance frequency. Setting $\varepsilon^* = \varepsilon' + i\varepsilon''$ in the Maxwell Garnett formula (3.4.6) and letting ε'' tend to infinity, we find the desired dependence to be

$$\omega \approx \omega_p \sqrt{\frac{1-P}{3}}.$$

Problem 3.38. Derive the two-particle distribution function for a sparse medium of hard spheres of radius a, assuming that the spheres fill the scattering volume uniformly on average and cannot approach each other closer than a sphere diameter (i.e. taking into account the hole correction).

Solution. From probability theory the two-particle distribution function is $P_{12} = P_{1|2}P_2$ where P_2 is the one-particle distribution function, and $P_{1|2}$ is the probability of finding particle 1 at $\mathbf{r}_1$ given that particle 2 is at $\mathbf{r}_2$. For hard spheres, it is natural to approximate $P_{1|2}$ as

$$P_{1|2} = \begin{cases} c\theta_v(\mathbf{r}_1), & \rho_{12} > 2a \\ 0, & \rho_{12} < 2a \end{cases} = c\theta_v(\mathbf{r}_1)[1-\theta(2a-\rho_{12})]. \tag{1}$$

Here, $\rho_{12} = |\mathbf{r}_1 - \mathbf{r}_2|$, $\theta(x)$ is the Heaviside step function, and $\theta_v(\mathbf{r})$ is the characteristic function of the scattering volume V_s which is unity inside and zero outside V_s. An unknown constant c appearing in eqn (1) can be found from the normalization condition

$$\int P_{12} d\mathbf{r}_1 d\mathbf{r}_2 = 1. \tag{2}$$

Taking into account that the particle's volume $v_1 = (4\pi/3)a^3$ is small compared to V_s, we have $c \approx 1/(V_s - v_1)$, so that

$$P_{12} = P_1 P_2 \frac{1-\theta(2a-\rho_{12})}{1-\beta}, \tag{3}$$

where $P_1 = \theta_{V_s}(\mathbf{r}_1)/V_s$, $\beta = v_1/V_s << 1$. The correlation function $q_{12} = P_{12} - P_1P_2$, corresponding to eqn (3), does not vanish in separation of the arguments:

$$q_{12} \underset{\rho_{12}>2a}{=} P_1 P_2 \frac{\beta}{1-\beta}.$$

This behavior owes its existence to residual long-range correlations due to expulsion by a particle a free volume which remains equal to $V_s - v_1$ for the other particles. Although such correlations are small ($\sim \beta = v_1 / V_s << 1$), in this model, they ensure that the correlation function q_{12} satisfies the condition

$$\int q_{12} d\mathbf{r}_2 = 0. \tag{4}$$

Problem 3.39. Show that, for a particle centered at $\mathbf{r}_l$, the scattering amplitude differs from that of a particle at the origin of coordinates by the factor $\exp(i\mathbf{q}\mathbf{r}_l)$ where $\mathbf{q} = k(\mathbf{n} - \mathbf{n}_0)$ is the vector of scattering.

Solution. The scattering amplitude is expressed in terms of the T-operator by eqn (3.3.62). Shifting the origin of coordinates by $\mathbf{a}$ shifts the kernel $T(\mathbf{r}, \mathbf{r}_0)$ of the operator T by $-\mathbf{a}$ in both arguments $\mathbf{r}$ and $\mathbf{r}_0$. Indeed, writing the action of T on u as $T_0 u_0$, $u_0 = u(\mathbf{r})$ in the initial coordinate system, and as $T_a u_a$, $u_a = u(\mathbf{r} - \mathbf{a})$ in the coordinate system shifted by a we obtain

$$T_a u_a = (T_0 u_0)_{\mathbf{r} \to \mathbf{r} - \mathbf{a}}. \tag{1}$$

Expressing this relationship in terms of the kernels of the operators we have

$$T_a(r, r_0) = T_0(r - a, r_0 - a). \tag{2}$$

Substituting eqn (2) into (3.3.62) for the scattering cross section yields

$$f_l(\mathbf{n} \leftarrow \mathbf{n}_0) = f_0(\mathbf{n} \leftarrow \mathbf{n}_0) \exp(-i\mathbf{q}\mathbf{r}_l), \tag{3}$$

where f_0 and f_l are the scattering amplitudes for a particle at the origin and at $\mathbf{r}_l$, respectively.

Problem 3.40. Derive the modulation function (3.4.43) for spherical particles uniformly distributed in a spherical scattering volume of radius R_s.

Solution. Express first the volume average quantities $\langle \xi_1 \rangle$ and $\langle \xi_1 \xi_2^* \rangle$. We have

$$\langle \xi_1 \rangle = \int_{V_s} \xi_1 \frac{d^3 r_1}{V_s} = \int_{V_s} \exp(i\mathbf{q}\mathbf{r}_1) \frac{d^3 r_1}{V_s} = 3 \frac{\sin z - z \cos z}{z^3} \equiv f(z), \tag{1}$$

where $z = |\mathbf{q}| R_s$, and

$$\langle \xi_1 \xi_2^* \rangle = \iint_{V_s} \exp(i\mathbf{q}(\mathbf{r}_1 - \mathbf{r}_2)) P_{12} d^3 r_1 d^3 r_2, \tag{2}$$

where P_{12} is the two-particle distribution function. In the case of a sufficiently sparse system of particles, $N\beta = Nv_1 / V_s << 1$, where v_1 is the particle volume, we may approximate P_{12} using eqn (2) of Problem 3.38, then

$$\langle \xi_1 \xi_2^* \rangle = \iint\limits_{V_s} \exp(i\mathbf{q}(\mathbf{r}_1 - \mathbf{r}_2)) \frac{1-\theta(2a-\rho_{12})}{1-\beta} d^3r_1 d^3r_2$$

$$= \frac{1}{1-\beta}\left[|f(z)|^2 - \iint\limits_{V_s} \exp(i\mathbf{q}(\mathbf{r}_1 - \mathbf{r}_2))\theta(2a-\rho_{12}) \frac{d^3r_1 d^3r_2}{V_s^2} \right]. \quad (3)$$

The integral in this expression can be easily estimated by changing the variables $\mathbf{r}_{1,2} = \mathbf{R} \pm \rho / 2$:

$$\iint\limits_{V_s V_s} \exp(i\mathbf{q}\rho)\theta(2a-\rho) \frac{d^3r_1 d^3r_2}{V_s^2} \approx \int\limits_{V_s} \frac{d^3R}{V_s} \int\limits_{V_s} \frac{d^3\rho}{V_s} \exp(i\mathbf{q}\rho)\theta(2a-\rho) = 4\beta f(2aq). \quad (4)$$

Finally, for the modulation factor (3.4.43) we have

$$\{.\} \approx N^2 |f(qR_s)|^2 + N\left(1 - |f(qR_s)|^2\right) - \frac{(N^2 - N)\beta}{1-\beta}\left(4f(2qa) - |f(qR_s)|^2\right). \quad (5)$$

Similarly, one can consider a system of particles with long-range correlation if a respective two-particle distribution function P_{12} is specified.

Problem 3.41. For a sparse discrete medium, derive the general expansions for the operators of effective inhomogeneities V^{eff} and K_{12} in powers of the T-operator of a single scatterer $t^{(l)}$.

Solution. In order to obtain the expansion for V^{eff}, we substitute eqn (3.3.60): $G = G_0 + G_0 T G_0$, into Dyson's equation $\langle G \rangle = G_0 + G_0 V^{\text{eff}} \langle G \rangle$. This substitution leads us directly to the Dyson equation for $\langle T \rangle$:

$$\langle T \rangle = V^{\text{eff}} + V^{\text{eff}} G_0 \langle T \rangle. \quad (1)$$

Thus, the expansion for V^{eff} in powers of $\langle T \rangle$ has the form

$$V^{\text{eff}} = \langle T \rangle (1 + G_0 \langle T \rangle)^{-1} = \sum_{n=0}^{\infty} \langle T \rangle (-G_0 \langle T \rangle)^n. \quad (2)$$

The desired expansion of V^{eff} results from eliminating $\langle T\rangle$ [obtained by averaging the series termwise over the order of scattering (3.4.33)] between this expression

$$\langle T\rangle = \sum\langle T^{(N,l)}\rangle = \sum\langle t^{(i)}\rangle + \sum_{i\neq j}\langle t^{(i)}G_0 t^{(j)}\rangle + \sum_{i\neq j\neq k}\langle t^{(i)}G_0 t^{(j)}G_0 t^{(k)}\rangle + \ldots \tag{3}$$

and eqn (2).

Similarly, from the Bethe–Salpeter equation (3.3.29) and from eqn (3.3.60), we obtain the expansion for K_{12} in powers of T:

$$\begin{aligned} K_{12} &= \langle G_1\rangle^{-1}\langle G_2^*\rangle^{-1} - \langle G_1 G_2^*\rangle^{-1} \\ &= \sum_{n=0}^{\infty}(-1)^n\langle G_{01}G_{02}^*\rangle^{-1}\Big\{\big(G_{01}\langle T_1\rangle + G_{02}^*\langle T_2^*\rangle + G_{01}G_{02}^*\langle T_1\rangle\langle T_2^*\rangle\big)^n \\ &\quad -\big(G_{01}\langle T_1\rangle + G_{02}^*\langle T_2^*\rangle + G_{01}G_{02}^*\langle T_1 T_2^*\rangle\big)^n\Big\}. \end{aligned} \tag{4}$$

Substituting here T expressed through the multiple scattering series (3.4.33), we obtain the desired expansion for K_{12}.

Problem 3.42. Given a medium of discrete scatterers in a free space, show that the procedure discussed in Section 3.4.10 leads to a representation for the effective permittivity in the form of the series (3.4.71).

Solution. For eqn (3.4.59), we introduce the T-operator of scattering T_F by the relation

$$\mathcal{F} = G^0 + G^{0\prime}\kappa\mathcal{F} \equiv G^0 + G^{0\prime}T_F G^0. \tag{1}$$

The operator T_F then satisfies an equation similar to eqn (3.4.28):

$$T_F = \kappa + \kappa G^{0\prime}T_F. \tag{2}$$

For the discrete medium under consideration, we may write κ as the sum of contributions from individual particles

$$\kappa = \sum\kappa_i\ , \quad \kappa_i = \frac{v_i}{1 - v_i/3k_0^2\varepsilon_0} = -3k_0^2\varepsilon_0\frac{\varepsilon_i - \varepsilon_0}{\varepsilon_i + 2\varepsilon_0}\ . \tag{3}$$

In calculations similar to those in Section 3.4.10, we replace V, T, and G^0 with κ, T_F, and $G^{0\prime}$, respectively, and $t^{(j)}$ with the operator t_F defined by

$$t_F^{(j)} = \kappa_j(1 + G_0' t_F^j) \tag{4}$$

to obtain the relationships

$$\kappa^{\text{eff}} = \sum_{n=0}^{\infty} \langle T_F \rangle \left(-G^{0'} \langle T_F \rangle\right)^n,$$
$$\langle T_F \rangle = \sum_{s=1}^{\infty} \sum_{(i,i_2 \dots i_s)}{}' \left\langle t_F^{(i)} G^{0'} t_F^{(i)} G^{0'} \dots t^{(i_s-1)} G^{0'} t^{(i_s)} \right\rangle, \tag{5}$$

which determine the desired expansion for κ^{eff}.

To show that the operator $t_F^{(j)}$ does coincide with the operator $t^{(j)}$ from Section 3.4.5, we use eqns (4), (3.4.58), (3.4.31), and the well known operator relation $(AB)^{-1} = B^{-1}A^{-1}$. Then

$$t_F^{(i)} = \frac{1}{1-\kappa_i G^{0'}} \kappa_i = \left[1 - \frac{v_i}{1 - v_i / 3k_0^2 \varepsilon_0} G^{0'}\right]^{-1} \frac{v_i}{1 - v_i / 3k_0^2 \varepsilon_0}$$
$$= \frac{1}{1 - v_i\left(G^{0'} + 1/3k_0^2 \varepsilon_0\right)} v_i = t^{(i)}. \tag{6}$$

4. Solution Techniques for Transfer Equations

In this chapter, we turn to the methods for solving the radiative transfer equation. We do not aim at giving a full review of all existent approaches, but restrict ourselves to a brief description of the most popular methods with a focus on their correlation and statistical content, ranges of validity, and limitations. At the same time, we tend to make this exposition sufficiently full in order that the reader could use the outlined methods for solving practical problems.

Section 4.1 is a brief characterization of the analytical methods used to solve the transfer equation. It describes two widely used methods — Fourier and invariant embedding. To demonstrate the application of the former, we obtain the exact solution to the transfer equation for an infinite scattering medium. The invariant embedding is stated in a very general form suitable not only to treat the reflection with the transfer equation formalism, but also to handle a variety of other wave problems.

Section 4.2 considers various forms of the small-angle approximation which is used to describe narrow beam waves. In addition, it presents the small-angle transfer equation for a composite medium containing continuous fluctuations and correlated discrete scatterers.

Section 4.3 is devoted to another limiting case of nearly isotropic scattering that allows a description in the context of the diffusion approximation. This approximation provides a basic solution for many situations, above all for media with small-scale inhomogeneities in the first place.

Section 4.4 discusses the solution of the transfer equation by iterations. This solution is obtained for the nonstationary problem with boundary and initial conditions in a general formulation. The applicability conditions for the first iteration, usually called the modified Born approximation, are estimated.

Section 4.5 outlines the simplest two-flux approximation allowing for multiple scattering in the diffusion range. This method is very simple, but it can provide a satisfactory accuracy in handling multiple scattering applications by a meaningful extraction of components of radiation with a wide radiation pattern.

Finally, Section 4.6 briefly describes the numerical methods for solving the transfer equation.

4.1 ANALYTICAL SOLUTIONS: THE FOURIER AND INVARIANT EMBEDDING METHODS

4.1.1 Rigorous Methods in Transfer Theory

Rigorous analytical solutions to the transfer equation have been obtained only for a few relatively simple problems. These are, for example, the isotropic scattering in an infinite medium, the problem of albedo in the reflection of a plane wave incident on a scattering half-space, the related Milne problem of sources radiating from a scattering medium, and problems with scattering layers. These solutions are of a limited applicability to practical problems, however, they are of a significant help in developing an initial intuition in multiple scattering problems and in verifying the accuracy of approximate numerical and analytical methods.

Since the transfer equation can be written in an integral form, methods used for its solution are directly related to the general theory of integral equations. Among the efficient analytical methods used in radiative transfer theory, we should first of all mention the invariance principle (invariant embedding method). This approach was proposed by Ambartsumyan (1943, 1944) and had a strong effect on the development of mathematical aspects of the theory.

A development of the invariance principle led to the discovery of important classes of special functions such as the Chandrasekhar function (Chandrasekhar, 1960) and the Sobolev function which arose in solving scattering layer problems and, as a special case, in tackling scattering half-space problems. Later, the method of invariant embedding was widely used in a variety of wave problems (Bellman and Wing, 1975; Klyatskin, 1986).

In 1960, a new approach to solving the transfer equation was devised by Case who proposed the method of expanding the transfer operator in terms of singular eigenfunctions. A detailed description of this method may be found in the book of Case and Zweifel (1967).

At present, the techniques employed to solve the transfer equation form a separate branch of mathematical physics containing many original approaches. A detailed presentation of this theory may be found in the cited above literature, including in the first place, the books by Chandrasekhar (1960) and Sobolev (1975), and the monographs by van de Hulst (1980), Davison (1957), and Bell and Glasstone (1970).

In order to provide insight into these analytical methods, we restrict our consideration to two approaches. These are the Fourier method, which is valid in the simplest case of an infinite medium with isotropic scattering (Section 4.1.2), and the method of invariant embedding, which is stated here for a scattering layer in a general operator form suitable not only for the equation of transport of scalar and electromagnetic radiation, but also for a class of wave problems (Section 4.1.3). The operator approach enables one to avoid cumbersome integral expressions at intermediate stages and, what is more important, reveals common merits in problems of transfer theory and wave theory that seem unlike at first glance.

4.1.2 An Exact Solution Obtained by the Fourier Method: Green's Function for an Infinite Isotropically Scattering Medium

In a statistically homogeneous and isotropically scattering medium, the scalar Green's function $G = G(\mathbf{n},\mathbf{r};\mathbf{n}_0,\mathbf{r}_0)$ obeys the transfer equation

$$(\mathbf{n}\nabla + \alpha)G(\mathbf{n},\mathbf{r};\mathbf{n}_0,\mathbf{r}_0) = \sigma\int G(\mathbf{n}',\mathbf{r};\mathbf{n}_0,\mathbf{r}_0)d\Omega_{\mathbf{n}'} + \delta(\mathbf{r}-\mathbf{r}_0)\delta_2(\mathbf{n}\mathbf{n}_0), \tag{4.1.1}$$

where α is the extinction coefficient. This function serves as a kernel of the operator

$$\hat{G} = (\mathbf{n}\nabla + \alpha - \hat{\sigma})^{-1}, \tag{4.1.2}$$

where the integral operator

$$\hat{\sigma}I(\mathbf{n},\mathbf{r}) = \sigma\int I(\mathbf{n}',\mathbf{r})d\Omega_{\mathbf{n}'}$$

acts on the angular variable $\mathbf{n}$ and describes isotropic scattering.

The operator $\hat{G}$ is uniform in the spatial coordinate $\mathbf{r}$, so that it is natural to use the Fourier transform with respect to $\mathbf{r}$ for evaluating $G(\mathbf{n},\mathbf{r};\mathbf{n}_0,\mathbf{r}_0) = G(\mathbf{r}-\mathbf{r}_0;\mathbf{n},\mathbf{n}_0)$. Using the results of Problem 4.1, we readily obtain for the kernel of $\hat{G}$:

$$G(\mathbf{n},\mathbf{r};\mathbf{n}_0,\mathbf{r}_0) = \int \frac{d^3\kappa}{(2\pi)^3} e^{i\kappa\mathbf{R}} \frac{1}{i\kappa\mathbf{n} + \alpha}\left[\frac{\sigma}{[1-\lambda(\kappa)](i\kappa\mathbf{n}_0 + \alpha)} + \delta_2(\mathbf{n}\mathbf{n}_0)\right], \tag{4.1.3}$$

where $\mathbf{R} = \mathbf{r} - \mathbf{r}_0$,

$$\lambda(\kappa) = \sigma\int\frac{d\Omega_{\mathbf{n}}}{i\kappa\mathbf{n} + \alpha} = \sigma\int_{-1}^{1}\frac{2\pi d\mu}{i\kappa\mu + \alpha} = \frac{4\pi\sigma}{\kappa}\arctan\frac{\kappa}{\alpha}. \tag{4.1.4}$$

The Green function (4.1.3) determines the radiance for a point directional source $\delta(\mathbf{r}-\mathbf{r}_0)\delta_2(\mathbf{n}\mathbf{n}_0)$. In the general case of an arbitrary distribution of sources $q(\mathbf{n},\mathbf{r})$, the radiance $I(\mathbf{n},\mathbf{r})$ is expressed through a superposition

$$I(\mathbf{n},\mathbf{r}) = \int G(\mathbf{n},\mathbf{r};\mathbf{n}',\mathbf{r}')q(\mathbf{n}',\mathbf{r}')d\Omega_{\mathbf{n}'}d^3r'. \tag{4.1.5}$$

In a special case of a point isotropic source with unit power when $q = \delta(\mathbf{r}-\mathbf{r}_0)/4\pi$, the total intensity of radiation takes the form

$$W(\mathbf{r}) = \int I(\mathbf{n},\mathbf{r})d\Omega_{\mathbf{n}} = \int \frac{d^3\kappa}{(2\pi)^3} e^{i\kappa\mathbf{R}} \frac{\lambda(\kappa)}{\sigma[1-\lambda(\kappa)]}$$

$$= -\frac{1}{(2\pi)^2 \sigma R} \partial_R \int_{-\infty}^{\infty} \frac{\lambda(\kappa)e^{i\kappa R}}{1-\lambda(\kappa)} d\kappa , \tag{4.1.6}$$

where again $\mathbf{R} = \mathbf{r} - \mathbf{r}_0$.

Thus, for an infinite homogeneous medium, the solution to the transfer equation (4.1.1) can be expressed through the integrals (4.1.2) or (4.1.6) which are not expressed in elementary functions and need a separate investigation. The behavior of these integrals depends substantially on the kind of integrand poles determined from the condition

$$\lambda(\kappa) = 1. \tag{4.1.7}$$

These integrals can be simplified by extracting the contribution due to the diffusion pole which is directly related to the solution of the transfer equation in the diffusion approximation. We will discuss this solution in Section 4.3.

Without going into a details of the expressions obtained, we note only that, in the absence of absorption, at small distances, the radiation intensity $W(\mathbf{r})$ of an isotropic source varies as R^{-2} which corresponds to the propagation in a free space. At large distances, when $R \to \infty$, the behavior of $W(\mathbf{r})$ is determined by the diffusion mechanism resulting in a slower dependence, R^{-1} (see, e.g., Ishimaru, 1978a).

4.1.3 Method of Invariant Embedding

This method was proposed by Ambartsumyan nearly half a century ago. The name "invariant embedding" owes its existence to a variety of the so-called *invariance principles*. A simple example of such a principle is associated with the description of reflection from a semi-infinite medium. The reflecting properties of such a medium are not changed when a scattering layer of an arbitrary thickness is added to this medium because the total thickness of the scattering medium remains infinite. These invariance principles enable one to obtain new (as a rule nonlinear) equations containing additional information on the solution of the problem.

In a wide sense, the method of invariant embedding consists in varying a certain parameter that controls the given system in a chosen class of systems. The initial value of this parameter corresponds to a simple system whose properties are well known. Depending on the statement of the problem, this parameter may be, for example, the depth of the scattering layer (this case will be treated below) or the radius of a scattering sphere. However, other geometrical or, even more general, physical parameters of the problem may be used. As a result, this approach enables one to move step by step from a simple system to a more complicated.

Formally, simplifications achieved with this approach are due to the "causality" of the invariant embedding equations with respect to the varying parameter, i.e. due to the switching from a boundary value problem to an initial value formulation.

In such a wide treatment, the invariant embedding proves to be applicable not only to the radiative transfer equation, but also to a more general class of linear and even nonlinear problems (Klyatskin, 1986). To illustrate the ideas underlying this method, we use the results of Apresyan (1987) where the invariant embedding equations were derived in general operator form. In this representation, final results can be obtained without intermediate cumbersome integral expressions. In addition, its results are suitable not only for the transfer equation but also for a variety of other wave equations describing the behavior of acoustic, electromagnetic, elastic, and other kinds of waves in inhomogeneous media.

Consider the general linear equation of the form (3.3.11)

$$G = G^0 + G^0 V G, \tag{4.1.8}$$

where G is the unknown operator, G^0 is the unperturbed Green operator, and V is the perturbation operator *local* with respect to the spatial argument $\mathbf{r}$, that is, the operator with a kernel containing a delta function $\delta(\mathbf{r} - \mathbf{r}_0)$. Note that, in the case of a plane-parallel layer, it is sufficient to require that the operator V be local only with respect to the z coordinate normal to the layer. In other words, the action of the operator V reduces to the multiplication by an operator $v(\mathbf{r})$.

The integral forms of a variety of linear problems, including the transfer equation (4.1.1), the scalar wave equation (3.3.3), and the electromagnetic wave equation (3.3.4), may serve as examples of equation (4.1.8).

As an illustrative example of the problem, consider the plane-parallel geometry assuming that the perturbation operator $v(\mathbf{r})$ is nonzero within the layer $0 \le z \le L_0$, so that

$$v(\mathbf{r}) = 0 \text{ for } z < 0 \text{ and for } z > L_0. \tag{4.1.9}$$

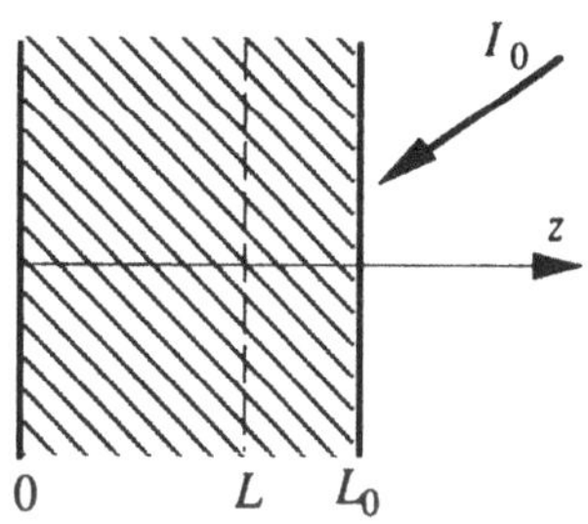

Figure 4.1. To scattering at a layer.

At the depth $L_0 = 0$, we come to the problem of a radiation propagating in a free space. Consider the possibility of going step by step from $L_0 = 0$ to a finite value of $L_0 > 0$. For this purpose we introduce a procedure truncating the perturbation V. It consists in replacing the operator $V = v(\mathbf{r})$ in eqn (4.1.8) with

$$V_L = \theta(L - z)v(\mathbf{r}), \tag{4.1.10}$$

where $\theta(L - z)$ is the Heaviside step function, and switching from eqn (4.1.8) to the equation

$$G = G^0 + G^0 V_L G, \tag{4.1.11}$$

where G depends implicitly on the parameter L.

Differentiating both sides of eqn (4.1.11) with respect to L, we have

$$\partial_L G = G^0 V_L' G + G^0 V_L \partial_L G,$$

where $V_L' = d_L V_L$. In view of eqn (4.1.8), we immediately obtain the *nonlinear* equation for G:

$$\partial_L G = G V_L' G, \tag{4.1.12}$$

which must be augmented by the obvious initial condition

$$G|_{L=0} = G^0. \tag{4.1.13}$$

Equation (4.1.12) with the initial condition (4.1.13) yields in principle the solution of the problem and enables us to determine the value of G at $L = L_0$ from its initial value G^0 at $L = 0$. It is important that this equation is valid not only for the layer under consideration, but also for any dependence of the perturbation V on the parameter L. Unfortunately, the nonlinear equation (4.1.12) is not simpler than the initial linear equation (4.1.8) and may be used only for a numerical analysis of the desired solution. Numerical calculations can be simplified using the causality of the problem (4.1.12) - (4.1.13) that differs it from the initial formulation which usually corresponds to a boundary value problem. For the layer and some other geometries of the scattering medium, one can advance a bit further and obtain a simpler invariant embedding equation for the reflection coefficient.

To prove this statement we write out the spatial arguments of the kernel of the operator G in explicit form, treating $G(\mathbf{r}, \mathbf{r}_0)$ as a kernel which can also act on other arguments, e.g., on time t in nonstationary wave problems, or on angle arguments $\mathbf{n}$ in the transfer equation. Consider the operator

$$H_L(\mathbf{r}_\perp,\mathbf{r}_{0\perp}) = G(\mathbf{r}_\perp,L;\mathbf{r}_{0\perp},L), \tag{4.1.14}$$

acting on the transverse arguments $\mathbf{r}_\perp$ on the surface $z = L$. This operator may be treated as a "restriction" of the operator G onto the layer surface $z = L$.

Differentiating H_L with respect to L and taking into account equation (4.1.12) where $V_L' = \partial_L v\theta(L - z) = v\delta(L - z)$, we have

$$\partial_L H_L = (\partial_z + \partial_{z_0})G(\mathbf{r}_\perp,z;\mathbf{r}_{0\perp},z_0)\Big|_{z=z_0=L} + H_L v(\mathbf{r}_\perp,L)H_L \tag{4.1.15}$$

In order that this equation be closed relative to H_L, we need to express the term involving G through H_L. The feasibility of this closing can be verified only with reference to the properties of the free propagation operator G^0 that are assumed known. In particular, it is natural to assume the operator G^0 to be uniform in $\mathbf{r}$, $G^0(\mathbf{r},\mathbf{r}_0) = G^0(\mathbf{r} - \mathbf{r}_0)$, so that

$$(\partial_z + \partial_{z0})G^0 = 0. \tag{4.1.16}$$

It can be readily shown (see Problem 4.2) that if the closure conditions

$$\begin{aligned} \partial_L G^0(r_\perp,L;r_{0\perp},z) &= B_+ G^0, \quad z < L, \\ \partial_L G^0(r_\perp,z;r_{0\perp},L) &= G^0 B_-, \quad z < L, \end{aligned} \tag{4.1.17}$$

where B_+ and B_- are some operators acting on the plane $z = L$ and independent of z for $z < L$, are satisfied for G^0, then the term with G in eqn (4.1.15) becomes closed with respect to H_L, and eqn (4.1.15) takes the form of a nonlinear equation for H_L:

$$\partial_L H_L = B_+(H_L - H_L^0) + (H_L - H_L^0)B_- + H_L v(r_\perp,L)H_L. \tag{4.1.18}$$

Here, $H_L^0 = G^0(\mathbf{r}_\perp,L;\mathbf{r}_{0\perp},L)$ is the restriction of the operator G^0 onto the plane $z = L$ similar to eqn (4.1.14). Equation (4.1.18) along with the obvious initial condition

$$H_L\big|_{L=0} = H_L^0\big|_{L=0} \equiv G^0(\mathbf{r}_\perp - \mathbf{r}_{0\perp}) \tag{4.1.19}$$

determines completely the behavior of H_L.

Equation (4.1.18) has a wide range of applications. This equation occurs in handling the reflection of waves of arbitrary nature and encompasses results obtained in various times in different branches of physics, including those for other problem geometries. The only difference between them is the explicit form of operators appearing in eqn (4.1.18). In addition to transfer theory, eqn (4.1.18) appears, for example, in the method of reduced potentials and in the method of phase functions in the theory of potential scattering (Taylor, 1972; Calojero, 1967); in the method of surface impedance describing surface acoustic waves (Biryukov et al., 1995), and in other wave problems (Klyatskin, 1986). In essence, eqn (4.1.18) may be thought of as a far reaching generalization of the Riccati equation which is well known in mathematics and in physical applications (see, e.g., Zakhar-Itkin, 1973).

The operator H_L can be related to the reflectance of a layer. Indeed, consider the operator

$$R = G - G^0. \tag{4.1.20}$$

Since the operator G describes the total field and G^0 describes an incident wave in scattering problems, then R corresponds to the total field minus the incident wave, i.e. to the scattered field. The restriction of the operator R onto the plane $z = L$ similarly to eqn (4.1.14) yields the operator

$$R_L = R(\mathbf{r}_\perp, L; \mathbf{r}_{0\perp}, L), \tag{4.1.21}$$

which enables us to express the scattered field emerging from the layer at $z = L$, i.e., the reflection from the layer, and has the meaning of the reflection coefficient. We therefore call R_L the *reflection operator*.

Expressing H_L through R_L as $H_L = H_L^0 + R_L$, we can write eqn (4.1.18) as an equation for the reflection operator R_L:

$$\partial_L R_L = B_+ R_L + R_L B_- + (H_L^0 + R_L)\nu(\mathbf{r}_\perp, L)(H_L^0 + R_L), \tag{4.1.22}$$

which is completed by the initial condition, following from eqn (4.1.19),

$$R_L\big|_{L=0} = 0. \tag{4.1.23}$$

In eqn (4.1.22) we took into account that $\partial_L H_L^0 = 0$.

Thus, we have shown that, in order to obtain the equations of invariant embedding (4.1.22) for linear problems of reflection from a layer, it is necessary:

(*a*) to formulate the initial problem in the form of the integral operator equation (4.1.8) with a spatially local perturbation $\nu(\mathbf{r})$;

(*b*) to find operators B_+ and B_-, i.e., to verify that the free-space propagation operator satisfies the closure conditions (4.1.17).

In practice, it is sufficient to verify only the first of the conditions (4.1.17) because, for problems satisfying the reciprocity theorem, the second condition follows from the first one, and the operator B_- is the transpose of B_+. In wave problems, B_+ and B_- prove to be integral operators (see Problem 4.3), and, in the case of the stationary transfer equation, they prove to be differential operators. We illustrate this simple scheme using the radiative transfer equation as an example (Apresyan, 1990a).

Let the scattering from the layer $0 \le z \le L$ be described by the transfer equation

$$(\mathbf{n}\nabla + \alpha)I = \hat{\sigma}I + q \equiv \int \sigma(\mathbf{n} \leftarrow \mathbf{n}')I(\mathbf{n}',\mathbf{r})d\Omega_{\mathbf{n}'} + q\,, \tag{4.1.24}$$

where q is the function of sources, and α is the extinction coefficient. The integral form of this equation is similar to eqn (4.1.8) where $V = \hat{\sigma}$, and the free-space propagation operator $G^0 = (\mathbf{n}\nabla + \alpha)^{-1}$ has the kernel

$$G^0(\mathbf{n},\mathbf{r};\mathbf{n}_0,\mathbf{r}_0) = \delta_2(\mathbf{n}\cdot\mathbf{n}_0)\frac{\exp\left(-\alpha\dfrac{z-z_0}{n_z}\right)}{|n_z|}\theta\left(\frac{z-z_0}{n_z}\right)\delta\left(\mathbf{r}_\perp - \mathbf{r}_{0\perp} - \mathbf{n}_\perp\frac{z-z_0}{n_z}\right), \tag{4.1.25}$$

where $\theta\big((z-z_0)/n_z\big)$ is the Heaviside step function (see Problem 4.4).

A direct substitution shows that the free propagation operator satisfies the conditions (4.1.17) where $B_+ = -B_- = -(\alpha + \mathbf{n}_\perp\nabla/n_z)$. In view of eqn (4.1.25) the operator H_L^0 in eqn (4.1.18) has the kernel $\delta_2(\mathbf{n}\mathbf{n}_0)\delta(\mathbf{r}_\perp - \mathbf{r}_{0\perp})/|n_z|$, therefore its action reduces to multiplying by $1/|n_z|$. As a result, we can write eqn(4.1.22) for the reflection operator as

$$\partial_L R_L = -\frac{\alpha + \mathbf{n}_\perp\nabla}{n_z}R_L + R_L\frac{\alpha + \mathbf{n}_\perp\nabla}{n_z} + \left(\frac{1}{|n_z|} + R_L\right)\hat{\sigma}\left(\frac{1}{|n_z|} + R_L\right). \tag{4.1.26}$$

This operator equation obtained by Apresyan (1990a) extends to the bounded incident beams a similar equation due to Chandrasekhar (1960) who considered plane waves [this case corresponds to eqn (4.1.26) with $\mathbf{n}_\perp\nabla$ equal to zero]. Changing from the compact operator form to kernels, we can readily write eqn (4.1.26) in the form of the integro-differential equation for the kernel $R_L(\mathbf{n}_\perp,\mathbf{n};\mathbf{r}_{0\perp},\mathbf{n}_0)$ of the operator R_L (Problem 4.5). The physical meaning of each term in eqn (4.1.26) is given in Problem 4.6.

We now consider the connection of eqn (4.1.26) with Chandrasekhar's results. The radiance of the incident wave I^0 is related to the source function q by $I^0 = G^0 q$. The similar radiance I_L^0 in the plane L is expressed in terms of the source distribution q_L in the same plane as

$$I_L^0 = H_L^0 q_L\,. \tag{4.1.27}$$

Taking this into account, we have for the radiance of reflected radiation I^r

$$I^r = R_L q_L = R_L (H_L^0)^{-1} I_L^0. \tag{4.1.28}$$

On the other hand, the radiance I^r is equal to $S(\mathbf{n},\mathbf{n}_0)/4\pi n_z$, where $S(\mathbf{n},\mathbf{n}_0)$ is the Chandrasekhar scattering function. Since the radiance of the incident plane wave $I^0 = \delta_2(\mathbf{n}\mathbf{n}_0)$ is independent of $\mathbf{r}_\perp$, it follows that

$$S(\mathbf{n},\mathbf{n}_0) = 4\pi n_z \int R_L(\mathbf{n},\mathbf{r}_\perp;\mathbf{n}_0,\mathbf{r}'_\perp) d^2 r'_\perp |n_{0z}|, \tag{4.1.29}$$

where $R_L(\mathbf{n},\mathbf{r}_\perp;\mathbf{n}_0,\mathbf{r}_{0\perp})$ is the kernel of the operator R_L. From this expression and eqn (4.1.26), we can easily obtain an equation for $S(\mathbf{n},\mathbf{n}_0)$ that coincides with the similar Chandrasekhar equation.

In the elementary case of an isotropically scattering half-space, for problems with axial symmetry, the scattering function $S(\mathbf{n},\mathbf{n}_0)$ becomes $S(\mu,\mu_0)$, where $\mu = n_z$ and $\mu_0 = |n_{0z}|$, and is expressed through the Chandrasekhar function $H(\mu)$. The latter is governed by a nonlinear integral equation (Problem 4.7) and was tabulated by Chandrasekhar (1960).

Thus, the general operator approach outlined in this section generalizes the known results of the invariant embedding method to the case of a beam wave. For other results of this method for the transfer equation the reader is referred to the literature quoted in this section.

4.2 RADIATIVE TRANSFER EQUATION IN THE SMALL-ANGLE APPROXIMATION

4.2.1 Scattering of Narrow Beams at Large-Scale Inhomogeneities

Large-scale inhomogeneities, whose size l_k is greater than the characteristic wavelength ($l_k >> \lambda$), scatter waves in a narrow cone around the forward direction. This circumstance appreciably simplifies the description of narrow beams by allowing one to use the small-angle approximation (Dolin, 1964b; 1966; 1968; Bremmer, 1964). We consider a transition to this approximation by taking the scattering of stationary monochromatic radiation in a time-invariant medium where no change in frequency occurs. We start with the transfer equation (4.1.1) in the absence of radiation sources:

$$(d_s + \alpha^{\text{ext}})I = \int \sigma(\mathbf{n} \leftarrow \mathbf{n}') I(\mathbf{n}',\mathbf{R}) d\Omega_{\mathbf{n}'} \equiv \hat{\sigma} I, \tag{4.2.1}$$

where $d_s = \mathbf{n}\nabla$ is the operator of differentiation along the ray path, $\alpha^{\text{ext}} = \alpha^{\text{abs}} + \alpha^{\text{sc}}$ is the extinction coefficient, α^{abs} is the absorption coefficient, and

$$\alpha^{\text{sc}} = \int \sigma(\mathbf{n}' \leftarrow \mathbf{n}) d\Omega_{\mathbf{n}'} \tag{4.2.2}$$

is the scattering coefficient.

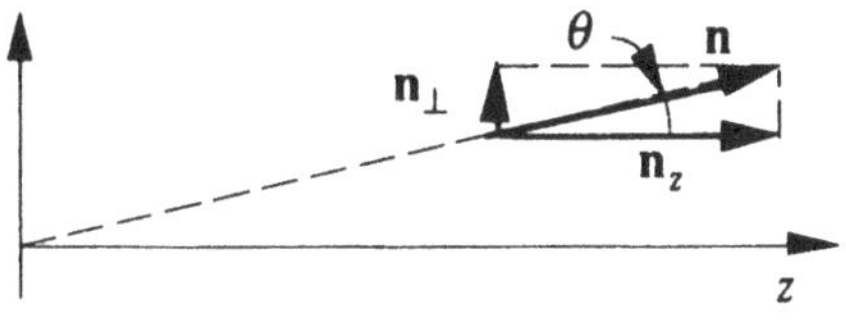

Figure 4.2. Scattering of a pencil beam at small angles.

Consider a narrow beam propagating along the z axis. Label by $\perp$ the transverse (relative to z) component of the respective vector, so that $\mathbf{n} = (\mathbf{n}_\perp, n_z)$ (Fig.4.2). In the small-angle approximation, the vector $\mathbf{n}$ is assumed to be close to the z axis, i.e.

$$|\mathbf{n}_\perp| \approx |\theta| << 1, \tag{4.2.3}$$

where θ is the angle between the vector $\mathbf{n}$ and the z axis, and the longitudinal component $n_z = (1 - n_\perp^2)^{1/2} \approx 1 - n_\perp^2 / 2$ differs from unity in the second order of $\mathbf{n}_\perp$.

Inequality (4.2.3) serves as the main necessary condition for the applicability of the small-angle approximation. When eqn (4.2.3) is satisfied, we may set $n_z \approx 1$, i.e.,

$$\mathbf{n} = (\mathbf{n}_\perp, n_z) \approx (\mathbf{n}_\perp, 1). \tag{4.2.4}$$

The initial value of radiance in the plane $z = 0$ is a function of $\mathbf{R}_\perp$ and $\mathbf{n}_\perp$ only, so that $I|_{z=0} = I^0(\mathbf{n}_\perp, \mathbf{R}_\perp)$, and the radiance itself depends in addition on distance $z = R_z$:

$$I = I(\mathbf{n}_\perp; \mathbf{R}) = I(\mathbf{n}_\perp; \mathbf{R}_\perp, z).$$

In the approximation (4.2.4) the operator of differentiation along the ray path is

$$d_s = \mathbf{n}\nabla = n_z \partial_z + \mathbf{n}_\perp \nabla_\perp \approx \partial_z + \mathbf{n}_\perp \nabla_\perp, \tag{4.2.5}$$

where $\nabla_\perp = \partial_{R_\perp}$.The elementary solid angle $d\Omega_\mathbf{n}$ may be replaced with $d^2\mathbf{n}_\perp$, and the integration with respect to $\mathbf{n}_\perp$ may be extended to infinity.

We restrict this consideration to the convolution type operator $\hat{\sigma}$ when the scattering cross section $\sigma(\mathbf{n} \leftarrow \mathbf{n}')$ depends only upon the difference $\mathbf{n} - \mathbf{n}'$. This assumption is exactly satisfied, for example, in a continuous medium in the Born approximation. In the general case, it may be used as an approximation. The right-hand side of eqn (4.2.1) may then be represented as

$$\hat{\sigma} I = \int \sigma(\mathbf{n} - \mathbf{n}') I(\mathbf{n}') d\Omega_{\mathbf{n}'} \approx \iint_{-\infty}^{\infty} \sigma(\mathbf{n}_\perp - \mathbf{n}'_\perp) I(\mathbf{n}'_\perp) d^2\mathbf{n}'_\perp. \tag{4.2.6}$$

Instead of eqn (4.2.1) we finally obtain the equation

$$(\partial_z + \mathbf{n}_\perp \nabla_\perp + \alpha^{\text{ext}}) I = \int \sigma(\mathbf{n}_\perp - \mathbf{n}'_\perp) I(\mathbf{n}'_\perp) d^2\mathbf{n}'_\perp \tag{4.2.7}$$

with the initial condition

$$I|_{z=0} = I^0(\mathbf{n}_\perp, \mathbf{R}_\perp). \tag{4.2.8}$$

Equation (4.2.7) is called the radiative transfer equation in the small-angle approximation. In this approximation, the scattering coefficient (4.2.2) transforms into

$$\alpha^{sc} = \int \sigma(\mathbf{n}' \leftarrow \mathbf{n}) d\Omega_{\mathbf{n}'} \approx \iint_{-\infty}^{\infty} \sigma(\mathbf{n}'_\perp - \mathbf{n}_\perp) d^2\mathbf{n}'_\perp . \tag{4.2.9}$$

4.2.2 Small-Angle Approximation and the Parabolic Equation

The small-angle transfer equation (4.2.7) is closely related to the parabolic-equation approximation (or quasi-optical approximation) for the wave equation and to the description of statistical moments in the Markovian approximation. The detailed treatment of these approximations is given in the book by Rytov et al.(1989b). We present here only some results directly connected with transfer theory.

In the case of continuous fluctuating media, the parabolic equation has the form

$$(2ik\partial_z + \Delta_\perp + k^2\tilde{\varepsilon})u' = 0 , \tag{4.2.10}$$

where k is the wave number, $\Delta_\perp = \partial_x^2 + \partial_y^2$ is the transverse Laplacian, $\hat{\varepsilon} = \varepsilon - \langle\varepsilon\rangle$ is the permittivity fluctuation, and the amplitude u' differs from the total amplitude of the wave field u by the multiplier e^{-ikz}, i.e. $u = u'e^{ikz}$.

Equation (4.2.10) is the simplest parabolic approximation to the wave equation $[\Delta + k^2(1+\tilde{\varepsilon})]u = 0$ and follows from it when the term with the second derivative ∂_z^2 is neglected. This one-wave approximation does not allow for backscattering and is close to diffraction treatments in the Fraunhofer approximation. Numerous attempts have been made in recent years to extend the approximation (4.2.10) to finite angles of scattering. They have resulted in various parabolic equations of higher order (see, e.g., Halpern and Trefethen, 1988).

Let us define the transverse coherence function as

$$\Gamma_\perp = \langle u'(\mathbf{r}_{1\perp}, z) u'^*(\mathbf{r}_{2\perp}, z)\rangle = \langle u(\mathbf{r}_{1\perp}, z) u^*(\mathbf{r}_{2\perp}, z)\rangle . \tag{4.2.11}$$

In the Markov approximation, from eqn (4.2.10) one can derive the equation for $\Gamma_\perp$

$$[2ik\partial_z + \Delta_1 - \Delta_2 + 2ikH_1(\mathbf{r}_1 - \mathbf{r}_2)]\Gamma_\perp = 0 . \tag{4.2.12}$$

Here, the function $H_1(\mathbf{r}_1 - \mathbf{r}_2)$ is expressed through the Born scattering cross section

$$\sigma(\mathbf{n}_\perp) = \frac{\pi k^4}{2}\Phi_\varepsilon(k\mathbf{n}_\perp) \equiv \frac{\pi k^4}{2}\int \langle\tilde{\varepsilon}(\rho)\tilde{\varepsilon}(0)\rangle \exp(-ik\mathbf{n}_\perp\rho_\perp)\frac{d^3\rho_\perp}{(2\pi)^3} \tag{4.2.13}$$

by

$$H_1(\rho_\perp) = \alpha^{abc} + \int \sigma(\mathbf{n}'_\perp)(1-\cos(k\mathbf{n}'_\perp \rho_\perp))d^2 n'_\perp , \tag{4.2.14}$$

where $\rho_\perp = \mathbf{r}_{1\perp} - \mathbf{r}_{2\perp}$.

On the other hand, from the general relation (3.5.22) it follows that, if we neglect the difference between the effective wave number k^{eff} and the free-space wave number k, the coherence function will be represented by the two-dimensional Fourier transform of the radiance

$$\begin{aligned} \Gamma &= \left\langle u\left(\mathbf{R}+\frac{\rho}{2}\right)u^*\left(\mathbf{R}-\frac{\rho}{2}\right)\right\rangle \\ &= \int I(\mathbf{n},\mathbf{R})e^{ik\mathbf{n}\rho}d\Omega_\mathbf{n} \\ &\approx \exp(ik\rho_z)\int I(\mathbf{n}_\perp,\mathbf{R})\exp(ik\mathbf{n}_\perp\rho_\perp)d^2n_\perp . \end{aligned} \tag{4.2.15}$$

Setting here $\rho_z = 0$,we have

$$\begin{aligned} \Gamma_\perp &= \Gamma|_{\rho_z=0} \\ &= \left\langle u\left(\mathbf{R}_\perp+\frac{\rho_\perp}{2},z\right)u^*\left(\mathbf{R}_\perp-\frac{\rho_\perp}{2},z\right)\right\rangle \\ &= \int I(\mathbf{n}_\perp;\mathbf{R}_\perp,z)\exp(ik\mathbf{n}_\perp\rho_\perp)d^2n_\perp . \end{aligned} \tag{4.2.16}$$

Taking the inverse Fourier transform of eqn (4.2.16), we find that the expression for radiance through the transverse coherence function $\Gamma_\perp$ also has the form of a two-dimensional Fourier transform:

$$I(\mathbf{n}_\perp,\mathbf{R}) = \left(\frac{k}{2\pi}\right)^2 \int \Gamma_\perp(\mathbf{R}_\perp,z;\rho_\perp)\exp(-ik\mathbf{n}_\perp\rho_\perp)d^2\rho_\perp . \tag{4.2.17}$$

If we now multiply the small-angle transfer equation (4.2.7) by $\exp(-ik\mathbf{n}_\perp\rho_\perp)$ and integrate the result over $\mathbf{n}_\perp$, we come to equation (4.2.12) written in the variables $\mathbf{R}_\perp = (\mathbf{r}_{1\perp}+\mathbf{r}_{2\perp})/2$ and $\rho_\perp = \mathbf{r}_{1\perp} - \mathbf{r}_{2\perp}$. Thus, equation (4.2.7) is equivalent to the equation for the coherence function $\Gamma_\perp$ which is obtained from the parabolic equation in the Markov approximation. This important result was obtained in the pioneering work of Dolin (1968) who proposed also a method for solving the small-angle transfer equation (1964).

It is important that in contrast to transfer theory dealing with the coherence function, that is, with the second moments of the radiant field, the Markov approximation of wave theory enables us to examine moments of arbitrary order which is especially important in the case of non-Gaussian fluctuations of the field.

4.2.3 General Solution of the Radiative Transfer Equation in the Small-Angle Approximation

The transfer equation in the small-angle approximation with the convolution integral operator (4.2.7) is remarkable in that it affords the exact solution (Problem 4.8):

$$I(\mathbf{n}_\perp,\mathbf{R}_\perp,z)=\int G(\mathbf{R}_\perp-\mathbf{R}'_\perp,z;\mathbf{n},\mathbf{n}')I^0(\mathbf{n}'_\perp,\mathbf{R}'_\perp)d^2R'_\perp d^2n'_\perp. \tag{4.2.18}$$

Here, the Green's function $G(\mathbf{R}_\perp,z;\mathbf{n}_\perp,\mathbf{n}'_\perp)$ is the kernel of the operator $(\partial_z+n_\perp\nabla_\perp+\alpha^{\text{ext}}-\hat{\sigma})^{-1}$:

$$G(\mathbf{R}_\perp,z;\mathbf{n}_\perp,\mathbf{n}'_\perp)$$
$$=\frac{k_0^2}{(2\pi)^4}\iint\exp\left\{ik_0(\mathbf{n}_\perp-\mathbf{n}'_\perp)\rho'_\perp+i\mathbf{k}'_\perp\mathbf{R}_\perp-z\overline{H_1\left(\rho'-\frac{z'\mathbf{k}'_\perp}{k_0}\right)}^z\right\}d^2\rho'_\perp d^2k'_\perp, \tag{4.2.19}$$

where $\overline{f(z')}^z$ denotes a function $f(z')$ averaged over the interval $[0,z]$:

$$\overline{f(z')}^z=\frac{1}{z}\int_0^z f(z')dz'=\int_0^1 f(zs)ds. \tag{4.2.20}$$

If we go from the radiance I to the coherence function $\Gamma_\perp$ in accordance with (4.2.17), the solution to the small-angle transfer equation (4.2.18) may then be written as

$$\Gamma_\perp(\mathbf{R}_\perp,z,\rho_\perp)=\int G(\mathbf{R}_\perp-\mathbf{R}'_\perp,z;\rho_\perp,\rho'_\perp)\Gamma^0(\mathbf{R}'_\perp,\rho'_\perp)d^2R'_\perp d^2\rho'_\perp, \tag{4.2.21}$$

where

$$G(\mathbf{R}_\perp,z;\rho_\perp,\rho'_\perp)=\left(\frac{k}{2\pi}\right)^2\exp\left[\frac{ik}{z}\mathbf{R}_\perp(\rho_\perp-\rho'_\perp)-z\int_0^1 H_\perp(\rho s+\rho'(1-s))ds\right], \tag{4.2.22}$$

The initial value

$$\Gamma_\perp^0(\mathbf{R}_\perp,\rho_\perp)=\left\langle u(\mathbf{R}_\perp+\frac{\rho_\perp}{2},0)u^*(\mathbf{R}_\perp-\frac{\rho_\perp}{2},0)\right\rangle \tag{4.2.23}$$

entering eqn (4.2.21) is related to I^0 by eqn (4.2.17) which is the small-angle limit ($n_z\sim 1$) of the general expression (3.2.19) for the boundary value of radiance in the diffraction in the free half-space. Expressions (4.2.18) - (4.2.21) give the general solution of the transfer equation for radiance and coherence function in the small-angle approximation. They are widely used in applications (Ishimaru, 1978a; Rytov et al., 1989b).

4.2.4 Diffusion in Angular Variables

The solution of the transfer equation obtained in the small-angle approximation is rather unwieldy and proves to be inconvenient for numerical calculations. A simpler approximate solution can be obtained by switching from the small-angle approximation (4.2.7) to the approximation corresponding to the diffusion of scattered radiation in the angular variables (Dolin, 1966).

For this purpose we represent the radiance as the sum

$$I = I^{\text{coh}} + I^{\text{incoh}}, \tag{4.2.24}$$

Here, I^{coh} corresponds to the coherent incident wave and satisfies the homogeneous transfer equation (4.2.1):

$$(d_s + \alpha^{\text{ext}})I^{\text{coh}} = 0, \tag{4.2.25}$$

(thus we assume the incident wave to be described in the approximation of geometrical optics). The term I^{incoh} describes the incoherent (scattered) wave and obeys the equation

$$(d_s + \alpha^{\text{ext}})I^{\text{incoh}} = \hat{\sigma}(I^{\text{ihcoh}} + I^{\text{coh}}). \tag{4.2.26}$$

Under small-angle scattering, the cross section $\sigma(\mathbf{n} \leftarrow \mathbf{n}')$ differs from zero only if $|\mathbf{n} - \mathbf{n}'| \leq \theta_0 << 1$, where θ_0 is some characteristic scattering angle. If the radiance of scattered radiation I^{incoh} is a smooth function of $\mathbf{n}$, that is, varies only slightly with a variation of $\mathbf{n}$ about θ_0, the term $\hat{\sigma}I^{\text{incoh}}$ appearing in eqn (4.2.26) may then be approximated by the differential expression:

$$\begin{aligned}\hat{\sigma}I^{\text{incoh}} &= \int d^2n'_\perp \sigma(\mathbf{n}'_\perp) I^{\text{incoh}}(\mathbf{n}_\perp - \mathbf{n}'_\perp) \\ &\approx \int d^2n'_\perp \sigma(\mathbf{n}'_\perp)(1 - \mathbf{n}'_\perp \nabla_{n_\perp} + \frac{1}{2}(\mathbf{n}'_\perp \nabla_{n_\perp})^2 + \ldots) I^{\text{incoh}}(\mathbf{n}_\perp) \\ &= (\alpha^{\text{sc}} + \mathcal{D}\nabla^2_{n\perp} + \ldots) I^{\text{incoh}}. \end{aligned} \tag{4.2.27}$$

Here $\nabla_{\mathbf{n}_\perp} = \partial_{\mathbf{n}_\perp}$,

$$\mathcal{D} = \frac{1}{4}\int n_\perp^2 \sigma(\mathbf{n}_\perp) d^2n_\perp = \frac{1}{4}\langle\theta^2\rangle\alpha^{\text{sc}} \tag{4.2.28}$$

is the diffusion coefficient, and the quantity

$$\langle\theta^2\rangle = \int n_\perp^2 \sigma(\mathbf{n}_\perp) d^2n_\perp \Big/ \int \sigma(\mathbf{n}_\perp) d^2n_\perp \tag{4.2.29}$$

has the meaning of average square of scattering angle in a single act of scattering. In

(4.2.27) we have assumed that the cross section σ depends only upon $|\mathbf{n}_\perp|$ $\sigma(\mathbf{n}_\perp) = \sigma(|\mathbf{n}_\perp|)$ and have used the relations

$$\alpha^{sc} = \int \sigma(\mathbf{n}_\perp) d^2 n_\perp,$$
$$\int \mathbf{n}_\perp \sigma(\mathbf{n}_\perp) d^2 n_\perp = 0, \tag{4.2.30}$$
$$\int n_{\perp i} n_{\perp j} \sigma(\mathbf{n}_\perp) d^2 n_\perp = \frac{1}{2}\delta_{ij} \int \sigma(\mathbf{n}_\perp) n_\perp^2 d^2 n_\perp .$$

Substituting the expansion terms written out in eqn (4.2.27) into eqn (4.2.26), we obtain the equation

$$(\partial_t + \mathbf{n}_\perp \nabla_\perp + \alpha^{abs} - \mathcal{D}\nabla^2_{\mathbf{n}\perp}) I^{incoh} = \hat{\sigma} I^{coh} , \tag{4.2.31}$$

which determines, together with (4.2.25) and the initial condition, the radiance (4.2.24) in the small-angle approximation.

Equation (4.2.31) describes the diffusion of scattered radiation over the transverse angle variables $\mathbf{n}_\perp$ with the diffusion coefficient $\mathcal{D}$, and may be easily solved, for example, using the Fourier transformation. Such a solution is significantly simpler than the more exact general solution of the small-angle transfer equation (4.2.18) (see, e.g., Dolin, 1966; Ishimaru, 1978). We do not intend to analyze this solution here and restrict ourselves to some comments on the applicability conditions for the diffusion approximation (4.2.27).

A necessary applicability condition can be obtained by requiring that the next nonvanishing term of the expansion (4.2.27) be small compared to the last retained term. As a result, we come to the inequality

$$|\langle\theta^4\rangle| \nabla^4_{\mathbf{n}\perp} I^{ihcoh} << |\langle\theta^2\rangle| \nabla^2_{\mathbf{n}\perp} I^{incoh} , \tag{4.2.32}$$

where

$$\langle\theta^4\rangle = \int n_\perp^4 \sigma(n_\perp) d^2 n_\perp \Big/ \int \sigma(\mathbf{n}_\perp) d^2 n_\perp \tag{4.2.33}$$

is the average of the forth power of the scattering angle. Inequality (4.2.32) represents mathematically the above requirement of smoothness of the scattered radiance with respect to the angular variables.

4.2.5 Nonstationary Radiation and Pulse Scattering

The small-angle approximation may be used to analyze the propagation and scattering of time-bounded pulse signals. In this case, the radiance depends not only on the spatial variables but also on time: $I = I_\omega = I(\mathbf{n}, \omega, x)$, $x = (\mathbf{r}, t)$. Following to the work of Furutsu (1979), we consider the transition to the small-angle approximation for the nonstationary transfer equation

$$(c^{-1}\partial_t + \mathbf{n}\nabla + \alpha^{\text{ext}})I(\mathbf{n},\omega,x)$$
$$= \int \sigma(\mathbf{n},\omega \leftarrow \mathbf{n}',\omega')I(\mathbf{n}',\omega',x)d\omega' d\Omega_{\mathbf{n}'} \equiv \hat{\sigma}I\,. \tag{4.2.34}$$

For the sake of simplicity the effective propagating medium is assumed close in its properties to a free space, i.e., $k^{\text{eff}} \approx k_0$. As before, we assume that the frequency changes in scattering are small, and the cross section $\sigma(\omega,\mathbf{n} \leftarrow \omega',\mathbf{n}')$ different from zero only for $|\omega - \omega'| \ll \omega$, $\omega' \sim \omega_0$, where ω_0 is some characteristic frequency of radiation. The extinction coefficient entering eqn (4.2.34) is $\alpha^{\text{ext}} = \alpha^{\text{abs}} + \alpha^{\text{sc}}$ where α^{abs} is the absorption coefficient, and

$$\alpha^{\text{sc}} = \int \sigma(\omega',\mathbf{n}' \leftarrow \omega,\mathbf{n})d\omega' d\Omega_{\mathbf{n}'} \tag{4.2.35}$$

is the scattering coefficient.

As an example we consider scattering from continuous fluctuations. In the Born approximation, the scattering cross section for the model considered in Section 3.5.13 has the form (3.5.60)

$$\sigma(\omega,\mathbf{n} \leftarrow \omega',\mathbf{n}')$$
$$= (\pi k_0^4/2)\int (2\pi)^{-4} d^3\rho d\tau \langle \tilde{\varepsilon}(\rho,\tau)\tilde{\varepsilon}(0,0)\rangle \exp[-i(\mathbf{q}\rho - \Delta\omega\tau)]\,, \tag{4.2.36}$$

where $\mathbf{q} = \mathbf{k} - \mathbf{k}'$ is the scattering vector, $\mathbf{k} = k\mathbf{n}$, $\mathbf{k}' = k'\mathbf{n}'$, $k = \omega/c$, $k' = \omega'/c$, and $\Delta\omega = \omega - \omega'$.

For the model of frozen fluctuations drifting with a velocity $\mathbf{V}$, we may write $\langle \tilde{\varepsilon}(\rho,\tau)\tilde{\varepsilon}(0,0)\rangle = \langle \tilde{\varepsilon}(\rho - \mathbf{V}\tau,0)\tilde{\varepsilon}(0,0)\rangle$. The drifting velocity $\mathbf{V}$ is always significantly smaller then the velocity of light c so that we have approximately $\mathbf{q} = \mathbf{k} - \mathbf{k}' \approx k_0(\mathbf{n} - \mathbf{n}')$. As a result, the cross section (4.2.36) takes the form

$$\sigma(\omega,\mathbf{n} \leftarrow \omega'\mathbf{n}') = \sigma(\mathbf{n} - \mathbf{n}')\delta\big(k_0\mathbf{V}(\mathbf{n} - \mathbf{n}') - \Delta\omega\big)$$
$$\equiv \sigma(\omega - \omega',\mathbf{n} - \mathbf{n}')\,. \tag{4.2.37}$$

Here, $\sigma(\mathbf{n} - \mathbf{n}')$ is the scattering cross section for time-independent fluctuations (4.2.15), and the scattering coefficient α^{sc} (4.2.35) transforms into eqn (4.2.2).

Consider now the small-angle approximation for equation (4.2.34). Furutsu (1979) has shown that, in the nonstationary case, it is advantageous to take a more accurate main approximation by retaining in $\mathbf{n}_\perp$ the term quadratic in $\mathbf{n}_z$, that is, by setting

$$\mathbf{n} = (\mathbf{n}_\perp, n_z) \approx (\mathbf{n}_\perp, 1 - \mathbf{n}_\perp^2/2).$$

The operator of differentiation along the ray path now becomes

$$d_s = c^{-1}\partial_t + \mathbf{n}\nabla \approx c^{-1}\partial_t + \mathbf{n}_\perp\nabla_\perp + (1 - \mathbf{n}_\perp^2/2)\partial_z\,.$$

By analogy with to eqn (4.2.6), in the small-angle approximation, the right-hand side of eqn (4.2.34) is represented as

$$\hat{\sigma}I = \int \sigma(\mathbf{n}-\mathbf{n}',\omega-\omega')I(\mathbf{n}',\omega',x)d\omega' d\Omega_{\mathbf{n}}$$
$$\approx \int \sigma(\mathbf{n}_\perp-\mathbf{n}'_\perp,\omega-\omega')I(\mathbf{n}'_\perp,\omega',x)d\omega' d^2 n'_\perp \,.$$

Thus, in place of eqn (4.2.34), we obtain the transfer equation in the small-angle approximation:

$$\left[c^{-1}\partial_t + \mathbf{n}_\perp \nabla_\perp + (1-n_\perp^2/2)\partial_z + \alpha^{\text{ext}}\right]I$$
$$= \int \sigma(\mathbf{n}_\perp-\mathbf{n}'_\perp,\omega-\omega')I(\mathbf{n}'_\perp,\omega',x)d\omega' d^2 n'_\perp \,. \tag{4.2.38}$$

This equation was used to describe the scattering of pulse signals from both continuous scatterers (Furutsu, 1979) and discrete scatterers (Ito, 1980).

The time-dependent radiance $I(\mathbf{n}',\omega,x)$ is related to the coherence function Γ by the general relationship (3.5.22) which takes the form in the small-angle approximation

$$\Gamma = \langle u(R+\rho/2)u^*(R-\rho/2)\rangle = \int I(\mathbf{n},\omega,R)\exp i(k_0 \mathbf{n}\mathbf{\rho} - \omega\tau)d\omega d\Omega_{\mathbf{n}}$$
$$\approx \int I(\mathbf{n}_\perp,\omega,R)\exp i[k_0 \mathbf{n}_\perp \mathbf{\rho}_\perp + k_0(1-n_\perp^2/2)\rho_z - \omega\tau]d\omega d^2 n_\perp \,, \tag{4.2.39}$$

where $R=(\mathbf{R},T)$ and $\rho=(\mathbf{\rho},\tau)$ are the space-time arguments. For simplicity, we set here $k^{\text{eff}} \approx k_0 = \omega/c$. By analogy with the monochromatic case (4.2.16), it is expedient to introduce the transverse coherence function

$$\Gamma_\perp = \Gamma|_{\rho_z=0} \approx \int I(\mathbf{n}_\perp,\omega,R)\exp i(k_0 \mathbf{n}_\perp \mathbf{\rho}_\perp - \omega\tau)d\omega d^2 n_\perp \,, \tag{4.2.40}$$

in terms of which the radiance I is expressed through

$$I(\mathbf{n}_\perp,\omega,R) = \left(\frac{k_0}{2\pi}\right)^2 \int \Gamma_\perp(R;\mathbf{\rho}_\perp,\tau)\exp i(k_0 \mathbf{n}_\perp \mathbf{\rho}_\perp - \omega\tau)\frac{d\tau d^2\rho_\perp}{2\pi} \,. \tag{4.2.41}$$

We assume that the pulse radiation is quasi-monochromatic. This implies that the time dependence of the field u is close to exponential: $u \approx \exp(-i\omega_0 t)$ where ω_0 is the characteristic frequency of the radiation. We may then replace $k_0 = \omega/c$ by ω_0/c, so that $k_0 \approx \omega_0/c$. Substituting eqn (4.2.41) in the small-angle transfer equation (4.2.38) yields the equation for the transverse coherence function:

$$\left[c^{-1}\partial_T + \frac{i}{k_0}\nabla_{\rho\perp}\nabla_{R\perp} + (1+\frac{1}{2k_0^2}\nabla_{\rho\perp}^2)\partial_z + P(\mathbf{\rho}_\perp,\tau)\right]\Gamma_\perp = 0 \,, \tag{4.2.42}$$

where

$$P(\rho_\perp, \tau) = \alpha^{\mathrm{abs}} + \int \sigma(\mathbf{n}'_\perp, \omega')(1 - \exp i(k_0 \rho_\perp \mathbf{n}'_\perp - \omega'\tau)) d^2 n'_\perp d\omega' .$$

In the model of frozen continuous fluctuations, where the cross section has the form (4.2.37), the last expression transforms into $P = H_1(\rho_\perp - \mathbf{V}_\perp \tau)$ where $H_1(\rho_\perp)$ is given by eqn (4.2.14).

Equation (4.2.42) enables us to compare the small-angle approximation with the widely used alternative approach to the description of quasi-monochromatic pulses that uses the so-called *two-frequency coherence function* (see, e.g., Ishimaru, 1978; Erukhimov et al., 1973; and the literature cited therein). In order to define this function, we consider the representation of the field $u(x)$, $x = (\mathbf{r}, t)$ in the form

$$u(x) = \int \tilde{u}(\omega, x) e^{-i\omega t} d\omega , \tag{4.2.43}$$

where $\tilde{u}(\omega, x)$ is an "incompletely expanded" frequency spectrum of $u(x)$. The representation (4.2.43) is generally ambiguous and acquires a physical meaning only when one can single out fast and slow time dependences of u. This incompletely expanded spectrum will then correspond to the Fourier transform with respect to the fast time at a fixed value of the slow argument. Such an approach is similar to the known asymptotic method of two-scale expansions (Nayfe, 1972).

Substituting eqn (4.2.43) in the expression for the coherence function yields

$$\begin{aligned}\Gamma &= \langle u(R + \rho/2) u^*(R - \rho/2) \rangle \\ &= \int \langle u(\omega_1, R + \rho/2) u^*(\omega_2, R - \rho/2) \rangle \exp(-i\omega_\tau \tau - i\omega_T T) d\omega_\tau d\omega_T \\ &= \int \Gamma^{(2)}_\omega (\omega_\tau, \omega_T, R, \rho) \exp(-i\omega_\tau \tau - i\omega_T T) d\omega_\tau d\omega_T .\end{aligned} \tag{4.2.44}$$

Here, $R = (x_1 + x_2)/2 = (\mathbf{R}, T)$, $\rho = x_1 - x_2 = (\rho, \tau)$, $\omega_\tau = (\omega_1 + \omega_2)/2$, and $\omega_T = \omega_1 - \omega_2$ are the frequencies corresponding to the arguments τ and T, and

$$\begin{aligned}\Gamma^{(2)}_\omega (\omega_\tau, \omega_T, R, \rho) &= \left\langle u(\omega_1, x_1) u^*(\omega_2, x_2) \right\rangle \\ &= \left\langle u\left(\omega_\tau + \frac{\omega_T}{2}, R + \frac{\rho}{2}\right) u^*\left(\omega_\tau - \frac{\omega_T}{2}, R - \frac{\rho}{2}\right) \right\rangle\end{aligned} \tag{4.2.45}$$

is the two-frequency coherence function.

As we have seen, in the small-angle approximation, it is sufficient to consider the transverse coherence function (4.2.40) which is expressed through the similar transverse two-frequency coherence function

$$\Gamma^{(2)}_\perp \equiv \Gamma^{(2)} |_{\rho_z = 0}$$

as

$$\Gamma_\perp = \int \Gamma^{(2)}_{\perp\omega} \exp(-i\omega_\tau \tau - i\omega_T T) d\omega_\tau d\omega_T .$$

The equation for the two-frequency coherence function $\Gamma_{\perp}^{(2)}$ can be derived from the parabolic equation (see, e.g., Erukhimov et al., 1973). For quasi-monochromatic pulses propagating in a medium with continuous fluctuations, this equation takes the form

$$\left[\partial_z + \frac{i}{k_0}\nabla_{\rho\perp}\nabla_{\perp} - i\Delta k\left(1 + (2k_0^2)^{-1}\nabla_{\rho\perp}^2\right) + H_1(\rho_1 - \mathbf{V}_{\perp}\tau)\right]\Gamma_{\perp\omega}^{(2)} = 0\,, \tag{4.2.46}$$

where $\Delta k = \omega_1 / c - \omega_2 / c$. Comparing this equation with the small-angle transfer equation (4.2.42), we see that these equations coincide when $c^{-1}\partial_\tau$ is replaced with $\partial_z$and ∂_z is replaced with $-i\Delta k$ in eqn (4.2.42). Therefore,

$$\Gamma_{\perp} = \int \Gamma_{\perp\omega}^{(2)}\Big|_{z=cT} \exp(-i\omega_\tau \tau - i\Delta k z)\, d\omega_\tau d\omega_T\,, \tag{4.2.47}$$

will satisfy equation (4.2.42). On the other hand, the representation (4.2.42) coincides with eqn (4.2.46) if we take $z \approx cT$ that is admissible for sufficiently short pulses whose length is much shorter than the path length z.

Thus, in the case of quasi-monochromatic pulses, the small-angle approximation for the nonstationary transfer equation (4.2.34) proves to be equivalent to the respective limiting case of the parabolic equation for the two-frequency coherence function. This result, obtained by Furutsu (1979), extends the relations between the small-angle approximation for the transfer equation and the parabolic approximation [considered in Section 4.2.2] to the case of nonmonochromatic radiation.

4.2.6 Small-Angle Transfer Equation for a Medium with Continuous Fluctuations and Discrete Scatterers in the Straight-Ray Approximation

The small angle transfer equation with an integral convolution-type operator (4.2.7) is applicable not only to continuous media with the Born scattering cross section (4.2.13) but also to media containing continuous fluctuations and large-scale discrete scatterers. Here, we intend to demonstrate this for a sparse medium with optically soft scatterers.

Such scatterers can be described in the straight-ray approximation which is also called the approximation of anomalous diffraction. It was devised for single particles by van de Hulst (1957). The physical meaning of the straight-ray approximation is quite clear: for optically soft particles with the refractive index close to that of a surrounding medium, one may neglect the refraction of rays on the particle and assume that the effect of particles reduces to adding (in the presence of complex absorption) a phase increment along straght rays crossing the particles.

We use the results of Apresyan (1991) who derived for optically soft medium (with permittivity close to unity) the Markov approximation equations for moments of arbitrary order on the basis of the parabolic wave equation (4.2.10). In particular, for the transverse coherence function $\Gamma_{\perp}$ (second moment), the respective equation takes the form similar to eqn (4.2.12):

$$\left(\partial_z + \frac{\nabla_{\rho\perp}\nabla_\perp}{ik_0} + H_{\mathrm{cont}}(\rho) + H_{\mathrm{part}}(\rho)\right)\Gamma_\perp = 0. \tag{4.2.48}$$

Here $H_{\mathrm{cont}}(\rho)$ describes the scattering from continuous fluctuations and is determined by eqn (4.2.14) with the Born cross section (4.2.13), and $H_{\mathrm{part}}(\rho)$ corresponds to the scattering by particles and is given by an infinite series expansion in terms of the correlation functions of scatterer positions:

$$H_{\mathrm{part}}(\rho) = \nu\langle\xi^{(1|z)}\rangle - \frac{\nu^2}{2}\langle\tilde{\xi}^{(1|z)}\tilde{\xi}^{(2)}\rangle + \ldots, \tag{4.2.49}$$

where ν is the concentration of particles, $\tilde{\xi} \equiv \xi - \langle\xi\rangle$ is the fluctuation of ξ,

$$\xi^{(s|z)} = \delta(z - z_s)\xi^{(s)}, \tag{4.2.50}$$

$$\xi^{(s)} = 1 - \exp\left[i\left(\psi_1^{(s)} - \psi_2^{(s)*}\right)\right], \tag{4.2.51}$$

and

$$\psi_j^{(s)} = \frac{k_0}{2}\int_{-\infty}^{\infty}\tilde{\varepsilon}_s(\mathbf{r}_{j\perp} - \mathbf{r}_{s\perp}, z')dz' \tag{4.2.52}$$

is the additional phase shift along the rectilinear ray crossing particle s and passing through the point $\mathbf{r}_{j\perp}$. (Here, $\mathbf{r}_s$ is the radius-vector, and $\tilde{\varepsilon}_s = \varepsilon_s - 1$ is the difference between the permittivity ε_s of particle s and the permittivity of the medium which, for simplicity, is assumed equal to unity.)

If we switch in eqn (4.2.48) from $\Gamma_\perp$ to the radiance I using eqn (4.2.17), we obtain a transfer equation of the form (4.2.7) where the extinction coefficient α^{ext} and the cross section $\sigma(\mathbf{n}_\perp)$ will split into the contributions from continuous fluctuations and discrete particles. In particular, for nonabsorbing and uncorrelated particles, when the right-hand side of eqn (4.2.49) contains the first term alone, the contributions into the scattering cross section and the extinction coefficient prove to be

$$\sigma_{\mathrm{part}}(\mathbf{n}_\perp) = \nu\left|\frac{k_0}{2\pi}\int(1 - \exp(i\psi_1^{(s)})\exp(ik\mathbf{n}_\perp\rho_{1\perp})d^2r_{1\perp}\right|^2, \tag{4.2.53}$$

$$\alpha_{\mathrm{part}}^{\mathrm{ext}} = \int\sigma_{\mathrm{part}}(\mathbf{n}'_\perp)d^2n'_\perp. \tag{4.2.54}$$

The cross section (4.2.53) is identical with the well-known approximation of the anomalous van de Hulst diffraction (van de Hulst, 1957) originally obtained for a single scatterer. Equation (4.2.49) extends the approximation of anomalous diffraction to the case of an ensemble of correlated particles.

The solution to the small angle transfer equation (4.2.7), discussed in Section 4.2.3, may be used directly also in the case of eqn (4.2.48) for a medium with continuous and discrete fluctuations (Problem 4.9). In practice, however, the approximation of uncorrelated particles is usually applied, because the allowance for correlations hampers by the lack of information about correlation functions of scatterers.

4.3 DIFFUSION APPROXIMATION

4.3.1 Elementary Derivation of Diffusion Approximation Equations

The small angle approximation discussed in the preceding section describes scattering of narrow beam waves. This approximation proves to be valid for numerous problems of scattering from large-scale inhomogeneities such as the propagation of laser radiation in a turbulent atmosphere. However, it is valid for not too long propagation paths over which the beam remains sufficiently narrow. The beam width increases with the path length and eventually the distribution of radiation over angles becomes nearly isotropic — the radiation "forgets" the direction of the initial wave (such a memory lapse is not total; for memory effects, see Section 5.3). The conversion of the radiation pattern to an isotropic form is especially significant for small-scale scatterers with an isotropic scattering pattern when even a single scattering smoothes strongly the directional pattern.

A convenient method to describe such a quasi-isotropic radiation at long distances from the boundaries of a scattering medium is the diffusion approximation which proves to be much simpler than the original integro-differential transfer equation because it reduces to a differential equation. As a consequence, the diffusion approximation is widely used in practical computations as a very efficient means of allowing for multiple scattering under the conditions where the characteristic dimension of a scattering medium, L, is large compared with the free path length L_{sc}, and the characteristic times of observation are much greater than the relaxation time L_{ext} / c (see Section 4.3.4). Below we discuss some general approaches to the diffusion approximation and estimate conditions for its validity.

Consider first the heuristic derivation of diffusion approximation equations for monochromatic radiation, suppressing the dependence on frequency. We start with the transfer equation for the radiance $I(\mathbf{n},x)$, $x=(\mathbf{r},t)$:

$$\left(c^{-1}\partial_t + \mathbf{n}\nabla + \alpha^{\text{ext}}\right)I(\mathbf{n},x) = \int \sigma(\mathbf{n} \leftarrow \mathbf{n}')I(\mathbf{n}',x)d\Omega_{\mathbf{n}'}\,, \tag{4.3.1}$$

which is different from eqn (4.2.34), discussed in the preceding section, in that it excludes frequency changes in scattering ($\sigma(\omega,\mathbf{n} \leftarrow \omega',\mathbf{n}') = \delta(\omega-\omega')\sigma(\mathbf{n} \leftarrow \mathbf{n}')$). Here, $\alpha^{\text{ext}} = \alpha^{\text{sc}} + \alpha^{\text{abs}}$, where α^{sc} is given by eqn (4.2.2). The scattering medium is deemed isotropic, so that $\sigma(\mathbf{n} \leftarrow \mathbf{n}') = \sigma(\mathbf{n} \cdot \mathbf{n}')$.

Similar to the above derivation of the angular diffusion equation in Section 4.2.4, we represent the radiance as the sum $I = I^{\text{coh}} + I^{\text{incoh}}$, where the term I^{coh}, associated with the coherent incident wave, satisfies the homogeneous transfer equation

$$\left(\frac{1}{c}\partial_t + \mathbf{n}\nabla + \alpha^{\text{ext}}\right)I^{\text{coh}} = 0, \tag{4.3.2}$$

and I^{incoh} describes the incoherent (scattered) wave and obeys the equation

$$\left(\frac{1}{c}\partial_t + \mathbf{n}\nabla + \alpha^{\text{ext}}\right)I^{\text{incoh}} = \int \sigma(\mathbf{n} \cdot \mathbf{n}')(I^{\text{incoh}}(\mathbf{n}', x) + I^{\text{coh}}(\mathbf{n}', x))d\Omega_{\mathbf{n}'}. \tag{4.3.3}$$

Assuming that I^{incoh} is nearly isotropic, we represent it as the sum

$$I^{\text{inc}}(\mathbf{n}, x) \approx \frac{c}{4\pi}W(x) + \frac{3}{4\pi}\mathbf{S}(x)\mathbf{n}, \tag{4.3.4}$$

where $W(x)$ is independent of $\mathbf{n}$ and corresponds to the isotropic radiation, and the term

$$\left|\frac{3}{4\pi}\mathbf{Sn}\right| << \frac{c}{4\pi}W(x)$$

is the correction for anisotropy. The numerical multipliers have been introduced in eqn (4.3.4) for the sake of convenience.

Integrating both sides of eqn (4.3.4) over all directions $\mathbf{n}$, we find

$$W(x) = \frac{1}{c}\int I^{\text{incoh}}(\mathbf{n}, x)d\Omega_{\mathbf{n}}, \tag{4.3.5}$$

that is, $W(x)$ is equal to the energy flux of scattered radiation. Similarly, multiplying both sides of eqn (4.3.4) by the unit vector $\mathbf{n}$ and integrating the results over the unit sphere, we have

$$\mathbf{S}(x) = \int I^{\text{incoh}}(\mathbf{n}, x)\mathbf{n}d\Omega_{\mathbf{n}}, \tag{4.3.6}$$

that is, $\mathbf{S}(x)$ is identical to the average flux of scattered energy.

Substitute now eqn (4.3.4) into eqn (4.3.3) and integrate the result over the directions $\mathbf{n}$. In view of eqns (4.3.5) and (4.3.6), we obtain the equation

$$\nabla\mathbf{S}(x) + (\partial_t + c\alpha^{\text{abs}})W(x) = \alpha^{\text{sc}}\int I^{\text{coh}}(\mathbf{n}', x)d\Omega_{\mathbf{n}'}, \tag{4.3.7}$$

which expresses the law of energy conservation for scattered radiation. Similarly, integrating equation (4.3.3) multiplied by $\mathbf{n}$ over all directions, we arrive at the equation

$$\frac{c}{3}\nabla W(x)+\left(\frac{1}{c}\partial_t+\sigma_{\mathrm{tr}}\right)\mathbf{S}(x)=\alpha^{\mathrm{sc}}\bar{\mu}\int I^{\mathrm{coh}}(\mathbf{n}',x)d\Omega_{\mathbf{n}'}. \tag{4.3.8}$$

Here, the parameter

$$\bar{\mu}=\int \mathbf{n}\cdot\mathbf{n}'\sigma(\mathbf{n}\cdot\mathbf{n}')d\Omega_{\mathbf{n}'}\Big/\int\sigma(\mathbf{n}\cdot\mathbf{n}')d\Omega_{\mathbf{n}'} \tag{4.3.9}$$

has the meaning of the average cosine of the scattering angle, $\mu=\mathbf{n}\cdot\mathbf{n}'=\cos\theta$, and vanishes in the case of isotropic scattering [$\sigma(\mathbf{n}\cdot\mathbf{n}')=$ const.]; and the quantity

$$\sigma_{\mathrm{tr}}=\alpha^{\mathrm{abs}}+\alpha^{\mathrm{sc}}(1-\bar{\mu})=\alpha^{\mathrm{abs}}+\int(1-\mathbf{n}\cdot\mathbf{n}')\sigma(\mathbf{n}\cdot\mathbf{n}')d\Omega_{\mathbf{n}'} \tag{4.3.10}$$

is called the *transport scattering cross section* per unit volume. The reciprocal quantity $\sigma_{\mathrm{tr}}^{-1}=L_{\mathrm{tr}}$ is called the *transport length of scattering*.

Equations (4.3.7) and (4.3.8) enable us, in principle, to determine the energy density W and the energy flux $\mathbf{S}$. If variations of the radiance I^{incoh} in time are sufficiently slow, so that the inequality

$$\left|\frac{1}{c}\partial_t\mathbf{S}\right|\ll\left|\sigma_{\mathrm{tr}}\mathbf{S}\right|, \tag{4.3.11}$$

is satisfied, then we may neglect the term $c^{-1}\partial_t\,\mathbf{S}$ and find from (4.3.8)

$$\mathbf{S}=-D\nabla W+\frac{3}{c}D\alpha^{\mathrm{sc}}\bar{\mu}\mathbf{S}^{\mathrm{coh}}, \tag{4.3.12}$$

where

$$\mathbf{S}^{\mathrm{coh}}=\int\mathbf{n}'I^{\mathrm{coh}}(\mathbf{n}',x)d\Omega_{\mathbf{n}'} \tag{4.3.13}$$

is the flux density of the coherent radiation, and the diffusion coefficient of radiation is

$$D=\frac{c}{3\sigma_{\mathrm{tr}}}=\frac{cL_{\mathrm{tr}}}{3}. \tag{4.3.14}$$

For isotropic scattering and in the absence of absorption, $\sigma_{\mathrm{tr}}=\alpha^{\mathrm{sc}}$ so that

$D = cL_{sc}/3$, where $L_{sc} = 1/\alpha^{sc}$ is the length of scattering.

In the absence of coherent radiance ($I^{coh} = 0$), eqn (4.3.12) simplifies and reduces to

$$\mathbf{S} = -D\nabla W; \tag{4.3.15}$$

that is, the flux is proportional to the gradient of the energy density W. This relationship is characteristic of the diffusion regime which differs appreciably from the free-space propagation regime when, for plane waves, the flux is proportional to the energy density itself ($\mathbf{S} = c\mathbf{n}W$).

Substituting eqn (4.3.12) into eqn (4.3.7), we obtain the diffusion equation for the intensity of scattered radiation:

$$(\partial_t + c\alpha^{abs} - D\Delta)W = \alpha^{sc}\int I^{coh}(\mathbf{n}',x)d\Omega_{\mathbf{n}'} - 3D\alpha^{sc}\overline{\mu}W^{coh}. \tag{4.3.16}$$

In contrast to the diffusion of incoherent radiance in the angular variables $\mathbf{n}$, discussed in Section 4.2.4, this equation describes the diffusion of the total intensity of scattered radiation $W(x)$ in the spatial variables $\mathbf{r}$.

In order to ensure that eqn (4.3.16) is uniquely solvable, we need some additional conditions. The choice of such conditions is ambiguous and is usually dictated by physical considerations. One of the possibilities consists in imposing an approximate boundary condition following from the requirement that a portion of the total flux of scattered radiation directed inside the scattering medium must vanish on the medium's boundary Σ:

$$\int I^{incoh}(\mathbf{n}',x)\mathbf{n}'\cdot\mathbf{n}_\Sigma d\Omega_{\mathbf{n}'}\Big|_\Sigma = 0. \tag{4.3.17}$$

Here $\mathbf{n}_\Sigma$ is the external normal to the boundary of the scattering medium. This constraint together with eqn (4.3.4) yield the condition

$$(W(x) - 2D\mathbf{n}_\Sigma\nabla W(x) + 6D\alpha^{sc}\mu W^{coh}_{(x)})_\Sigma = 0. \tag{4.3.18}$$

This kind of boundary conditions was discussed, for example, by Ishimaru (1978) and Caze and Zweifel (1967) (see also the paper of Barabanenkov and Ozrin, 1991).

4.3.2 Diffusion Approximation for Green's Function of an Unbounded Medium

As a simple example of using the diffusion approximation, we consider the radiation of a point source in an unbounded scattering medium. To describe such a source, we introduce a source function q in the starting equation (4.3.1). In the absence of the incident wave, $I^{coh} = 0$, and, repeating the reasoning of the preceding section one can readily demonstrate that the diffusion equation (4.3.16) becomes

$$(\partial_t + c\alpha^{abs} - D\Delta)W = \int q(\mathbf{n}', x)d\Omega_{\mathbf{n}'} - \nabla\frac{1}{\sigma_{tr}}\int q(\mathbf{n}', x)\mathbf{n}'d\Omega_{\mathbf{n}'}. \tag{4.3.19}$$

For comparison with the exact result, we restrict this consideration to the case of the stationary radiation of a point isotropic source in an isotropically scattering medium when $q = \delta(\mathbf{r} - \mathbf{r}_0)/4\pi, \partial_t W = 0$ and $D = c/3\alpha^{ext}$. Then eqn (4.3.19) becomes

$$\left(\alpha^{abs} - \frac{1}{3\alpha^{ext}}\Delta\right)W = \frac{1}{c}\delta(\mathbf{r} - \mathbf{r}_0). \tag{4.3.20}$$

This equation can be easily solved using the Fourier transformation (Problem 4.12). As a result we have

$$W = \frac{3\alpha^{ext}}{4\pi c}\exp\frac{|\mathbf{r} - \mathbf{r}_0|}{l_a}, \tag{4.3.21}$$

where the quantity

$$l_a = \sqrt{\frac{L_{abs}L_{sc}}{3}} = \frac{1}{\sqrt{3\alpha^{abs}\alpha^{sc}}} \tag{4.3.22}$$

has the meaning of the absorption length in the diffusion regime (insight into its physical meaning is given in Problem 4.13).

This expression for W may be compared with the exact solution of the transfer equation (4.1.6). Since the diffusion approximation is assumed to be applicable to the description of the large-scale behavior of the solution in a region where multiple scattering is significant, it is natural to compare eqn(4.3.21) with the asymptotic of the exact solution far from the source. This asymptotic representation can be obtained by evaluating the integral in eqn (4.1.6) as $|\mathbf{r} - \mathbf{r}_0| \to \infty$. We then have (Ishimaru, 1978)

$$W \approx \frac{\alpha^{ext}}{2\pi c}\frac{\alpha_0^2(1 - \alpha_0^2)}{w_0(\alpha_0^2 + w_0 - 1)}\frac{\exp(-\alpha^{ext}\alpha_0|\mathbf{r} - \mathbf{r}_0|)}{|\mathbf{r} - \mathbf{r}_0|}, \tag{4.3.23}$$

where $w_0 = \alpha^{sc}/(\alpha^{sc} + \alpha^{abs})$ is the single-scattering albedo, and α_0 is a parameter related to the solution of eqn (4.1.7) by

$$\Lambda(i\alpha^{ext}\alpha_0) \equiv 1 - \frac{w_0}{2\alpha_0}\ln\frac{1 + \alpha_0}{1 - \alpha_0} = 0. \tag{4.3.24}$$

[Note that this equation determines the poles of the integrand in eqn (4.1.6).]
The solution to eqn (4.3.24) can be easily estimated in the limiting cases of weak ($\alpha^{abs} << \alpha^{sc}$) and strong ($\alpha^{abs} >> \alpha^{sc}$) absorption (Ishimaru, 1978). In the first case, when $\alpha^{abs} << \alpha^{sc}$, $w_0 \to 1$, and we have

$$\alpha_0 \approx \sqrt{3\frac{1-w_0}{w_0}} = \sqrt{3\frac{\alpha^{abs}}{\alpha^{sc}}}\,. \tag{4.3.25}$$

In the case of the strong absorption, $\alpha^{abs} >> \alpha^{sc}$, $w_0 \to 0$, and

$$\alpha_0 \approx 1 - 2\exp(-2/w_0)\,. \tag{4.3.26}$$

A comparison of eqns (4.3.21) and (4.3.23) shows that they are consistent in the limit of weak absorption (4.3.25). Thus, the solution of the diffusion equation (4.3.20) correctly describes the large-scale asymptotics for the total intensity in the weak absorption case when scattering is the main mechanism for a change of radiance. This conclusion, drawn for a point isotropic source, is true also for more complex applications.

4.3.3 Diffusion Approximation and the Model of Random Walks

The diffusion approximation admits a simple physical interpretation in terms of random walks of wave trains (or photons). The simplest version of this model assumes that, in a scattering medium, the intensity is proportional to the number of photons arriving at the observation point. The paths of photons are interpreted as the sum of independent random walks $\mathbf{l}_i$ between sequential random scattering events:

$$\mathbf{r} = \sum_{i=1}^{N} \mathbf{l}_i\,, \quad \langle \mathbf{l}_i \rangle = 0\,. \tag{4.3.27}$$

The velocity of the photons is assumed to be equal to c and the length of each step $|\mathbf{l}_i|$ is assumed to be the free path length L_{sc}. This length serves as some microscopic scale relative to macroscopic observation scales $L >> L_{sc}$. It is convenient to consider the free path time $\tau_{sc} = L_{sc} / c$ along with the length L_{sc}. The number of walks is then expressed in terms of the observation time t as $N = t / \tau_{sc} = ct / L_{sc}$.

For isotropic scattering in an unbounded scattering medium, the problem may be reduced to describing random walks along an arbitrarily chosen axis x. In view of isotropy, $\overline{x^2} = \overline{r^2}/3$.

This simple and clear model explains almost all the qualitative features of the diffusion regime of wave propagation and gives correct order-of-magnitude estimates.

For example, the diffusion coefficient D may be determined by the known relationship $\overline{x^2} = Dt$. In accordance with eqn (4.3.27),

$$\overline{x^2} = \frac{1}{3}\overline{r^2} = \frac{1}{3}NL_{sc}^2 = \frac{cL_{sc}}{3}t,$$

whence it follows

$$D = cL_{sc}/3$$

which coincides with the above expression (4.3.14) for the isotropic scattering. One more example of using the random walks model is presented in Problem 4.13.

4.3.4 Justification of Diffusion Approximation: Methods of Spherical Harmonics and Asymptotic Expansion. The Chain of Equations for Moments

The above derivation of the diffusion approximation was based on an approximate representation of the radiance as the sum (4.3.4). This is a heuristic approach that does not permit estimation of corrections to this approximation. There exist more rigorous methods that are free of this drawback. The best known of them is the expansion in terms of spherical harmonics that was developed in neutron transport theory (Case and Zweifel, 1967; Sobolev, 1975; Davison, 1957). This method converts the transfer equation into an infinite system of equations for coefficients of the expansion of the radiance I in a series in terms of spherical harmonics. Truncations of this system lead to the so-called P_L-approximations. The simplest of them is the set of equations for four functions, which is equivalent to the diffusion approximation. Such an approach yields only a stationary diffusion equation, proves to be rather cumbersome, and in addition does not lead to unambiguous boundary conditions.

A more consistent technique of asymptotic expansion in terms of a small parameter, equal to the ratio of the extinction length to the characteristic size of the scattering volume, has been discussed by Larsen and Keller (1974, 1976) and Habetler and Matkovsky (1975). This technique represents the solution to the transfer equation as a sum of terms. One of these terms satisfies the diffusion equation in the small absorption limit. Far from boundaries of the scattering volume and from the initial time, this term becomes principal, whereas the others allow for boundary effects and differ noticeably from zero only in the boundary layer of width on the order of the extinction length or else at small time.

Following the works of Furutsu (1980a, b), we describe a different and simpler way of deriving the diffusion approximation.[1] For this purpose we consider the moments of the radiance of scattering radiation:

[1] A comparison of Furutsu's results with another approach to the diffusion approximation, proposed by Ishimaru (1978), has been given by Ishimaru (1984).

$$J(x) = \int I^{\text{incoh}}(\mathbf{n},x)d\Omega_{\mathbf{n}}\,, \qquad J_i(x) = \int n_i I^{\text{incoh}}(\mathbf{n},x)d\Omega_{\mathbf{n}}\,,$$
$$J_{ij}(x) = \int n_i n_j I^{\text{incoh}}(\mathbf{n},x)d\Omega_{\mathbf{n}}\,, \quad J_{ijk}(x) = \int n_i n_j n_k I^{\text{incoh}}(\mathbf{n},x)d\Omega_{\mathbf{n}}\,, \tag{4.3.28}$$

and so forth. (The first two moments $J(x)$ and $J_i(x) = S_i(x)$ were used in the preceding section.) Similar moments related to the coherent radiance I^{coh} will be labeled by primes $J', J_i', J_{ij}', \ldots$.

If the radiance I^{incoh} is concentrated within the region where $n_i \geq 0$, the moments (4.3.28) form a non-increasing sequence:

$$J \geq J_i \geq J_{ij} \geq \ldots$$

Multiplying the equation of transfer (4.3.3) by 1, n_i, $n_i n_j, \ldots$ and integrating the results over the directions $\mathbf{n}$ we obtain a chain of equations for the moments (4.3.28):

$$(c^{-1}\partial_t + \alpha^{\text{abs}})J + \nabla_i J_i = \alpha_s J'\,, \tag{4.3.29}$$

$$\left[c^{-1}\partial_t + \alpha^{\text{abs}} + \alpha^{\text{sc}}(1-\overline{\mu})\right]J_i + \nabla_l J_{il} = \alpha_s \overline{\mu} J_i'\,, \tag{4.3.30}$$

$$\left[c^{-1}\partial_t + \alpha^{\text{abs}} + \frac{3\alpha^{\text{sc}}}{2}\left(1-\overline{\mu^2}\right)\right]J_{ij} + \nabla_l J_{ijl}$$
$$= \frac{\alpha^{\text{sc}}}{2}\delta_{ij}\left(1-\overline{\mu^2}\right)(J+J') - \frac{\alpha^{\text{sc}}}{2}\left(1-3\overline{\mu^2}\right)J_{ij}'\,, \tag{4.3.31}$$

and so forth. The summation is performed by recurring indices, and the parameter

$$\overline{\mu^2} = \int (\mathbf{n}\cdot\mathbf{n}')^2 \sigma(\mathbf{n}\cdot\mathbf{n}')d\Omega_{\mathbf{n}'} \Big/ \int \sigma(\mathbf{n}\cdot\mathbf{n}')d\Omega_{\mathbf{n}'} \tag{4.3.32}$$

is the mean square of the cosine of the scattering angle. For isotropic scattering, $\overline{\mu^2} = 1/3$.

The infinite chain of equations for the moments J, J_i, $J_{ij}, \ldots$ may be truncated by setting any of these moments J_{ijk} equal to zero. A truncation of the chain at the second step, made by putting $J_{ij} = 0$ in eqn (4.3.30), proves to be very crude and does not yield nontrivial results. Therefore, we truncate the the chain at the third step putting J_{ijl} equal to zero in eqn (4.3.31). This approximation will be allowable provided that

$$\left|\left(\alpha^{\text{abs}} + \frac{3\alpha^{\text{sc}}}{2}(1-\overline{\mu})\right)J_{ij}\right| \sim \left|\alpha^{\text{ext}}J_{ij}\right| >> \left|\nabla_l J_{ijl}\right|, \tag{4.3.33}$$

or more roughly

$$\left|\alpha^{\text{ext}} J_{ij}\right| \sim \alpha^{\text{ext}} I^{\text{incoh}} = I^{\text{incoh}} / L^{\text{ext}} >> \left|\nabla_l J_{ijl}\right| \sim \left|\nabla I^{\text{incoh}}\right| \sim I^{\text{incoh}} / L_I . \tag{4.3.34}$$

Here, $L_I \sim \left|\nabla \ln I^{\text{incoh}}\right|^{-1}$ is the characteristic spatial variation scale of the incoherent radiance I^{incoh}, and $L^{\text{ext}} = (\alpha^{\text{ext}})^{-1}$ is the extinction length. Thus, eqn (4.3.33) roughly reduces to the condition $L_I >> L^{\text{ext}}$ that expresses the requirement of smooth spatial variations of the scattered radiance I^{incoh} over the extinction length L^{ext}.

As a result of this truncation we come to the set of three equations (4.3.29) - (4.3.31) for three moments J, J_i, and J_{ij}. This set, however, does not enable us to derive a diffusion equation for J of the first order in time t. To derive such an equation, we require additionally that the radiance I^{incoh} should change rather slowly:

$$\left|c^{-1}\partial_t I^{\text{incoh}}\right| << \left|\left(\alpha^{\text{abs}} + \frac{3\alpha^{\text{sc}}}{2}\left(1-\overline{\mu^2}\right)\right) I^{\text{incoh}}\right| \sim \alpha^{\text{ext}} I^{\text{incoh}} \tag{4.3.35}$$

which simultaneously means switching to large time intervals

$$t >> (\alpha^{\text{ext}} c)^{-1}. \tag{4.3.36}$$

We now assume that satisfying the condition (4.3.35) for I^{incoh} satisfies similar inequalities for the moments J_i and J_{ij}. Therefore, satisfying eqn (4.3.35), we may omit the term $c^{-1}\partial_t J_{ij}$ in eqn (4.3.31) and the term $c^{-1}\partial_t J_i$ in eqn (4.3.30). After these manipulations eqns (4.3.30) and (4.3.31) become algebraic equations for the moments J_i and J_{ij}. Solving this equation yields

$$J_i = -\left[\alpha^{\text{abs}} + \alpha^{\text{sc}}(1-\overline{\mu})\right]^{-1} \nabla_l J_{ij} + O(J'), \tag{4.3.37}$$

$$J_{ij} = J\delta_{ij}\alpha^{\text{sc}}\left(1-\overline{\mu^2}\right)\left[2\alpha^{\text{abs}} + 3\alpha^{\text{sc}}\left(1-\overline{\mu^2}\right)\right]^{-1} + O_1(J'), \tag{4.3.38}$$

where $O(J')$ and $O_1(J')$ stand for the terms dependent on I^{coh}

Substituting first eqn (4.3.38) in eqn (4.3.37) and then eqn (4.3.37) into eqn (4.3.29), we come to the diffusion equation for J:

$$\left(c^{-1}\partial_t + \alpha^{\text{abs}} - \frac{D_1}{c}\Delta\right) J = \alpha^{\text{sc}} J' + O_2(\nabla J', \Delta J'/\alpha^{\text{ext}}), \tag{4.3.39}$$

where $O_2(a,b)$ stands for the sum of terms on the order of a and b, and

$$\frac{D_1}{c} = \alpha^{sc}\left(1-\overline{\mu^2}\right)\left[\left(\alpha^{ext} - \bar{\mu}\alpha^{sc}\right)\left(2\alpha^{abs} + 3\alpha^{sc}\left(1-\overline{\mu^2}\right)\right)\right]^{-1} \tag{4.3.40}$$

is the diffusion coefficient. If the smoothness condition $L_I >> L_{sc} = (\alpha^{sc})^{-1}$, where $L_I \sim |d_s \ln I^{coh}|^{-1}$, is satisfied for the coherent radiance, we may omit the term O_2 on the right-hand side of eqn (4.3.39).

The diffusion coefficient (4.3.40) is different from the diffusion coefficient (4.3.14), obtained in elementary theory, and refines the latter, as one might expect. This conclusion is borne out by the following example. For an unbounded medium with isotropic scattering, where an exact solution of the transfer equation is known, the intensity of a stationary isotropic point source contains a diffusion component that varies as

$$\frac{1}{r}\exp\left[-\left(3\alpha^{abs}(\alpha^{ext})^2/\alpha^{sc}\right)^{1/2} r\right]$$

in the small absorption limit $\alpha^{abs} << \alpha^{sc}$ according to eqns (4.3.23) and (4.3.25).

The diffusion equation (4.3.39) then takes the form

$$(\alpha_{abs} - \frac{D_1}{c}\Delta)J \propto \delta(r), \tag{4.3.41}$$

where numerical multipliers are omitted for simplicity, and the delta function on the right-hand side corresponds to the point source. In accordance with eqn (4.3.41), the intensity I varies as $r^{-1}\exp[-(\alpha^{abs}/D_1)^{1/2} r]$. Substituting in this expression the diffusion coefficient (4.3.40) which is equal to $D_1 = \alpha^{sc}c/3(\alpha^{ext})^2$ in the isotropic case, we obtain $J \propto r^{-1}\exp\{-[3\alpha^{abs}(\alpha^{ext})^2/\alpha^{sc}]^{1/2} r\}$, that is, the same law as from the solution of the transfer equation. At the same time, according to eqn (4.3.14), we have $D = c/3\alpha^{ext}$ which yields the dependence $J \propto r^{-1}\exp(-(3\alpha^{abs}\alpha^{ext})^{1/2} r)$, that proves to be less accurate.

Other derivations of a diffusion approximation may result in somewhat different diffusion coefficients. However, all these differences disappear in passing to the small absorption limit $\alpha^{abs} << \alpha^{sc}$.[2] This transition corresponds to the physical meaning of the diffusion approximation: a description of scattering as a diffusion process becomes appropriate only if the scattering effects dominate over the absorption effects ($\alpha^{abs} << \alpha^{sc}$), whereas in the opposite case ($\alpha^{abs} \geq \alpha^{sc}$), we essentially have one wave strongly attenuated because of absorption.

[2] See the paper of Pomraning (1990) where the differences in the diffusion coefficients are noted for the anisotropic scattering only.

Similarly, the diffusion approximation does not describe the boundary effects which manifest themselves near boundaries and at the initial time t_0 when the radiation has not had enough time to "forget" the boundary and initial conditions. Mathematically, this means that the diffusion approximation yields a nonuniform asymptotic for the solution to the transfer equation in the small absorption limit. To approximately describe the radiative transfer in boundary regions, it is preferable to use asymptotic methods discussed by Nayfe (1972), Habelter and Matskovsky (1975), and Larsen and Keller (1974, 1976).

4.4 ITERATIVE SOLUTION TO THE EQUATION OF RADIATIVE TRANSFER

4.4.1 Integral Form and Iterative Series for the Transfer Equation

The iteration of the equation of transfer in integral form is one of the most spread approximate methods for its solution. This method is applicable in case of the weak scattering when a scattering medium is dilute and a scattering volume is not too large.

In the first approximation, the iterative solution to the equation of transfer yields the result similar to the conventional Born approximation of wave theory and known as "modified Born approximation" or as "the first order approximation of multiple scattering theory" (Ishimaru, 1978).

Let us consider the iterative solution to the equation of transfer in the most general statement treating on equal terms the spatial and temporal arguments. We write the original equation of transfer in the form

$$(d_s + \alpha)I = \hat{\sigma} I, \tag{4.4.1}$$

where d_s is the differentiation operator along the space-time ray. The meaning of the parameter α and the operator $\hat{\sigma}$ was discussed above. In (4.4.1), both the scalar radiance and the four-component Stokes vector (in case of electromagnetic radiation) may serve as $I = I(\mathbf{n}, \omega, x)$, where $x = (\mathbf{R}, t)$. Note that an explicit form of the parameters of the equation of transfer (4.4.1) is still nonessential.

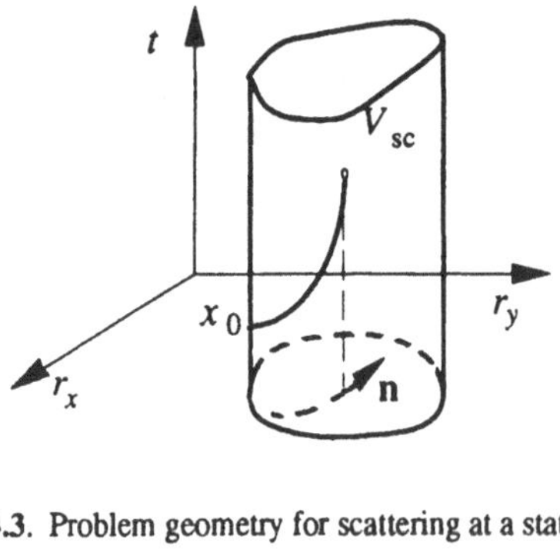

Figure 4.3. Problem geometry for scattering at a stationary object.

Let us assume that one needs to determine I inside of a hypersurface Σ confining a region $V_{sc}^{(4)}$ in the 4D space $x=(\mathbf{R},t)$. Such a statement, in principle, enables one to describe the scattering of radiation from moving objects. The conventional scattering problem from a stationary volume V_{sc} for $t \geq 0$ corresponds to Σ degenerated into the surface of a semi-cylinder schematically shown in Fig. 4.3, where the path of a space-time ray passing from x_0 to x and its spatial projection are indicated.

Generalizing the known results of linear transfer theory (Case and Zweifel, 1967), one may infer that in order to solve this problem it is necessary to know the distribution of radiance I over the hypersurface Σ for rays entering Σ, that is, to complement (4.4.1) by the condition

$$I(\mathbf{n},\omega,x)|_{\Sigma,\mathbf{n}=\text{in}} = I^{\Sigma}(\mathbf{n},\omega,x), \tag{4.4.2}$$

where $\mathbf{n}=\text{in}$ implies that (4.4.2) is specified only for $\mathbf{n}$ corresponding to rays entering Σ. For definiteness, we consider the boundary condition for a coherent incident wave which is described in the geometrical optics approximation. This means that the right-hand side of (4.4.2) contains the corresponding delta function of directions of $\mathbf{n}$.

As before, it is convenient to represent the radiance I as a sum $I = I^{\text{coh}} + I^{\text{inc}}$, where I^{coh} satisfies the homogeneous equation (4.4.1):

$$(d_s+\alpha)I^{\text{coh}} = 0, \tag{4.4.3}$$

with a boundary condition like (4.4.2), and I^{incoh} describes the scattered radiation and is governed by the equation

$$(d_s+\alpha)I^{\text{incoh}} = \hat{\sigma}(I^{\text{incoh}} + I^{\text{coh}}). \tag{4.4.4}$$

This component obeys the boundary conditions like (4.4.2) with $I^{\Sigma}=0$.

Let us assume that only one ray passes through any two points x and x_0 thus eliminating caustics from our consideration (Kravtsov and Orlov, 1990, 1993). The solution of the homogeneous equation (4.4.3) can then be written as

$$\begin{aligned} I^{\text{coh}}(\mathbf{n},\omega,x) &= I^{\text{coh}}(\mathbf{n}(s),\omega(s),x(s)) = I^{\text{coh}}(s) \\ &= \exp[-\int_{s_0}^{s} \alpha(s')ds']I^{\Sigma}(s_0) \\ &\equiv (d_s+\alpha)^{-1} I^{\Sigma}(s_0)\delta(s-s_0). \end{aligned} \tag{4.4.5}$$

Henceforth, all the integrals are evaluated along the ray $x(s)$ emanating from a point $x_0 = x(s_0)$ on the boundary Σ and arriving at an observation point $x = x(s)$ in the direction $\mathbf{n}$ for a frequency ω as it is shown in Fig. 4.3. Note that the parameters $\mathbf{n}$, s, and ω which are specified for the observation point x determine *implicitly* the parameter of ray emanating s_0, i.e., at a given s we have $s_0 = s_0(x, \mathbf{n}, \omega)$ in (4.4.5).

The expression (4.4.5) describes the propagation of a coherent wave in the geometrical-optics approximation. This wave is attenuated because of the extinction according to the factor $\exp(-\int_{s_0}^{s} \alpha ds')$. For electromagnetic radiation, this factor will give also the change in the wave polarization under shifting along the ray if α is interpreted as the 4×4 matrix. For an inhomogeneous medium the computation of the factor with a matrix exponent appearing in (4.4.5) represents a nontrivial problem which is equivalent to solving a set of fourth order linear differential equations with varying coefficients.

In full analogy to (4.4.5), the equation of transfer (4.4.4) for the incoherent radiance of scattered radiation I^{incoh} may be written in the integral form as

$$I^{\text{incoh}} = (d_s + \alpha)^{-1}\hat{\sigma}(I^{\text{coh}} + I^{\text{incoh}}) \equiv \hat{A}\hat{\sigma}(I^{\text{coh}} + I^{\text{incoh}}). \tag{4.4.6}$$

Here, the operator $\hat{A} = (d_s + \alpha)^{-1}$ acts on a function $\varphi = \varphi(\mathbf{n}, \omega, x)$ according to the rule

$$\hat{A}\varphi = (d_s + \alpha)^{-1}\varphi = \int_{s_0}^{s} ds' \exp[-\int_{s'}^{s} \alpha(s'')ds'']\varphi(s'), \tag{4.4.7}$$

where $\varphi(s) = \varphi(\mathbf{n}(s), \omega(s), x(s))$.

The iterating of the integral equation of transfer (4.4.6) yields the infinite series for the scattered radiance I^{incoh}

$$I^{\text{incoh}} = \hat{A}\hat{\sigma}I^{\text{coh}} + \hat{A}\hat{\sigma}\hat{A}\hat{\sigma}\hat{A}\hat{\sigma}I^{\text{coh}} + \ldots . \tag{4.4.8}$$

Here the operators $\hat{\sigma}$ describe "scattering acts", and the operators $A = (d_s + \alpha)^{-1}$ describe the propagation of radiation between the "scattering acts". According to this, the n term of the series (4.4.8) is due to the n-fold scattered field.

4.4.2 Modified Born Approximation

In practice, one rarely retains more than two terms of the series in the expansion (4.4.8) and commonly uses only the first term which corresponds to the so-called *modified Born approximation*. In this approximation the radiance of scattered radiation is

$$I^{\text{incoh}} \approx I_{\text{sc}}^{(1)} = \hat{A}\hat{\sigma}I^{\text{coh}} = (d_s + \alpha)^{-1}\hat{\sigma}I^{\text{coh}}$$

$$= \int_{x_0}^{x} ds' \exp[-\int_{x_0}^{x} \alpha(s'')ds''](\hat{\sigma}I^{\text{coh}})(s'), \tag{4.4.9}$$

where I^{coh} is given by (4.4.5), and x_0 is the point of ray entering which lies on the boundary Σ and corresponds to the values of finite ray parameters x, $\mathbf{n}$, ω.

The expression (4.4.9) is applicable in the most general case of an inhomogeneous and nonstationary scattering medium. Let us consider in more detail the case of stationary radiation where medium parameters are time-independent, and the scattering without change in frequency occurs. In this case the dependence of radiance upon time and frequency is insignificant, so that we may use the argument $x = \mathbf{R}$ in place of $x = (\mathbf{R}, t)$ and omit the argument ω in the above formulas thus dealing with stationary (time-independent) rays.

If an incident wave is scattered from a volume V_{sc}, the integration in (4.4.9) is then performed along the ray which crosses the scattering volume V_{sc} and passes through the observation point $\mathbf{R}$ (Fig. 4.4). All the quantities of the integrand in (4.4.9) are functions of ray arclength s'. The initial value $s' = s_0$ corresponds to the boundary of the scattering region Σ, current values of s' are related to the scattering points $\mathbf{R}'$, and the final value $s' = s$ respects to the observation point $\mathbf{R}$.

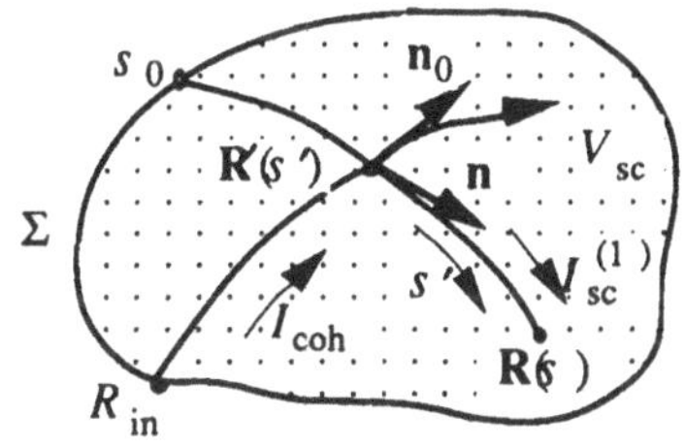

Figure 4.4. Transformation of the coherent field component I^{coh} in the singly scattered (incoherent) component $I_{sc}^{(1)}$.

The coherent component of radiance $I_{coh}(\omega, \mathbf{n}, \mathbf{R}')$ is related to the local intensity of the field $\mathcal{I}_{coh}(\mathbf{R}', \omega) = |u(\mathbf{R}', \omega)|^2$ by (3.5.38):

$$I_{coh}(\mathbf{n}, \omega, \mathbf{R}') = \mathcal{I}_{coh}(\omega, \mathbf{R}')\delta_2(\mathbf{n}\mathbf{n}_0). \tag{4.4.10}$$

where the vector $\mathbf{n}_0 = \mathbf{n}_0(\mathbf{R}')$ indicates the direction of an incident wave at the scattering point $\mathbf{R}'$, and the vector $\mathbf{n} = \mathbf{n}(\mathbf{R}')$ is tangent to the ray directed at the observation point $\mathbf{R}$ (Fig. 4.4). The intensity $\mathcal{I}_{coh}$ at the scattering point $\mathbf{R}'$ is related in turn to the "input" value of the intensity on the boundary Σ by

$$\mathcal{I}_{coh}(\omega, \mathbf{R}') = \mathcal{I}_{in} \exp[-\int_{s_{in}}^{s_1'} \alpha(s_1) ds_1] \tag{4.4.11}$$

where the exponential factor describes the attenuation of the intensity of the incident wave field due to its scattering and absorption along the path from the input point $\mathbf{R}_{in}$ to the scattering point $\mathbf{R}'$.

The "angular" delta function $\delta_2(\mathbf{n}\mathbf{n}_0)$ in (4.4.10) enables us to perform integration over angles in the expression for $\hat{\sigma}I_{\text{coh}}$ thus reducing the action of the operator $\hat{\sigma}$ on I_{coh} to the integration with respect to frequency only:

$$\hat{\sigma}I_{\text{coh}}(\mathbf{n}_0,\omega,\mathbf{R}') = \int d\omega' \sigma(\mathbf{R}';\mathbf{n}(\mathbf{R}'),\omega \leftarrow \mathbf{n}_0(\mathbf{R}'),\omega')\mathcal{I}_{\text{coh}}(\omega';\mathbf{R}) \tag{4.4.12}$$

Here the dependence σ upon $\mathbf{R}'$ corresponds to the case of an inhomogeneous scattering medium.

By virtue of (4.4.12), the radiance of the singly scattered field (4.4.9) becomes

$$I_{sc}^{(1)}(\mathbf{n}(\mathbf{R}),\omega,\mathbf{R}) = \int_{s_0}^{s} ds' \exp[-\int_{s'}^{s} \alpha(s'')ds'']$$
$$\times \int d\omega' \sigma(\mathbf{R}';\mathbf{n}(R'),\omega \leftarrow \mathbf{n}_0(R'),\omega')\mathcal{I}_{\text{coh}}(\mathbf{R}',\omega'). \tag{4.4.13}$$

According to (4.4.13), the modified Born approximation allows for both the attenuation of an incident wave along the path from the input point $\mathbf{R}_{\text{in}}$ to the scattering point $\mathbf{R}'$ [exponential factor in (4.4.11)] and the attenuation of the scattered field from the scattering point $\mathbf{R}'$ to the observation point $\mathbf{R}$ (exponential factor of the integrand in (4.4.13)). Specifically these factors, which are due to the propagation of incident and scattered waves in the effective medium, make the modified Born approximation different from the conventional Born approximation. The latter is derived by integrating the original wave equation and does not allow for the wave attenuation due to the scattering.

4.4.3 Coherence Function of the Scattered Field

The coherence function of a singly scattered field $\Gamma^{(1)}$ can be obtained by applying the Fourier transformation to $I_{sc}^{(1)}$:

$$\Gamma_{\text{sc}}^{(1)}(\mathbf{R},\rho,\tau) = \oint d\Omega_{\mathbf{n}} \int d\omega I_{\text{sc}}^{(1)}(\mathbf{n},\omega,\mathbf{R})\exp(ik\mathbf{n}\rho - i\omega\tau)$$
$$= \oint d\Omega_{\mathbf{n}} \int d\omega \int_{s_0}^{s} ds' \exp[-\int_{s'}^{s} \alpha(s'')ds'' + ik\mathbf{n}\rho - i\omega\tau]$$
$$\times \int d\omega' \sigma(\mathbf{R}';\mathbf{n}(\mathbf{R}'),\omega \leftarrow \mathbf{n}_0(\mathbf{R}'),\omega')\mathcal{I}_{\text{coh}}(\omega',R') . \tag{4.4.14}$$

For the sake of brevity the vectors $\mathbf{n}$ without argument are actually attributed to the observation point $\mathbf{R}$, i.e. $\mathbf{n} = \mathbf{n}(\mathbf{R})$, and the wavenumber k is in fact the real part of the effective wavenumber k_{eff}: $k = \operatorname{Re} k_{\text{eff}}$

In order to relate these expressions to the results of the Born scattering theory we switch in eqn (4.4.14) from the integration over solid angle and arclength s' to the integration over scattering volume. For this purpose, we take into account that

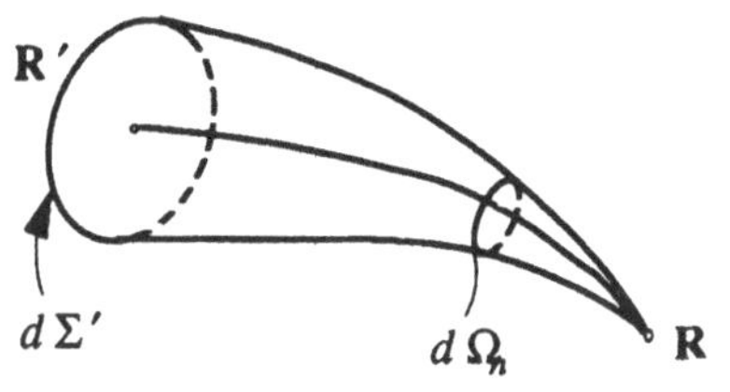

Figure 4.5. Solid angle $d\Omega_{\mathbf{n}}$ near the observation point **R** corresponds to an area $d\Sigma'$ near the scattering point $\mathbf{R}'$.

$$d\Omega_{\mathbf{n}}ds' = |d\Sigma'/d\Omega_{\mathbf{n}}|^{-1} d\Sigma' ds' = |d\Sigma'/d\Omega_{\mathbf{n}}|^{-1} d^3R', \tag{4.4.15}$$

where $d\Sigma'$ is the area of foundation of a curvilinear cone with a solid angle $d\Omega_{\mathbf{n}}$ (Fig. 4.5). The vertex of the cone is at the observation point $\mathbf{R}$, and its foundation is attributed to the scattering point $\mathbf{R}'$.

In view of (4.4.15), (4.4.14) becomes

$$\Gamma_{\text{sc}}^{(1)}(\mathbf{R},\rho,\tau) = \int d\omega \int d^3R' |d\Sigma'/d\Omega_{\mathbf{n}}|^{-1} \exp[-\int_{s'}^{s} \alpha(s'')ds'')]\exp(ik\mathbf{n}\rho - i\omega\tau)$$
$$\times \int d\omega' \sigma(\mathbf{R}',\mathbf{n}(\mathbf{R}'),\omega \leftarrow \mathbf{n}_0(\mathbf{R}'),\omega') \mathcal{I}_{\text{coh}}(\omega'\mathbf{R}'). \tag{4.4.16}$$

Letting $\rho = 0$ and $\tau = 0$ in (4.4.16), we obtain the intensity of the singly scattered field

$$\mathcal{I}_{\text{sc}}^{(1)}(\mathbf{R}) = \Gamma_{\text{sc}}^{(1)}(\mathbf{R},0,0) = \int d\omega \int d^3R' |d\Sigma'/d\Omega_{\mathbf{n}}|^{-1}$$
$$\times \exp[-\int_{s'}^{s} \alpha(s'')ds''] \int d\omega' \sigma(\cdot) \mathcal{I}_{\text{coh}}(\omega',R'). \tag{4.4.17}$$

If we set $\alpha = 0$, that is, if we neglect the effect of the scattering along the portions of propagation path from the source to the scattering point $\mathbf{R}'$ and from $\mathbf{R}'$ to the observation point $\mathbf{R}$, then (4.4.17) will be exactly equivalent to the result of the conventional Born theory (see Rytov, Kravtsov, and Tatarskii, 1989). This implies that (4.4.17) allows for the diffraction effects under scattering similarly to the conventional Born approximation. For various examples of using the modified Born approximation see, e.g., the book by Ishimaru (1978) (see also problems 4.14 and 4.15).

4.4.4 Conditions for Convergence of Iteration Series

Let us turn now to examining conditions for convergence of the iteration series (4.4.8). For the sake of simplicity, we restrict ourselves to the case of a stationary, homogeneous, and isotropic scattering medium, so that the rays are rectilinear in this medium and the frequency ω remains constant along the rays.

One can treat the convergence of (4.4.8) in different ways. The simplest point-by-point convergence (i.e. convergence at every fixed argument's value) is usually of little interest because only integrals of the radiance I which express energy characteristics of a field have a physical meaning rather than the radiance I itself. Taking this into account, consider the convergence of (4.4.8) in norm. (For mathematical properties of norms see, e.g., Kato, 1966, and Richtmyer, 1978)

The norm $\|a\|$ of a function (operator) a is a positive number that estimates the "magnitude" of this function (operator) and has the properties: $\|a+b\| \le \|a\| + \|b\|$, $\|ca\| \le |c|\,\|a\|$, where c is a number. It is clear that one can differently define the "magnitude" of a function varying with its arguments using only a number. In this case, for radiance I, we use the following definition for the norm:

$$\|I\| = \|I(\mathbf{n},\omega,x)\| = \max_x \int d\Omega_{\mathbf{n}} d\omega I(\mathbf{n},\omega,x), \tag{4.4.18}$$

that is, the norm $\|I\|$ is equal to the maximum value of the field energy density.

Estimating $\hat{A}\hat{\sigma}I$ in the norm (4.4.18), in view of the nonnegativity of all the integrand factors, one can easily get

$$\begin{aligned}
\left\|\hat{A}\hat{\sigma}I\right\| &= \max_x \int d\Omega_{\mathbf{n}} d\omega \int_{s_0}^{s} ds' \exp[-\int_{s_0}^{s} \alpha(s'')ds''] \\
&\quad \times \int d\omega' d\Omega_{\mathbf{n}'} \sigma(\omega,\mathbf{n} \leftarrow \omega',\mathbf{n}') I(\mathbf{n}',\omega',x(s')) \\
&\le \max_x \int d\Omega_{\mathbf{n}} d\omega \int_{s_0}^{s} ds' \exp[-\int_{s'}^{s} \alpha(s'')ds''] \max_{\omega_0,\mathbf{n}_0} \sigma(\omega,\mathbf{n} \leftarrow \omega_0,\mathbf{n}_0) \|I\| \\
&\equiv C\|I\| .
\end{aligned} \tag{4.4.19}$$

It follows that $\left\|(\hat{A}\hat{\sigma})^n I\right\| \le C^n \|I\|$, and estimating the terms of (4.4.8) in the norm, we get for $C < 1$

$$\left\|I^{\mathrm{incoh}}\right\| \le (C + C^2 + C^3 + \ldots)\left\|I^{\mathrm{coh}}\right\| = \frac{C}{1-C}\left\|I^{\mathrm{coh}}\right\|. \tag{4.4.20}$$

This means that, for $C < 1$, (4.4.8) converges in the norm (4.4.18). The physical meaning of this convergence consists in that the maximum value of the density of the field energy related to every subsequent term of the expansion (4.4.8) will be smaller than the preceding one. If we keep n first terms of (4.4.8) and assume that $\left\|(\hat{A}\hat{\sigma})^n I\right\| \sim C^n \|I\|$, then the contribution from all omitted terms will be on the order of $C^{n+1}\left\|I^{\mathrm{coh}}\right\|$, i.e., this will be smaller than the contribution from the last retained term. It is this circumstance that justifies the use of the modified Born approximation in which only one term is retained in (4.4.8).

Even though the condition $C < 1$ is sufficient and, strictly speaking, is not necessary, from physical considerations one may expect that the modified Born approximation will be of little efficiency under violation of this condition.

Let us obtain a more simple estimate of the conditions for convergence in the case of isotropic scattering without frequency alteration. In (4.4.19) one may then assume $\alpha = \text{const}$, $\sigma(\omega, \mathbf{n} \leftarrow \omega', \mathbf{n}') = (4\pi)^{-1}\alpha_s\delta(\omega - \omega')$ where α_s is the scattering coefficient. As a result we have

$$\begin{aligned} C &= \max_x \int d\Omega_{\mathbf{n}} d\omega \int_{s_0}^{s} \exp(-\alpha s')ds'(4\pi)^{-1}\alpha^{sc}\delta(\omega - \omega_0) \\ &\le \int d\Omega_{\mathbf{n}} \int_0^{s(\mathbf{n})} \exp(-\alpha s')ds'(4\pi)^{-1}\alpha^{sc} \\ &= \int d\Omega_{\mathbf{n}} \alpha^{-1}\left[1 - e^{-\alpha s(\mathbf{n})}\right](4\pi)^{-1}\alpha^{sc} \\ &= w_0\left(1 - \langle e^{-\alpha s(\mathbf{n})}\rangle_{\mathbf{n}}\right), \end{aligned} \tag{4.4.21}$$

where $w_0 = \alpha^{sc} / \alpha$ is the albedo of the single scattering,

$$\langle \cdot \rangle_{\mathbf{n}} = \int (4\pi)^{-1} d\Omega_{\mathbf{n}} \ldots \tag{4.4.22}$$

is the operation of averaging over directions $\mathbf{n}$, and $s(\mathbf{n}) = \max(s - s_0)$ is the maximum depth of the scattering volume along the direction $\mathbf{n}$. It follows that in the case of isotropic scattering the following condition is sufficient for applicability of the modified Born approximation:

$$\left(1 - \langle e^{-\alpha s(\mathbf{n})}\rangle_{\mathbf{n}}\right)w_0 << 1, \tag{4.4.23}$$

which provides the fast convergence for the series (4.4.8). The conventional Born approximation is valid under satisfaction of the much more severe condition $\alpha s << 1$, where s is the path of the initial wave in the scattering medium.

Equation (4.4.23) suggests that three essentially different situations exist in which the modified Born approximation is applicable:

(i) The case of strongly absorbing medium ($w_0 << 1$, $\alpha^{sc} << \alpha^{abs}$) for which the scattering effects are ignorable compared to the absorption effects.

(ii) The case of a weakly absorbing ($w_0 \sim 1$) and small scattering volume, so that $\alpha s(\mathbf{n}) << 1$ for any $\mathbf{n}$, i.e., the optical depth is small in all directions.

(iii) The case of a weakly absorbing ($w_0 \sim 1$) unbounded scattering volume, such that though $\alpha s(\mathbf{n}) > 1$ at some $\mathbf{n}$, the factor $1 - \langle e^{-\alpha s(\mathbf{n})}\rangle_{\mathbf{n}}$ is small compared to unity. The latter case relates to special problem geometries where with a high probability the singly scattered radiation leaves the scattering volume before it can be scattered twice,

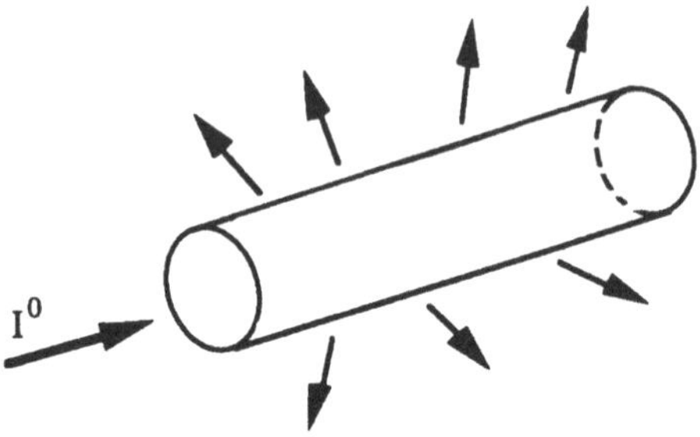

Figure 4.6. Scattering of an axially incident wave at a semi-infinite cylinder of diameter $D \ll \alpha^{-1}$. Here, the Born approximation is inapplicable because of the excluded extinction.

as for example, in the case of an infinite scattering cylinder (or layer) with a radius (or depth) far smaller than the scattering length (Fig. 4.6).

In eqn (4.4.21), $C \leq 1$. In the presence of an arbitrary weak absorption, $w_0 < 1$ and hence $C < 1$, so that, in this case, the iteration series (4.4.8) converges in the norm (4.4.21) for any geometry of the scattering volume including an unbounded medium. In the case of a weak absorption ($w_0 \sim 1$), however, for $C \sim 1$, this series will converge very slowly thus requiring a large number of terms from (4.4.8) and rendering the modified Born approximation almost inapplicable.

4.5 TWO-FLUX APPROXIMATION

4.5.1 General

The theory of radiative transfer seem to have been the first occasion when physicists were faced with the necessity to solve integro-differential equations which differ substantially from habitual ordinary or partial differential equations by nonlocal integral terms depending on the behavior of functions not only in the neighborhood of a point but also within a certain finite range of argument values. In the initial approach to such equations, it was natural to simplify the mathematical aspect of a problem so far as possible by adopting rough phenomenological idealizations. For the equation of radiative transfer, in the simplest approximation, this approach yields a model of two light fluxes propagating in opposite directions which is usually referred to as *two-flux theory*. This theory is usually associated with the efforts of Kubelka and Munk who presented it in 1931 (see Ishimaru, 1978a), although Schuster (1905) had used similar equations.

The two-flux theory is applicable in the case of *diffuse radiation* which has a wide directional pattern. Narrow paraxial beams are excluded from consideration. To describe them one needs to construct a more complicated four-flux approximation.

A vast literature is devoted to modifications of the two-flux and four-flux theories. Among these papers it is worth noting the instructive paper of Bohren (1987) where the two-flux theory was used to explain some qualitative features of multiple scattering in natural media such as clouds, snow, and sand.

4.5.2 Equations of Two-Flux Theory

The simplest approach used to derive equations of two-flux theory is based on heuristic considerations of the conservation of energy applied to the total fluxes of radiation propagating in the direct and opposite directions. This approach brings about free parameters that generally are in no way related to the parameters of the transfer equation and must be determined experimentally. A more consistent way is to derive equations of two-flux theory directly from the transfer equation. This approach reveals limitations of two-flux theory and expresses its parameters through the respective scattering cross sections (Ishimaru, 1978a).

We choose a simplified approach which, on the one hand, is adjacent to the heuristic derivation of the two-flux approximation and, on the other hand, uses the exact transfer equation directly. For a statistically homogeneous medium, this equation has the form

$$\left(\frac{1}{v_g}\partial_t + \mathbf{n}\nabla + \alpha\right)I = \int \sigma(\mathbf{n}\mathbf{n}')I(\mathbf{n},\mathbf{r},t')d\Omega_{\mathbf{n}} \equiv \hat{\sigma}I, \qquad (4.5.1)$$

where v_g is the group velocity of propagation, and the extinction coefficient $\alpha \equiv \alpha^{\text{ext}}$ is related to the absorption coefficient α^{abs} and the scattering coefficient α^{sc} through conventional relationships

$$\alpha = \alpha^{\text{abs}} + \alpha^{\text{sc}}, \qquad \alpha^{\text{sc}} = \int \sigma(\mathbf{n}\mathbf{n}')d\Omega_{\mathbf{n}'}. \qquad (4.5.2)$$

Suppose that a scattering layer $[0, L]$ is hit by a radiation characterized by a unit intensity, a uniform, spatial distribution, and a wide axially symmetrical angular distribution (diffuse irradiation, see Fig. 4.7). This is a plane symmetric problem, therefore all the quantities under consideration depend only on the z component of $\mathbf{r} = (\mathbf{r}_{\perp}, z)$ and on the angular variable $\mu = n_z$, so that eqn (4.5.1) takes the form

$$(\partial_{v_g t} + \mu\partial_z + \alpha)I(\mu, z, t) = \hat{\sigma}I. \qquad (4.5.3)$$

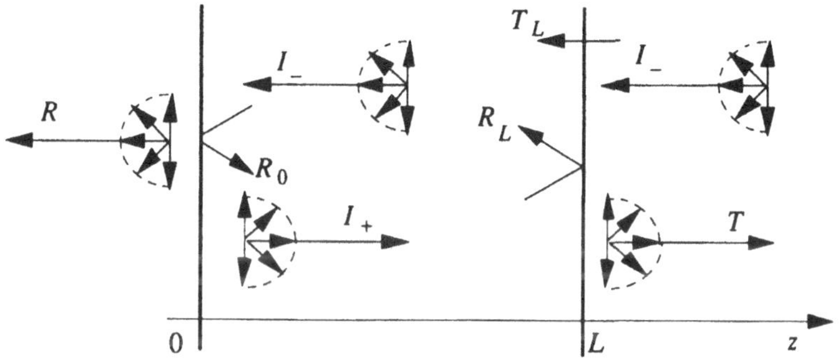

Figure 4.7. Illustrating the two-flux theory of diffuse scattering at a layer.

In the two-flux approximation, the total fluxes, traveling in the directions $\pm z$, are considered rather than the distribution of radiance over angles $\mathbf{n}$. Strictly speaking, such fluxes are expressed in terms of the z component of the flux vector $\mathbf{S} = \mathbf{n}I$ integrated over two hemispheres. It is this definition for fluxes that is used in attempting to rigorously justify the two-flux theory on the basis of eqn (4.5.1) (Ishimaru, 1978a). We adopt a simpler approach. We let the direction vector $\mathbf{n}$ take only two values $\pm z$, so that $\mu = \pm 1$. Then the radiance I transforms into the set of two quantities I_+ and I_- which describe the fluxes, and the integrals over directions $\mathbf{n}$ transform into the respective sums. As a result, eqn (4.5.3) becomes the set of two differential equations for $I_\pm$

$$\left(\frac{1}{v_g}\partial_t \pm \partial_z + \alpha\right) I_n = \sum_{n'=\pm} \sigma_{nn'} I_{n'}, \quad n = \pm. \tag{4.5.4}$$

These equations form the basis of the two-flux theory.

A more consistent integral derivation of the equations of two-flux theory, discussed in Problem 4.16, substitutes in eqn (4.5.4) the parameters $\pm\mu^*\partial_z$ for $n\partial_z = \pm\partial_z$, where $\mu^* = 1/2$ is the average cosine of the inclination angle over the hemisphere. This distinction does not alter the form of eqns (4.5.4) and has an effect only on the numerical values of their coefficients.

It is convenient to write eqn (4.5.4) in the form

$$(\partial_{\tilde{t}} + n\partial_\tau + 1) I_n = \sum_{n'} w_0 P_{nn'} I_{n'}, \tag{4.5.5}$$

where $n = \pm$, and $n' = \pm 1$.

$$w_0 = \frac{\alpha^{\text{sc}}}{\alpha^{\text{ext}}} = \frac{\alpha^{\text{sc}}}{\alpha^{\text{abs}} + \alpha^{\text{sc}}} \tag{4.5.6}$$

is the single-scattering albedo, $\tilde{t} = v_g t \alpha^{\text{ext}}$ is the reduced time, $\tau = \alpha^{\text{ext}} z$ is the optical path, and the parameters

$$P_{nn_0} = \frac{\sigma_{nn_0}}{\alpha^{\text{sc}}} = \frac{\sigma_{nn_0}}{\sum_{n'} \sigma_{n'n_0}}, \qquad \sum_n P_{nn_0} = 1 \tag{4.5.7}$$

may be treated as the probabilities that the radiation is scattered from the direction n_0 into the direction n.

In an isotropic medium, $P_{nn_0} = P_{n_0 n}$ and $P_{nn} = P_{n_0 n_0}$, which will be assumed henceforth. Note in passing that *the isotropy of the medium* should not be confused with *isotropic scattering*. In terms of scattering cross section, the former implies that σ depends only on the scalar product $\mathbf{n}\mathbf{n}_0$: $\sigma = \sigma(\mathbf{n} \cdot \mathbf{n}_0)$, and the latter means that σ depends neither on $\mathbf{n}$ nor on $\mathbf{n}_0$.

Following Bohren (1978), it is convenient to introduce the quantity

$$q = P_{nn} - P_{n_0 n} = P_{n_0 n_0} - P_{n n_0}, \quad n_0 \neq n, \tag{4.5.8}$$

which may be treated as the average cosine $\bar{\mu}$ of the scattering angle in an act of single scattering. This parameter describes the anisotropy of scattering, namely, for isotropic scattering, $q = 0$; for forward scattering, $q = 1$; and for backscattering, $q = -1$. From eqn (4.5.8) it follows that

$$P_{n_0 n} = P_{n n_0} = \frac{1-q}{2}, \quad P_{nn} = P_{n_0 n_0} = \frac{1+q}{2}, \tag{4.5.9}$$

where $n \neq n_0$.

4.5.3 Boundary Conditions

The system of two-flux equations should be supplemented with initial and boundary conditions chosen from physical considerations. We restrict our consideration to the stationary case ($\partial_t I_\pm = 0$) of wave scattering on a layer with an average refractive index $m \neq 1$. Taking then into account the reflection on layer boundaries, it is natural to impose the following boundary conditions:

$$\begin{aligned} I_-(z = L) &\equiv I_{-L} = T_L I_L^0 + R_L I_{+L}, \\ I_+(z = 0) &\equiv I_{+0} = R_0 I_{-0}, \end{aligned} \tag{4.5.10}$$

where R_L is the intensity reflection coefficient from the boundary $z = L$, T_L the similar transmission coefficient through the boundary $z = L$, R_0 the reflection coefficient from the boundary $z = 0$, and I_L^0 the flux of radiation incident upon the plane $z = L$ (see Fig. 4.7).

The reflection and transmission coefficients are expressed through the Fresnel reflection (r) and transmission (t) coefficients as follows [see eqn (2.2.22)]

$$\begin{aligned} T_L &= m^2(1 - |r|^2), \\ R_0 &= |r|^2 = R_L, \end{aligned} \tag{4.5.11}$$

where r and t are taken for the normal incidence of radiation upon a layer with the effective refractive index m:

$$\begin{aligned} r &= \frac{1-m}{1+m}, \\ t &= \frac{2}{1+m}. \end{aligned} \tag{4.5.12}$$

When the effective refractive index of the layer m is close to unity, so that the reflection from boundaries may be neglected, then $r \approx 0$ and the conditions (4.5.10) simplify:

$$I_{-L} = I_L^0 ,$$
$$I_{+0} = 0 . \tag{4.5.13}$$

We will say that the conditions (4.5.13) correspond to an optically "soft" layer in contrast to the general case of an optically "stiff" layer (4.5.12).

4.5.4 Estimation of Parameters of Two-Flux Approximation Equations

The above approach to deriving equations of the two-flux approximation on the basis of the transfer equation enables us to directly estimate parameters of this approximation. For the simplest model of sparse medium with discrete scatterers, these parameters can be expressed through the characteristics of individual particles using the relationships:

$$\alpha^{\text{abs}} = \nu_0 \sigma_{\text{abs}}^{(1)} , \quad \alpha^{\text{sc}} = \nu_0 \sigma_{\text{sc}}^{(1)} , \quad \sigma_{\text{sc}}^{(1)} = \int \sigma^{(1)}(\mathbf{n}\mathbf{n}_0) d\Omega_{\mathbf{n}} \tag{4.5.14}$$

where ν_0 is the concentration of particles $[\text{m}^{-3}]$, and $\sigma_{\text{abs}}^{(1)}$, $\sigma_{\text{sc}}^{(1)}$, and $\sigma^{(1)}(\mathbf{n}\mathbf{n}_0)$ are the absorption cross section, scattering cross section, and differential scattering cross section for an individual particle, respectively. The total albedo w_0 is identical with the albedo of an individual particle:

$$w_0 = \frac{\alpha^{\text{sc}}}{\alpha^{\text{abs}} + \alpha^{\text{sc}}} = \frac{\sigma_{\text{abs}}^{(1)}}{\sigma_{\text{abs}}^{(1)} + \sigma_{\text{sc}}^{(1)}} \equiv w_1 .$$

Henceforth it is natural to identify q with the mean cosine of scattering angle

$$q = \mu = \frac{\int n_z \sigma^{(1)}(\mathbf{n} \cdot \mathbf{n}_0) d\Omega_{\mathbf{n}}}{\sigma_{\text{sc}}^{(1)}} . \tag{4.5.15}$$

Naturally, all these estimates are valid only accurate to factors on order unity. However, these are quite suitable for rough estimations and for the construction the general pattern of the effects of parameters of the scattering layer.

When a more exact quantitative relationship with the parameters of the original transfer equation (4.5.1) is required, then one may invoke the integral approach to the two-flux approximation (Problem 4.16). This switching brings about, first, the factor $\mu^* = 1/2$ on ∂_z and, second, replaces the forward and backward scattering cross sections σ_{nn_0} by the quantities

$$\tilde{\sigma}_{nn_0} \equiv \frac{1}{2\pi} \int \sigma(\mathbf{n}' \leftarrow \mathbf{n}_0') \theta(n n_z') \theta(n_0 n_{0z}') d\Omega_{\mathbf{n}'} d\Omega_{\mathbf{n}_0'} , \tag{4.5.16}$$

which describe the scattering within the front and back hemispheres (here $\theta(x)$ is the Heaviside step function).

4.5.5 Optically Soft Scattering Layer

The two-flux theory and its modifications [the four-flux approximation (Ishimaru, 1978a; Maheu et al., 1984, 1989)] are widely used in various applications, especially in situations where the *integral* reflection and transmission coefficients are required, and it is sufficient to estimate the effect of multiple scattering to moderate accuracy. As a simple example we consider an optically soft scattering layer which has no reflection from the boundaries. (A more complicated case of using the two-flux approximation is considered in Problem 4.16)

For a unit incident flux with $I_L^0 = 1$, the stationary solution of the system (4.5.5) with boundary conditions (4.5.13) gives the reflection and transmission coefficients as

$$\begin{aligned} R &= I_+(z=L) = \frac{(P_2^2 - P_1^2)(e^d - e^{-d})}{(P_2+P_1)^2 e^d - (P_2-P_1)^2 e^{-d}}, \\ T &= I_-(z=0) = \frac{4P_1P_2}{(P_2+P_1)^2 e^d - (P_2-P_1)^2 e^{-d}}, \end{aligned} \tag{4.5.16}$$

$$P_1 = \sqrt{1-w_0}, \quad P_2 = \sqrt{1-w_0 q}, \quad d = \alpha^{\text{ext}} P_1 P_2 L. \tag{4.5.17}$$

In the particular case of a nonabsorbing layer, $\alpha_a = P_1 = 0$, the expressions for R and T contain indeterminate forms of the type $0/0$ which are treated to yield

$$\begin{aligned} R &= \frac{\tau^*}{2+\tau^*}, \\ T &= \frac{2}{2+\tau^*}, \end{aligned} \tag{4.5.18}$$

where

$$\tau^* = (1-q)\alpha^{\text{ext}} L. \tag{4.5.19}$$

It follows that an increase in the layer thickness L decreases the transmission coefficient as $1/L$.

In another limiting case of a scattering half-space ($L = \infty$) the reflection coefficient (4.5.16) reduces to the simple expression (Bohren, 1987)

$$\begin{aligned} R = R_\infty &= \frac{P_2 - P_1}{P_2 + P} \\ &= \frac{\sqrt{1-w_0 q} - \sqrt{1-w_0}}{\sqrt{1-w_0 q} + \sqrt{1-w_0}}, \end{aligned} \tag{4.5.20}$$

which enables one to estimate the effect of absorption on the reflection from an optically thick layer.

4.6 NUMERICAL AND STATISTICAL SOLUTION TECHNIQUES

4.6.1 Numerical Methods for Solving the Transfer Equation

Relevant numerical methods were initially developed under the influence of the needs of astrophysics that gave birth to the theory of radiative transfer in scattering media. Beginning from the 1940s, and particularly from the 1950s, applied nuclear physics, in which the transport theory of neutrons forms the basis of reactor design, added a strong stimulus. This theory also uses the transfer equation in which the neutron velocity distribution plays the role of radiance. A renewed interest in the numerical methods of transfer theory has been inspired by opticians.

The numerical methods used in radiative transfer theory can be divided into two large groups: mesh methods and the methods based on expansions of the solution in terms of base functions. The radiative transfer equation written in its original integro-differential form or transformed into its integral formulation reduces to a set of linear algebraic equations of a rather high order. This set of algebraic equations is solved by conventional techniques, as a rule, based on iterative procedures.

For the general information on the numerical methods used in transfer theory we refer the reader to the books by Greenspan (1968), Marchuk and Lebedev (1986), Lewis and Miller (1984), and the special sections of the classical treatise by Chandrasekhar (1960).

4.6.2 Expansion in Terms of Base Functions

This method develops the Galerkin technique and uses various sets of functions. The most popular approach is based on *spherical harmonics*. The angular dependence of radiance $I(\mathbf{n},\mathbf{r})$is described by a set of spherical harmonics $Y_{lm}(\mathbf{n})$. The coefficients $I_{lm}(\mathbf{r})$ of the expansion of the radiance in a series in terms of spherical harmonics are point functions to be determined.

A truncation of set of equations for $I_{lm}(\mathbf{r})$ leads to the so-called P_L approximations where L is the number of angular harmonics retained. The P_1 approximation proves to be equivalent to the diffuse approximation. A compact description of the fundamentals of the spherical harmonic approach may be found in the book by Davison (1957).

4.6.3 The Discrete Ordinate Method

This method is a modification of the mesh technique in the angular representation. In this method, the angular dependence of radiance is approximated by a set of values at discrete angles. The integral term in the transfer equation is simultaneously approximated by a quadrature formula. This discretization replaces the transfer equation by a system of linear differential equations.

Fundamentals of this method are outlined in the book by Marchuk and Lebedev (1986); the state of the art is given in the book by Bass, Voloshchenko, and Germogenova (1986).

A promising modification of this approach is the nodal methods. These methods sew the local solutions to the transfer equation obtained in the neighborhood of discrete nodes with computational schemes of high accuracy. Advances of nodal methods have been described by Badruzzman (1990).

4.6.4 Statistical Modeling Methods

The mesh methods and the methods of expansion in a set of base functions are referred to deterministic methods. Neutron transport theory frequently resorts to the Monte Carlo method based on the statistical modeling of neutron paths. It deals with a set of large number of individual neutron paths. The distribution of neutrons over spatial and angular coordinates is then determined by calculating the number of paths corresponding to the values of $\mathbf{r}$ and $\mathbf{n}$ in some neighborhood $\Delta\mathbf{r}\Delta\mathbf{n}$ of the point under consideration.

The method of statistical modeling is used when the precision specified requires very fine meshes, and also the integral, rather than local, characteristics of the solution are needed. In principle, a combined approach realizing strong features of the mesh method and statistical modeling is also possible.

PROBLEMS

Problem 4.1. Using the Fourier transformation show that the solution of the transfer equation (4.1.1) has the form (4.1.3).

Solution. Define the Fourier transform of $G = G'(\mathbf{r},\mathbf{n};\mathbf{r}_0,\mathbf{n}_0)$ by

$$\tilde{G}(\kappa,\mathbf{n}) = \int G e^{-i\kappa\mathbf{r}} d^3\mathbf{r} / (2\pi)^3 \tag{1}$$

(insignificant arguments are omitted). This function satisfies an equation that follows from eqn (4.1.1):

$$(i\kappa\mathbf{n} + \alpha)\tilde{G}(\kappa,\mathbf{n}) = F(\kappa) + \frac{\exp(-i\kappa\mathbf{r}_0)\delta^{(2)}(\mathbf{n}\cdot\mathbf{n}_0)}{(2\pi)^3}, \tag{2}$$

where

$$F(\kappa) = \sigma \int \tilde{G}(\kappa,\mathbf{n}) d\Omega_{\mathbf{n}}. \tag{3}$$

Substituting $\tilde{G}$ expressed from eqn (2) through $F(\kappa)$, we obtain for

$$F(\kappa) = \sigma \int \frac{d\Omega_{\mathbf{n}}}{i\kappa\mathbf{n} + \alpha} F(\kappa) + \frac{\sigma}{(2\pi)^3} \frac{\exp(-i\kappa\mathbf{r}_0)}{(i\kappa\mathbf{n}_0 + \alpha)}.$$

Solving this equation for $F(\kappa)$, we obtain from eqn (2) the expression for $\tilde{G}(\kappa,\mathbf{n})$ which immediately yields eqn (4.1.3).

Problem 4.2. Using the closure conditions (4.1.17), we switch from relation (4.1.15) containing the operator G to eqn (4.1.18) which is closed for H_L.

Solution. Using eqn (4.1.11), we write the first term on the right-hand side of (4.1.15) as

$$\left|\partial_z G(\mathbf{r}_\perp, z; \mathbf{r}_{0\perp}, z_0) = \right|\partial_z G^0 + \partial_L \int G^0(\mathbf{r}_\perp, L; \mathbf{r}')V(\mathbf{r}')G(\mathbf{r}', \mathbf{r}_0)d^3r', \tag{1}$$

where vertical bars denote the substitution $z = z_0 = L$. According to eqn (4.1.17), the operator ∂_L may be replaced with B_+. Using again eqn (4.1.1), we obtain

$$\left|\partial_z G = \right|\partial_z G^0 + \left|B_+(G - G^0) = \partial_z G^0 + B_+(H_L - H_L^0).\right. \tag{2}$$

Transforming the second term on the right-hand side of eqn (4.1.15) in a similar manner and taking into account eqn (4.1.16), we arrive at eqn (4.1.18).

Problem 4.3. Write an explicit form of operators entering the equation of invariant embedding for the reflection coefficient (4.1.22) for the scalar wave problem.

Solution. In the case of a scalar wave problem (3.3.3), the direct differentiation of the kernel of free-space propagation operator $G^0 = (\Delta + k^2)^{-1}$ does not enable us to find the operators B_+ and B_- which satisfy the condition (4.1.17). Therefore, we resort to the representation of the kernel of G^0 in the form of the Rayleigh plane-wave expansion

$$G^0(\mathbf{r}, \mathbf{r}_0) = -\frac{\exp(ik|\mathbf{r} - \mathbf{r}_0|)}{4\pi|\mathbf{r} - \mathbf{r}_0|} = \int \frac{\exp\left[i\sqrt{k^2 - \kappa_\perp^2}|z - z_0| + i\kappa_\perp(\mathbf{r}_\perp - \mathbf{r}_{0\perp})\right]}{8\pi^2 i\sqrt{k^2 - \kappa_\perp^2}} d^2\kappa_\perp. \tag{1}$$

We see that the Fourier representation in the transverse coordinates reduces the action of the operators B_+ and B_-, satisfying the conditions (4.1.17), to multiplication by $i\sqrt{k^2 - \kappa_\perp^2}$, or, in operator form, $B_+ = B_- = i\sqrt{k^2 + \Delta_\perp}$. In the r-representation, these operators have integral form explicitly derived by Klyatskin (1986) [see this book for many other wave problems].

In our case, the kernel of the operator H_L^0 is

$$H_L^0(\mathbf{r}_\perp, \mathbf{r}_{0\perp}) = -\frac{\exp(ik|\mathbf{r}_\perp - \mathbf{r}_{0\perp}|)}{4\pi|\mathbf{r}_\perp - \mathbf{r}_{0\perp}|}, \tag{2}$$

and the action of v reduces to a multiplication by $-k^2\varepsilon(\mathbf{r})$.

Problem 4.4. Find the Green function for the scalar transfer equation (4.1.2) in the absence of scattering.

Solution. In the absence of scattering, $\sigma = 0$, and the transfer equation (4.1.24) takes the form

$$(\mathbf{n}\nabla + \alpha)I(\mathbf{n},\mathbf{r}) = q(\mathbf{r},\mathbf{n}), \tag{1}$$

where $\alpha = \alpha^{\text{abs}}$, and the vector $\mathbf{n}$ is a parameter. Chose the s axis along $\mathbf{n}$ and write the vector $\mathbf{r}$ as $\mathbf{r} = (\rho, s)$ where ρ is the component normal to $\mathbf{n}$. Then eqn (1) will be converted to the ordinary differential equation

$$(\partial_s + \alpha)I = q, \tag{2}$$

with a solution

$$I = \int_{-\infty}^{s} e^{-\alpha(s-s')} q(\rho, s'; \mathbf{n}) ds' = \int_{0}^{\infty} e^{-\alpha s'} q(\rho, s - s'; \mathbf{n}) ds'. \tag{3}$$

Since $(\rho, s - s') = \mathbf{r} - s'\mathbf{n}$, we have

$$I = \int_{0}^{\infty} e^{-\alpha s'} q(\mathbf{r} - s'\mathbf{n}; \mathbf{n}) ds' = \int_{0}^{\infty} ds' \int d\mathbf{r}' e^{-\alpha s'} \delta(\mathbf{r} - \mathbf{r}' - s'\mathbf{n}) q(\mathbf{r}', \mathbf{n}). \tag{4}$$

Comparing eqn (4) with the expression for I written in terms of the Green's function

$$I = \int G^0(\mathbf{n},\mathbf{r};\mathbf{n}',\mathbf{r}') q(\mathbf{n}',\mathbf{r}') d\mathbf{r}' d\Omega_{\mathbf{n}'}, \tag{5}$$

we obtain

$$G^0(\mathbf{n},\mathbf{r};\mathbf{n}',\mathbf{r}') = \delta_2(\mathbf{n}\cdot\mathbf{n}') \int_{0}^{\infty} e^{-\alpha s'} \delta(\mathbf{r} - \mathbf{r}' - s'\mathbf{n}) ds'. \tag{6}$$

This integral is easily evaluated, and the result can be represented in the form reflecting problem symmetry. For example, in the case of a scattering layer, it is convenient to derive longitudinal (||) and lateral ($\perp$) vector components (relative to the normal to the layer) thus writing the delta function in eqn (6) as

$$\delta(\mathbf{R} - \mathbf{n}s') = \delta(\mathbf{R}_\perp - \mathbf{n}_\perp s') \delta(R_\perp - n_z s'). \tag{7}$$

As a result, the Green's function is expressed in the form (4.1.25).

One more form for the Green's function can be obtained by substituting another expression for the delta function:

$$\delta(\mathbf{R}-\mathbf{n}s')=\frac{\delta(R-|\mathbf{n}s'|)}{R^2}\delta_2(\hat{\mathbf{R}}\cdot\mathbf{n}), \tag{8}$$

where $\hat{\mathbf{R}}=\mathbf{R}/R$. Then

$$\begin{aligned} G^0(\mathbf{n},\mathbf{r};\mathbf{n}',\mathbf{r}') &= \delta_2(\mathbf{n}\cdot\mathbf{n}')\int_0^\infty e^{-\alpha s}\frac{\delta(R-s)}{R^2}\delta_2(\hat{\mathbf{R}}\cdot\mathbf{n})ds \\ &= \delta_2(\mathbf{n}\cdot\mathbf{n}')\delta_2(\hat{\mathbf{R}}\cdot\mathbf{n})\frac{\exp(-\alpha R)}{R^2}. \end{aligned} \tag{9}$$

This expression has a simple physical meaning if we observe that $1/R^2$ is the factor of divergence corresponding to free-space propagation, and the factor $e^{-\alpha R}$ describes absorption.

Problem 4.5. Change from equation (4.1.26) for the operator $\hat{R}_L$ to the equation for the kernel of this operator $R_L(\mathbf{n},\mathbf{r}_\perp;\mathbf{n}_0,\mathbf{r}_{0\perp})$.

Solution. The action of this operator on a test function $f(\mathbf{n},\mathbf{r}_\perp)$ is expressed as

$$\hat{\mathbf{R}}_L f=\int R_L(\mathbf{n},\mathbf{r}_\perp;\mathbf{n}',\mathbf{r}_\perp)f(\mathbf{n}',\mathbf{r}_\perp')d^2r_\perp' d\Omega_{\mathbf{n}'}. \tag{1}$$

Assuming that the function f tends to zero as $r_\perp\to\infty$ and integrating by parts, we have

$$\hat{R}_L\nabla f=\int R_L\nabla' f d^2r_\perp' d\Omega_{\mathbf{n}'}=-\int(\nabla_{r'}R_L)fd^2r_\perp' d\Omega_{\mathbf{n}'}. \tag{2}$$

It follows that the operator $\hat{R}_L\nabla$ in the equation for the kernel R_L is converted to $-\nabla_{r'}R_L$. Taking this fact into account and expressing the action of operators in eqn (4.1.26) in explicit form, we obtain the equation for the kernel R_L:

$$\begin{aligned} &\partial_L R_L(\mathbf{n},\mathbf{r}_\perp;\mathbf{n}_0,\mathbf{r}_{0\perp}) \\ &=\left(-\frac{\alpha+\mathbf{n}_\perp\nabla}{\mu}-\frac{\alpha-\mathbf{n}_{0\perp}\nabla_{r_0}}{\mu_0}\right)R_L(\mathbf{n}_\perp,\mathbf{r}_\perp;\mathbf{n}_0,\mathbf{r}_{0\perp}) \\ &+\int\left[\frac{\delta(\mathbf{r}_\perp-\mathbf{r}_\perp')\delta_2(\mathbf{n}\cdot\mathbf{n}')}{\mu}+R_L(\mathbf{n}_\perp,\mathbf{r}_\perp;\mathbf{n}_\perp',\mathbf{r}_\perp')\right]\sigma(\mathbf{n}'\leftarrow\mathbf{n}'') \end{aligned}$$

$$\times\left[\frac{\delta(\mathbf{r}'_\perp-\mathbf{r}_{0\perp})\delta_2(\mathbf{n}'\cdot\mathbf{n}_0)}{\mu_0}+R_L(\mathbf{n}',\mathbf{r}'_\perp;\mathbf{n}_0,\mathbf{r}_{0\perp})\right]d^2r'_\perp d^2r''_\perp d\Omega_{\mathbf{n}'}d\Omega_{\mathbf{n}''}\,. \qquad (3)$$

Here $\mu=|n_z|$, $\mu_0=|n_{0z}|$ and we observed that the kernel $R_L(\mathbf{n},\mathbf{r}_\perp;\mathbf{n}_0,\mathbf{r}_{0\perp})$ is other than zero only for $n_{0z}\le 0$ (because $\mathbf{n}_0$ is the direction of the wave incident on the layer from the right, see Fig. 4.1), and $n_z\ge 0$ ($\mathbf{n}$ is the direction of the wave reflected to the left).

Problem 4.6. Establish the physical meaning of different terms in equation (4.1.26). For this purpose consider the increment δR_L caused by an infinitesimal change in layer's thickness $h\to 0$ and take into account the result of Problem 4.5 (Apresyan, 1990a).

Solution. The reflection coefficient R_L may also be written as $R_L=R_{+-}$ where the index "-" corresponds to the primary wave propagating in the direction $-z$, and "+" to the reflected wave propagating in the direction $+z$ (see Fig. 4.1). Taking into account the property of the operator R_L, discussed in Problem 4.5, and introducing similar index notations σ_{++}, σ_{+-}, σ_{-+}, and σ_{--} for the scattering events $+z\to+z$, $-z\to+z$, $+z\to-z$, $-z\to-z$, we write eqn (4.1.26) in the form

$$\partial_L R_{+-}=-\frac{\alpha+\mathbf{n}_\perp\nabla}{\mu}R_{+-}-R_{+-}\frac{\alpha+\mathbf{n}_\perp\nabla}{\mu}$$
$$+\frac{1}{\mu}\sigma_{+-}\frac{1}{\mu}+\frac{1}{\mu}\sigma_{++}R_{+-}+R_{+-}\sigma_{--}\frac{1}{\mu}+R_{+-}\sigma_{-+}R_{+-}\,. \qquad (1)$$

This form has a simple physical meaning. The first two terms on the right-hand side describe the contribution into $\delta_L R_{+-}$ due to a change of L (disregarding the scattering) which is made up of the changes in extinction and lateral displacements of rays, the third term gives the contribution from single scattering $-z\to+z$, the forth and fifth terms give contributions from the forward scattering in the directions $+z\to+z$ and $-z\to-z$, and the last nonlinear term corresponds to backscattering $+z\to-z$.

Problem 4.7. Given an isotropically scattering half-space where the $\mathbf{n}$-dependencies reduce to the dependencies on $n_z=\mu$, derive the equation of invariant embedding for the scattering function $S=S(\mu,\mu_0)$ [see eqn (4.1.29)].

Solution. The axial symmetry implies that the dependence on spatial arguments $\mathbf{r}_\perp$ becomes insignificant, and we may omit the terms with gradients. In addition, for the half-space, $L\to\infty$ and $\partial_L R\to 0$. Accordingly, we have from eqn (4.1.26)

$$\frac{\alpha}{n_z}R_L-R_L\frac{\alpha}{n_z}=\left(\frac{1}{|n_z|}+R_L\right)\hat{\sigma}\left(\frac{1}{|n_z|}+R_L\right) \qquad (1)$$

Switching here from operators to kernels, quite similarly to the preceding problem, we express the kernel R_L as $R_L(n,n_0)=(4\pi\mu\mu_0)^{-1}S(\mu,\mu_0)$, where $\mu=n_z$ and $\mu_0=|n_{0z}|$. Recognizing that $\sigma(\mathbf{n}\leftarrow\mathbf{n}_0)=\sigma$ is independent of $\mathbf{n}$ and $\mathbf{n}_0$ in isotropic scattering, we obtain the equation for $S(\mu,\mu_0)$:

$$\left(\frac{1}{\mu}+\frac{1}{\mu_0}\right)S(\mu,\mu_0)=w_0\left(1+\frac{1}{2}\int_0^1 S(\mu,\mu')\frac{d\mu'}{\mu'}\right)\left(1+\frac{1}{2}\int_0^1 S(\mu'',\mu)\frac{d\mu''}{\mu''}\right), \tag{2}$$

where $w_0=\sigma/\alpha$ is the albedo of single scattering.

In view of the symmetry relation $S(\mu,\mu_0)=S(\mu_0,\mu)$ the right-hand side of this equation may be written as the product $w_0H(\mu)H(\mu_0)$, where

$$H(\mu)\equiv 1+\frac{1}{2}\int_0^1 S(\mu,\mu')\frac{d\mu'}{\mu'} \tag{3}$$

is the Chandrasekhar H-function (not to be confused with the kernel of the operator H_L). If we now express $S(\mu,\mu')$ in terms of $H(\mu)$:

$$S(\mu,\mu')=w_0H(\mu)H(\mu')\frac{\mu\mu'}{\mu+\mu'}, \tag{4}$$

we obtain the nonlinear integral equation for $H(\mu)$

$$H(\mu)=1+\frac{w_0}{2}\mu H(\mu)\int_0^1\frac{H(\mu')}{\mu+\mu'}d\mu', \tag{5}$$

which was discussed in detail by Chandrasekhar (1960).

Problem 4.8. Find the general solution to the small-angle transfer equation (4.2.7) with the initial condition (4.2.8).

Solution. We first pass over from eqn (4.2.7) for I to eqn (4.2.12) for $\Gamma_\perp$. Then it is convenient to perform the Fourier transformation with respect to transverse coordinates of the center of gravity $R_\perp$, that is, to change from $\Gamma_\perp$ to the following function

$$\tilde{\Gamma}(\mathbf{k}_\perp,z,\rho)=\int\Gamma_\perp(\mathbf{R}_\perp,z;\rho_\perp)e^{-i\mathbf{k}_\perp\mathbf{R}_\perp}\frac{d^2R_\perp}{(2\pi)^2}. \tag{1}$$

This function satisfies the linear partial differential equation that follows from eqn (4.2.12):

$$\left[\partial_z + \frac{\mathbf{k}_\perp \nabla_{\rho\perp}}{k_0} + H_1(\rho)\right]\tilde{\Gamma}_\perp = 0 \tag{2}$$

subject to the initial condition following from eqn (4.2.8). The solution for $\tilde{\Gamma}$ can be easily found using the standard method of characteristics:

$$\tilde{\Gamma}(\mathbf{k}_\perp, z; \rho) = \tilde{\Gamma}^0\left(\mathbf{k}_\perp, \rho_\perp - \frac{\mathbf{k}_\perp z}{k_0}\right)\exp\left(-\int_0^z H_1(\rho_\perp - \frac{z'\mathbf{k}_\perp}{k_0})dz'\right). \tag{3}$$

Returning now from $\tilde{\Gamma}$ to the radiance I by the inverse Fourier transformation, we finally come to the solution (4.2.18), (4.2.19).

Problem 4.9. Using the small-angle approximation of transfer theory, consider the joint effect of continuous fluctuations and discrete scatterers (particles) on the coherence functions and the radiance of a plane incident wave (Borovoi et al., 1988).

Solution. For a plane incident wave we may omit the term with gradients in equation (4.2.48) for the coherence function Γ. This omission yields a simple equation whose solution may be written in the form

$$\Gamma_\perp = \Gamma_{\text{cont}}(\rho)\Gamma_{\text{part}}(\rho), \tag{1}$$

where Γ_{cont} and Γ_{part} are the coherence functions corresponding to the presence of only continuous or only discrete fluctuations in the medium:

$$\Gamma_{\text{cont}} = \exp[-zH_{\text{cont}}(\rho)], \tag{2}$$

$$\Gamma_{\text{part}} = \exp[-zH_{\text{part}}(\rho)]. \tag{3}$$

According to eqn (4.2.17) the generalized radiance corresponding to eqn (1) is expressed as a convolution with respect to the angular argument $\mathbf{n}_\perp$:

$$I(\mathbf{n}_\perp, z) = I_{\text{cont}}(\mathbf{n}_\perp, z) * I_{\text{part}}(\mathbf{n}_\perp, z)$$
$$\equiv \int I_{\text{cont}}(\mathbf{n}'_\perp, z) I_{\text{part}}(\mathbf{n}_\perp - \mathbf{n}'_\perp, z) d^2\mathbf{n}'_\perp, \tag{4}$$

where I_{cont} relates to continuous fluctuations and I_{part} to discrete fluctuations, both I_{cont} and I_{part} being the Fourier transforms (4.2.17) of Γ_{cont} and Γ_{part}, respectively.

Problem 4.10. If the absorption by particles is so large that the imaginary part of the phase along the ray passing through the particle significantly exceeds unity, $\operatorname{Im}\psi_j^{(s)} >> 1$, then the anomalous diffraction approximation naturally transforms into the model of black screens in which particles are replaced by opaque screens consistent with the contour of their shadows. Formally this transition implies the replacement of the phase $\psi_j^{(s)}$ in eqn (4.2.51) with $i\infty$, so that $\exp(i\psi_j^{(s)})$ is transformed into $1-\eta(\mathbf{r}_{j\perp}-\mathbf{r}_{s\perp})$ where $\eta(\mathbf{r}_\perp)$ is the characteristic function of particle shadow which is is unity within the shadow and zero outside the shadow. Using this substitution for spherical particles of radius a, evaluate explicitly the first term on the right-hand side of eqn (4.2.49), which corresponds to the approximation of independent particles.

Solution. Let particles be uniformly distributed over the scattering volume. Then setting $\exp(i\psi_j^{(s)}) \sim 1-\eta_{sj}$, where $\eta_{sj}=\eta(\mathbf{r}_{j\perp}-\mathbf{r}_{s\perp})$, in eqn (4.2.51), we have from eqn (4.2.49)

$$\begin{aligned} H_{\text{part}}(\rho) &= \nu\left\langle \xi^{(1|z)}\right\rangle \\ &= \nu\left\langle \delta(z-z_s)\left[1-(1-\eta_{s1})(1-\eta_{s2})\right]\right\rangle \\ &= \nu\int \delta(z-z_s)(\eta_{s1}+\eta_{s2}-\eta_{s1}\eta_{s2})d^3r_s \\ &= \nu\left(2\Sigma-\int \eta_{s1}\eta_{s2}d^2r_{s\perp}\right). \end{aligned} \tag{1}$$

Here, $\Sigma=\pi a^2$ is the area of the particle shadow. For spherical particles, the integral in eqn (1) is easily evaluated to yield

$$H_{\text{part}}(\rho)=\nu 2a^2\left[\pi-2\left(\arcsin\frac{\rho}{2a}-\frac{\rho}{2a}\sqrt{1-\left(\frac{\rho}{2a}\right)^2}\right)\theta(2a-\rho)\right], \tag{2}$$

where $\theta(x)$ is the Heaviside step function.

Problem 4.11. Calculate the contribution of the incoherent radiance (4.3.4) to the coherence function.

Solution. Evaluating the integral for the coherence function corresponding to eqn (4.3.4) yields

$$\begin{aligned} \Gamma^{\text{incoh}} &= \int I^{\text{incoh}}(\mathbf{n},\mathbf{R})e^{ik\mathbf{n}\rho}d\Omega_{\mathbf{n}} \\ &= cW(\mathbf{R})\frac{\sin k\rho}{k\rho}+\frac{3\mathbf{S}(\mathbf{R})\rho}{i\rho}\left(d_x\frac{\sin x}{x}\right)_{x=k\rho}. \end{aligned} \tag{1}$$

Here $k = \mathrm{Re}\, k^{\mathrm{eff}} \approx k_0$, and the superscript "incoh" implies that this function describes the scattered field rather than the total field.

Problem 4.12. Find the solution to the diffusion equation (4.3.20).

Solution. The desired solution is easily obtained by the Fourier transformation. Substituting

$$W = \int W_k e^{i\mathbf{kR}} d^3k \tag{1}$$

into eqn (4.3.20) yields

$$\left(\alpha^{\mathrm{abs}} + \frac{k^2}{3\alpha^{\mathrm{ext}}}\right) W_{\mathbf{k}} = \frac{1}{(2\pi)^3 c}, \tag{2}$$

whence

$$\begin{aligned} W &= \int \frac{e^{i\mathbf{kR}}}{\alpha^{\mathrm{abs}} + \dfrac{k^2}{3\alpha^{\mathrm{ext}}}} \frac{d^3k}{(2\pi)^3 c} \\ &= \frac{3\alpha^{\mathrm{ext}}}{(2\pi)^3 c} \int_0^\infty k^2 dk \int_{-1}^{1} d\mu \int_0^{2\pi} d\varphi \frac{e^{ikR\mu}}{k^2 + 3\alpha^{\mathrm{ext}}\alpha^{\mathrm{abs}}} \\ &= -\frac{3\alpha^{\mathrm{ext}}}{(2\pi)^2 cR} \partial_R \int_{-\infty}^{\infty} \frac{e^{ikR}}{k^2 + 3\alpha^{\mathrm{ext}}\alpha^{\mathrm{abs}}} dk. \end{aligned} \tag{3}$$

This integral is evaluated by residues at the point $k = i\sqrt{3\alpha^{\mathrm{ext}}\alpha^{\mathrm{abs}}}$ and the final result is

$$\begin{aligned} W &= -\frac{3\alpha^{\mathrm{ext}}}{c(2\pi)^2 R} \partial_R 2\pi i \underset{k=i\sqrt{3\alpha^{\mathrm{ext}}\alpha^{\mathrm{abs}}}}{\mathrm{Res}} \frac{\exp(ikR)}{k^2 + 3\alpha^{\mathrm{ext}}\alpha^{\mathrm{abs}}} \\ &= \frac{3\alpha^{\mathrm{ext}}}{4\pi c} \frac{\exp(-\sqrt{3\alpha^{\mathrm{ext}}\alpha^{\mathrm{abs}}} R)}{R} \end{aligned} \tag{4}$$

Problem 4.13. Using the random walk model estimate the characteristic absorption length in the diffusion regime.

Solution. In the model of random walks, the attenuation of intensity due to absorption is obviously proportional to the factor $\exp(-S / L_{\mathrm{abs}})$ where L_{abs} is the absorption length, and S is, in the general case, the fluctuating length of the quantum path. To estimate the diffusive absorption length, this factor must be averaged over S. If we neglect fluctuations of S, then we can set $S \sim NL_{sc}$, where N is the number of steps, and L_{sc} is the step length. On the other hand, the root-mean-square photon path is estimated as

$$\bar{L} = \left\langle x^2 \right\rangle^{1/2} \sim \sqrt{\frac{NL_{sc}^2}{3}},$$

whence we have $S \sim 3\bar{L}^2 / L_{sc}$, so that $\exp(-S / L_{abs}) = \exp-(L / l_a)^2$, where $l_a = \sqrt{L_{abs} L_{sc} / 3}$ has the meaning of the absorption length in the diffusion regime.

This estimate yields a correct value for l_a which coincides with the result (4.3.22). However, the absorption factor so obtained proves to be valid only for $\bar{L} \leq l_a$. For larger $\bar{L}$, the correct absorption factor has the form $\exp(-\bar{L} / l_a)$ [see eqn (4.3.21)] and can be obtained only with a more complicated model that takes into account the fluctuations of S.

Problem 4.14. Using the modified Born approximation, write the intensity of the scattered field for a plane wave incident on a statistically uniform scattering layer (Fig. 4.8).

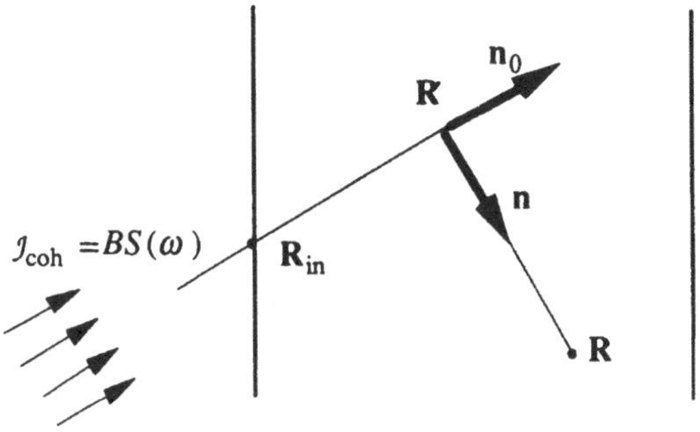

Figure 4.8. Incidence of a plane wave on a scattering layer.

Solution. Let us write an incident wave intensity as

$$\mathcal{I}_{coh} = BS(\omega'), \tag{1}$$

and neglect reflection on the boundary. Within the layer we have

$$\mathcal{I}_{coh}(R', \omega') = BS(\omega')Q(\mathbf{R}_{in} - \mathbf{R}'), \tag{2}$$

where $Q(R)$ is the attenuation factor

$$Q(R) = e^{-\alpha|R|}, \tag{3}$$

and $\mathbf{R}_{in}$ is the point where the ray enters the layer (Fig. 4.8). In a homogeneous layer, rays are not curved, therefore the factor $|d\Sigma'/d\Omega_{\mathbf{n}}|^{-1}$ in eqn (4.4.17) reduces to $|\mathbf{R}-\mathbf{R}'|^{-2}$, and we obtain the expression for the intensity of the field scattered at $\mathbf{R}$

$$\mathcal{I}_{sc}^{(1)}(\mathbf{R}) = B\int d\omega \int \frac{d^3R'}{|\mathbf{R}-\mathbf{R}'|^2} Q(\mathbf{R}_{in}-\mathbf{R}')Q(\mathbf{R}'-\mathbf{R})$$
$$\times \int d\omega' \sigma(\mathbf{R}';\mathbf{n}(\mathbf{R}'),\omega \leftarrow \mathbf{n}_0,\omega')S(\omega'). \tag{4}$$

Problem 4.15. In the single-scattering approximation, evaluate the Doppler-like shift of spectral lines in scattering from random inhomogeneities with the general Gaussian space-time correlation function (James and Wolf, 1990). Assume that the scattering volume is small compared to the extinction length and distances from the source and the receiver.

Solution. Let the scattering occur in a compact scattering volume (Fig. 4.9) and $S(\omega')$ be the spectral density of the incident wave. Since we are interested in the spectral composition of the scattered field $S_{sc}(\omega')$, we take only the inner part of the double integral (4.4.13)

$$S_{sc}^{(1)}(\omega) = c\int d\omega' \sigma(\mathbf{n},\omega \leftarrow \mathbf{n}_0,\omega')S(\omega'). \tag{1}$$

The scattering cross section σ can be expressed through the spectral density of the fluctuation of permittivity $\Phi_\varepsilon(\mathbf{q},\omega-\omega',\omega)$:

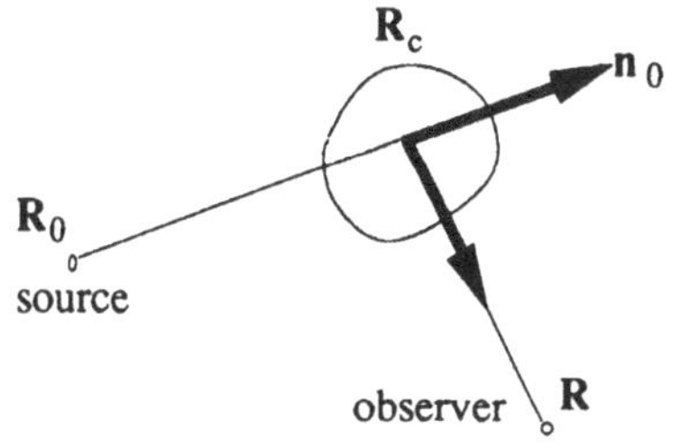

Figure 4.9. Scattering at space-time inhomogeneities within a small scattering volume leads to Doppler-like shift of spectral lines (after James and Wolf, 1990).

$$\sigma(\mathbf{n},\omega \leftarrow \mathbf{n}_0,\omega') = \frac{\pi\omega^4}{2c^4}\Phi_\varepsilon(\mathbf{q},\omega-\omega',\omega).$$

$\Phi_\varepsilon(\mathbf{q},\Delta\omega,\omega)$ is in turn related to the correlation function of the permittivity fluctuation at a frequency ω through the space-time Fourier transform:

$$\Phi_\varepsilon(\mathbf{q},\Delta\omega,\omega) = \frac{1}{(2\pi)^4}\int d^3R\int dT\,\psi_\varepsilon(\mathbf{R},T,\omega)\exp[i(\mathbf{qR}-\Delta\omega T)].$$

Here, the scattering vector $\boldsymbol{q}$ is the difference between the wave vector of the incident wave $\mathbf{k}^{\mathrm{inc}} = (\omega'/c)\mathbf{n}'$ and the wave vector of the scattered wave $\mathbf{k}^{\mathrm{sc}} = (\omega/c)\mathbf{n}$: $\mathbf{q} = \mathbf{k}^{\mathrm{sc}} - \mathbf{k}^{\mathrm{inc}} = (\omega\mathbf{n} - \omega'\mathbf{n}')/c$.

In the case of a monochromatic signal, when $S(\omega') = \delta(\omega'-\omega_0)$, formula (1) takes a simple form:

$$S_{\mathrm{sc}}^{(1)}(\mathbf{r},\omega) = A\omega^4\Phi_\varepsilon(\mathbf{q},\Delta\omega,\omega_0), \tag{2}$$

where $A = \mathrm{const}$. For the anisotropic Gaussian correlation function

$$\psi_\varepsilon(\mathbf{R},T,\omega) = \langle\tilde{\varepsilon}^2\rangle\exp\left[-\frac{1}{2}\left(\frac{X^2}{\sigma_x^2}+\frac{Y^2}{\sigma_y^2}+\frac{Z^2}{\sigma_z^2}+\frac{c^2T^2}{\sigma_\tau^2}\right)\right],$$

we have

$$\Phi_\varepsilon(\mathbf{q},\Delta\omega,\omega) = \Phi_0\exp\left[-\frac{1}{2}(\sigma_x^2q_x^2+\sigma_y^2q_y^2+\sigma_z^2q_z^2+\frac{\sigma_\tau^2(\Delta\omega)^2}{c^2}\right],$$

where $\Phi_0 = \dfrac{\langle\tilde{\varepsilon}^2\rangle c}{(2\pi)^2\sigma_x\sigma_y\sigma_z\sigma_\tau}$.

We derive the terms containing ω^2, $\omega\omega'$, and ω'^2 in the exponent to represent the spectrum Φ_ε in the form

$$\Phi_\varepsilon(q,\Omega,\omega) = B\exp\left[-\frac{1}{2}\left(\alpha'\omega'^2 - 2\beta\omega\omega' + \alpha\omega^2\right)\right]. \tag{3}$$

Here

$$\alpha = \frac{1}{c^2}(\sigma_x^2n_x^2+\sigma_y^2n_y^2+\sigma_z^2n_z^2+\sigma_\tau^2),$$

$$\alpha' = \frac{1}{c^2}(\sigma_x^2 n_x'^2 + \sigma_y^2 n_y'^2 + \sigma_z^2 n_z'^2 + \sigma_\tau^2), \tag{4}$$

$$\beta = \frac{1}{c^2}(\sigma_x^2 n_x n_x' + \sigma_y^2 n_y n_y' + \sigma_z^2 n_z n_z' + \sigma_\tau^2).$$

Finally, substituting eqn (3) into eqn (2) and rearranging the terms in the exponent, we obtain

$$S_{\mathrm{sc}}^{(1)}(\mathbf{r},\omega) = A\Phi_0\omega^4 \exp\left(-\frac{1}{2}\omega_0^2 \frac{\alpha\alpha' - \beta^2}{\alpha}\right)\exp\left[-\frac{\alpha}{2}\left(\omega - \frac{\beta}{\alpha}\omega_0\right)^2\right]. \tag{5}$$

We see that the scattered field is a Gaussian peak with central frequency $\overline{\omega} = (\beta/\alpha)\omega_0$ and width $\Gamma = 1/\sqrt{\alpha}$.

The coefficient β may take on either positive or negative values. When the spectrum is interpreted in terms of only positive frequencies, $|\beta|$ should be substituted into eqn (5) for β: $\overline{\omega} = (|\beta|/\alpha)\omega_0$.

It follows from eqn (5) that the relative frequency shift $z = (\omega_0 - \overline{\omega})/\omega_0$ is equal to $z = (\alpha/|\beta|) - 1$. As in the case of the Doppler effect, the relative shift z is independent of frequency and may take on values from -1 to ∞. The similarity with the Doppler effect enabled James and Wolf (1990) to speak about the Doppler-like shift of spectral lines. We show some other features of the *Wolf effect*.

The pre-exponent factor ω^4 is responsible for a frequency-dependent change in the line intensity that shifts the source to the blue side. In the first approximation, ω may be taken equal to $\overline{\omega}$.

For an isotropically scattering medium, when $\sigma_x = \sigma_y = \sigma_z = \sigma$, eqns (4) become $\alpha = \alpha' = (1/c^2)(\sigma^2 + \sigma_\tau^2)$. In the general case if the forward scattering is excluded ($\mathbf{n}'$ is not parallel to $\mathbf{n}$), the quantity $\beta = (1/c^2)(\sigma^2 \mathbf{n}\cdot\mathbf{n}' + \sigma_\tau^2)$ does not exceed α. The relative shift z is therefore always positive; that is, the scattering necessarily produces a *red shift*. Although this effect is not related to the expansion of the universe it should be kept in mind when interpreting data of astrophysical experiments, for example, the scattering of light by interstellar clouds of hot gases.

The magnitude of the relative shift z depends on the relationship between σ_τ and σ. At a slow motion of inhomogeneities ($v << c$) when $\sigma_\tau \approx c\sigma/v >> \sigma$, the shift z is small and comparable in the order of magnitude with $(v/c)^2$. Conversely, at relativistic velocities $v \to c$, the shift z is comparable to unity.

Problem 4.16. With reference to Problem 4.7, find the scattering function $S(\mu,\mu_0)$ which describes the diffuse reflection from a half-space with isotropic scatterers (Hapke, 1981). We recommend to split the radiance into the coherent and scattered components and use the two-flux approximation to describe the scattered component.

Solution. The scattering function $S(\mu,\mu_0)$ describes the diffuse reflection of a radiation with radiance $J_0\delta(n_z+\mu_0)2\pi^{-1}$ incident on a scattering half-space; it is defined by the expression

$$I^{\text{refl}}(\mu)=\frac{1}{4\pi\mu}S(\mu,\mu_0)J_0, \tag{1}$$

where I_μ^{refl} is the radiance of the reflected wave, $\mu=n_z$, $\mu_0=|n_0^z|$, and the problem geometry is similar to that shown in Fig. 4.1 (the origin must be put at $z=L_0$ and L allowed to go to infinity).

In the case of isotropic scattering, the attenuated incident component is sharply anisotropic. However, even the singly scattered radiation has a significantly smoother indicatrix enabling us to use the two-flux approximation to describe the scattered component.

Write the radiance I as the sum $I=I^{\text{coh}}+I^{\text{incoh}}$, where I^{coh} and I^{incoh} satisfy the equations

$$(\mu\partial_z+\alpha)I^{\text{coh}}=0, \tag{2}$$

$$(\mu\partial_z+\alpha)I^{\text{incoh}}=\sigma J. \tag{3}$$

Here

$$J=\int(I^{\text{coh}}+I^{\text{incoh}})d\Omega_{\mathbf{n}}=J^{\text{coh}}+J^{\text{incoh}} \tag{4}$$

is the total intensity and J^{coh} and J^{incoh} correspond to contribution from I^{coh} and I^{incoh}, respectively.

Equation (2) for I^{coh} completed by the initial condition

$$I^{\text{coh}}\Big|_{z=0}=J_0\delta(n_z+\mu_0)2\pi^{-1}$$

is easily solved:

$$I^{\text{coh}}=\frac{J_0}{2\pi}\delta(n_z+\mu_0)\exp(\alpha z/\mu_0). \tag{5}$$

Thus we obtain the attenuated coherent component

$$J^{\text{coh}}=\int I^{\text{coh}}d\Omega_{\mathbf{n}}=\int_{-1}^{1}\int_{0}^{2\pi}I^{\text{coh}}d\mu d\varphi=J_0\exp\left(\frac{\alpha z}{\mu_0}\right). \tag{6}$$

To solve the transfer equation (3) we use the two-flux approximation. Integrating both sides of this equation over two hemispheres $n_z > 0$ and $n_z < 0$ and neglecting the anisotropy of I^{incoh}, that is, the dependence of I^{incoh} on μ, we obtain a system of two-flux equations similar to eqn (4.5.4):

$$\begin{aligned} \left(\frac{1}{2}\partial_z + \alpha\right)J_+ &= 2\pi\sigma(J_+ + J_- + J^{\text{coh}}), \\ \left(-\frac{1}{2}\partial_z + \alpha\right)J_- &= 2\pi\sigma(J_+ + J_- + J^{\text{coh}}), \end{aligned} \tag{7}$$

for

$$J_\pm = \int_{0<n_z>0} I^{\text{inc}} d\Omega_{\mathbf{n}}. \tag{8}$$

The set (7) is augmented by the evident boundary conditions

$$\begin{aligned} J_{-(z=0)} &= 0, \\ J_{+(z=\infty)} &= 0, \end{aligned} \tag{9}$$

and is easily solved using eqn (6). As a result, we find the total energy density

$$\begin{aligned} J &= J_+ + J_- + J^{\text{coh}} \\ &= \frac{J_0}{1-\lambda^2\mu_0^2}\left((2-\lambda)(1+2\mu_0)\mu_0 e^{\lambda\alpha z} + (1-4\mu_0^2)e^{\alpha z/\mu_0}\right), \end{aligned} \tag{10}$$

where $z = 0$, $\lambda = 2\sqrt{1-w_0}$, and $w_0 = 4\pi\sigma/\alpha$ is the albedo.

On the other hand, the total radiance I satisfies the equation

$$(\mu\partial_z + \alpha)I = \sigma J, \tag{11}$$

equivalent to eqns (2) and (3), and is expressed as

$$I(\mu, z) = \int_{-\infty}^{z} \frac{\sigma}{\mu}\exp\left(-\frac{\alpha(z-z')}{\mu}\right)J(z')dz'. \tag{12}$$

Substituting here eqn (10) for $J(z)$ and using eqn (1), we obtain finally

$$S(\mu,\mu_0) \equiv \frac{4\pi\mu}{J_0} I(\mu, z=0) = \frac{4\pi\sigma}{J_0} \int_{-\infty}^{0} \exp\left(\frac{\alpha z'}{\mu}\right) J(z')dz'$$

$$= \frac{w_0 \mu \mu_0}{\mu + \mu_0} H(\mu) H(\mu_0), \tag{13}$$

where

$$H(\mu) = \frac{1+2\mu}{1+2\sqrt{1-w_0}\mu}. \tag{14}$$

This expression for $S(\mu,\mu_0)$ differs from the exact relationship (2) of Problem 4.7 only in a form of functions $H(\mu)$ which are determined by a nonlinear integral equation in the exact theory. A comparison of eqn (4) with the exact $H(\mu)$, tabulated by Chandrasekhar (1960), shows the approximation (14) is in error by not more than 3%.

5. Radiative Transfer and Coherent Effects

In this chapter, we consider the most important corollaries of the statistic, wave approach to radiative transfer theory. They stem from the correlation approach of radiance and lie outside the scope of the simple energy-balance interpretation of the radiation transfer equation.

Section 5.1 deals with the coherent backscatter enhancement effect (sometimes called the weak localization effect), which occurs when radiation is scattered back to the source. This effect corresponds to the region of weak disorder, where the wavelength λ is small in comparison with the mean free path of radiation, $\lambda << L_{sc}$, and owes its existence to the symmetry of wave equations with respect to time reversal. Mathematically, the description of a weak localization reduces to augmenting the solution of the transfer equation by the contribution from the so-called cyclic diagrams, that, in turn, are expressed through the solution of the transfer equation.

Weak localization is not the only possible cause of backscattering enhancement. Other enhancement mechanisms are described in Section 5.2, which is focused on the non-coherent mechanisms that do not involve directly the phase relationships of scattered waves.

Section 5.3 deals with the new correlation effects of radiative transfer theory that occur in the region of weak disorder where $\lambda << L_{sc}$. It elucidates various forms of memory effects, including intensity correlation, scalar, and polarization effects.

Section 5.4 is devoted to the description of scattering in the region of high disorder, where the Ioffe–Rigel condition $\lambda \sim L_{sc}$ holds. Here, the ordinary transfer theory fails, and a strong localization — the so-called Anderson's transition — may take place. No rigorous theory for such a transition has been developed, but a promising approach to its description seems to be the self-consistent theory based on a nonlinear approximation of the Bethe–Salpeter equation, which will be considered in this chapter.

5.1 COHERENT BACKSCATTERING ENHANCEMENT. WEAK LOCALIZATION

5.1.1 Coherent Channels of Backscattering Enhancement

Placing the receiver near the source of radiation brings about specific coherent backscattering channels not covered by radiative transfer theory. The origin of these channels may be explained as follows.

Consider static discrete scatterers S_1, S_2, ...,S_j that scatter one by one a monochromatic wave, radiated by a source at point O, so that the wave eventually returns to the receiver placed at the same point O, as shown in Fig. 5.1.

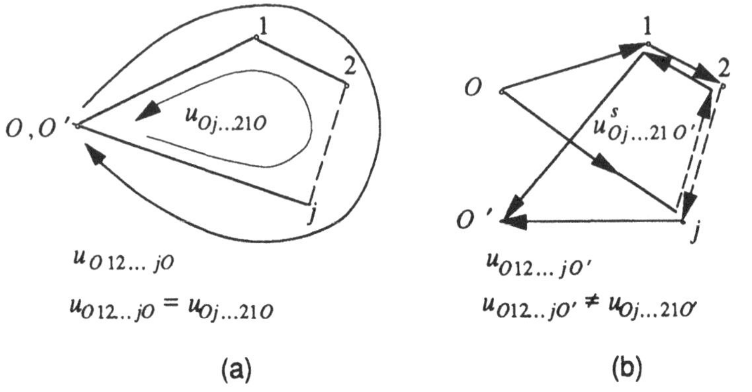

Figure 5.1. (a) Watson's coherent backscattering channels occur when the receiver and the source are at one point. (b) When the receiver and the source are spaced apart, the fields passed over direct (*12...j*) and inverse (*j...21*) sequences are no longer mutually coherent.

In a continuous scattering medium, points $s_1, s_2, ..., s_j$ may be viewed as centers of elementary scattering volumes dV_1, dV_2, $...dV_j$.

Let the scattered wave $u_{012...j0}$ correspond to the scattering sequence $Os_1s_2...s_jO$. By the reciprocity theorem, a wave passing the same scatterers in the inverse order $s_j, s_{j-1}, ..., s_2, s_1$ produces the same field:

$$u_{012...j0} = u_{0j...210} \,. \tag{5.1.1}$$

We will call a sequence of scattering events at $s_1 \to s_2, ..., \to s_j$ a *scattering channel.* With this convention, the relationship (5.1.1) tells us that the direct and inverse channels contribute equally in the scattered field, or that these scattering channels are mutually coherent.

Coherent backscattering channels will be referred to as Watson's channels, in deference to Watson (1969) who was the first to demonstrate their existence in radiation transfer theory. It is important to emphasize that mutual coherence of the fields $u_{012...j0}$ and $u_{0j...210}$ holds for any arrangement of scatterers however complex.

Equation (5.1.1) holds in view of the reciprocity theorem, i.e. in view of the system symmetry relative to time reversal. All factors destroying such symmetry (motion of scatterers, gyrotropy of the medium, etc.) invalidate the relationship (5.1.1) and, consequently, the mutual coherence of Watson's channels.

The coherence of fields scattered forward and backward is characteristic only of situations where the positions of the receiver and the source coincide. If receiver O' and source O are separated in space, as in Fig. 5.1b, then, beginning from a certain separation OO', the fields $u_{O12...jO'}$ and $u_{Oj...21O'}$ loose mutual coherence.

$$u_{O12...jO'} \neq u_{Oj...21O'} \,. \tag{5.1.2}$$

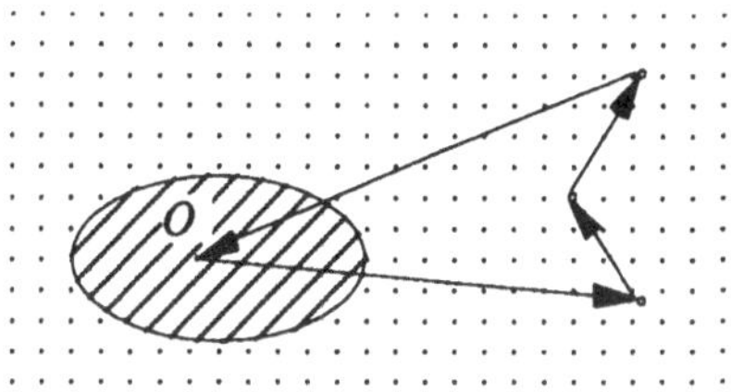

Figure 5.2. Zone of coherence of backscattering channels is formed around the source O. Within this zone, transfer theory is not applicable directly.

Thus, the source is surrounded by a certain zone where coherent backscattering channels can occur, as shown in Fig. 5.2.

Let us estimate the size of the mutual coherence zone assuming all scatterers and observation points are in the far (Fraunhofer) zone relative each other. Digressing for a moment from the slow-varying amplitude factors, we consider each scattering event as a change in the direction of propagation of the plane wave. The mutual coherence is associated with the coincidence of the phase increments for the waves passing a sequence of scatterers in forward and backward directions (Akkerman et al., 1986; Kaveh et al., 1986).

For simplicity, we consider the case of point scatterers (Fig. 5.3).

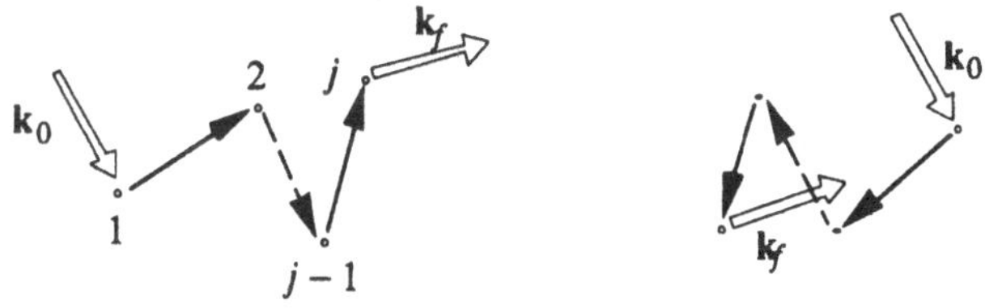

Figure 5.3. Determining the phase increments of a plane wave passed through Watson's scattering channels with point scatterers.

The phase increment of the wave $u_{O12...jO'}$ is

$$\psi_+ = \mathbf{k}_f(\mathbf{r} - \mathbf{r}_j) + \varphi_+ + \mathbf{k}_0\mathbf{r}_1, \tag{5.1.3}$$

where $\mathbf{k}_0$ and $\mathbf{k}_f$ are the wave vectors of the incident and scattered waves, and φ_+ is the arbitrary phase depending upon the properties of scatterers. Similarly, for the phase of the wave $u_{Oj...21O'}$ passed this sequence in the inverse order, we have

$$\psi_- = \mathbf{k}_f(\mathbf{r} - \mathbf{r}_1) + \varphi_- + \mathbf{k}_0\mathbf{r}_j, \tag{5.1.4}$$

where $\varphi_- = \varphi_+$ is the same phase as in eqn (5.1.3), in accordance with the reciprocity theorem. From these relationships we have for the phase difference

$$\delta\psi = \psi_+ - \psi_- = (\mathbf{k}_f + \mathbf{k}_0)(\mathbf{r}_1 - \mathbf{r}_j). \tag{5.1.5}$$

Here, $\mathbf{r}_1$ and $\mathbf{r}_j$ are the vectors of the first and last scatterers in the sequence under consideration.

For backward scattering, $\mathbf{k}_f = -\mathbf{k}_0$, the phase difference (5.1.5) becomes zero, and both considered waves found themselves in phase; i.e., they are mutually coherent for any arrangement of scatterers 1, 2, ..., j. For other directions, $\mathbf{k}_f \neq -\mathbf{k}_0$, the phase difference $\delta\psi$ depends only on the difference $\mathbf{l} = \mathbf{r}_1 - \mathbf{r}_j$ of the vectors of the initial and final scatterers.

From the point of view of classic transfer theory, it is unusual that in backscattering ($\mathbf{k}_f = -\mathbf{k}_0$) Watson's pairs are in phase whatever the length of the spatial loop $\mathbf{r}_1$, $\mathbf{r}_2$, ...$\mathbf{r}_j$, along which the wave can acquire an arbitrary phase increment (5.1.3) or (5.1.4). Thus, in this case, we speak of the mutual coherence of random waves $u_{012...jO}$ and $u_{Oj...210}$, rather than of more conventional coherence treated as the absence of elements of randomness.

This phenomenon of the mutual coherence of Watson's pairs is usually referred to as the *weak localization* of scattered waves, in contrast to the strong, or Anderson, localization which we shall touch briefly in Section 5.4. The term "weak localization" originates from solid-state physics, where the mutual coherence of Watson's channels was investigated for electrons interacting with impurities in metals. Inclusion of weak localization in the analysis results in a decrease in the metal conductivity at low temperatures as compared to that predicted by classic kinetic theory (Altschuler et al., 1982).

If we regard the vector $\mathbf{l} = \mathbf{r}_1 - \mathbf{r}_j$, completing the loop 1, 2, ..., j, as a random variable — what is almost always true in scattering media — then estimating the coherence region from the condition

$$|\delta\psi| = |(\mathbf{k}_f + \mathbf{k}_0)\mathbf{l}| \le 1 \tag{5.1.6}$$

and neglecting an occasional equal-phase situation of orthogonal vectors $\mathbf{k}_j = \mathbf{k}_0$ and $\mathbf{l}$ we have

$$k \cdot l \cdot 2\sin\frac{\vartheta}{2} \le 1. \tag{5.1.7}$$

Here, $k = 2\pi/\lambda$ is the wave number, ϑ is the angle between the wave vector of the scattered wave and the direction toward the source, and $l = |\mathbf{r}_1 - \mathbf{r}_j|$. According to eqn (5.1.7) the angle of backscatter coherence ϑ_{coh}, within which Watson's pairs remain coherent, is estimated by

$$\vartheta_{coh} \sim 1/kl. \tag{5.1.8}$$

This relationship gives simple estimates of the coherence angle for many applications without having to solve the scattering problem.

We now give a physical description of the formation of angular coherence by contributions of coherent channels with different characteristic dimensions (scales).

According to eqn (5.1.8), extended scattering loops, i.e. large scales l, correspond to small coherence angles, and vise versa: short scattering loops (small l) correspond to large angles ϑ_{coh}. One may estimate characteristic values of l from physical considerations.

As an example, we consider the reflection of a plane wave from a weakly scattering half-space whose extinction length L_{ext} far exceeds the wavelength: $L_{ext} >> \lambda$. It is quite natural to estimate the characteristic distance l^* between the points where the ray enters and leaves the medium by the relation $l^* \sim L_{ext}$, whence for the angle of coherence ϑ_{coh} we have $\vartheta_{coh} \sim 1/kL_{ext}$.

In the other limiting case, when the characteristic size of a scattering volume L^* is small compared to the extinction length, the scale l^* is obviously of the order of L^*, and so $\vartheta_{coh} \sim 1/kL^*$.

These simple considerations may be detailed to some extent by estimating the contributions from the multiple scattering loops. For example, for double scattering in an extended medium, $l \sim L_{ext}$, so that the coherence angle $\vartheta_{coh} \sim 1/(kL_{ext})$ may be associated mainly with double scattering (Barabanenkov, 1973). The contributions due to multiple scattering (with large order $j >> 1$) may be estimated in the diffusion approximation (Section 4.3.3) assuming that $l_j^2 = (\mathbf{r}_1 - \mathbf{r}_j)^2$ increases in accordance with the diffusion spreading law $l_j^2 \sim DL_j/c$, where the diffusion coefficient $D \sim L_{ext}c/3$ is given by eqn (4.3.14), and the length of the loop is estimated as $L_j \sim jL_{ext}$. These scattering events lead to

$$\vartheta_{coh,j} \sim \frac{1}{k\sqrt{L_{ext}L_j}} \sim \frac{1}{kL_{ext}\sqrt{j}}.$$

Bounding the scattering volume breaks down first of all the channels of high scattering multiplicity, which affects mainly the coherent properties at small angles ϑ_{coh}.

Naturally, these simple estimates tell us nothing about the mechanism in which the mutual coherence affects the observables. We look into this topic in more detail.

5.1.2 Backscattering Enhancement

The coherent backscattering enhancement caused by the mutual coherence of Watson pairs may be observed using relatively simple techniques and is the most striking manifestation of weak localization in the field of classical waves. To consider the origin of this effect, we will adhere, for the sake of definiteness, to the model of discrete scatterers.

We represent the scattered field u^{sc} as a series expansion in scattering orders

$$u^{sc}(\mathbf{r}) = \sum u^{(1)}(\mathbf{r}) + \sum u^{(2)}(\mathbf{r}) + \sum u^{(3)}(\mathbf{r}) + \ \dots , \tag{5.1.9}$$

where the summation covers all possible combinations of scatterers responsible for singly, doubly, and multiply scattered fields. When the point of observation $\mathbf{r}$ approaches the source $\mathbf{r}_0$, the sum (5.1.9) acquires coherent Watson's channels.

We now combine all scattering channels in pairs of the form (5.1.1) and denote the respective fields as u_γ for the forward channel $Os_1s_2...s_jO'$ and $u_{-\gamma}$ for backward channel $Os_js_{j-1}...s_1O'$. The right-hand side of (5.1.9) contains not only such mutually adjoint (coupled) pairs but also single (uncoupled) channels, whose field will be denoted by $u_{\gamma'}$. The channels whose forward propagation coincides with backward propagation:

$$s_1,\ s_2,\ ...,s_j = s_j,\ s_{j-1},\ ...,s_2,\ s_1. \tag{5.1.10}$$

will be also categorized as uncoupled channels. For $j > 1$, these sequences are typical of the model of discrete scatterers. The uncoupled channels always consist of an odd number of scattering events $j = 2m+1$, because scatterer s_{2m+1} coincides with s_1, s_{2m} with s_2, and so on, as illustrated in Fig. 5.4.

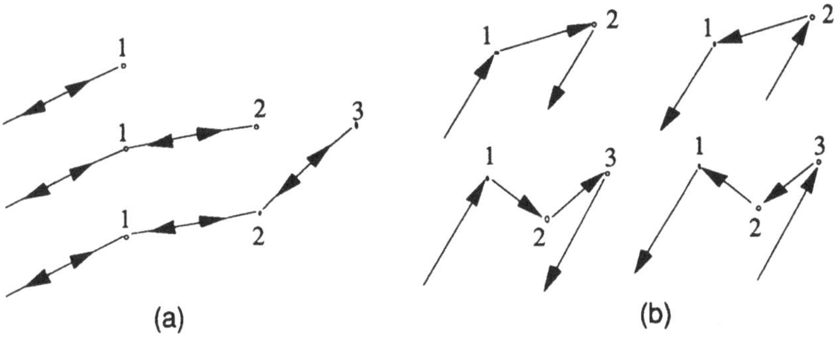

Figure 5.4. Examples of (*a*) uncoupled and (*b*) coupled channels for the model of discrete scatterers.

Scattering sequences of type *O1O, O121O, O12321O*, and so on refer to uncoupled channels. Single scattering plays the dominant role among uncoupled channels, therefore

$$\sum u_{\gamma'} \approx \sum u^{(1)}. \tag{5.1.11}$$

We note that the model of a continuous medium with smooth distributions of fluctuations has no uncoupled channels of the order $j \geq 2$, because, here, diagrams $1 \to 2 \to 1$ give way to diagrams $1 \to 2 \to 3$ (see Fig. 5.4). A transition from the continuous model to the discrete model is possible but calls for singular distributions with delta functions enabling reversible channels of the type $1 \to 2 \to 1$.

We now represent the scattered field as the sum of uncoupled and coupled components

$$u^{sc} = \sum u_{\gamma'} + \sum \left(u_\gamma + u_{-\gamma}\right). \tag{5.1.12}$$

Averaging the squared module of this expression yields

$$\left\langle I^{\mathrm{sc}}\right\rangle = \left\langle \left|u^{\mathrm{sc}}\right|^2\right\rangle = \langle I'\rangle + \langle I_+\rangle + \langle I_-\rangle + 2\,\mathrm{Re}\sum\langle u_\gamma u_{-\gamma}^*\rangle, \tag{5.1.13}$$

where

$$\langle I'\rangle = \sum\langle |u_{\gamma'}|^2\rangle, \quad \langle I_\pm\rangle = \sum\langle |u_{\pm\gamma}|^2\rangle. \tag{5.1.14}$$

Here we took into account that the uncoupled components $u_{\gamma'}$ are incoherent with respect to all coupled components u_γ and $u_{-\gamma}$, so that

$$\left\langle u_{\gamma'} u_\gamma^*\right\rangle = \left\langle u_{\gamma'} u_{-\gamma}^*\right\rangle = 0.$$

If the receiver O' is in the coherent zone near the source O, $\mathbf{r} \approx \mathbf{r}_0$, then $\langle I_+\rangle = \langle I_-\rangle = \mathrm{Re}\sum\left\langle u_\gamma u_\gamma^*\right\rangle$ and eqn (5.1.13) takes the form

$$\left\langle I^{\mathrm{sc}}\right\rangle \approx \left\langle I_{\mathrm{bsc}}\right\rangle = \langle I'\rangle + 2\langle I_+\rangle + 2\,\mathrm{Re}\sum\left\langle u_\gamma u_\gamma^*\right\rangle = \langle I'\rangle + 4\langle I_+\rangle. \tag{5.1.15}$$

Beyond this coherence zone, Watson's pair coherence breaks down, and $\langle u_\gamma u_{-\gamma}^*\rangle \approx 0$. If the receiver and the source are within a comparatively small distance $\mathbf{r} - \mathbf{r}_0$ from one another, the intensities $\langle I_+\rangle$ and $\langle I_+\rangle$ may, as before, be considered approximately identical: $\langle I_+\rangle \approx \langle I_-\rangle$. Then eqn (5.1.13) yields

$$\left\langle I^{\mathrm{sc}}\right\rangle \approx \left\langle I_{\mathrm{sep}}\right\rangle = \langle I'\rangle + 2\langle I_+\rangle. \tag{5.1.16}$$

For monostatic observation ($\mathbf{r} = \mathbf{r}_0$), the intensity (5.1.15) is always greater than the intensity (5.1.16) for bistatic observation:

$$K_{\mathrm{bsc}} = \frac{\left\langle I_{\mathrm{bsc}}\right\rangle}{\left\langle I_{\mathrm{sep}}\right\rangle} = \frac{\langle I'\rangle + 4\langle I_+\rangle}{\langle I'\rangle + 2\langle I_+\rangle} > 1. \tag{5.1.17}$$

This inequality describes backscattering enhancement quantitatively, and K_{bsc} stands for the enhancement factor. This effect owes its existence to Watson's channels that form coherent pairs near the source. Moving the receiver away from the source breaks down the coherence of Watson's channels and decreases the average intensity of the scattered field.

It is important to note that inequality (5.1.17) relates quantities, averaged over a statistical ensemble, rather than separate realizations. Therefore, a well pronounced maximum in the backscattering direction may be observed only after statistical averaging, whereas, in separate realizations, the scattered intensity fluctuates strongly and results in speckle structures with maxima and minima, as shown in Fig. 5.5

(Etemad et al., 1986; Kaveh et al., 1986).

In the general case, the positions of these maxima vary randomly. The statistical averaging smoothes the pattern leaving a single maximum in the backscattering direction. In scattering from moving inhomogeneities, e.g. Brownian particles in suspensions, this averaging proceeds automatically because of averaging over observation time and the presence of temporal ergodicity. However, for scattering from static scatterers, an additional averaging should be made.

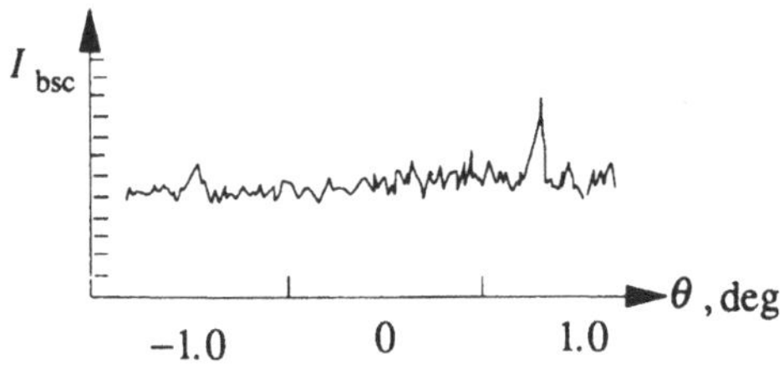

Figure 5.5. Intensity of backscattering from a rigid fluff of submicron SiO_2 beards in air (after Etemad, Thompson, and Andreico, 1986).

When the effect of uncoupled scattering channels is negligible $\langle I' \rangle << \langle I_+ \rangle$, or, what is the same, when the contribution of single scattering is small compared with multiple scattering, the enhancement factor K_{bsc} tends to 2. Thus, the enhancement factor depends upon the proportion between uncoupled and coupled channels and varies from unity, when single scattering dominates and $\langle I' \rangle >> \langle I_+ \rangle$, to two in the opposite case $\langle I' \rangle << \langle I_+ \rangle$:

$$1 < K_{bsc} < 2. \tag{5.1.18}$$

Let the source of radiation and the receiver be far from the scattering volume and the approximation (5.1.5) be valid in describing the phase difference between the direct and inverse scattering channels. Neglecting variations in the slow amplitude factors, we can rewrite (5.1.13) as

$$\langle I^{sc} \rangle \approx \langle I' \rangle + 2\langle I_+ \rangle \left(1 + \left\langle \cos\left[\left(\mathbf{k}_f + \mathbf{k}_0\right)\left(\mathbf{r}_1 - \mathbf{r}_j\right)\right]\right\rangle_\gamma\right), \tag{5.1.19}$$

where

$$\langle \ldots \rangle_\gamma = \sum P_\gamma \ldots . \tag{5.1.20}$$

Here, the quantity

$$P_\gamma = \langle |u_\gamma|^2 \rangle \big/ \sum \langle |u_\gamma|^2 \rangle \tag{5.1.21}$$

satisfies the conditions

$$P_\gamma \geq 0, \quad \sum P_\gamma = 1, \tag{5.1.22}$$

which allow us to treat P_γ as the probability to meet a Watson's pair with index γ.

We define the enhancement factor $K_{\text{enh}}(\theta)$ by

$$K_{\text{enh}}(\theta) = \frac{\left\langle I^{\text{sc}} \right\rangle}{\left\langle I_{\text{sep}} \right\rangle}, \tag{5.1.23}$$

where θ is the angle between $\mathbf{k}_f$ and the backscattering direction, and $\left\langle I_{\text{sep}} \right\rangle$ is given by eqn (5.1.16). In the approximation (5.1.19) we have

$$K_{\text{enh}}(\theta) = 1 + \frac{\left\langle \cos\left[\left(\mathbf{k}_f + \mathbf{k}_0\right)\left(\mathbf{r}_1 - \mathbf{r}_j\right)\right]\right\rangle_\gamma}{1 + \Delta}, \tag{5.1.24}$$

where

$$\Delta = \langle I' \rangle / 2\langle I_+ \rangle. \tag{5.1.25}$$

By virtue of eqn (5.1.11), $\langle I' \rangle$ is very close to the single scattering intensity $\left\langle I^{\text{single}} \right\rangle$, so that $2\langle I_+ \rangle$ represents the multiple scattering intensity calculated without allowance for the mutual coherence of Watson's pairs. Thus, we may rewrite (5.1.25) symbolically as

$$\Delta = \frac{\text{single scattering}}{\text{multiple scattering}}. \tag{5.1.26}$$

The formula (5.1.24) allows us to trace out the transition from the bistatic reception (5.1.16) to the monostatic reception (5.1.13). Indeed, for large values of θ, the factor $K(\theta) \approx 1$; whereas, for backscattering, $K(\theta) = 2 - \Delta/(1+\Delta)$. In the definition (5.1.23), the unknown quantity P_γ may be estimated using, e.g., the diffusion approximation (Akkerman, Wolf, and Maynard, 1986). A more rigorous approach, which we shall discuss in the ensuing section, leads to similar expressions for the enhancement factor.

5.1.3 Weak Localization and Radiative Transfer Theory

Radiative transfer theory does not allow for the coherence of Watson's channels because the radiative transfer equation is based on the incoherent summation of wave intensities. Watson (1969) noted the inapplicability of this equation near the source. Howerer, transfer theory may be modified to include the backscattering enhancement due to weak localization. In a simple form, such a modification is as follows.

Let $I_{\text{RTE}}(\mathbf{n}, \mathbf{R})$ be a solution of the radiative transfer equation for the radiance in the vicinity of a point source or in the direction opposite to the incidence of the primary wave. The contribution of Watson's pairs into the radiance may be taken into account by adding the intensity of multiple scattering $I_{\text{multiple}} = I_{\text{RTE}} - I_{\text{single}}$, i.e. by substituting:

$$I_{\text{RTE}} \to I_{\text{RTE}} + I_{\text{multiple}} = 2I_{\text{RTE}} - I_{\text{single}}, \tag{5.1.27}$$

where I_{single} is the radiance in the single scattering approximation.

In situations with predominantly multiple scattering and a small amount of single scattering, $I_{\text{single}} << I_{\text{RTE}}$, the relationship (5.1.27) becomes

$$I_{\text{RTE}} \to 2I_{\text{RTE}}, \tag{5.1.28}$$

that is, the inclusion of backscatter enhancement reduces to doubling the backscattering radiance calculated by transfer theory.

This result can be treated as an increase of the probability for the radiation to return back to the source, or, in terms of the diffusion approximation, as a hampering of the diffusion of radiation. The last phenomenon may be explained qualitatively as a decrease in the effective radiation diffusion coefficient D in comparison with its classical Boltzman's value (4.3.14) $D_{\text{B}} = cL_{\text{tr}}/3$. A simple estimation of this reduction, considered in Problem 5.1, gives

$$D = D_{\text{B}}\left[1 - c'(\lambda/L_{\text{sc}})^2\right], \tag{5.1.29}$$

where c' is a constant factor of order unity.

This estimator is valid under the condition of weak disorder $\lambda << L_{\text{sc}}$, when ordinary transfer theory is applicable. It has important corollaries in solid-state physics, where the diffusion of electrons is considered in place of the diffusion of photons (see, e.g., Lee and Ramakrishnan, 1985). In the field of classical waves, however, the decrease in the diffusion coefficient due to weak localization is more difficult to evaluate directly than the backscattering enhancement because the former effect is of the second order in the small parameter $\lambda / L_{\text{sc}} << 1$ and the latter is of the first order in this parameter.

A need to modify transfer theory for the case of backscattering follows from the early works of Barabanenkov (1973, 1975). His results are based on revising the Bethe–Salpeter equation with respect to backscattering and deserve a separate consideration. We trace out the transition from the enhanced radiance (5.1.27) to the nonenhanced radiance I_{RTE} outside the coherence zone using the diagram technique.

5.1.4 Diagram Interpretation of Backscattering Enhancement. Maximal Crossed Diagrams

Barabanenkov (1973, 1975) observed that, in the vicinity of the source, a new class of diagrams (which he called cyclic and which are known today as maximal crossed diagrams) appreciably contributed to the scattering along with ladder diagrams.

Maximal crossed diagrams reflect the appearance of Watson's coherent channels. In the vicinity of the source, the contribution of these diagrams coincides with the contribution of all ladder diagrams except for the first diagram which corresponds to single scattering. This result proves the correctness of the substitution (5.1.27). Far from the source, the contribution of cyclic diagrams becomes negligible, thus reflecting a breakdown of the coherence of Watson's scattering channels, and the backscattering enhancement vanishes.

All these qualitative considerations may be endowed with an exact quantitative meaning. To simplify the problem, we consider an isotropic point source of radiation embedded in a continuous scattering medium with Gaussian fluctuations. In wave terms, such a source is described by the delta function $\delta(\mathbf{r}-\mathbf{r}_0)$, whereas in transfer theory, it is described by the function $\delta(\mathbf{r}-\mathbf{r}_0)/4\pi$, which is independent of the direction $\mathbf{n}$ of the radiation flux. In this case, the solution to the transfer equation (Section 3.5) corresponds approximately to an infinite sum of the ladder sequence. This sum is the solution of the Bethe–Salpeter equation without an infinite set of diagrams with assumingly little contribution to the solution.

In the conditions of strong multiple scattering near the source, the ladder diagrams must be augmented with cyclic diagrams that have the form (Barabanenkov, 1973)

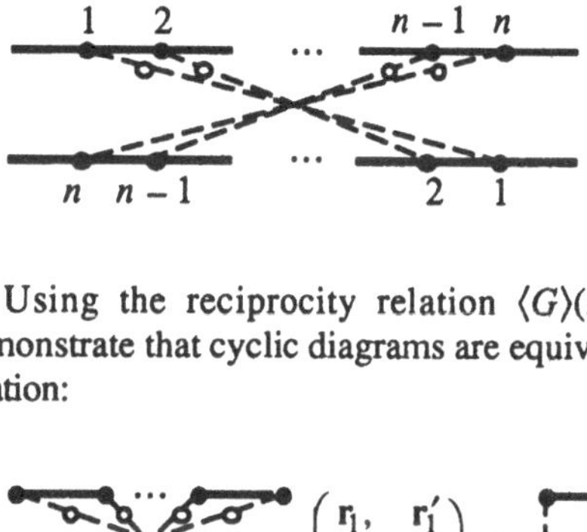

. (5.1.30)

Using the reciprocity relation $\langle G\rangle(r_1,r_2)=\langle G\rangle(r_2,r_1)$ it is not difficult to demonstrate that cyclic diagrams are equivalent to ladder diagrams in the sense of the relation:

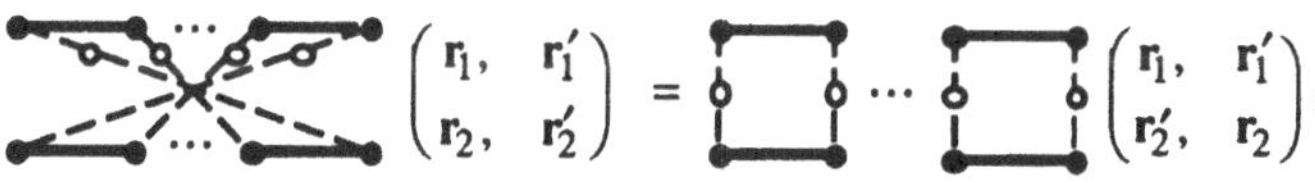

(5.1.31)

which shows the arguments corresponding to the end points of the diagrams. In other words, the propagation lines in these diagrams may be reversed so that the cumulant ties be retained, and this is achieved by interposing the respective arguments in the kernel.

In terms of operators, the relationship (5.1.31) means that if C is an operator corresponding to the left-hand diagram of (5.1.31) and L is an operator corresponding to the right-hand diagram, then

$$\mathrm{C}\begin{pmatrix}\mathbf{r}_1, & \mathbf{r}_1' \\ \mathbf{r}_2, & \mathbf{r}_2'\end{pmatrix}=\mathrm{L}\begin{pmatrix}\mathbf{r}_1, & \mathbf{r}_1' \\ \mathbf{r}_2', & \mathbf{r}_2\end{pmatrix}. \tag{5.1.32}$$

This property must obviously hold for the sum of all cyclic diagrams:

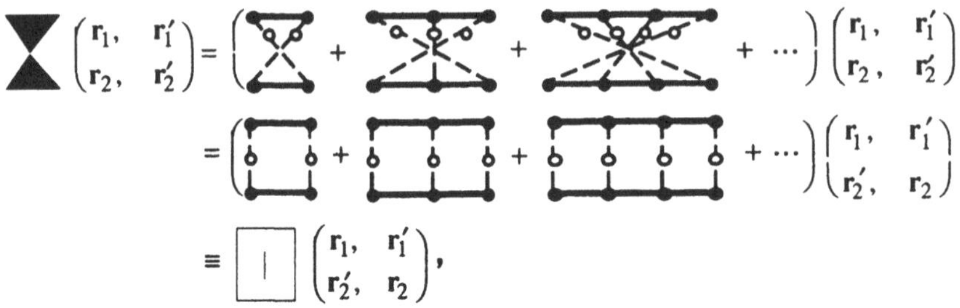

$$\blacktriangledown\!\!\blacktriangle\begin{pmatrix} \mathbf{r}_1, & \mathbf{r}_1' \\ \mathbf{r}_2, & \mathbf{r}_2' \end{pmatrix} = \left(\bowtie + \cdots + \cdots \right)\begin{pmatrix} \mathbf{r}_1, & \mathbf{r}_1' \\ \mathbf{r}_2, & \mathbf{r}_2' \end{pmatrix}$$

$$= \left(\;\square\; + \;\square\;\square\; + \;\square\;\square\;\square\; + \cdots \right)\begin{pmatrix} \mathbf{r}_1, & \mathbf{r}_1' \\ \mathbf{r}_2', & \mathbf{r}_2 \end{pmatrix}$$

$$\equiv \;\boxed{\ |\ }\;\begin{pmatrix} \mathbf{r}_1, & \mathbf{r}_1' \\ \mathbf{r}_2', & \mathbf{r}_2 \end{pmatrix},$$

(5.1.33)

where the black triangles denote the sum of cyclic diagrams and the final rectangle denotes the sum of ladder diagrams.

Using eqn (5.1.33), it is not difficult to demonstrate that the sum of cyclic diagrams of the form (5.1.30)

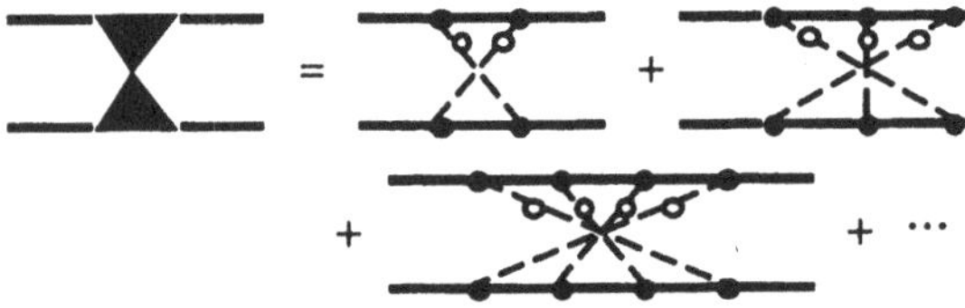

(5.1.34)

that appears in the expansion of the second moment, in the Fraunhofer approximation, is expressed in terms of the solution to the transfer equation. Indeed, in view of (5.1.33) we write

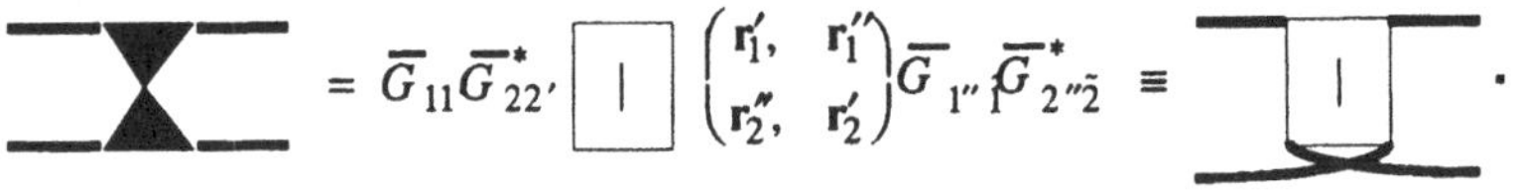

(5.1.35)

In the sum of ladder diagrams, the transfer to the Fraunhofer approximation may be represented schematically as follows:

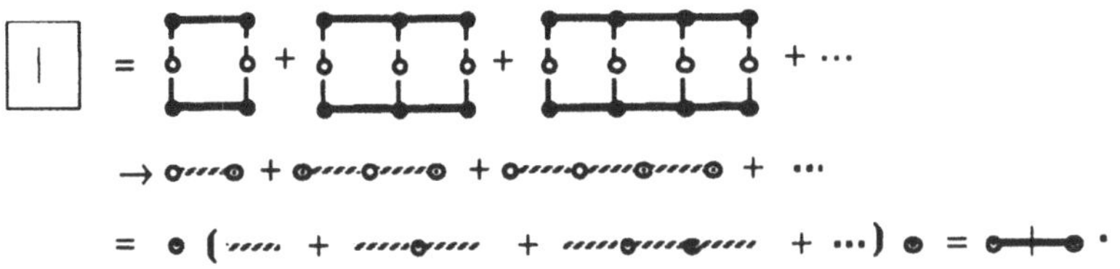

(5.1.36)

Here, we used the diagram representation of the solution to the transfer equation (3.6.27). In the Fraunhofer approximation, the right-hand side of (5.1.35) takes the form

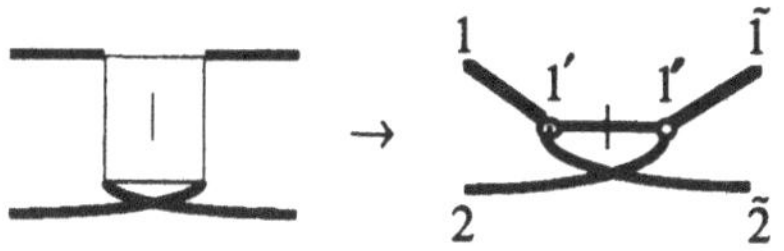

$$=\bar{G}_{11'}\bar{G}^*_{21''}$$

$$\times(4\pi)^2\sigma_{r_1'}(\mathbf{k}_{11'},-\mathbf{k}_{1'\tilde{2}}\leftarrow k^{\mathrm{eff}}\mathbf{n}',k^{\mathrm{eff}}\mathbf{n}')F(\mathbf{r}_1',\mathbf{n}';\mathbf{r}_1'',\mathbf{n}'')$$

$$\times(4\pi)^2\sigma_{r_1''}(k^{\mathrm{eff}}\mathbf{n}'',k^{\mathrm{eff}}\mathbf{n}''\leftarrow\mathbf{k}_{1''\tilde{1}},-\mathbf{k}_{21''})\bar{G}_{1''\tilde{1}}\bar{G}^*_{1'\tilde{2}}, \qquad (5.1.37)$$

where the integration is with respect to the primed variables.

To include cyclic diagrams in addition to ladder diagrams, we add the sum (5.1.34) into the expression for the second moment $\langle G_1G_2^*\rangle$. Then, instead of (3.6.25) and (3.6.26) we obtain (Apresyan, 1981a)

$\langle G_1G_2^*\rangle =$

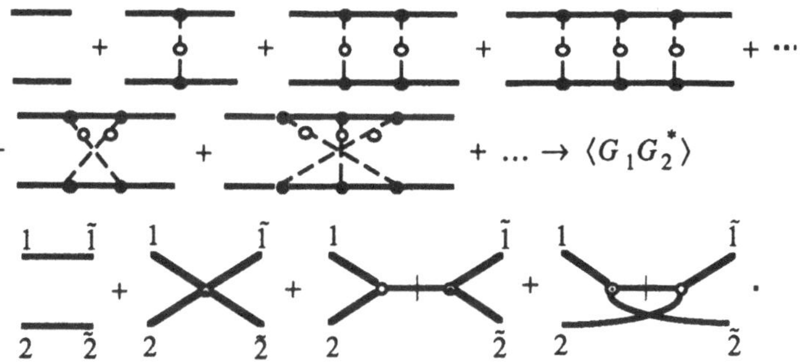

(5.1.38)

For convenience we have given the indices of the arguments in the operator kernel $\langle G_1G_2^*\rangle$.

Equation (5.1.38) shows the relation between the second moment of Green's function, that contains the contributions from ladder and cyclic diagrams, and the Green's function of the radiative transfer equation. This relation describes the general case of extended sources and separated points of observation.

We now illustrate the transition from (5.1.38) to a simpler calculation of intensity near a point source. When the source positions $\tilde{\mathbf{r}}_1$, $\tilde{\mathbf{r}}_2$ and observation points $\mathbf{r}_1$, $\mathbf{r}_2$ coincide, $\mathbf{r}_1=\mathbf{r}_2=\tilde{\mathbf{r}}_1=\tilde{\mathbf{r}}_2=\mathbf{r}$, which corresponds to the description of the backward scattered intensity, the lower branches in the last diagram of (5.1.38) may be first turned apart and then joined:

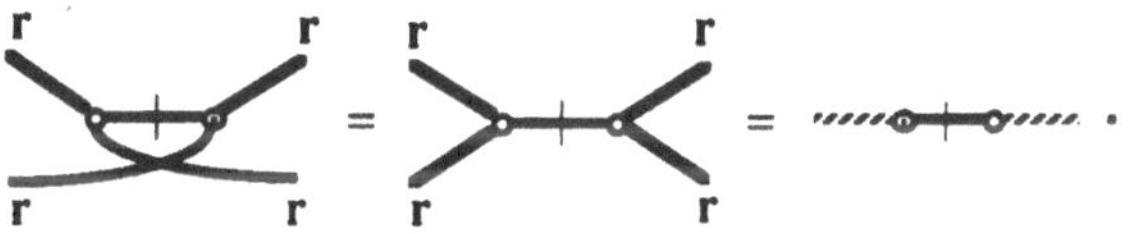

(5.1.39)

As a result we have

$$\langle G_1 G_2^* \rangle_{r_1 = r_2 = \tilde{r}_2 = \tilde{r}_2} =$$

(5.1.40)

Using eqn (5.1.32) we may redraw the right-hand side of this relation as

(5.1.41)

which is equivalent to the right-hand side of eqn (5.1.27) if we observe here the coherent (non-scattered) component of the radiation. The second, third, and forth components on the right-hand side of eqn (5.1.40) yield a diagrammatic analog of eqn (5.1.15) for the scattered intensity.

Equation (5.1.41) may be interpreted as follows. The intensity of backward radiation is equal to twice the intensity found from the transfer equation (first term on the right-hand side) minus the single-scattered field (third term). The second term of eqn (5.1.41) describes the propagation of a coherent wave and, for the backscattering direction, may be neglected in most problems. If the multiple scattered field exceeds the singly scattered field, one may neglect the two last terms in eqn (5.1.41); in this case eqn (5.1.41) gives the known result: the backscattered intensity is twice the intensity found from the transfer equation.

Thus, in the conditions of strong multiple scattering the solution of transfer equation is inapplicable near the source because of the backscattering enhancement effect. However, the results of transfer theory may be saved by modifying the solution to include the contributions of cyclic diagrams. This result of Barabanenkov (1973) is widely used, as well as his proposal to pass over, for this purpose, from the radiative transfer equation to the diffusion approximation.

5.1.5 Observation of Weak Localization and Other Results

In his pioneering work Watson (1969) discussed the effect of multiple scattering on the effective backscattering cross section of electrons in a randomly inhomogeneous plasma. De Wolf (1971) used this idea to estimate the backscattering enhancement for light and microwaves scattered by inhomogeneities of a turbulent atmosphere. However, the backscattering intensity was found to be very small and practically inaccessible for experimental measurements.

An allied enhancement effect of backscattering from bodies in a turbulent medium is much stronger. This effect was analyzed by Belenkii and Mironov (1972) and Vinogradov, Kravtsov, and Tatarskii (1973). It were the last authors who introduced the term backscattering enhancement. We will discuss this effect in more detail in Section 5.2.

Almost simultaneously, Barabanenkov (1973, 1975) considered the backscattering from a half-space filled with random inhomogeneities and showed the presence of a peak of scattering in the direction opposite to the incidence of the primary wave. Many results of the early Barabanenkov's papers on coherent backscattering enhancement were reiterated later by other researchers.

Enhanced backscattering from bodies in a turbulent medium was experimentally measured by Kashkarov and Gurvich (1977). As to the peak of backscattering from suspensions and emulsions predicted by Barabanenkov, its observation was retarded by almost ten years. It was observed only in 1984-1985 simultaneously by several research groups including Kuga and Ishimaru (1984); Kuga, Tsang, and Ishimaru (1985); Wolf and Maret (1985); Van Albada and Lagendijk (1985). These authors used the term *weak localization* by analogy with the related effect in solid state physics.

At present, the literature abounds in theoretical and experimental studies dealing with investigations of coherent optical reflection. The results cover the reflection from semi-infinite medium solved using in the diffusion approximation (Akkerman et al., 1986; Edrei and Kaveh, 1987) and with a more rigorous radiative transfer theory (Gorodnichev et al., 1989; Ozrin, 1992), an allowance for moving scatterers (Golubentsev, 1984a), the effect of a magnetic field in a gyrotropic medium (Golubentsev, 1984b), the effect of reflection from interfaces (Freund and Berkovitz, 1990), polarization effects (Wolf and Maret, 1985), to list but a few. The reader may found a detailed, though not comprehensive, list of pertinent references in a review by Barabanenkov, Kravtsov, Ozrin, and Saichev (1991).

We will not dwell on all the mentioned results and, to close the topic, we present only a simple estimator for the enhancement factor (5.1.24). This formula was derived by Akkerman, Wolf, and Maynard (1986) for reflection from a half-space with point isotropic scatterers using the diffusion approximation in the limiting case of $\Delta \cong 0$:

$$K(\theta) \approx 1 + \frac{1 - \exp(-3.4kL_{\text{ext}}|\theta|)}{3.4kL_{\text{ext}}|\theta|} . \tag{5.1.42}$$

This expression has a maximum at $\theta = 0$ that corresponds to the backscattering enhancement factor of two. At the tip, this peak has a form of a triangle with a characteristic width of about $1/kL_{\text{ext}}$ (Fig. 5.6). This triangular peak in the curve $K(\theta)$ is a characteristic feature of the reflecting half-space. Factors that limit the length of scattering loops and smooth out the peak include the finite thickness of the scattering slab, absorption losses, and such.

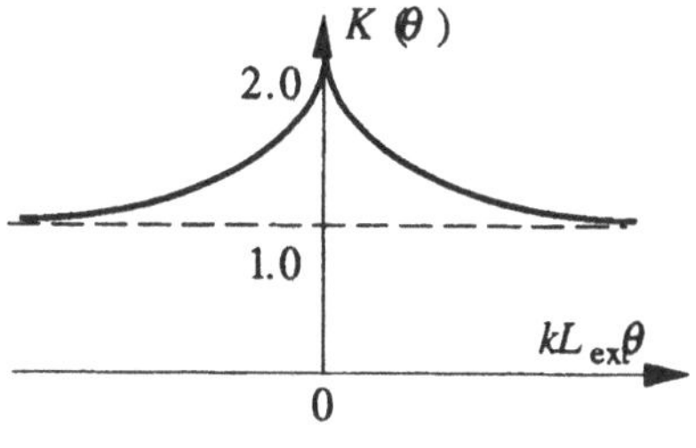

Figure 5.6. Triangular tip of a peak near the backscattering direction for a half-space with isotropic scatterers.

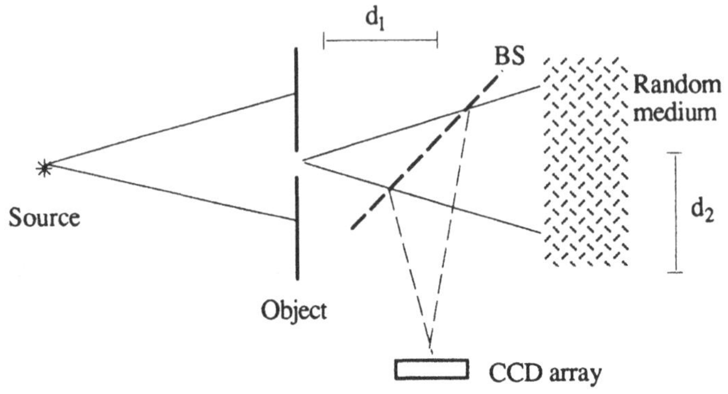

Figure 5.7. Experimental configuration for the observation of an image due to backscattering without any focusing optics. The image is best focused when $d_1 \approx d_2$ (after Rochon and Bissonnette, 1990).

Among experimental works on backscatter enhancement we mention an interesting paper by Rochon and Bissonnette (1990), where the backscattering enhancement effect was used to image a transparency without a lens (Fig. 5.7). In this experiment, a weak image of a transparency was recovered from a relatively uniform background of backscattered light (512 x 256 array of charge coupled devices was used as a detector). We should note, however, that the experimental configuration in this work cannot uniquely relate the observed backscattering enhancement to weak localization as it does in the above experiments. Other enhancement mechanisms may work here as will be described in Section 5.2.

5.2 INCOHERENT MECHANISMS OF BACKSCATTERING ENHANCEMENT

5.2.1 General Description of Physical Mechanisms in Backscattering Enhancement

The backscatter enhancement mechanism, based on the mutual coherence of Watson's channels as outlined in the preceding section, is not unique. Long before the coherent phenomenon of backscattering enhancement was discovered, the backscattering enhancement effect had been observed in astronomy, where it was called the opposition effect. For the first time it was observed as far back as the last century (Seeliger, 1895). Later it was observed for the reflection of solar radiation from the moon (Barabashev, 1922) and from other space objects. The opposition effect has long been described in astronomy textbooks in purely geometrical terms without such notions as coherence, phase, or wavelength. It is natural to call this enhancement mechanism *incoherent* in contrast to *coherent* effects such as weak localization.

Subsequent research demonstrated that backscattering enhancement has been a frequent phenomenon encountered in different forms and in different fields, and, therefore, it may be deemed universal in a certain sense. Following astronomy and optics of turbulent and turbid media, similar effects were observed in a laser sounding of the upper ocean layer (Vlasov, 1985), in reflections of light from different random surfaces (Oetking, 1966; Egan and Hilgeman, 1977), and in an ion sounding of the surface layers of solids (Pronko et al., 1979). Various mechanisms (both coherent and incoherent) were suggested to explain these effects. Some of them will be considered in this section following the approach of Apresyan (1993).

5.2.2 Statistical and Dynamic Enhancement Mechanisms

As a matter of terminological convention, we will use the term backscattering enhancement with reference to a robust peak around the direction to the source in the pattern of the scattered field, which persists when the position of the source is changed within wide limits. As a rule, this maximum denies description by conventional radiative transfer theory in which the multiple-scattered radiation "forgets" the direction of propagation of the primary wave.

We will call a backscattering enhancement mechanism *dynamic* if it gives rise to a pronounced backscattering maximum for each individual realization. If such a maximum appears only as a result of ensemble averaging, then the respective mechanism will be called *statistical*. Statistical enhancement is a weaker concept than dynamic enhancement because the latter holds in averaging and, hence, is necessarily statistical, but not vice versa.

A typical example of statistical enhancement is the case of weak localization discussed in the preceding section. Near the backscattering direction, in each realization, its pattern of scattered radiation has the form of a speckle structure, i.e. a sequence of alternating maxima and minima (Fig. 5.5). A single maximum in the direction of backscattering appears only after statistical averaging.

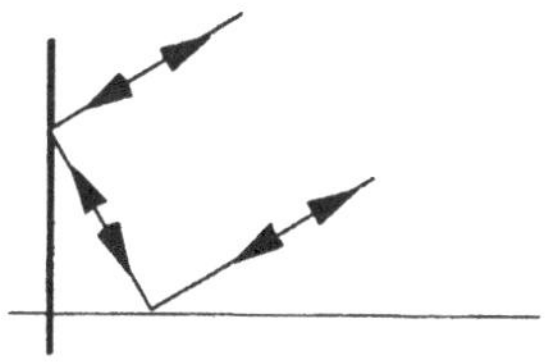

Figure 5.8. Corner reflector is a characteristic example of dynamic enhancement, which occurs in each realization

Examples of dynamic enhancement will be given later. Here, for the sake of illustration, we confine ourselves to an artificial case of a medium containing cube-corner reflectors (Fig. 5.8). From simple raytracing geometry it is evident that the reflection maximum of each retroreflector is in the direction to the source. Accordingly, the total scattered field will also have a maximum in each realization.

5.2.3 Single Particle Mechanisms

The example of cube-corner reflectors suggests the existence of a group of single particle mechanisms among the mechanisms of backscattering enhancement. In this group, the backscattering enhancement is hidden within the directivity pattern of elementary scattering events which must be extended in the backward scattering direction. Such a directivity pattern can be produced by scattering spheres which can be large or small compared to the wavelength (small spheres must be of large permittivity material, see, e.g., van de Hulst, 1957).

The single particle mechanism gives the sole example of backscattering enhancement described by conventional transfer theory, or, more precisely, by the single scattering approximation of this theory. When the contribution of multiple scattering increases, the importance of this mechanism decreases, because, for multiple scattering events of higher order, the form of the directivity pattern of an individual scatterer is no longer significant. On the other hand, an increase of the absorption in the medium decreases the significance of multiple scattering thus increasing the enhancement.

By the single particle mechanism, the magnitude of the backscattering maximum (enhancement factor K_{enh}) and its angular width (enhancement angle θ_{enh}) are governed by the shape of the single scattering pattern. The same is valid for polarization, which is completely controlled by the properties of an elementary scattering event. For example, depolarization in backscattering from scattering spheres is completely determined by multiple scattering, therefore, the effect of single particle enhancement is absent in the depolarized component of the reflected field.

Single particle enhancement is more often dynamic, though a possibility of a statistical single particle mechanism cannot be excluded. To illustrate this statement, we give a hypothetical example of a system of a few freely orientable particles whose scattering pattern exhibits a maximum in the backward direction after averaging over the orientation of scatterers. If the scattering cloud has many such particles, then a measurement would automatically involve ensemble averaging and the enhancement becomes dynamic.

5.2.4 Collective Mechanisms

If a backscattering enhancement occurs but is not predicted with the single-scattering approximation, then we understand that collective mechanisms come into play. These mechanisms are characteristic of ensembles of particles or scattering inhomogeneities and may manifest themselves in different forms. Some of them will be discussed below.

5.2.5 Geometrical Mechanisms

Collective mechanisms may be subdivided into *geometric*, called sometimes *incoherent*, and *phase*, or *coherent*, mechanisms. Note that in the literature, the term coherent is sometimes used — and without sufficient justification — for all the mechanisms of backscattering enhancement.

We will classify the mechanisms with little sensitivity to the phase of wave, which occur mainly because variations of intensity associated with the deformation of rays in inhomogeneous media, as geometric mechanisms, and the mechanisms like weak localization, which are related to the coincidence of phases of waves propagating in forward and backward directions (contribution of cyclic diagrams), as phase mechanisms.

Both the geometric and phase mechanisms of backscattering enhancement are similar in their nature and result from double passage of back-scattered radiation through the same inhomogeneities. In the final analysis this nature comes to time reversibility of equations governing the propagation of radiation. The main difference between geometric and phase mechanisms is as follows: in first case the question is about the passing a wave through separate, large-scale inhomogeneities, and in second case it is about the passing through a number $N \geq 2$ of different inhomogeneities of scattering medium.

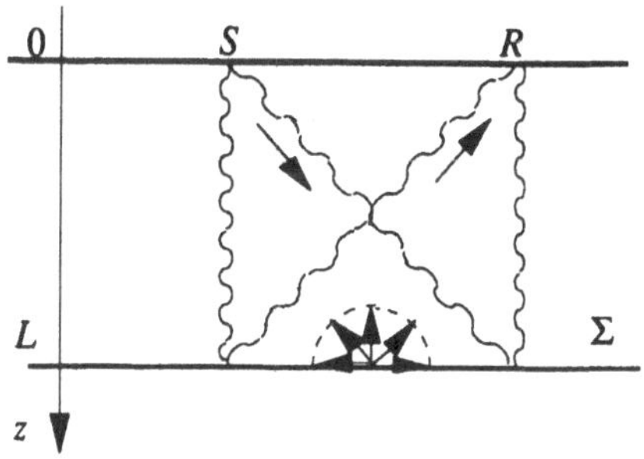

Figure 5.9. Geometry of the problem.

To illustrate geometric mechanisms of enhancement, we consider an experimental scheme shown in Fig. 5.9. A wave launched by a radiation source S propagates through a large-scale medium and after reflection from an obstacle Σ is detected by a receiver R. For simplicity, we choose Σ to be a homogeneous, diffusely reflecting plane with a uniform scattering pattern and assume that S and R have identical directivity patterns. Then, using the reciprocity theorem, the received power P may be represented, accurate to a constant factor, as

$$P = \int_{\Sigma} \mathcal{J}^{(+)}(\mathbf{r}_\perp)\mathcal{J}^{(-)}(\mathbf{r}_\perp)d^2r_\perp, \tag{5.2.1}$$

where

$$\mathcal{J}^{(\pm)}(\mathbf{r}_\perp) = \int I^\pm(\mathbf{n},\mathbf{r}_\perp)d^2n \tag{5.2.2}$$

are the total intensities at the reflecting plane: due to the source, I^+, and due to the receiver, I^-, acting as a source.

As noted in Section 4.2, waves traveling in large-scale medium may be described with the small-angle approximation for the radiative transfer equation that may be easily derived by averaging the parabolic equation of wave theory. Taking the parabolic equation as the basis, we can treat eqn (5.2.1) as nonaveraged dynamic relation, where $\mathcal{J}^{(+)}$ and $\mathcal{J}^{(-)}$ are the fluctuating intensities of the incident and reflected waves, which depend on random fluctuations in the scattering medium. The step of going to radiative transfer theory consists in averaging the both sides of (5.2.1) without regard to correlation between direct and inverse waves. This gives

$$\langle P\rangle_{\mathrm{RT}} = \int \langle \mathcal{J}^{(+)}(\mathbf{r})\rangle\langle \mathcal{J}^{(-)}(\mathbf{r})\rangle d^2r, \tag{5.2.3}$$

where $\langle \mathcal{J}^{(\pm)}\rangle$ are the mean intensities that occur in transfer theory.

Equation (5.2.1) gives a quantitative description and simple qualitative explanation of the statistical effects of backscatter enhancement in an inhomogeneous medium with single reflection, i.e., with single backscattering (Vinogradov, Kravtsov, and Tatarskii, 1973). In accordance with the reciprocity theorem, for monostatic reception, $\mathcal{J}^{(+)} = \mathcal{J}^{(-)} = \mathcal{J}$, so that the received power $\langle P\rangle$ is proportional to the integral of the squared intensity $\langle \mathcal{J}^2\rangle$. This is the formal reason of the enhancement effect. As a result, in monostatic reception, eqn (5.2.1) includes the ensemble-average quantity $\langle \mathcal{J}^{(+)}\mathcal{J}^{(-)}\rangle = \langle \mathcal{J}^2\rangle$, while, in bistatic reception, the quantity $\langle \mathcal{J}^{(+)}\rangle\langle \mathcal{J}^{(-)}\rangle = \langle \mathcal{J}\rangle^2$, corresponding to the transfer theory (5.2.3), arises, because, in this case, the direct and inverse waves become uncorrelated. In view of the inequality $\langle \mathcal{J}^2\rangle \geq \langle \mathcal{J}\rangle^2$ we have the relation $\langle P\rangle > \langle P\rangle_{\mathrm{RT}}$ that implies an enhancement of the backscattered radiation as compared to the side scattered radiation.

This explanation is applicable to all statistical enhancement mechanisms with single backward reflection but it tells us nothing about the nature of processes and gives us no possibility to single out the dynamic enhancement.

The wavelength λ does not represent a characteristic parameter in geometric mechanisms; therefore, these mechanisms do not disappear in the limit of geometrical optics as $\lambda \to 0$. In this case, the characteristic value of the enhancement angle θ_{cnh} does not depend explicitly upon the wavelength and is given by $\theta_{\mathrm{cnh}} \sim l_0 / L$, where L is the characteristic path length (extinction length L_{ext} for a scattering half-space) and l_0 is some internal scale of inhomogeneities in the medium (for example, the mean distance between particles, or the mean size of focusing inhomogeneities). Below we consider various geometric mechanisms in some more detail.

5.2.6 Focusing Mechanism

The *focusing* (or *lens*) *mechanism* is a geometric mechanism governing variations of the cross section of a ray tube passing through large-scale phase inhomogeneities that act like focusing or defocusing lenses.

Qualitatively, the contribution of the focusing mechanism may be estimated as follows. Define the dynamic backscattering enhancement factor K_d as the ratio of the random power P, given by eqn (5.2.1), to the unperturbed power P_0, which would be received in a medium without fluctuations:

$$K_d = \frac{P}{P_0}. \tag{5.2.4}$$

For the monostatic reception of a narrow beam passing through a single phase inhomogeneity that acts as a weak lens, we can easily estimate K_d if we observe that the total flux of beam energy is conserved (Problem 5.2), viz.,

$$K_d \sim \frac{\Sigma_0}{\Sigma_0 + \delta\Sigma} \sim 1 - \frac{\delta\Sigma}{\Sigma_0} + \left(\frac{\delta\Sigma}{\Sigma_0}\right)^2 + \ldots, \tag{5.2.5}$$

where Σ_0 and $\Sigma_0 + \delta\Sigma$ are the cross sections of the beam in the region of reflection with and without the inhomogeneity, respectively.

The dynamic enhancement factor K_d may be greater or lower than unity depending on the sign of the random quantity $\delta\Sigma / \langle\delta\Sigma\rangle$. However, the contribution of focusing inhomogeneities dominates the average over the ensemble:

$$\langle K_d \rangle = 1 + \left\langle \left(\frac{\delta\Sigma}{\Sigma_0}\right)^2 \right\rangle \geq 1. \tag{5.2.6}$$

This expression suggests that the focusing mechanism is, generally speaking, statistical.

In the strong focusing region, the quantity $\Sigma_0 + \delta\Sigma$ is small ($\Sigma_0 + \delta\Sigma < \Sigma_0$) and its magnitude is limited by diffraction and aberrations. As a result, the mean enhancement factor may be much greater then unity. Such strong enhancement effects may occur, for example, behind phase screens, when $\langle K_d \rangle$ is proportional to $\ln \sigma_\psi$, where σ_ψ is the dispersion of the random phase behind the screen (we note the paper of Jakeman, 1988 among recent works on the phase screen). A laser sounding the subsurface layer in the ocean gives a real example of a phase screen, which is formed by the random boundary between the air and the water (Apresyan and Vlasov, 1988).

5.2.7. Autoadaptive Mechanism

The autoadaptive mechanism is one of geometric mechanisms. An insight into this mechanism is provided by Fig. 5.10. A backscattering enhancement occurs in monostatic reception due to the fact that the source beam $I^{(+)}$ and the observation beam $I^{(-)}$ propagate through the same inhomogeneities. Moreover, these beams completely coincide as if the medium were automatically adjusted to the reflected ray. Rye (1981) used the term autoadaptive precisely in this sense. For bistatic reception, these specified beams can overlap only partially, thus decreasing the overlapping integral (5.2.1). In other words, in the monostatic case, the observer sees only illuminated reflectors, whereas, in the bistatic case, he sees both illuminated and unilluminated areas.

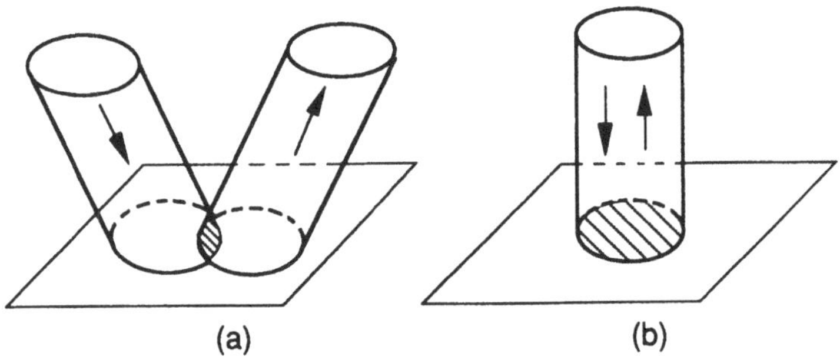

Figure 5.10. Autoadaptive mechanism.

The autoadaptive mechanism is not related with any focusing or variations in the cross-section of the beam. It manifests itself in each given realization and is a dynamic mechanism in this sense. However, since the autoadaptive behavior is observed for refracting (phase) inhomogeneities, it is always accompanied by an enhancement due to focusing. As a result, the total effect may manifest itself as a purely statistical effect.

5.2.8 Shadow Mechanism

This mechanism is similar to autoadaptive with that difference that the shadows behind absorbing or opaque particles replace the ray bending at phase inhomogeneities (Fig. 5.11). Similar shadows may also be produced by transparent, highly refractive particles, which give rise to a refracted field with a broad directional pattern and the small angles diffraction depending mainly on the diffraction at the rim of the particle.

It is the shadow, purely amplitude mechanism that is used in astronomy to describe the opposition effect. Fuks (1976) has discussed the shadow mechanism analyzing the reflection from rough surfaces.

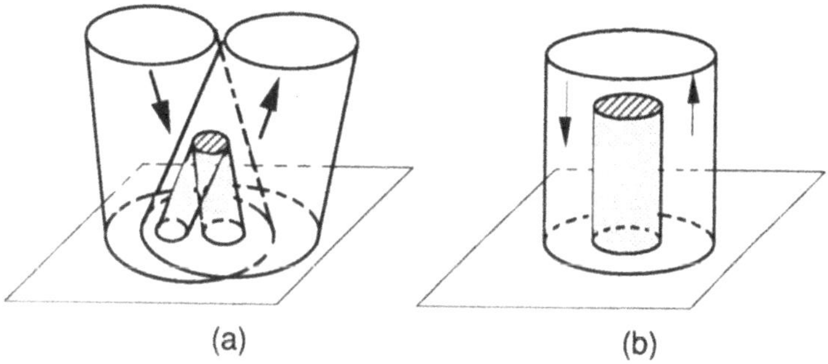

Figure 5.11. Shadow mechanism.

Hapke (1963) has analyzed the popular astronomic estimate of the magnitude and shape of backscattering peaks. For the enhancement factor, this estimate reduces to the replacement of the integrand in eqn (5.2.3) for backscattering case; namely, $\langle \jmath^{(+)} \rangle \langle \jmath^{(-)} \rangle \sim e^{-2\tau}$, where τ is the optical depth of reflecting plane, is replaced with $\langle \jmath^{(+)} \rangle \langle \jmath^{(-)} \rangle \sim e^{-\tau}$. This replacement is justified by simple heuristic reasoning about the probability for a ray to arrive at the reflector: the transfer theory considers the arrivals of a ray from the source and a ray from the observation point to the reflector as independent events, whereas, in backscattering, they totally coincide. Integrating the corrected quantity $\langle \jmath^{(+)} \rangle \langle \jmath^{(-)} \rangle$ with respect to τ, we obtain 2 for the enhancement factor of a half-space in the single-scattering approximation. This value approximately corresponds to results of astronomic measurements. A formal correction (suggested by Hapke) for the backscattering enhancement reduces to a modification of the term, corresponding to single scattering in the solution of the transfer equation. In multiple scattering, the effect of the shadow mechanism is proved to be of low significance.

5.2.9 Phase Mechanisms

Mechanisms similar to weak localization, which are due to the mutual phase coherence of waves passed the sequence of scatterers 1, 2, ..., j, $j \geq z$, in direct and inverse directions, fall into phase mechanisms.

Phase mechanisms are generally statistic mechanisms. There is a complicated set of maxima and minima in given realizations, and a clear maximum in backscattering direction appears only as a result of statistical averaging.

In Section 5.1 we have considered the mechanism of weak localization in detail. Here, we dwell shortly on an alternative phase mechanism, which is connected with the reflection from scatterers embedded into large-scale random media (Vinogradov, Kravtsov, and Tatarskii, 1973; Kravtsov and Saichev, 1982). When an incident plane wave propagates trough such a medium, there exists one-ray region at small enough paths $L << F$, where F is the characteristic focus length of the medium. Only one ray

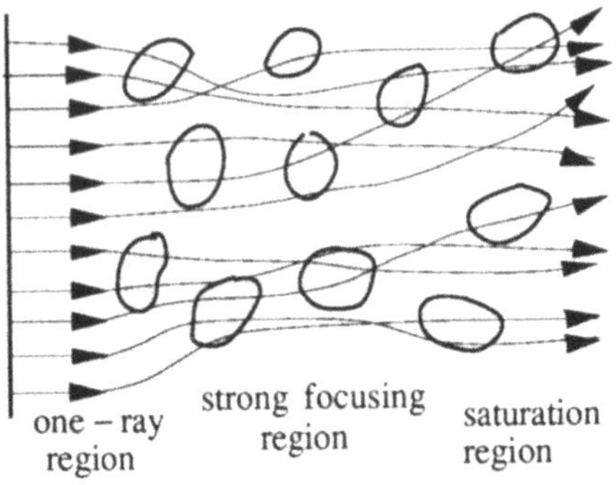

Figure 5.12. Regions for wave propagation in a large-scale medium.

runs through each point of that region (Fig. 5.12), and all aforementioned geometric mechanisms (like lens or autoadaptive mechanisms) are applicable to scatterers in that region. A greater $L >> F$ corresponds to a multiple ray region, or a region with saturated fluctuations of intensity.

Here, one may asymptotically consider the incident on a scatterer field as Gaussian field. Then, in spite of multiplicative effect of medium, the coherence function of reflected wave is represented by the sum of two terms: $\Gamma = \Gamma_1 + \Gamma_2$, where Γ_1 conforms to radiation with wide angular spread about λ / ρ_{c1} and coherence length ρ_{c1}, which corresponds to propagation through a path L in direct direction, and Γ_2 describes the enhancement effect.

The coherence length of Γ_2 is of the order of transverse displacement of a ray $\sigma_d \sim \lambda L/\rho_{c1}$ (Fig. 5. 13). Hence, for the enhancement angle, we have the estimate $\theta_{enh} \sim \lambda/\rho_{c2} \sim \rho_{c1} / \bar{L}$, where $\rho_{c1} = \rho_{c1}(L)$ depends upon the fluctuations in medium. In particular, if the length of path $L \sim L_{ext}$, then $\rho(L_{ext}) \sim l_c$, where l_c is the characteristic scale of inhomogeneous medium, and we have $\theta_{enh} \sim l_c / L_{ext}$. This estimate is appropriate to the reflection from a half-space because L_{ext} represents a natural length of path in this case. In contrast to weak localization, the enhancement angle does not depend explicitly upon the wavelength that makes this example close to the above considered geometric mechanisms. One may found details of this example and numerous references in the review by Kravtsov and Saichev (1982).

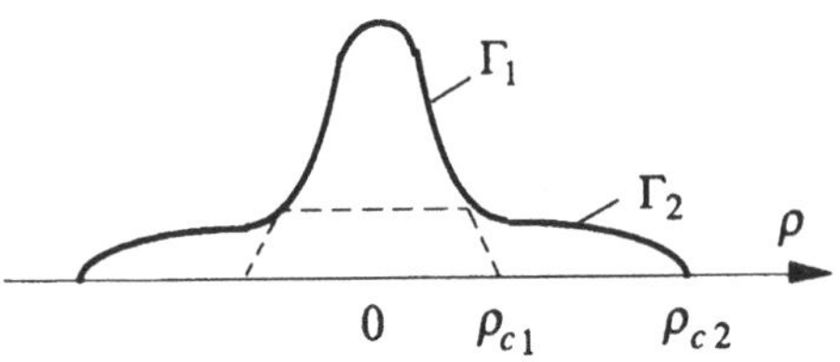

Figure 5.13. Coherence function versus ρ in the region of saturated intensity fluctuations.

This example is very close in many features to that of weak localization. The term Γ_1 is similar to the ladder diagram and Γ_2 to the cyclic diagram. The analogy becomes total if one represents the field of reflected wave as $u_{\text{ref}} = G_- r G_+ = \ominus\!\!-\!\!-\!\!\oplus$, where $G_\pm = \textcircled{\pm}$ are the random operators of propagation in an inhomogeneous medium for direct and reflected waves and $r = $ ▬ is the operator of reflection. Then

$\Gamma_1 =$ 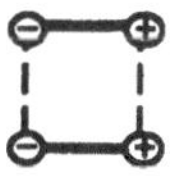 , $\Gamma_2 =$ 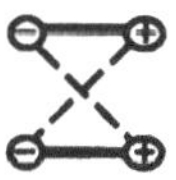 .

The dashed connections denote here the operation of averaging in pairs.

A mechanism of the same kind occurs also in the aforementioned example of backscattering enhancement which arises as a result of reflection of ions from surface layers of a substance. Modeling computations showed (Holland and Barrett, 1987) that the enhancement here is also determined by single backscattering with contribution to the enhancement from the paths which run through the same scatterers and the paths, which run through the sequences of different scatterers in opposite directions.

In the end of this section, we emphasize once more that, in all the above mechanisms save for the one-particle mechanism, the backscattering enhancement originates from the correlation between the waves propagating in forward and backward directions. Backscattering enhancement forces one to modify the classic radiative transfer theory that does not include such correlations.

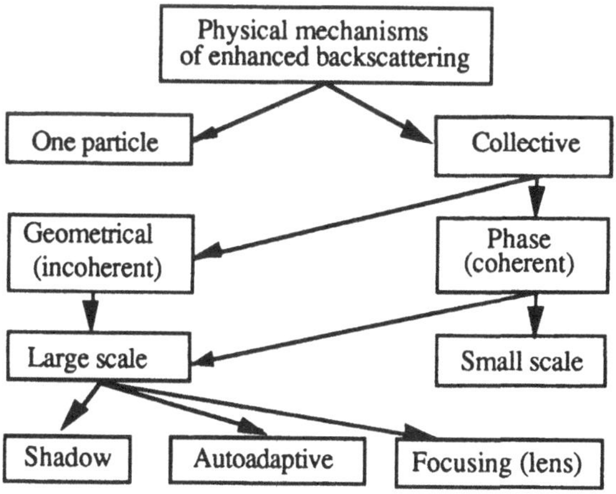

Figure 5.14. Classification of backscattering enhancement mechanisms.

5.2.10 Classification of Backscattering Enhancement Mechanisms

Figure 5.14 summarizes the considered mechanisms of backscattering enhancement. In accord with this figure, these enhancement mechanisms may be primarily classified as one-particle and collective. The last, in turn, are subdivided into two main groups: geometric (incoherent) and phase (coherent) mechanisms. Geometric mechanisms are connected mainly with the passage through large-scale inhomogeneities, while phase mechanisms may be determined by both small-scale and large-scale inhomogeneities. There are three similar but different in some features geometric mechanisms: shadow, autoadaptive, and focusing (or lens).

This classification is rather broad, especially because the considered mechanisms usually contribute collectively rather than individually. Nevertheless, the diagram of Fig. 5.14 may be found useful in clarifying the description of particular examples of backscattering enhancement.

5.3 NEW EFFECTS IN STATISTICAL RADIATIVE TRANSFER

5.3.1 Background

The statistical approach to radiation transfer brought to light a number of new phenomena that cannot be understood without a wave interpretation of transfer theory. These phenomena may be divided into two classes. One class includes effects which occur under the conditions of weak disorder $\lambda << L_{sc}$, where linear transfer theory is applicable. The other class deals with effects under the conditions of strong disorder $\lambda \sim L_{sc}$, where this theory fails. In this section we consider effects from the first class. The range of high disorder and related strong localization phenomena will be briefly outlined in Section 5.4.

The effects of weak localization considered in Section 5.1 are related to the range of weak disorder $\lambda << L_{sc}$. However, recently new effects were evaluated that can be understood only in the context of the statistical wave formalism. Usually, these phenomena do not require qualitative changes of the theory and are often a simple correlation treatment of the classical phenomenological approach. Later we consider some of these effects, including phenomena related to a non-Gaussian distribution of field fluctuations (Sections 5.3.2 and 5.3.3), scalar memory effect (Section 5.3.4), time-reversed memory effect (Section 5.3.5), long-correlation effects for intensity (Section 5.3.6), polarization effects (Section 5.3.7), and, finally, the effects related to image transfer through turbid media (Section 5.3.8).

Intending primarily to describe the possibilities of transfer theory, we focus our attention on the phenomena that can be described using for this theory the diffusion approximation characteristic of media with small-scale inhomogeneities. To handle the important case of large-scale media, including turbulent media, it is more convenient to use the parabolic equation and related Markov's approximation for moments. Detailed expositions of this approach may be found, for example, in books by Rytov, Kravtsov, and Tatarskii (1989), and Banakh and Mironov (1987).

Classic transfer theory was a set of concepts constructed for sparse media. These treatments fail completely in handling effects occurring under the conditions of strong disorder $L_{sc} \sim \lambda$. This fact is clear even from a phenomenological standpoint. Indeed, for $L_{sc} \sim \lambda$, the scattered field loses the form of a quasi-plane propagating wave, and this point breaks down the grounds of the heuristic approach to radiative transfer theory. The range of strong disorder is the hardest for both theory and experiment. The most interesting achievements in this region are connected with the prediction of strong, or Anderson's, localization. Investigations in the field of strong localization are still far from completion. Despite numerous efforts and a large body of literature published since the first prediction of this phenomenon in the theory of solids (Anderson, 1958), clarity as to how the condition of strong localization $L_{sc} \leq \lambda$ may be realized in practice for electromagnetic and acoustic fields is still at large (see Section 5.4).

5.3.2 Complex Gaussian Fields

Analysis of higher moments and complete statistics for scattered radiation is known to be a complex problem whose solution is far from completion. The Gaussian statistics of the field may be a single exception. Such a statistics is usually formed asymptotically when the field is treated as a sum of many independent, approximately equal contributions (central limit theorem). This situation often occurs in problems of transfer theory. Let us examine the Gaussian statistics on the basis of rough physical considerations.

Assume the complex amplitude u of a scattered field in the form of a sum of a large number $N >> 1$ of independent random contributions u_l:

$$u = \sum_{l=1}^{N} u_l . \tag{5.3.1}$$

We assume the mean value of the field is zero by setting $\langle u_l \rangle = 0$.

Let us consider the sequence of field cumulants and, estimate the variance $\sigma_u^2 = \langle |u|^2 \rangle$. Since u_l and u_m are mutually independent for $l \neq m$, we have

$$\sigma_u^2 = \langle uu^* \rangle_c = \sum_{l,m} \langle u_l u_m^* \rangle = \sum_{l} \langle u_l u_l^* \rangle = N\sigma_1^2. \tag{5.3.2}$$

It follows that the variance of individual contributions $\sigma_1^2 = \langle |u_e|^2 \rangle = \sigma_u^2 / N$ tends to zero as N increases for a fixed value of σ_u^2. Using this fact, it is convenient to re-normalize the contributions u_l, replacing them with the amplitudes $a_l = \sqrt{N} u_l$. The variance of the amplitude $\sigma_a^2 = \langle |a_l|^2 \rangle = \sigma_u^2$ is already independent of N. Therefore, in order of magnitude estimations it is expedient to assume $u_l \sim \sigma_1 \sim \sigma_u / \sqrt{N}$ and $a_l \sim \sigma_a \sim u$.

Thus, in place of eqn (5.3.1) we will consider the sum

$$u = \frac{1}{\sqrt{N}} \sum_{l=1}^{N} a_l , \tag{5.3.3}$$

whence, for the second order cumulants $\langle uu^* \rangle_c$ and $\langle uu \rangle_c$, we have

$$\langle uu^* \rangle_c = \sigma_u^2 = \langle a_l a_l^* \rangle_c = \sigma_a^2 \tag{5.3.4}$$

and

$$\langle uu \rangle_c = \langle a_l a_l \rangle_c \tag{5.3.5}$$

that are independent of N.

Similar expressions for cumulants of higher orders $l \geq 3$ were found to be proportional to $N^{-(l-2)/2}$, so that they vanish as $N \to \infty$. For example, the third order cumulant $\langle uuu \rangle_c$ is expressed as

$$\langle uuu \rangle_c = N^{-3/2} \sum \langle a_l a_l a_l \rangle_c = \frac{\langle a_l a_l a_l \rangle_c}{\sqrt{N}} . \tag{5.3.6}$$

As a result, the second order cumulants (5.3.4) and (5.3.5) may be considered as the only non-zero cumulants of the sum (5.3.3) for large N. This is what means that the sum (5.3.3) is the (complex) Gaussian random quantity.

The statistical description of a complex random quantity u is equivalent to the statistical description of a pair of real random quantities, which may be, for example, the amplitude and the phase or the real and imaginary parts of u. In practice, the amplitude and the phase of the components u_l are often statistically independent and, in addition, the phase of u_l fluctuates strongly and is distributed uniformly over the interval $[0, 2\pi]$. In this case, the second cumulant (5.3.5) vanishes, $\langle uu \rangle_c = 0$, and we arrive at the model of a circular Gaussian random quantity. This model is widely used in describing the statistics of speckles produced by reflections of a coherent laser radiation from rough surfaces (see, e.g., Dainty,.1984; Goodman, 1985).

For an circular Gaussian quantity, the only non-zero cumulant is $\langle uu^* \rangle_c$. The components u_l satisfy the relationships

$$\langle u_l u_l^* \rangle_c = \langle I_l \rangle , \quad \langle u_l u_m \rangle = 0 , \quad \text{for } l \neq m . \tag{5.3.7}$$

Using these relations it is not difficult to show that the intensity of the field $I = |u|^2$ obeys the Rayleigh distribution

$$P(I) = \frac{e^{-I/\langle I \rangle}}{\langle I \rangle}, \tag{5.3.8}$$

whereas the phase of the field is distributed uniformly. The real and imaginary parts prove to be uncorrelated Gaussian quantities (see Problems 5.3 and 5.4).

For a circular process, the joint distribution of the intensity $I = |u|^2$ and the phase $\varphi = \arg u$ is expressed as

$$P(I, \varphi) = P(I) \frac{1}{2\pi}. \tag{5.3.9}$$

This distribution defines completely all the one-point statistics of the field u.

Multipoint statistical moments of a complex Gaussian field $u(x)$ are expressed through the covariances $\langle u(x_1) u^*(x_2) \rangle$ and $\langle u(x_1) u(x_2) \rangle$, of which the first can be found from transfer theory, and the second may be neglected in many cases. Thus, the solution of the transfer equation in the form of a circular Gaussian process defines almost completely the statistics of radiation.

5.3.3 Sources of Non-Gaussian Scattered Field

In investigating the statistics of a scattered field, it is reasonable to ask what conditions render this statistics Gaussian. The inverse problem is also legitimate: what can prevent a scattered field from being Gaussian?

There are two main reasons for non-Gaussian statistics. One is due to the different scattering order of the terms in the series (3.3.17). Some of these terms may give the predominant contribution to the scattered field, so that the effective number N^{eff} of terms in the sum (5.3.1) may be insufficient to normalize the process.

The other possible reason is the presence of functional relationships (correlations) between individual terms of the series with respect to the scattering order. These relations may also violate the applicability conditions of the central limit theorem.

Unfortunately, it is usually difficult to rigorously check the validity of using the Gaussian statistics for a scattered field. Nevertheless, the hypothesis of normal distribution is often used as a point of departure. Deviations from the Gaussian statistics are considered as certain "anomalies" which, however, are often of great importance and give useful information on the properties of the medium (see, e.g., the review of Jakeman and Tough, 1988; and a paper of Shnerb and Kaven, 1991).

As we have already mentioned, the Gaussian statistics is completely defined by the two first moments of radiation. If the field is not strictly Gaussian, the differences from the Gaussian statistics are more significant in higher moments, whereas lower moments, in particular forth moments, may often be calculated as if the field were Gaussian. In this "quasi-Gaussian" approximation, higher moments give no information in addition to that given by second moments, but experimental measurement of higher moments — for example, measurement of intensity correlations (Section 5.3.5) — may be a simpler problem than a measurement of field correlations, which are also more sensitive to noise.

5.3.4 Scalar Memory Effect

In classic radiative transfer theory it is assumed usually that, in multiple scattering, radiation forgets its initial characteristics such as the direction of propagation or the angular spectrum and the initial frequency if the scattering is accompanied by changes in frequency. However, this assumption is only partially correct. The information on the initial characteristics of radiation disappears not completely, rather it is "encoded" in the process of multiple scattering.

Consider a standard problem of multiple scattering of a plane monochromatic wave by a statistically uniform layer of a scattering medium. For the sake of definiteness, we will consider the transmitted wave, though similar conclusions may be drawn for the reflected radiation. To simplify the problem, we will assume that the inhomogeneities in the layer are small-scale. Then, if the layer thickness L by far exceeds the mean free path of radiation L_{sc}, the transmitted radiation will be nearly isotropic, i.e., the radiation pattern behind the layer will be nearly isotropic, no matter what the direction of propagation of the incident wave was at the moment of incidence on the layer. It is in this sense that one may speak of "forgetting" of the initial wave characteristics. In reality, however, the propagating wave memorizes partially the incident wave. Let us consider some of such "memories", and, above all, the *scalar memory effect.*

The scalar memory effect was first predicted in the theoretical paper of Feng et al. (1988) and was borne out experimentally by Freund et al. (1988). The essence of this effect may be formulated as follows.

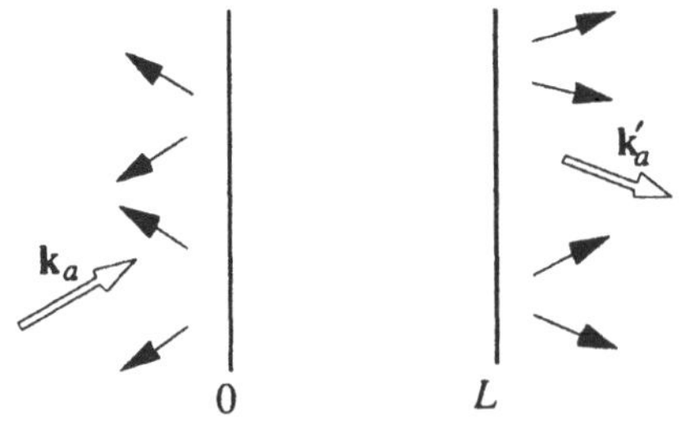

Figure 5.15. Scattering of a plane wave by a statistically uniform layer (wave vectors for scattering process $\mathbf{k}_a \to \mathbf{k}'_a$ under consideration are indicated by double-shafted arrows).

Consider a plane wave with a wave vector $\mathbf{k}_a$ incident on a scattering layer. This wave is transformed in transmitted and reflected waves, which have generally wide angular spectra (Fig. 5.15). Keeping in mind that the radiation is usually observed in the statistical Fresnel zone relative to scatterers, we consider the transmitted wave with wave vector $\mathbf{k}'_a$. We will refer to such a scattering process $\mathbf{k}_a \to \mathbf{k}'_a$ as process "a".

Let us look now at another scattering process $\mathbf{k}_b \to \mathbf{k}'_b$, process "b", which differs from process "a" in that the directions of the wave vectors $\mathbf{k}_a$ and $\mathbf{k}'_a$ are replaced with $\mathbf{k}_b$ and $\mathbf{k}'_b$, respectively. If the scattering layer thickness L is large compared to the scattering length L_{sc} and the layer is statistically uniform in plane $z = const$, then the correlation of the two processes under consideration, "a" and "b", prove to be not very small only if

$$\mathbf{q}_{a_\perp} = \mathbf{q}_{b_\perp} \tag{5.3.10}$$

and

$$|\mathbf{k}_{a_\perp} - \mathbf{k}_{b_\perp}| \leq \frac{1}{L}. \tag{5.3.11}$$

Here $\mathbf{k}_\perp$ denotes the component of any vector $\mathbf{k}$ parallel to the layer, and $\mathbf{q}_a = \mathbf{k}'_a - \mathbf{k}_a$ and $\mathbf{q}_b = \mathbf{k}'_b - \mathbf{k}_b$ are the scattering vectors for processes "a" and "b", respectively. A quantitative formulation of this statement is given in Problems 5.5 and 5.6. The qualitative sense of the memory effect is as follows.

In the case of a medium with continuous fluctuation of permittivity ε, the first condition (5.3.10) may be rearranged in the Bragg form. Indeed, applying the Fourier transformation over coordinates $\mathbf{r}_\perp$ transverse with respect to z, we can represent the field in the layer as a superposition of all possible interacting waves of opposite directions $u^{(+)}(\mathbf{k}'_\perp, z)$ and $u^{(-)}(\mathbf{k}''_\perp, z)$, where $\mathbf{k}'_\perp$ and $\mathbf{k}''_\perp$ are wave vectors transverse with respect to z (see, for example, Malakhov and Saichev, 1979). The strongest resonant interaction occurs for waves which satisfy the condition of synchronism: $\mathbf{k}'_\perp - \mathbf{k}''_\perp = \mathbf{q}_\perp$, where $\mathbf{q}_\perp$ is the wave vector of fluctuations in the medium $\varepsilon(\mathbf{q}_\perp, z)$.

Let us assume that the resonant fluctuations of the medium are, on average, most important for the scattering process. This means that the fluctuations $\varepsilon(\mathbf{q}_{a\perp}, z)$ with $\mathbf{q}_{a\perp} = \mathbf{k}'_{a\perp} - \mathbf{k}_{a\perp}$ are essential for the process $\mathbf{k}_a \to \mathbf{k}'_a$ and, similarly, $\varepsilon(\mathbf{q}_{b\perp}, z)$ with $\mathbf{q}_{b\perp} = \mathbf{k}'_{b\perp} - \mathbf{k}_{b\perp}$ for the process $\mathbf{k}_b \to \mathbf{k}'_b$. Therefore, the correlation of amplitudes of the two processes is proportional — with allowance for the statistical uniformity of fluctuations in the layer — to the quantity $\langle \varepsilon(\mathbf{q}_{a\perp}, z')\varepsilon^*(\mathbf{q}_{b\perp}, z'')\rangle \propto \delta(\mathbf{q}_{a\perp} - \mathbf{q}_{b\perp})$, whose singularity explains the sense of the condition (5.3.10).

Figure 5.16 illustrates these considerations and shows the analogy between this problem and the ordinary refraction of a plane wave in a homogeneous layer. In the latter problem, the layer permittivity ε is constant, so that $\varepsilon(\mathbf{q}_\perp) \propto \delta(\mathbf{q}_\perp)$ is nonzero only for $\mathbf{q}_\perp = \mathbf{k}'_\perp - \mathbf{k}_\perp = 0$. This condition corresponds to the conservation of the wave vector component transverse to the normal to the layer. For the boundary between two homogeneous media, this condition gives the Snell law.

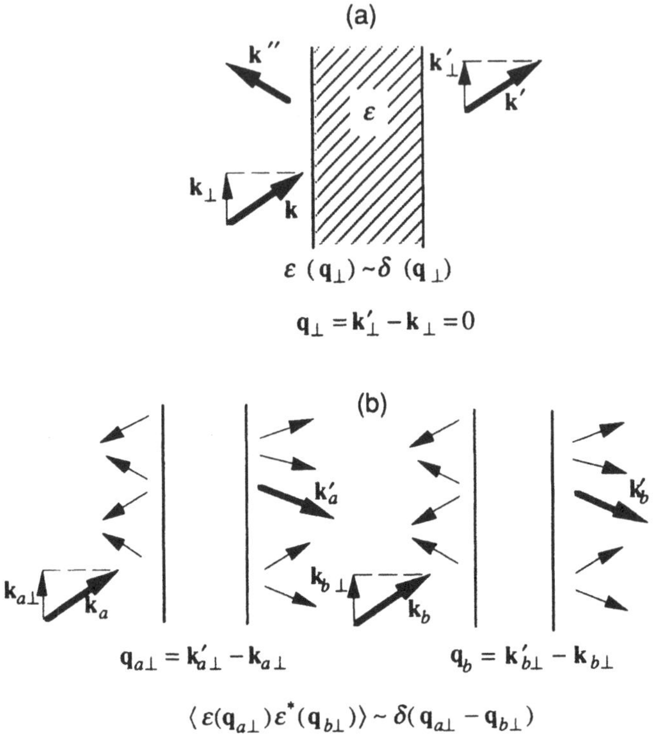

Figure 5.16. Analogy between the reflection from a homogeneous layer and the correlation of the scattering processes in a statistically uniform layer. (a) For the *homogeneous* layer, $\varepsilon(\mathbf{q}_\perp) = 0$ only at $\mathbf{q}_\perp = 0$ and the transverse component of the wave vector $\mathbf{k}_\perp$ is conserved. (b) For a *statistically uniform* layer, the correlation between the two processes is non-zero only if $\mathbf{q}_{a\perp} = \mathbf{q}_{b\perp}$.

To comprehend the second condition (5.3.11), we slightly move the incident wave vector $\mathbf{k}_a$ to find out from what amplitude this motion becomes noticeable for the complex speckle structure of the propagating wave. Consider an arbitrary point R at the far side of the layer at $z = L$. To this point there corresponds an "influence region" A at the near side of the layer ($z = 0$) such that a variation of the incident wave in A appreciably changes the magnitude of $u(\mathbf{R})$. This region is similar to the Fresnel zone for a wave in a free space. If L_A denotes the characteristic size of A, then changing from the wave vector $\mathbf{k}_a$ to its new value $\mathbf{k}_b$ is of little effect until the phase difference on the edges of this region remains small compared to unity, i.e. until $|\mathbf{k}_{a\perp} - \mathbf{k}_{b\perp}| L_A < 1$. In the problem under consideration, with its thick layer and small-scale fluctuations of scatterers, the wave evolution in the layer has a diffuse behavior, which results in L_A of the order of the layer thickness L : $L_A \sim L$. Thus, we have arrived at the condition (5.3.11).

These reasoning is valid not only for the transmitted wave but also for the reflected wave with the difference that, for a thick layer $L >> L_{\text{ext}}$, the "influence region" size L_A is about the extinction length L_{ext} and is independent of the layer thickness L. This means that, for the reflected wave, condition (5.3.10) keeps its form and condition (5.3.11) transforms into

$$|\mathbf{k}_{a\perp} - \mathbf{k}_{b\perp}| \leq \frac{1}{L_{\text{ext}}}. \tag{5.3.12}$$

We emphasize that the memory effect bears substantially on the assumption of statistical uniformity of the layer in the plane $z = \text{const}$. Therefore any factor destroying this uniformity (e. g., a finite size of the layer in the plane z = const.) breaks down the memory effect (Eliyahu et al., 1991).

If the incident wave propagates near the normal to the layer, then the memory effect admits a simpler interpretation. Indeed, in this case, $\mathbf{k}_{a\perp} - \mathbf{k}_{b\perp} \approx k_0\theta_{\text{ab}}$, where θ_{ab} is the angle between $\mathbf{k}_a$ and $\mathbf{k}_b$ so that the condition (5.3.11) denotes $\theta_{ab} = \theta'_{ab}$ where θ'_{ab} is the angle between $\mathbf{k}'_a$ and $\mathbf{k}'_b$ (see Fig. 5.15). In other words, deflection of the incident wave on a small angle $\theta_{\text{ab}} < \lambda / L$ results on average in deflection of the speckle structure of propagating wave as a whole on the same angle. For the reflected wave, the angle θ_{ab} is limited by the condition $\theta_{\text{ab}} < \lambda / L_{\text{ext}}$ that follows from (5.3.12). Note that, in the single-scattering approximation, such a deflection occurs not only on average, but in each particular realization (Problem 5.7).

5.3.5 Time Reversed Memory Effect

The scalar memory effect considered in the foregoing section defines the region of correlation of scattering processes "a" and "b" in the simplest case when difference between the wave vectors of these processes is small ($|\mathbf{k}_{a\perp} - \mathbf{k}_{b\perp}| \leq 1/L$ for transmitted wave, and $|\mathbf{k}_{a\perp} - \mathbf{k}_{b\perp}| \leq 1/L_{\text{ext}}$ for reflected wave). There is an additional, more complex mechanism of such a correlation — the *time-reversed memory effect* that is connected with symmetry under time reversal, i.e. with the reciprocity theorem.

If scatterers satisfy the conditions of this theorem (excluding from consideration gyrotropy, nonstationarity of medium, and other effects violating reciprocity), then the transmission factor through the layer $t(\mathbf{k}' \leftarrow \mathbf{k})$satisfies the reciprocity condition $t(\mathbf{k} \leftarrow \mathbf{k}') = t(-\mathbf{k}' \leftarrow -\mathbf{k})$ in the scalar approximation. A similar condition holds for the reflection factor. Hence, equal transmission factors correspond to the scattering processes $\mathbf{k}_a \rightarrow \mathbf{k}'_a$ and $-\mathbf{k}'_a \rightarrow -\mathbf{k}_a$, i.e. these processes are strongly correlated (Berkovitz and Kaveh, 1990a).

In the case of transmission, this new correlation can be observed only if scattering process "b" is due to the wave incident on the layer from the opposite side (Fig. 5.17a). In the case of reflection, there is no need to change the direction of propagation

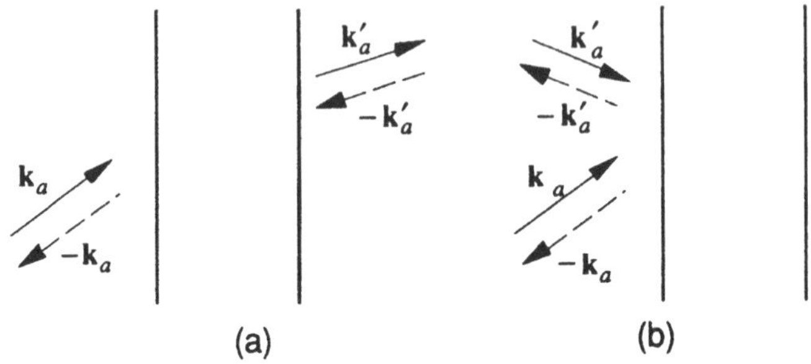

Figure 5.17. Strongly correlated direct and time-reversed waves for a scattering process symmetric with respect to time reversal. (a) transmission; (b) reflection.

of the incident wave because the wave vectors of direct and time-reversed processes lie on the one side of the scattering layer. In the last case, the time-reversed memory effect gives rise to an additional correlation peak that occurs when process "b" coincides with time-reversed process "a".

Polarization complicates the time-reversed memory effect. In this case, the transmission and reflection factors become tensor quantities, and the scalar reverse of time discussed above should be completed by the transposition of related tensors. As a result, the pattern of correlations becomes substantially dependent upon the polarizations of incident and scattered waves. These vector memory effects were described in Bercovitz and Kaveh (1990c).

5.3.6 Long Correlation Effects for Intensities

The most interesting results in this range are associated with the difference in the statistics of the scattered field from Gaussian. Even if scattering has a diffuse character and the average intensity can be described in the diffusion approximation of transfer theory, the correlations of intensities may include appreciable non-Gaussian corrections. These corrections result in a slower decay of correlations as compared to the results of the simple "quasi-Gaussian" approximation. We outline some results in this field avoiding complicated calculations.

In the range of weak disorder $\lambda << L_{sc}$, the standard perturbation theory is applicable in the form of diagram expansions. The diagram technique gives the forth moments in the form of four-deck diagrams similar to two-deck diagrams used to calculate second moments. Such diagrams can be partially summed by extracting internal two-deck blocks that can be estimated using the conventional "correlation" transfer theory. With this approach, Feng et al. (1988) showed that the normalized correlation function $C_I = \langle \tilde{I}(\mathbf{k}_1)\tilde{I}(\mathbf{k}_2)\rangle$ of intensity fluctuations $\tilde{I}(\mathbf{k})$ of the transmitted wave in a waveguide is representable as a sum of three terms:

$$C_I \approx C^{(1)} + C^{(2)} + C^{(3)} . \tag{5.3.13}$$

These terms are given in the diminishing order with respect to small parameter $(\lambda \setminus L_{\perp})^2 << 1$, i.e. $C^{(j)} \propto (\lambda / L_{\perp})^{2(j-1)}$, where $L_{\perp}$ is the transverse size of the waveguide. Wang and Feng (1989) extended this result to describe the reflected waves.

The first term $C^{(1)}$ corresponds to the "quasi-Gaussian" approximation, where the forth moment is represented in the form of the sum of pair products of second moments, as in the case with a circular random Gaussian quantity. This term describes the memory effect considered in the foregoing section. In an infinite medium, the term $C^{(1)}$ decays exponentially with the separation of the observation points. The characteristic scale of this decay is of the order of the extinction length L_{ext} (Shapiro, 1986).

The second and third terms in eqn (5.3.13) correspond to non-Gaussian corrections. The term $C^{(2)}$ is characterized by a long-range power-law behavior and describes slowly decaying correlations of intensities. These correlations were first found by Stephen and Cwilich (1987) who interpreted them as a result of an interference interaction of diffusion modes. Later it was shown that the term $C^{(2)}$ can be described using the Langeven approach (Zuzin and Spivak, 1987) describing the long-range correlations if the close-range correlations are considered as extraneous fluctuation currents (Pnini and Shapiro, 1989).

Finally, the smallest term $C^{(3)}$ corresponds to the so-called universal or permanent correlations. In k-space, this term turns out to be independent of the difference $\mathbf{k}_a - \mathbf{k}_b$ between the wave vectors of the two scattering processes "a" and "b". The diagrams, associated with this term, have a fairly complicated structure (Feng et al., 1988; Wang and Feng, 1989) and may be explained as a result of a more complicated interaction of diffusion modes than in the case of $C^{(2)}$. In the physics of solids, such correlations are related to "universal fluctuations of conductivity" that have been observed at low temperatures in "mesoscopic" samples of size $L << L_{\text{sc}}$ (Altshuler et al., 1986).

5.3.7 Polarization Effects

In addition to scalar correlation effects the literature describes also specific polarization effects associated with polarization of radiation (Freund, 1990a, 1991; Freund et al., 1990). Some of these effects can be obtained simply by extending the results of scalar theory to polarization characteristics. For instance, the scalar memory effect deals with the correlation of intensities of two scattering processes with different directions of incident and scattered waves. Extending these results to polarization characteristics, it is reasonable to explore how the correlations of intensities would change if we alter the polarization of the incident and scattered waves.

We present only the result of Freund et al., 1990 (see also Problem 2.17). Consider two scattering processes "a" and "b" with identical directions of incidence and scattering, but with orientations of the polarization vectors differing by an angle

$\Delta\theta$. Freund at al. (1990) showed that the normalized correlation function of intensities is represented as $C(\Delta\theta) = \cos^2(\Delta\theta)$ in the diffusion regime, i.e under the conditions of scattering with strong depolarization. It follows that the scattered waves turn out to be uncorrelated if the incident waves had orthogonal polarizations, whereas the correlation of scattered waves is nonzero for nonorthogonal polarizations of the incident waves.

5.3.8 Correlation Effects of Imaging in Diffuse Scattering Media

The problem on image transfer through highly turbid media, where the multiple scattering is predominant, is an interesting and practically important problem that involves correlation effects. This problem is of special importance in astronomy, where the effect of turbulent atmosphere limits considerably the possibilities of ground based telescopes. Special techniques of stellar speckle interferometry and adaptive optics were developed to compensate for this effect (Dainty, 1984; Beichman and Ridgway, 1991). Unfortunately, these techniques have been developed for large-scale media, which mainly produce phase distortions, and the model of phase distortions appears to be unable to describe the diffuse scattering in small-scale media considered in this section.

It is common lore that nothing can be seen through turbid media. In physical terms this means that multiple scattering prevents us from transferring clear optical images. It blurs the image and makes indistinguishable separate points of an object.

However, this does not mean that multiple scattering annihilates all the information on the primary wave. It would be more correct to say that this information is "encoded" in the process of multiple scattering into a complex speckle structure of the scattered wave. The fine details of this structure contain the imprint of both the incident wave and parameters of the medium. This fact suggests an idea how to "decode" the speckle structure to extract the useful information on the properties of the primary wave. This idea can be realized, in particular, by analyzing the higher statistical moments and multiple-point correlations of the scattered field.

The study of this problem is at its infancy. Nevertheless, there are some results that are worth mentioning. We give a simple example that itself is of little practical significance, but is useful to prove the possibility to transfer optical images through turbid media by input-output matched filtering of the transmitted signal (V.V. Karavaev, private communication).

Suppose that we need to image a transparency through a thick layer of a turbid medium. This difficult problem at first glance is easily solved if the transmission is point by point rather than for the whole object at once. The simplest way to effect this sequential transmission is to use two identical screens with small holes one of which is put on the transparency and the other in front of the photographic film right behind the scattering layer. The wave incident on the transparency passes through the hole in the screen. Because of multiple scattering in the layer, the transmitted wave becomes diffuse and almost uniformly illuminates the screen obscuring the film. This arrangement images a point to the film, the image intensity being proportional to the transmittance of the slide at the respective point. Thus identically moving the holes over the film and slide and repeating the registration process yields a correct image of the transparency on the film. This image can be either enlarged or reduced by either increasing or decreasing the amplitude of the motion of the second hole.

This example seem to be trivial. However, it includes the basic features of optical information transfer through turbid media. To be more specific (1) it is impossible to directly transfer an optical image without distortions and (2) it is necessary to establish an additional coupling between the input and output. This coupling is set up by a reference signal.

Speaking of the proposed methods of information transfer through turbid media, we should note above all the possibilities of memory effects considered above. Using these effects, one may estimate, for example, a variation in the direction of propagation or in the polarization state of the incident wave (Freund, 1990a). However, the possibilities of memory effects are rather limited. A development of the theory in this direction may be found in the work of Freund (1990b) who considered some characteristic properties of optical data transfer through scattering media by using the correlation analysis with reference signals. Berkovitz and Kaveh (1990c) reported the effect of scattering at an isolated scatterer on the mean transmission factor. In his later work, Berkovitz (1991) found that fluctuations of transmission factor are highly sensitive to motion of isolated scatterers.

Most of the results in this field are yet tentative in nature, so that the possibilities of progress in studies of the correlation phenomena are far from being exhausted. Note that, because of limited potentialities of measuring instruments, the object of investigations in classic transfer theory was for long time the energy characteristics alone, i. e., one-point second moments of the field. Such a description is not complete in the statistical sense and does not include higher and and many-point correlations of the field. However, these correlations may carry a definite additional information on the properties of the medium.

It is not a simple problem to measure the correlation characteristics of a field (especially in the optical range), but these difficulties are step by step overcome with the development of experimental instruments. Clearly it is the study of higher moments and statistical distributions of scattered radiation from which a significant advancement in both theory and experiment should be expected.

5.4 STRONG LOCALIZATION AND NONLINEAR TRANSFER THEORY

5.4.1 Effect of Strong Localization

Up to this point we have considered linear radiative transfer theory whose principal validity is restricted by the condition $\lambda << L_{ext}$, or, in the absence of losses, $\lambda << L_{sc}$.

We have already pointed out that, if this condition fails, i.e. in the region of high disorder, where the Ioffe–Rigel condition $\lambda \sim L_{sc}$ holds, a qualitatively new phenomenon of strong localization arises, whose description is beyond the scope of the theory under consideration. This phenomenon was predicted by Anderson (1958) in his classic work on the behavior of a quantum particle in a random potential.

Anderson demonstrated that a new localized regime arises for wave field in a scattering medium beginning with a certain level of disorder. In this regime, wave packets do not spread out, i.e. the eigenfunctions of the scattering system that describe free propagation of the wave function are standing waves exponentially vanishing at infinity in contrast to traveling quasi-plane waves considered in transfer theory. This is the reason why the description of Anderson's localization is far beyond the classic transfer theory.

Anderson's results are of importance for quantum theory of solids, particularly after Mott used the Anderson ideas to describe the metal-to-dielectric transition that occurs in systems with electric conductivity at low temperatures. Further development of the theory revealed that Anderson's transition to strong localization regime is a continuous phase transition that can be described in terms of the general theory of phase transitions such as diverging length, critical exponent, and order parameters (see, e.g., Ma, 1976). The importance of Anderson's results was confirmed by a 1977 Nobel award in physics.

Later it has been found that Anderson's transition is not a peculiar quantum effect and is related directly to the use of the Schrödinger equation. This transition owes its existence to wave interference phenomena and can appear in descriptions of different kinds of waves. Moreover, for one-dimensional and, probably, two-dimensional systems, the strong localization turns out to be a universal property that occurs at an arbitrary low level of disorder.

Linear transfer theory is not applicable to such systems when scattering occurs without frequency variations, thus forcing the researchers to use other, more complicated methods. For linear transfer theory, it is likely that, in the three-dimensional case, the localization occurs only under the condition of strong disorder $\lambda \sim L_{sc}$. For systems with weak fluctuations, the eigenfunctions are found to be delocalized, which relates them to familiar plane waves. A curve showing the transition to localized states in the space of disorder-describing parameters plus frequency is called the *mobility edge* (Mott, 1974).

For electrons in a solid, the condition of strong localization $\lambda \sim L_{sc}$ is realized relatively easy either by a small scattering length, or by a large wavelength λ. However, it is not a simple matter to construct the region of strong disorder for classic waves of different nature. Notwithstanding repeated attempts, no Anderson's transition has yet been observed for classic (acoustic or electromagnetic) waves in a three-dimensional system. Simple estimates of the scattering length in the independent-particle approximation ($L_{sc} \approx 1/\nu_0\sigma_1$; σ_1 is the cross section of an isolated scatterer, and ν_0 is the concentration of scatterers) show that the condition $\lambda \sim L_{sc}$ can be satisfied only under resonant conditions where the scattering cross section is large enough.

Some investigators have cast doubt on the principal feasibility of of Anderson's transition for classic waves because of pronounced losses in the medium that are absent for the wave function of electrons in a solid. Nevertheless, we consider it useful to elucidate certain features of strong localization to provide an insight into the limited possibilities of linear transfer theory.

The development of the theory of strong localization turned out to be a challenging problem. It has been approached by a variety of methods including numerical analysis, different forms of self-consistent theories, computational schemes based on perturbation theory; theories that describe localization by calculating Lyapunov's exponents, methods of percolation theory; and methods of the field theory such as the σ nonlinear model, and supersymmetric theories. An extensive literature is devoted to attempts of describing strong localization. We refer the reader to reviews of Kirkpatrick and Dorfman (1985) and Lee and Ramakrishnan (1985); to multiauthor volumes edited by Nagaoka and Fukuyama (1982) and by Kramers et al. (1985) that deal with strong localization of electrons; and to a multiauthor volume edited by Sheng et al. (1990) devoted to the search of localized states for classic waves.

Despite much attention drawn to the problem rigorous results are scarce. They are mainly related to one-dimensional problems that may be described in terms of 2×2 transition matrices (Erdos and Herndon, 1982). The best progress in the general theory of strong localization is usually associated with the work of Abrahams et al. (1979) who suggested a scaling theory.

The results of the self-consistent theory of Vollhardt and Wolfle (1980, 1982) are also widely accepted. Starting from the earlier self-consistent theories (Goetze, 1978) and using a perturbation theory in the diagram representation together with some auxiliary heuristic assumptions, these authors derived a self-consistent, nonlinear equation for a generalized diffusion coefficient that offers a reasonable pattern for Anderson's localization in one- and two-dimensional systems and allows the boundaries of the localization region to be estimated in the case of three-dimensional problems. The results of that theory are widely used to estimate the conditions for an incipient Anderson's transition in both quantum and classic systems (see, e.g., Kirkpatrick, 1985; Zhang et al., 1990).

In this section we briefly outline only one possible approach to the self-consistent theory of strong localization (Apresyan, 1989, 1990b). Using this approach, we show how the results of Vollhardt and Wolfle can be obtained without auxiliary heuristic assumptions accepted ad hoc.

The prime objective of this section is not to describe the phenomenon of strong localization, but rather to show how the conventional transfer theory should be revised to make possible analysis in the range of strong disorder $\lambda \sim L_{sc}$. A rigorous theory is not available yet for strong localization, and all methods known to us, including the one presented below, use hypotheses that extrapolate the estimates from the region of weak disorder to the region of high disorder.

5.4.2 Nonlinear Bethe–Salpeter Equation

An equation for second statistical moments, i. e. the Bethe–Salpeter equation (3.3.28), is a reasonable choice for describing Anderson's localization because this localization originates from phase correlations and results above all in a re-distribution of energy over the space. Indeed, on the one hand, second moments carry certain information on phases and, on the other hand, they are closely related to average energy characteristics of the field.

The strong localization effect is non-perturbative, so that it cannot be covered with the perturbation theory of any finite order in the denominator. Consequently a need exists for in a method beyond the scope of conventional linear theory. On the other hand, this effect, like the weak localization, is primarily caused by phase correlations of waves propagating in opposite directions. These correlations, as we have already seen, are taken into account by means of contributions from cyclic diagrams. However, in contrast to the weak localization, a simple addition of these contributions to the results of conventional transfer theory is inadequate here. It turns out that, for the Bethe–Salpeter equation, a nonlinear approximation can be constructed whose solution automatically includes the contributions from cyclic diagrams and gives the results of Vollhardt and Wolfle.

According to Apresyan (1990), the intensity operator K in the Bethe–Salpeter equation may be related to the required second moment of the Green's function $\Gamma = \langle G_1 G_2^* \rangle$ by the relationship

$$K = K^{\text{inv}} + SK\langle G_1 G_2^* \rangle K \,. \tag{5.4.1}$$

Here the operator S turns around the lower propagation lines (if any) in the diagram representation of K:

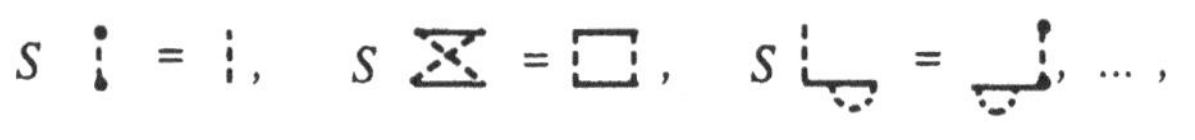

(5.4.2)

and K^{inv} is the initializing value of operator K defined as part of K invariant with respect to S. In the case of a medium with continuous Gaussian fluctuations, the diagram representation for K has the form:

(5.4.3)

Equation (5.4.1) together with the Bethe–Salpeter equation (3.3.28) rewritten as

$$(D_1 - D_2^*)\langle G_1 G_2^* \rangle = (\langle G_2^* \rangle - \langle G_1 \rangle)(K\langle G_1 G_2^* \rangle + 1) \tag{5.4.4}$$

sets up a system that defines K and $\langle G_1 G_2^* \rangle$. Equation (5.4.4) is derived by premultiplying eqn (3.3.28) by the operator $\langle G_2^* \rangle - \langle G_1 \rangle$ and is known as the generalized Boltzmann's equation. This form for the Bethe–Salpeter equation is best suited to radiative transfer theory. Indeed, starting with the scalar wave equation and taking the Fourier transformation with respect to the difference coordinate $\rho = x_1 - x_2$ at a constant center-of-gravity coordinate $R = x_1 - x_2 / 2$, we find that the operator $D_1 - D_2^*$ on the left-hand side of eqn (5.4.4) will contain the term $(\omega / c)\partial_T + K\nabla$ coinciding in form with the free transfer operator.

Assuming that the fluctuations are small, equation (5.4.1) may be solved for K by iterations. Retaining only the first term on the right-hand side of eqn (5.4.3) and introducing the notation

$$\delta\Gamma = \langle \tilde{G}_1 \tilde{G}_2^* \rangle - \langle G_1 G_2^* \rangle =$$ 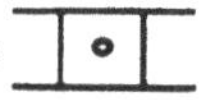 (5.4.5)

for the correlations of Green's operators G_1 and G_2^*, we arrive at an approximate expression for K in terms of $\delta\Gamma$:

$K \approx$  . (5.4.6)

Substituting this expression into the equation

$$(D_1 - D_2^*)\delta\Gamma = (\langle G_2^* \rangle - \langle G_1 \rangle) K (\delta\Gamma + \langle G_1 \rangle \langle G_2 \rangle), \tag{5.4.7}$$

following from the generalized Boltzmann's equation (5.4.4), we obtain a *nonlinear* equation for $\delta\Gamma$.

5.4.3 Selfconsistent Equation for Diffusion Coefficient

Solve equation (5.4.7) with the approximation (5.4.6) is a rather difficult task. However, it may be simplified if one would analyze only the asymptotic behavior of the solution for large distances from sources and for long times and pass to the diffusion approximation. We present some final results of such an approach omitting intermediate calculations.

In deriving the diffusion approximation for the transfer equation in Section 4.3, we have shown that the diffusion coefficient for radiation is given by the relation $D_B = c / (3\sigma_{tr})$ It can be rewritten in the form

$$\frac{1}{D_B} = \frac{3}{c} \int \sigma(1 - \mathbf{nn}') d\Omega_{n'} \ . \tag{5.4.8}$$

If we shall directly derive the diffusion approximation for the Bethe–Salpeter equation retaining only the first and the last diagrams in the approximation (5.4.6) and assume that the condition of weak disorder $\lambda < L_{sc}$ is met, we find that eqn (5.4.8) recasts into a nonlinear equation for the generalized diffusion coefficient $D = D(\omega_T, \omega)$ derived by Vollhardt and Wolfle (1980):

$$\frac{1}{D} = \frac{1}{D_B} + c_0 \frac{\lambda^2}{L_{sc}} \int_0^{q_s} \frac{q^2 dq}{q^2 - i\omega_T D}. \tag{5.4.9}$$

Here $D_B = cL_{sc}/3$ is the "Boltzmann" diffusion coefficient, c_0 is the factor about unity, and ω_T is the frequency due to nonstationarity and conjugated with the time argument $T = 1/2(t_1 + t_2)$ of the second moment of the field $\langle u_1 u_2^* \rangle$ (sometimes this frequency is called "external" in contrast to "internal" frequency ω corresponding to the difference argument $\tau = t_1 - t_2$). The cut-off parameter $q_{sc} \sim 1/L_{sc}$ arising in eqn (5.4.9) owes its existence to the fact that this equation is valid only for small $q < 1/L_{sc}$ when $\omega_T \to 0$.

5.4.4 Anderson Transition and Scattering Regimes in the Region of Localization

To give a general outlook of the modern ideas underlying the mechanism of Anderson's transition from delocalized to localized states, we first describe the physical pattern of this transition and then consider some consequences of eqn (5.4.9).

The scattering length $L_{sc} = L_{sc}(\omega)$, that depends on the "internal" frequency ω, is the simplest parameter characterizing "scattering strength". It is expected that Anderson's transition will appear when the Ioffe–Rigel condition $L_{sc} \sim \lambda$ holds; however, there exist certain phase correlations even in the range of weak disorder $L_{sc} >> \lambda$. These correlations are associated with weak localization that results in a coherent backscattering enhancement and in a rise of the probability for the wave packet to be returned back to the radiation source. Then, if the contribution from cyclic diagrams is allowed for, the effective coefficient of diffusion for radiation $D^{eff} = D(\omega_T \to 0)$ turns out to be smaller than its classic counterpart $D_B = cL_{sc}/3$ and is given by

$$D^{eff} = D_B\left(1 - c_1\left(\frac{\lambda}{L_{sc}}\right)^2\right), \tag{5.4.10}$$

where c_1 is the numerical factor [see, e.g., the review of Kaveh (1987) and Problem 5.1].

According to eqn (5.4.10), the diffusion coefficient decreases monotonically as the disorder increases (with growth of L_{sc}), so that the localized regime occurs at some critical value $L_{sc}(\omega^*) \sim \lambda$, where ω^* is the corresponding critical frequency. On transition D^{eff} equals zero, i.e. the diffusion of radiation vanishes and the waves fine themselves localized in a region of characteristic size about a certain *localization length* ξ. In the region of strong localization $L_{sc} \leq \lambda$, the formula (5.4.10) is no longer valid and the diffusion coefficient does not take negative values but remains equal to zero.

The localization length ξ depends on frequency, $\xi = \xi(\omega)$, and tends to infinity following the power law

$$\xi = \xi_0 \left| \frac{\omega^* - \omega}{\omega^*} \right|^{-1} \tag{5.4.11}$$

at the boundary of the localization region. This means that, in transition to the localized regime, localization tells first on large scales which decrease as the disorder increases deeper into the region of strong localization.

If we attempt to qualitatively describe the behavior of a wave in the region of strong disorder near the boundary of Anderson's transition, then, for $L_{sc} < \xi$ such a description might appear approximately as follows. At small distances from the source, or else for short paths $L < L_{sc}$, the radiation has no time to notice scattering. Here, the wave propagates much as if it were no scattering, and we face the regime of a *quasi-free propagation.*

Along paths $L_{sc} < L < \xi$, multiple scattering becomes significant, but the wave has still no time to notice that it is captured in some localized state. In this region, the behavior of the wave resembles a usual diffusion with the difference that the effective diffusion coefficient is now dependent on the characteristic dimension L, $D^{eff} = D^{eff}(L)$. Therefore, $L_{sc} < L < \xi$ may be called the region of *diffusion.*

Finally, on long tracks $L > \xi$, we deal with the *regime of localization* where the wave "recalls" that it is entrained. Beginning with the scale $L \sim \xi$, the effective diffusion coefficient $D^{eff}(L)$ tends to zero as $L \to \infty$, which means that there is no diffusion at infinity.

This simple qualitative pattern of propagation regimes in a medium with strong localization has not been rigorously proved but it has been confirmed in general terms by the results of different approximate treatments. It refers to the case of media without losses. The presence of even small losses significantly complicates the wave regimes both inside and outside the localization region (see, e.g., Sornette, 1989).

A considerable role of losses may be made clear with the aid of the following argument. In the absence of losses, the localized states may play the role of resonators with an infinitely high Q-factor and accumulate an infinite energy from a monochromatic source. Therefore losses, however small, are necessary for the steady-state regime could exist in the case of a monochromatic source in an infinite medium. In practice, the Q-factor of these resonators will be limited not only by losses but also by a finite size of real physical systems implying losses by radiation.

5.4.5 Diffusion with Memory and Localization Criteria

At present a number of criteria have been devised that make it possible to identify localized and delocalized regimes. For example Thouless (1974) has considered six such criteria. One of the most illustrative criteria for localization deals with the evolution of an initial disturbance. If an initial disturbance dies out with the passage of time, then the situation corresponds to a usual diffusion of radiation in a scattering medium, i. e. to the delocalized regime. On the contrary, if a spatial disturbance of a field does not vanish with time, then we deal with the localized regime that precludes a diffusion of radiation to infinity.We apply this simple criterion to the conventional diffusion equation

$$(\partial_T - D\Delta)\bar{I} = \delta(T)I^0(\mathbf{r}), \tag{5.4.12}$$

where $\bar{I}$ is the average intensity of radiation, D is the diffusion coefficient, and $I^0(\mathbf{r})$ is the initial disturbance. Taking the Fourier transform with respect to time and space coordinates yields

$$\begin{aligned}\bar{I}(\mathbf{k},\infty) &\equiv \lim_{T\to\infty} \bar{I}(k,T) = \lim_{\omega_T\to 0}\left(-i\omega_T I(k,\omega_T)\right) \\ &= \lim_{\omega_T\to 0}\frac{-i\omega_T I^0(k)}{-i\omega_T + Dk^2} \to 0 .\end{aligned} \tag{5.4.13}$$

in view of the limit theorem. This behavior corresponds to the usual diffuse spreading of an initial impulse signal.

In the case of strong localization the generalized diffusion coefficient D is determined by the solution of the self-consistent equation (5.4.9) and appears to be dependent on the "external" frequency ω_T that describes the transient processes, $D = D(\omega_T)$. In a time-domain representation like eqn (5.4.12) this fact is consistent generally with a nonlocal integral operator, which means that the wave possesses "temporal memory" abut the past states. Such a memory qualitatively changes the behavior of the solution to eqn (5.4.12).

Vollhardt and Wolfle (1980) have demonstrated that the asymptotic behavior of $D(\omega_T)$ as $\omega_T \to 0$ is of the form

$$D(\omega_T) = -i\omega_T\xi^2 + O(\omega_T^2) , \tag{5.4.14}$$

where ξ is the length of localization. This asymptotic specifies the behavior of the solution to eqn (5.4.12) for large times $T \to \infty$ and λ / L_{sc} exceeding some critical value corresponding to the Ioffe–Rigel criterion $\lambda \sim L_{sc}$. Substituting eqn (5.4.14) into eqn (5.4.13) yields

$$\bar{I}(k,\infty) = \lim_{\omega_T \to 0} \frac{-i\omega_T I^0(k)}{-i\omega_T (1+k^2\xi^2)}$$
$$= \frac{I^0(k)}{1+k^2\xi^2} \neq 0 \,, \tag{5.4.15}$$

which implies that the initial pulse does not spread out with time but is "locked" in a finite space region with dimensions approximately equal to ξ.

For one-dimensional problems, the length of localization ξ is about the free-path length of radiation L_{sc} for any level of disorder. For two-dimensional problems, the value of ξ is estimated as exponentially large compared to L_{sc}, but it remains finite for any level of disorder. In contrast, for the three-dimensional model considered here, the expression (5.4.14) for $D(\omega_T)$ is valid only beginning with a sufficiently high level of disorder $\lambda \geq L_{sc}$. The case of weak scattering fluctuations is not accompanied by localization so that the effective diffusion coefficient is nonzero, and we find ourselves in the validity region of conventional transfer theory.

Thus, the phenomenon of strong localization that breaks down the validity of conventional transfer theory may be described using nonlinear transfer theory with spatially and temporarily nonlocal scattering events. It might seem strange how linear initial equations have led to a *nonlinear* transfer equation without violation of the linearity of the initial problem. The answer to this doubt is that the nonlinear equation has been derived for the moments of the Green's function rather for the function of the problem under consideration, and this does not violate the linear nature of the field with respect to the sources of radiation.

To close his topic, we emphasize once again that the existent notions in the field of strong localization are incomplete. In particular, it is not yet clear how a scattering medium can be constructed for which the condition of strong disorder $\lambda \sim L_{sc}$ would be realized and Anderson's localization may be observed for classic waves.

PROBLEMS

Problem 5.1. Estimate the variation of the effective diffusion coefficient for a medium with an isotropic scattering pattern assuming that the effect of weak localization reduces to doubling the effective scattering cross section in a narrow range of angles $|\theta| < \theta_{coh} << 1$ near the backscattering direction.

Solution. For isotropic scattering, the diffusion coefficient (4.3.14) is related to the scattering cross section by

$$\frac{c}{3D} = \alpha_{sc} = \int \sigma(\mathbf{nn}') d\Omega_{\mathbf{n}'} = 4\pi\sigma \,. \tag{1}$$

Assuming that this relationship is valid also for the effective parameters D_{eff} and σ_{eff}, we have

$$\frac{c}{3D_{\text{eff}}} = \int \sigma_{\text{eff}} d\Omega_{\mathbf{n}'} = \int \sigma d\Omega_{\mathbf{n}'} + \int\limits_{|\theta| \le \theta_{\text{coh}}} \sigma d\Omega_{\mathbf{n}'}$$

$$= 4\pi\sigma + \pi\theta_{\text{coh}}^2 \sigma = \frac{c}{3D}\left(1 + \frac{\theta_{\text{coh}}^2}{4}\right). \quad (2)$$

We have

$$D_{\text{eff}} \approx D\left(1 - \frac{\theta_{\text{coh}}^2}{4}\right) \sim D\left(1 - c_1\left(\frac{\lambda}{L_{\text{sc}}}\right)^2\right), \quad (3)$$

where we estimated the enhancement angle as $\theta_{\text{coh}} \sim \lambda / L_{\text{sc}}$ and $c_1 = 1/4$.

The estimate (3) obtained for the effective diffusion coefficient coincides with the result of a more rigorous approach based on diagram expansions accurate to the numerical constant c_1. However, this estimate relies on an oversimplification of the physical pattern of weak localization. Indeed, in contrast to the conventional cross section σ that is formed in the far zone with respect to the scatterers, the scattering pattern due to weak localization is formed by multiple scattering in larger volumes of characteristic dimension $L^* > L_{\text{sc}}$. In estimations of the diffusion coefficient, this difference appears to be insignificant because the diffusion regime refers to the large-scale behavior of radiation with characteristic dimensions $L >> L_{\text{sc}}$.

Problem 5.2. Find the estimate (5.2.5) for the dynamic enhancement factor K_{d} in the case of a narrow beam propagating through an isolated phase inhomogeneity.

Solution. For a pencil incident beam, the phase inhomogeneity may be approximately replaced by an equivalent lens. We use the fact that the total energy flux is conserved in any section of the beam:

$$\int \mathcal{I} d^2 r_1 = \mathcal{I}\Sigma = \mathcal{I}_0 \Sigma_0, \quad (1)$$

where $\mathcal{I}$ and Σ are the characteristic values of the intensity and cross section of the beam, and $\mathcal{I}_0$ and Σ_0 are similar quantities of the incident beam. Assuming that the incident beam is paraxial and the diffraction is negligible, we may ignore the variation of incident beam that occurs in the absence of inhomogeneity. As a result, for the monostatic reception ($\mathcal{I}_+ = \mathcal{I}_- = \mathcal{I}$), we have

$$\int \mathcal{I}_+ \mathcal{I}_- \mathrm{d}^2 r_\perp \sim \mathcal{I}^2 \Sigma. \quad (2)$$

Substituting this estimate in (5.2.4) and (5.2.1) for K_{d} and using (1) yields eqn (5.2.5).

Problem 5.3. Suppose that a random quantity u can be represented as the sum (5.3.1) of an arbitrary number N of independent terms u_l each of which has only one nonzero cumulant $\langle u_l u_l^* \rangle$. Demonstrate that the intensity $I = |u|^2$ is then distributed with the Rayleigh law (5.3.8) and satisfies the condition $\langle I^m \rangle = m!\langle I \rangle$.

Solution. If the only nonzero cumulant of u_l is $\langle u_l u_l^* \rangle$, then u_l is a Gaussian random quantity, and, for any N, the quantity u, being a sum of independent Gaussian quantities, is also Gaussian. This basically proves the problem statement. However, it may be easily verified directly.

With this purpose we write the mean

$$\langle I \rangle = \sum_{l,m} \langle u_l u_m^* \rangle = \sum_l \langle |u_l|^2 \rangle = N\langle I_l \rangle \tag{1}$$

and other moments of intensity

$$\langle I^n \rangle = \sum_{l_1 \dots l_{2n}} \langle u_{l_1} u_{l_2} \dots u_{l_n} u_{l_{n+1}}^* u_{l_{n+2}}^* \dots u_{l_{2n}}^* \rangle. \tag{2}$$

The nonzero terms in this sum correspond to the partition of these moments into products of $\langle u_{l_i} u_{l_j}^* \rangle$

$$\begin{aligned}\langle I^n \rangle &= n! \sum \langle u_{l_1} u_{l_{n+1}}^* \rangle \langle u_{l_2} u_{l_{n+2}}^* \rangle \dots \langle u_{l_n} u_{l_{2n}}^* \rangle \\ &= n! \sum \langle I_{l_1} \rangle \langle I_{l_2} \rangle \dots \langle I_{l_n} \rangle = n! N^n \langle I_l \rangle^n,\end{aligned} \tag{3}$$

which means $\langle I^n \rangle = n!\langle I^n \rangle$. Using the integral representation for the delta function and taking the integral by the method of residues, we obtain the probability distribution

$$\begin{aligned}P(\vartheta) &= \langle \delta(\vartheta - I) \rangle = \int_{-\infty}^{\infty} \left\langle e^{ik(\vartheta - I)} \right\rangle \frac{dk}{2\pi} \\ &= \int_{-\infty}^{\infty} \frac{dk}{2\pi} e^{ik\vartheta} \sum_{n=0}^{\infty} \frac{(-ik)^n}{n!} n! \langle I^n \rangle \\ &= \int_{-\infty}^{\infty} \frac{dk}{2\pi} \frac{e^{ik\vartheta}}{1 - ik\langle I \rangle} = \frac{e^{-\vartheta/\langle I \rangle}}{\langle I \rangle}, \quad \vartheta \ge 0,\end{aligned} \tag{4}$$

which coincides with the Rayleigh law (5.3.8).

Problem 5.4. Show that, for a complex random field $u(x)$ with zero mean $\langle u \rangle = 0$ and only two nonzero cumulants

$$\langle u(x_1) u^*(x_2) \rangle_c = \langle u(x_1) u^*(x_2) \rangle \text{ and } \langle u(x_1) u(x_2) \rangle_c = \langle u(x_1) u(x_2) \rangle,$$

the real part $u' = \operatorname{Re} u$ and the imaginary part $u'' = \operatorname{Im} u$ are Gaussian, and, in the general case, they are mutually correlated fields.

Solution. Representing u' and u''as

$$u' = \frac{1}{2}(u + u^*) \text{ and } u'' = \frac{1}{2i}(u - u^*), \tag{1}$$

we note that the nth cumulants in $u'$and u'' can be linearly expressed in terms of the cumulants of the same order in u and u^*. Hence, u' and u'' are Gaussian random fields with only second order cumulants

$$\left.\begin{array}{l}\langle u'(x_1)u'(x_2)\rangle \\ \langle u''(x_1)u''(x_2)\rangle\end{array}\right\} = \frac{1}{2}\mathrm{Re}\langle u(x_1)u^*(x_2) \pm u(x_1)u(x_2)\rangle,$$
$$\langle u'(x_1)u''(x_2)\rangle = \frac{1}{2}\mathrm{Im}\left\langle u(x_1)u(x_2) - u(x_1)u^*(x_2)\right\rangle \tag{2}$$

being nonzero. If, in particular, $\langle u(x_1)u(x_2)\rangle = 0$, then we arrive at a circular Gaussian process with equal autocorrelations of the real and imaginary parts $\langle u'(x_1)u'(x_2)\rangle = \langle u''(x_1)u''(x_2)\rangle$ and with zero cross-correlation of u' and u'' taken at one point $\langle u'(x)u''(x)\rangle = 0$.

Problem 5.5. Consider a monochromatic plane wave with wave vector $\mathbf{k}_a$ incident on a static scattering layer, and the associated transmitted wave with wave vector $\mathbf{k}'_a$ (Fig. 5.18). If the amplitudes of these waves are $v(\mathbf{k}_a)$ and $u(\mathbf{k}'_a)$, then the scattering process $\mathbf{k}_a \to \mathbf{k}'_a$ will be completely described by the amplitude transmission factor $t(\mathbf{k}'_a \leftarrow \mathbf{k}_a)$ such that $u(\mathbf{k}'_a) = t(\mathbf{k}'_a \leftarrow \mathbf{k}_a)v(\mathbf{k}_a)$.

Determine the form of the correlation function of two transmission coefficients $\langle t(\mathbf{k}'_a \leftarrow \mathbf{k}_a)t^*(\mathbf{k}'_b \leftarrow \mathbf{k}_b)\rangle$ for a statistically uniform layer. (Naturally, the statistical uniformity in the plane z = const is implied, whereas the layer of *finite* thickness is inhomogeneous along the z axis by definition).

Solution. Let $G(\mathbf{r}, \mathbf{r}_0)$ be the response function describing the transmission through the scattering layer so that the amplitude $u(\mathbf{r})$ of the wave leaving the layer (at $z = L$) is related to the amplitude $v(\mathbf{r}_o)$ of the incident wave by

$$u(\mathbf{r}) = \int G(\mathbf{r}, \mathbf{r}_0)v(\mathbf{r}_0)d^2r_0\,. \tag{1}$$

The transmission factor is expressed — accurate to a numerical factor that for simplicity we set equal to unity — in terms of $G(\mathbf{r}, \mathbf{r}_0)$ as

$$t_a \equiv t(\mathbf{k}'_a \leftarrow \mathbf{k}_a) = \int e^{-i\mathbf{k}'_a\mathbf{r}}G(\mathbf{r}, \mathbf{r}_0)e^{i\mathbf{k}_a\mathbf{r}_0}d^2r_0d^2r. \tag{2}$$

From the statistical uniformity of the layer it follows that the second moment of the response function

$$\left\langle G(\mathbf{R}+\frac{\rho}{2},\mathbf{R}_0+\frac{\rho_0}{2})G^*(\mathbf{R}-\frac{\rho}{2},\mathbf{R}_0-\frac{\rho_0}{2})\right\rangle = G^{(2)}(\mathbf{R}-\mathbf{R}_0,\rho,\rho_0) \quad (3)$$

depends only on the difference $\mathbf{R}-\mathbf{R}_0$. Using eqns (2) and (3), after some straightforward algebra we obtain the desired moment

$$\langle t_a t_b^*\rangle = (2\pi)^2 \delta_\Sigma(\mathbf{k}'_{a\perp}-\mathbf{k}'_{b\perp}-\mathbf{k}_{a\perp}+\mathbf{k}_{b\perp})T(\mathbf{k}_{a\perp}-\mathbf{k}_{b\perp},\frac{\mathbf{k}'_{a\perp}+\mathbf{k}'_{b\perp}}{2},\frac{\mathbf{k}_{a\perp}+\mathbf{k}_{b\perp}}{2}). \quad (4)$$

Here, $\mathbf{k}_\perp$ denotes the component of an arbitrary vector $\mathbf{k}$ that is orthogonal to the z axis, the function

$$\delta_\Sigma(\mathbf{q}_\perp) = \int_\Sigma \exp(i\mathbf{q}_\perp\mathbf{r}_\perp)\frac{d^2r_\perp}{(2\pi)^2} \quad (5)$$

is expressed by the integral over the surface Σ of the layer and satisfies the conditions

$$\delta_\Sigma(\mathbf{q}_\perp) = \begin{cases} \delta(\mathbf{q}_\perp), & \text{as } \Sigma\to\infty, \\ \Sigma, & \text{for } q_\perp = 0, \end{cases} \quad (6)$$

and, finally, the quantity

$$T(\mathbf{k}'_\perp,\mathbf{k}''_\perp,\mathbf{k}'''_\perp) = \int \exp(-i\mathbf{k}'_\perp\mathbf{R}_\perp - i\mathbf{k}''_\perp\rho_\perp + i\mathbf{k}'''_\perp\rho_{0\perp})G^{(2)}(\mathbf{R}_\perp,\rho_\perp,\rho_{0\perp})d^2R_\perp d^2\rho_\perp d^2\rho_{0\perp} \quad (7)$$

is the Fourier-transform of the correlation coefficient (3).

Relationship (4), which defines the memory effect quantitatively, is valid for a layer with arbitrary scatterers. The meaning of this relationship may be given following Berkovitz et al. (1989). Let $\mathbf{R}_0$ be a point of incidence on and $\mathbf{R}$ be a point of emanation from the layer (Fig. 5.18). The respective transmission coefficient may be represented as the sum of contributions from all possible paths from $\mathbf{R}_0$ to $\mathbf{R}$. We assume that different propagation paths imply different phase increments, therefore, inside the layer, paths may be deemed mutually uncorrelated. Then, for an isolated

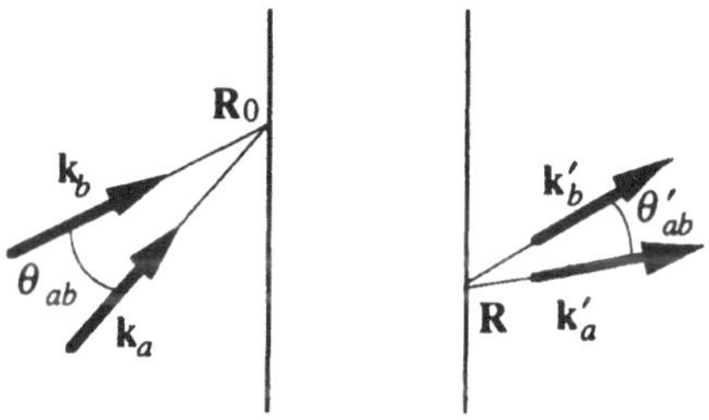

Figure 5.18

path connecting $\mathbf{R}_0$ and $\mathbf{R}$, the phase difference for two processes $\mathbf{k}_a \to \mathbf{k}'_a$ and $\mathbf{k}_b \to \mathbf{k}'_b$ depends only on the positions of points $\mathbf{R}_0$ and $\mathbf{R}$ and equals

$$\delta\psi = (\mathbf{k}'_a - \mathbf{k}'_b)\mathbf{R}_\perp - (\mathbf{k}_a - \mathbf{k}_b)\mathbf{R}_{0\perp}. \tag{8}$$

To calculate the correlation between the scattering processes under consideration, the phase factor $e^{i\delta\psi}$ must be averaged over the positions of entrance and exit points $\mathbf{R}_0$ and $\mathbf{R}$. In the general case, for a statistically nonuniform layer, the probability of transition $\mathbf{R}_0 \to \mathbf{R}$ depends on $\mathbf{R}_{0\perp}$ and $\mathbf{R}_\perp$ in a complicated manner, therefore the correlation noticeably differs from zero only in the trivial case of $a = b$. For a statistically homogeneous layer, this probability depends only on the difference $\mathbf{R}_{0\perp} - \mathbf{R}_\perp$ and does not depend on the position of the center of gravity $\mathbf{R}_c \equiv (\mathbf{R}_{0\perp} + \mathbf{R}_\perp)/2$. An averaging over $\mathbf{R}_c$ brings about a memory-effect-caused $\delta_\Sigma(\cdot)$ in eqn (4).

Problem 5.6. Assuming that the statistics of the radiation transmitted through a scattering layer is nearly circular Gaussian, find the normalized correlation function between the intensities of two scattering processes $\mathbf{k}_a \to \mathbf{k}'_a$ and $\mathbf{k}_b \to \mathbf{k}'_b$.

Solution. Let $I(\mathbf{k}'_a) = |u(\mathbf{k}'_a)|^2$ and $I(\mathbf{k}'_b) = |u(\mathbf{k}'_b)|^2$ be the radiation intensities of the two processes under consideration. The normalized correlation function of these quantities is defined as

$$C(\mathbf{k}_a, \mathbf{k}'_a, \mathbf{k}_b, \mathbf{k}'_b) = \frac{\langle \tilde{I}(\mathbf{k}'_a)\tilde{I}(\mathbf{k}'_b)\rangle}{\langle I(\mathbf{k}'_a)\rangle\langle I(\mathbf{k}'_b)\rangle} = \frac{\langle I(\mathbf{k}'_a)I(\mathbf{k}'_b)\rangle}{\langle I(\mathbf{k}'_a)\rangle\langle I(\mathbf{k}'_b)\rangle} - 1, \tag{1}$$

where $\tilde{I} = I - \langle I\rangle$ is the fluctuation of I. By assumption, the transmission factor $t_a = t(\mathbf{k}'_a \leftarrow \mathbf{k}_a)$ is a Gaussian random quantity. Substituting $I(\mathbf{k}'_a) = |t_a|^2 |v(\mathbf{k}_a)|^2$ and $I(\mathbf{k}'_b) = |t_b|^2 |v(\mathbf{k}_b)|^2$ in eqn (1) and taking into account the properties of Gaussian quantities, we obtain

$$C(\mathbf{k}_a, \mathbf{k}'_a, \mathbf{k}_b, \mathbf{k}'_b) = \frac{\left|\langle t_a t_b^* \rangle\right|^2}{\langle |t_a|^2\rangle\langle |t_b|^2\rangle}. \tag{2}$$

The statistical moments of the transmission factor involved may be expressed using eqn. (4) of Problem 5.5. We have finally

$$C(\mathbf{k}_a, \mathbf{k}'_a, \mathbf{k}_b, \mathbf{k}'_b) = \left| \frac{\delta_\Sigma(\mathbf{k}'_{a\perp} - \mathbf{k}'_{b\perp} - \mathbf{k}_{a\perp} + \mathbf{k}_{b\perp}) T\left(\mathbf{k}_{a\perp} - \mathbf{k}_{b\perp}, \frac{\mathbf{k}'_{a\perp} + \mathbf{k}'_{b\perp}}{2}, \frac{\mathbf{k}_{a\perp} + \mathbf{k}_{b\perp}}{2}\right)}{\Sigma\sqrt{T(0, \mathbf{k}'_{a\perp}, \mathbf{k}_{a\perp}) T(0, \mathbf{k}'_{b\perp}, \mathbf{k}_{b\perp})}} \right|^2. \tag{3}$$

Problem 5.7. Suppose that a plane wave undergoes a single scattering at a cluster of point scatterers. Show that if the direction of the incident wave is changed by a small angle θ, then the entire speckle structure, formed by the scattering field in the far field, is rotated through the same angle θ.

Solution. In the single scattering approximation, in far field, the amplitude of the scattered wave may be written as

$$u = \sum_j e^{-i(\mathbf{k}'-\mathbf{k})\mathbf{r}_j}, \tag{1}$$

where $\mathbf{k}'$ and $\mathbf{k}$ are the wave vectors of the incident and scattered waves, respectively, and $\mathbf{r}_j$ are the radius vectors of scatterers. This relationship depends only on the scattering vector $\mathbf{q} = \mathbf{k}' - \mathbf{k}$. In the case of propagation and scattering near the z axis (small angle approximation), vectors $\mathbf{k}'$ and $\mathbf{k}$ in the expression for $\mathbf{q}$ may be approximated by $\mathbf{k}'_\perp \approx k_0\theta_s$ and $\mathbf{k}_\perp \approx k_0\theta_i$, where $\mathbf{k}_0$ is the wave number, and θ_s and θ_i are the respective inclination angles (Fig. 5.19). We see that a simultaneous rotation of both $\mathbf{k}'$ and $\mathbf{k}$ through a small angle θ does not change the scattering vector $\mathbf{q} \approx k_0(\theta_s - \theta_i)$, thus proving the problem statement.

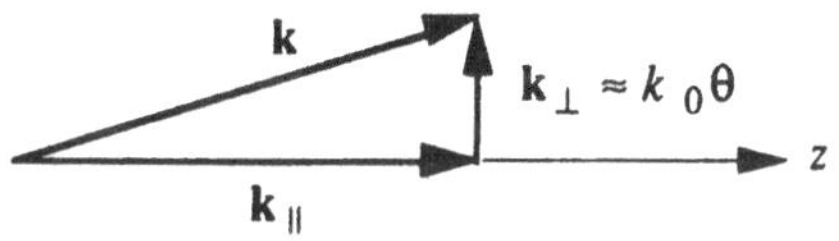

Figure 5.19. Components of the wave vector $\mathbf{k} = \mathbf{k}_\perp + \mathbf{k}_\parallel$ in the small angle scattering near the z axis.

Problem 5.8. In the case of a statistically homogeneous medium, the second moment of the Green's function Γ given in the *$R\rho$-representation* depends only on the difference $R - R_0$. Derive the generalized Boltzmann equation (5.4.4) for the complete Fourier transform of Γ in the model of the scalar wave equation $\mathrm{L} = \partial^2_{ct}(1+\mu(x)) - \Delta$, where a random function $\mu(x)$ describes fluctuation, using the Fourier transformation with respect to all arguments $R - R_0$, ρ and ρ_0.

Solution. Substituting the kernel of the second moment $\Gamma = \langle G_1 G_2 \rangle$ in the form

$$\Gamma = \left\langle G(R+\frac{\rho}{2}, R_0 + \frac{\rho_0}{2})G^*(R-\frac{\rho}{2}, R_0 - \frac{\rho_0}{2})\right\rangle = \Gamma(R - R_0, \rho, \rho_0)$$
$$= \int \Gamma(k_R, k_\rho, k_{\rho_0})\exp\left[i(k_R(R-R_0) + k_\rho\rho - k_{\rho_0}\rho_0)d^4k_R d^4k_\rho d^4k_{\rho_0}\right], \tag{1}$$

where $k_R = (\mathbf{k}_R, \omega_R)$, $k_\rho = (\mathbf{k}_\rho, \omega)$, $k_{\rho_0} = (\mathbf{k}_{\rho_0}, \omega_0)$, and the scalar product is defined as $kx = \mathbf{k}\cdot\mathbf{r} - \omega t$, into eqn. (5.4.4) and performing the Fourier transformation we obtain

$$\left[-i\frac{2\omega_T\omega}{c^2}+2i\mathbf{k}_R\mathbf{k}_\rho-i\left(V^{\mathrm{eff}}(k_\rho+\frac{k_R}{2})-V^{\mathrm{eff}*}(k_\rho-\frac{k_R}{2})\right)\right]\Gamma(k_R,k_\rho,k_{\rho_0})$$
$$=-i\left[G(k_\rho+k_R/2)-G^*(k_\rho-k_R/2)\right]$$
$$\times\left\{\int K(k_R,k_\rho,k'_\rho)\Gamma(k_R,k'_\rho,k_{\rho_0})d^4k'_\rho+\delta(k_\rho-k_{\rho_0})\right\}. \quad (2)$$

Here ω and ω_0 are the "internal" frequencies associated with the difference arguments τ and τ_0, and ω_T is the "external" frequency associated with the center of gravity T, $V^{\mathrm{eff}}(k)=e^{-\mathrm{i}kx}V^{\mathrm{eff}}e^{ikx}$,

$$G(k)=e^{-\mathrm{i}kx}\langle G\rangle e^{\mathrm{i}kx}=\left(-\left(\frac{\omega}{c}\right)^2+\mathbf{k}^2-V^{\mathrm{eff}}(k)\right)^{-1},$$

$(2\pi)^8K(k_R,k_\rho,k_{\rho_0})$ is the kernel of the intensity operator in the representation (1), and $\delta(k_\rho-k_{\rho_0})$ is the four-dimensional delta function.

The exact equation (2) is identical in form with the transfer equation for scattering events that are nonlocal in space and time. If the medium is also stationary and $\mu(x)=\mu(\mathbf{r})$, then scattering occurs without a change in frequency, and both K and Γ in eqn (2) contain delta functions of difference frequencies, so that equation (2) somewhat simplifies.

6. Related Problems

Numerous branch-offs develop hand in hand with the fundamental, main-line avenues of research in transfer theory. In this chapter we attempt to elucidate several of such side lines that were nearly omitted in traditional treatises on transfer theory.

In the first place, this is the radiative transfer through turbid media with random parameters (Section 6.1). In the circumstances, it is expedient to consider parameters in the transfer equation as random quantities with "secondary" averaging.

Section 6.2 describes the "warming" and "cooling" effects for random media and rough surfaces, whose thermal radiation depends on the nature of random inhomogeneities. Inclusion of these effects appears to be necessary in many problems related to passive location of natural objects.

Section 6.3 dwells on the application of the methods of transfer theory to acoustical problems and Section 6.4 covers wave propagation in multimode waveguides.

Finally, Section 6.5 takes a quick look at nonlinear problems of transfer theory and some related problems. This chapter encloses very dissimilar problems and thus looks necessarily like a review. For a more detailed presentation of the covered problems, the reader is referred to references cited in this chapter.

6.1 TURBID MEDIA WITH RANDOM PARAMETERS. SECONDARY AVERAGING OF THE TRANSFER EQUATION

6.1.1 Transfer Equation with Random Parameters

The concept of turbid media with random parameters may look strange at first sight because turbid media themselves are the objects with random characteristics. Nevertheless, the concept of *random turbid media* becomes justified if one takes into account the *scale hierarchy* in natural media.

The scattering cross-section σ and extinction coefficient α are introduced into the transfer equation as local parameters, i.e. averaged over a certain volume V. Dimensions of the averaging volume $l_{av} \sim V^{1/3}$ must evidently be larger than the correlation length l_{micro} of microinhomogeneities:

$$l_{av} >> l_{micro} \, . \tag{6.1.1}$$

Consider a turbid medium, whose local parameters α and σ exhibit slow variations with macroscales l_{macro} much larger than the averaging scale l_{av}. If these variations are of irregular nature and follow a statistical pattern, then it is appropriate to consider σ and α as random functions of coordinates in the range

$$l_{macro} >> l_{av} >> l_{micro}. \tag{6.1.2}$$

This switching from micro- to macro-scales in transfer theory has its analogy in hydrodynamics. A passage from microscopic to macroscopic (hydrodynamic) description implies consideration of quantities averaged over a certain region satisfying the condition (6.1.1). In this way macroscopic characteristics of fluid, such as pressure, density, and velocity, evolve. Nevertheless, averaged, i.e. macroscopic, characteristics may, in turn, become the subject of statistical analysis when large-scale processes satisfying the condition (6.1.2) are considered. The statistical theory of turbulence is build in this way.

Taking this into account, consider the random values of extinction coefficient α and scattering cross-section σ by setting

$$\begin{aligned} &\alpha(\mathbf{r}) = \bar{\alpha} + \tilde{\alpha}, \\ &\sigma(\mathbf{r},\mathbf{n} \leftarrow \mathbf{n}') = \bar{\sigma} + \tilde{\sigma}. \end{aligned} \tag{6.1.3}$$

We write the transfer equation with random parameters in the symbol form

$$\frac{dI}{ds} = AI, \tag{6.1.4}$$

where $dI / ds = \mathbf{n}\nabla I$ and A is the operator, introduced for brevity, that operates in accordance with the rule

$$AI = -\alpha(\mathbf{r})I + \int \sigma(\mathbf{r},\mathbf{n} \leftarrow \mathbf{n}')I(\mathbf{n}')d\Omega'. \tag{6.1.5}$$

The radiance I in eqn (6.1.4) becomes now a random quantity, whose fluctuations are caused by the fluctuations of the operator A. Considering eqn (6.1.4) as an initial *stochastic* equation, we could follow the procedure used in Section 3.3.3 for deriving the Dyson equation and immediately write the equation for the mean $\bar{I}$. However, we repeat simple calculations leading to equation for $\bar{I}$ to alleviate the perception.

The operator A is characterized by its mean value $\bar{A}$ and the fluctuation component $\tilde{A}$

$$A = \bar{A} + \tilde{A}, \tag{6.1.6}$$

that are expressed through $\bar{\alpha}$, $\tilde{\alpha}$, $\bar{\sigma}$, and $\tilde{\sigma}$ in an obvious manner. The radiance I can also be represented as the sum of its mean and fluctuating parts:

$$I = \bar{I} + \tilde{I}. \tag{6.1.7}$$

Substituting eqns (6.1.6) and (6.1.7) into the transfer equation (6.1.4), we single out the mean component

$$\frac{d\bar{I}}{dS} = \bar{A}\bar{I} + \langle \tilde{A}\tilde{I} \rangle, \tag{6.1.8}$$

and the fluctuating component with zero mean

$$\frac{d\tilde{I}}{dS} = \tilde{A}\bar{I} + \bar{A}\tilde{I} + \tilde{A}\tilde{I} - \langle \tilde{A}\tilde{I} \rangle. \tag{6.1.9}$$

The system of equations (6.1.8) and (6.1.9) cannot be decoupled into independent equations for $\bar{I}$ and $\tilde{I}$, therefore the evaluation of the mean radiance $\bar{I}$, the correlation function $\langle \tilde{I}_1 \tilde{I}_2 \rangle$, and higher moments $\langle \tilde{I}_1 ... \tilde{I}_m \rangle$ is possible only by using some approximate methods. We now consider a method that reduces the transfer equation with random parameters to an effective transfer equation for the mean radiance $\bar{I}$, i.e. to a transfer equation with effective parameters. However, as an introduction, we consider the bleaching effect in turbid media that predetermined interest to radiative transfer through media with fluctuating parameters.

6.1.2 Bleaching Effect in Turbid Media with Random Parameters

This effect might be first discussed by Dolin (1984) and Borovoi (1984). Here we will follow a recent paper by Kleeorin et al. (1989).

The bleaching effect manifests itself in that the scattering medium with random variations of its parameters becomes more transparent on average than the same medium without fluctuations. To put this another way, the medium exhibits is lowest transparency when its scattering inhomogeneities are distributed uniformly.

This peculiarity of radiative transport in a random turbid medium may be related to the physically obvious principle of least attenuation. According to this principle, in the presence of many competitive paths (or propagation channels) in the medium, radiation comes to the observer over the path with the lowest attenuation, because this path contributes to the observed intensity most effectively.

Figure 6.1 illustrates the least attenuation principle. The dark areas represent sites of high attenuation. Among all possible paths from the source S to the observer O, the path A yields the maximum intensity contribution because it is mainly through light regions, whereas other virtual paths cause higher attenuation. The enhancement of intensity at the observation point due to the contributions from paths with lower attenuation is greater than the loss of intensity due to paths with higher attenuation.

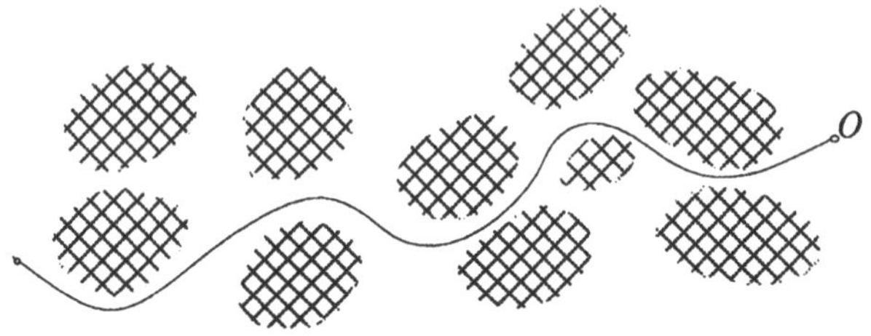

Figure 6.1. Illustration of the least attenuation principle: the radiation arrives at the observation point over a variety of propagation channels, but the channel corresponding to the least attenuation prevails.

To make this point clear, we consider the following simple model. Let the extinction coefficient α be a fluctuating quantity that can take on two values: $\langle\alpha\rangle+\Delta$ and $\langle\alpha\rangle-\Delta$ with equal probability $P_+=P_-=1/2$. Then the average value of the respective transmittance (i.e. the attenuation factor) $e^{-\alpha}$ is

$$\left\langle e^{-\alpha}\right\rangle=\frac{1}{2}e^{-\langle\alpha\rangle}\left(e^{\Delta}+e^{-\Delta}\right)$$

what is always greater than its unperturbed value $e^{-\langle\alpha\rangle}$. The least attenuation principle was first formulated heuristically (without mentioning radiative transfer theory) by Kravtsov, Tinin, and Cherkashin (1979) for problems of radio wave propagation in the ionosphere.

We support the above qualitative considerations by a simple calculation that can provide insight into the physical mechanism of the bleaching effect, but cannot give its complete universal description.

We now neglect the scattering and intrinsic radiation of the medium and consider a special case with fluctuations of the absorption factor $\alpha=\alpha_{\text{abs}}$ only. The transfer equation (6.1.4) then takes the form

$$\frac{dI}{ds}=-\alpha I\,, \qquad (6.1.10)$$

because $A=-\alpha$ under the above assumptions.

Assuming that the fluctuations $\tilde{\alpha}$ are small compared to $\overline{\alpha}$ and the fluctuations of radiance $\tilde{I}$ are small compared to $\overline{I}$, we neglect the difference $\tilde{\alpha}\tilde{I}-\langle\tilde{\alpha}\tilde{I}\rangle$in eqn (6.1.9) as a quantity of the second order of smallness with respect to the main term $\overline{\alpha}\overline{I}$.

The neglect of $\tilde{\alpha}\tilde{I}-\langle\tilde{\alpha}\tilde{I}\rangle$ is equivalent to the Bourret approximation for the Dyson equation for $\overline{I}$. True, here all the quantities are energy rather than field variables. Note that the derivation of the Dyson equation for $\overline{I}$ from eqns (6.1.8) and (6.1.9) was considered by Vardanyan (1988) and Manning (1989).

Neglecting difference $\tilde{\alpha}\tilde{I} - \langle \tilde{\alpha}\tilde{I} \rangle$ caries eqn (6.1.9) to the form (with allowance for $A = -\alpha$)

$$\frac{d\tilde{I}}{ds} = -\tilde{\alpha}\bar{I} - \bar{\alpha}\tilde{I} ,$$

that admits an explicit solution

$$\tilde{I} = -\int_{-\infty}^{s} \tilde{\alpha}\bar{I} \exp[\bar{\alpha}(s' - s)]ds' - \int_{s}^{\infty} \tilde{\alpha}\bar{I} \exp[\bar{\alpha}(s - s')]ds' , \tag{6.1.11}$$

which satisfies the reasonable requirement $\tilde{I} \to 0$ as $\tilde{\alpha} \to 0$. Substituting eqn (6.1.11) into eqn (6.1.8), we obtain a closed integro-differential equation for the mean radiance $\bar{I}$ — the Dyson equation in the Bourret approximation:

$$\frac{d\bar{I}}{ds} = -\bar{\alpha}\bar{I} + \int_{-\infty}^{\infty} \langle \tilde{\alpha}(\mathbf{r})\tilde{\alpha}(\mathbf{r}') \rangle \exp(-\bar{\alpha}|s - s'|)\bar{I}(r')ds' . \tag{6.1.12}$$

Here, s corresponds to the point $\mathbf{r}$, and s' corresponds to the point $\mathbf{r}'$.

Define the correlation length of the absorption factor by the integral formula

$$l_c = \int_{-\infty}^{\infty} \frac{\langle \tilde{\alpha}(\mathbf{r})\tilde{\alpha}(\mathbf{r}') \rangle}{\langle \tilde{\alpha}^2 \rangle} d\xi', \quad \xi' = |\mathbf{r} - \mathbf{r}'| ,$$

and assume that the correlation length l_c of its inhomogeneities is small compared with the absorption length $l_{abs} = 1/\bar{\alpha}$ coinciding with the characteristic variation length of the mean radiance $\bar{I}$. Thus, we assume that

$$\bar{\alpha} l_c << 1 . \tag{6.1.13}$$

Then the exponential factor in eqn (6.1.12) may be replaced with unity, and all the integral terms may be replaced with the local value

$$\int \langle \tilde{\alpha}(\mathbf{r})\tilde{\alpha}(\mathbf{r}') \rangle \exp(-\bar{\alpha}|s - s'|)\bar{I}(\mathbf{r}')ds' \approx \langle \tilde{\alpha}^2 \rangle l_c \bar{I}(\mathbf{r}) .$$

As a result, eqn (6.1.2) takes the form

$$\frac{d\bar{I}}{ds} = -\alpha_{\text{eff}} \bar{I} , \tag{6.1.14}$$

where the quantity

$$\alpha_{\text{eff}} = \overline{\alpha} - \left\langle \tilde{\alpha}^2 \right\rangle l_c \tag{6.1.15}$$

represents the effective absorption factor in the medium with fluctuating parameters. This quantity is always smaller than the mean value $\overline{\alpha} : \alpha_{\text{eff}} < \overline{\alpha}$, i.e. the turbid medium with fluctuating parameters appears more transparent than the nonfluctuating medium with the mean absorption factor $\overline{\alpha}$.

The simple analysis given above shows that the bleaching of turbid media with fluctuating parameters is associated with the correlation, or more correctly, with the anticorrelation between the fluctuations of the radiance $\tilde{I}$ and the absorption $\tilde{\alpha}$: the higher the absorption α the lower the radiance $\tilde{I}$. It is due to this anticorrelation that radiation tends to select a path corresponding to the smallest absorption.

6.1.3 Transfer Equation with Effective Parameters

If fluctuating parameters include not only the absorption factor α_{abs} but also the cross-section per unit volume σ, then mean-radiance calculations become more complicated. The equation for the mean radiance $\overline{I}$ does not reduce, in general case, to a radiative transfer equation. It becomes, however, a transfer equation with effective parameters if the condition (6.1.13) holds, i.e. if the optical thickness of an isolated inhomogeneity is small.

Suppose that the fluctuations in absorption and scattering are brought about by the same cause, namely, by the fluctuations in the concentration of absorbing and scattering particles. In this case, we may write

$$\begin{aligned} \tilde{\alpha} &= \mathrm{B}\tilde{n}\,, \quad \tilde{\sigma} = \mathrm{D}\tilde{n}\,, \\ \overline{\alpha} &= \mathrm{B}\overline{n}\,, \quad \overline{\sigma} = \mathrm{D}\overline{n}\,, \end{aligned} \tag{6.1.16}$$

where $\overline{n}$ is the average concentration of particles and $\tilde{n}$ is the fluctuating part of the concentration. Calculations identical with those of previous section with the substitution of $-\tilde{A} = \tilde{\alpha} - \tilde{\sigma} = (\mathrm{B}-\mathrm{D})\tilde{n}$ for $\tilde{\alpha}$ and $-\overline{A} = (\mathrm{B}-\mathrm{D})\overline{n}$ for $\overline{\alpha}$ yield the transfer equation (6.1.14), where the operator

$$A_{\text{eff}} = -(\mathrm{B}-\mathrm{D})\overline{n} - (\mathrm{B}-\mathrm{D})^2 \left\langle \tilde{n}^2 \right\rangle l_c \tag{6.1.17}$$

replaces $-\alpha_{\text{eff}}$.

Taking into account the notations (6.1.16), we may represent the operator relationship (6.1.17) in the equivalent form $A_{\text{eff}} = -\alpha_{\text{eff}} + \sigma_{\text{eff}}$, where $\alpha_{\text{eff}} = \overline{\alpha}(1 - \overline{\alpha}\varepsilon l_c)$, and the operator σ_{eff} has the kernel

$$\sigma_{\text{eff}}(\mathbf{n} \leftarrow \mathbf{n}') = (1 - 2\overline{\alpha}\varepsilon l_c)\overline{\sigma}(\mathbf{n} \leftarrow \mathbf{n}')$$
$$+ \varepsilon l_c \int \sigma(\mathbf{n} \leftarrow \mathbf{n}'')\sigma(\mathbf{n}'' \leftarrow \mathbf{n}')d\Omega''. \qquad (6.1.18)$$

Here $\varepsilon = \tilde{n}^2 / \overline{n}^2$ denotes relative fluctuations of concentration.

We see that the fluctuations of medium parameters result in a familiar decrease of the extinction factor and in some change of the scattering cross-section (6.1.18). First, in the expression for the average cross-section $\overline{\sigma}$, the factor $1-2\varepsilon l_c \overline{\alpha}$ appears that is even smaller than the factor $1-\varepsilon l_c \overline{\alpha}$ in the expression for α_{eff}. Second, in the relation (6.1.18) for the kernel, the integral term, that corresponds to "double scattering" of radiance by inhomogeneities of the turbid medium, appears. In the general case, the scattering pattern of that term is broader than that of the first term. The first and second terms taken together produce weaker scattering as compared to the case of a homogeneous medium. Thus, the above trend to decrease the absorption finds its analog in a decrease of the effective scattering.

6.1.4 Manifestations of the Bleaching Effect

This effect may manifest itself first in astrophysics. Inhomogeneities in plasma density and in the density of interstellar gas clouds may be the reason for an enhancement of brightness temperature of space objects. Vardanyan (1988) and Manning (1989) estimated the magnitude of this effect. Of course, in this case, it is required that equation (6.1.4) included an additional term describing the strength of sources.

The inhomogeneous snow cover is another physical system of practical interest. The brightness temperature of the earth's surface covered with snow is mainly decided by the transparency of the snow cover that shields the relatively warm earth from the air. Inhomogeneities increase the transparency of the snow cover and thereby rise its brightness temperature. This problem was theoretically considered by Kleeorin et al. (1989), whereas experimental evidence in support of the bleaching effect may be found in the paper of Wen et al. (1990). In parallel with the trend to increase the brightness temperature due to large-scale inhomogeneities, there exist another trend in the case of the thermal radiation of snow and ice: a reduction of the brightness temperature because of scattering. We will dwell on this effect in the ensuing section.

We mention also the optical effect connected with inhomogeneities in a continuous cloud cover: inhomogeneities, even small, may considerably increase the radiance of the earth's surface.

Finally, we notice one implication for systems of protection from neutron fluxes: the homogeneity of screening layers may be a more important factor than the total mass of the protection cover. This recommendation may be deduced from the heuristic principle of least attenuation: any inhomogeneities of the protecting screen, not necessarily random, act as "loop-holes" for neutrons leaving the reactor.

6.2 THERMAL RADIATION OF SCATTERING MEDIA AND SURFACES

6.2.1 "Warming" and "Cooling" Effects in Scattering Media

Remote sensing is now widely accepted to study the atmosphere, ocean, and land using receivers of thermal radiation aboard airplane and satellites. Brightness temperature measured with such receivers depends on the parameters of these media and above all on their refraction and absorption indexes. Recently it has been found that random volume inhomogeneities and irregularities of the interface can considerably contribute to the brightness temperature. Therefore, studies of the thermal radiation of natural media are an important source of information on regular and random parameters of these media.

A quantitative analysis of the effect of random inhomogeneities on thermal radiation can be carried out in principle using transfer theory. For scattering media, this was done, for example, by England (1974) and Tsang and Kong (1975). For random boundaries, calculations on transfer theory are technically difficult especially if the dimensions of surface irregularities are small compared to the wavelength.

In the circumstances, it turns out to be more convenient to resort to the electromagnetic theory of thermal fluctuations especially using the diffraction generalization of Kirchhoff's law. According to this generalization, the power of thermal radiation from a body or a medium at a given frequency and at a given point (or in a given direction) is proportional to the power of an auxiliary wave absorbed by the body or the medium. The auxiliary wave is assumed to be transmitted from the desired point (or in desired direction) and has the desired polarization and frequency.

In this formulation, the Kirchhoff statement — that the radiating and absorbing powers of a body are proportional to each other — as was initially derived in the limit of geometrical optics, is valid also for bodies with dimensions below or about the wavelength. This extension of the Kirchhoff law is given, in particular, in the book of Rytov, Kravtsov, and Tatarskii (1989a).

A general statement can be derived from the diffraction formulation of the Kirchhoff law: this statement concerns the conditions for "warming" and "cooling"of radiating bodies and media with random inhomogeneities. Let T_0 be the physical temperature of a material medium, and $T_b(\mathbf{n})$ be its brightness temperature in the direction $\mathbf{n}$ (Fig. 6.2). In accordance with the extended Kirchhoff law, the brightness temperature $T_b(\mathbf{n})$ differs from T_0 by the factor $\chi_{\text{eff}}(\mathbf{n}_0)$ that characterizes the effective absorbing power, i.e. the proportion of absorbed power arrived from the auxiliary source in the direction $\mathbf{n}_0 = -\mathbf{n}$:

$$T_b(\mathbf{n}) = T_0 \chi_{\text{eff}}(\mathbf{n}_0). \tag{6.2.1}$$

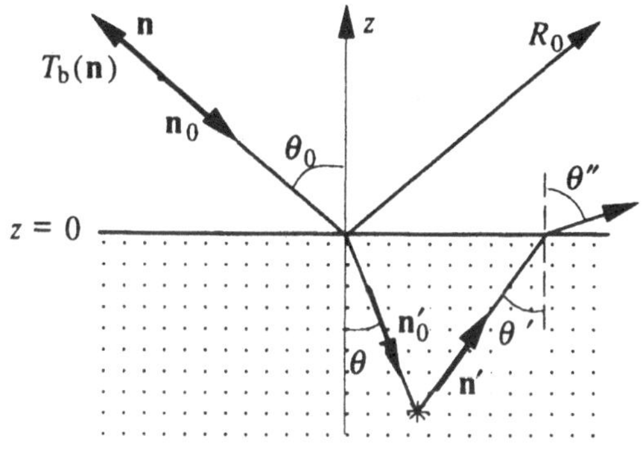

Figure 6.2. Illustration to calculation of thermal radiation of scattering half-space $z < 0$.

The absorption power, which is also called the degree of darkening, may be expressed in terms of the effective reflection factor R_{eff} that characterizes the proportion of reflected and scattered power, i.e. the proportion of the power, passed in the free upper half-space $z > 0$ (Fig. 6.2). In this case, $\chi_{\text{eff}}(\mathbf{n}_0) = 1 - R_{\text{eff}}(\mathbf{n}_0)$ and formula (6.2.1) takes the form

$$T_b(\mathbf{n}) = T_0\left[1 - R_{\text{eff}}(\mathbf{n}_0)\right]. \tag{6.2.2}$$

Using this formula, we may formulate the conditions for "warming" or "cooling" the medium when random inhomogeneities arise. Let $R_0(\mathbf{n}_0)$ be the energy reflection factor of a wave from the plane interface between a vacuum and a material medium. If inhomogeneities cause an increase in the reflectance, i.e. if

$$R_{\text{eff}}(\mathbf{n}_0) > R_0(\mathbf{n}_0), \tag{6.2.3}$$

then the brightness temperature decreases, and this may be interpreted as a "cooling" of the medium.Conversely, if

$$R_{\text{eff}}(\mathbf{n}_0) < R_0(\mathbf{n}_0), \tag{6.2.4}$$

then a "warming" of the medium occurs. Both effects may occur in actual media depending upon the nature of inhomogeneities. Let us consider several examples.

6.2.2 Reduction of the Brightness Temperature of the Thermal Radiation of Antarctic Glaciers

Airplane and satellite experiments have revealed that, in the centimeter range, the brightness temperature of antarctic glaciers is considerably lower than their physical temperature T_0 and, moreover, it is lower than the temperature $T_0(1-R_0)$ corresponding to a plane air-ice interface. Gurvich at al. (1973) and Gurvich and Krasilnikova (1977) showed that the reduction of the temperature may be naturally attributed to the scattering at random inhomogeneities of glacier snow that exhibits a clear-cut layered structure.

With bulk random inhomogeneities, the effective reflectivity $R_{\text{eff}}(\mathbf{n}_0)$ may be represented as the sum of the unperturbed reflectivity R_0 and of the term Q that characterizes the proportion of radiation returned to the upper half-space $z > 0$ after scattering in the lower half-space $z < 0$: $R_{\text{eff}} = R_0 + Q$. Then

$$T_b = T_0\left[1-(R_0+Q)\right]. \tag{6.2.5}$$

If the scattering is not very strong, one may use a perturbation method (i.e. the Born approximation) to estimate Q. Denote by

$$\sigma(\mathbf{n}') = \frac{\pi k^4}{2}\Phi_\varepsilon(\mathbf{q})\sin^2\chi' \tag{6.2.6}$$

the scattering cross-section per unit volume per unit solid angle in the direction $\mathbf{n}'$, where $k = 2\pi n/\lambda_0$ is the wave vector in the medium, χ' is the angle between the vector $\mathbf{n}'$ and the electric strength of the auxiliary wave $\mathbf{E}_0'$ in the lower medium, $\Phi_\varepsilon(\mathrm{k})$ is the spectral density of inhomogeneities, and $\mathbf{q} = k(\mathbf{n}_0' - \mathbf{n}')$ is the scattering vector.

The absorption in the lower medium $z < 0$ may be accounted for by the factor $P(z,\theta) = \exp(\alpha z/\cos\theta)$, where α is the energy absorption factor and $|z|/\cos\theta$ is the oblique length of the ray in the media from the boundary to the scattering point (Fig. 6.2). The angle θ is defined by Snell's law $n\sin\theta = \sin\theta_0$, where θ_0 is the angle of incidence of the auxiliary wave. When the auxiliary wave crosses the interface, the proportion $1-R_0$ of the power is reflected, and the energy flux density of the transmitted wave is changed by refraction. Both effects may be accounted for by the factor $(1-R_0)\cos\theta_0/\cos\theta$ of σ. As a result, the effective scattering cross-section takes the form

$$\sigma_{\text{eff}}(\mathbf{n}_{\text{sc}}') = (1-R_0)\frac{\cos\theta_0}{\cos\theta}\exp\left(\frac{\alpha z}{\cos\theta}\right)\sigma(\mathbf{n}'). \tag{6.2.7}$$

In order to calculate Q, i.e. the part of power that is scattered into the air through the unit surface of the boundary, we need to integrate eqn (6.2.7) with respect to z over the entire thickness H of the scattering medium and over the solid angle 2π corresponding to the upper hemisphere. We have

$$Q = \int_{-H}^{0} dz \int_{0}^{\pi/2} \sin\theta'' d\theta'' \int_{0}^{2\pi} d\varphi'' \frac{\cos\theta'}{\cos\theta''} (1 - R') P(z,\theta) \sigma_{\mathrm{eff}}(\mathbf{n}'), \tag{6.2.8}$$

where the factor P describes the absorption. The reflectivity factor R' is the function of the angle of incidence θ' of the scattered waves on the boundary and of their polarization, which is defined by the vector $\mathbf{n}' \times (\mathbf{E}_0' \times \mathbf{n}')$. The angle θ'' of the wave after its emanation from the medium is related to θ' by Snell's law.

Substituting eqn (6.2.8) into eqn (6.2.5), we may in principle analyze the brightness temperature as a function of all problem parameters : the sight angle θ_0, wavelength λ_o, polarization $\mathbf{E}_0$, and the thickness H of the scattering medium. The calculations for Q are simplified if the thickness of the scattering layer H is larger than the absorption length $1/\alpha$ and if inhomogeneities are "oblate" in the direction of the z-axis, i.e. the correlation length of inhomogeneities in the horizontal plane is much larger than that in the vertical direction. The layered structure of the medium is associated with the forming processes of the glacier snow.

In this case, it is possible to integrate eqn (6.2.8) over the angles. As a result, we obtain

$$Q = \frac{\pi k_0^2 n^2 \cos\theta_0}{4\alpha} V(2k\cos\theta)(1 - R_o)^2 \sigma_\varepsilon^2 \tag{6.2.9}$$

for the horizontal polarization (vector $\mathbf{E}_0$ is perpendicular to the plane of incidence) and

$$Q = \frac{\pi k_0^2 n^2 \cos\theta_0}{4\alpha} \cos^2(2\theta)\cos\theta \, V(2k\cos\theta)(1 - R_o)^2 \sigma_\varepsilon^2 \tag{6.2.10}$$

for the vertical polarization. The quantity $V(\kappa)$ is defined as the "vertical" one-dimensional spectrum of inhomogeneities, i.e. the Fourier transform of the normalized correlation function of inhomogeneities $b(0,0,z)$ with respect to z:

$$V(\kappa) = \frac{1}{2\pi} \int_{-\infty}^{\infty} b(0,0,z) e^{-i\kappa z} dz.$$

From the relationships (6.2.9) and (6.2.10) if follows that the only spectral component of inhomogeneities with the wave number $\kappa = 2k\cos\theta$ scatters effectively, and it corresponds to the space scale $\lambda / 4\pi(n^2 - \sin^2\theta_0)^{1/2}$.

For observations of brightness temperature in the nadir direction the formula for Q simplifies and we have

$$Q = \frac{\pi k_0^2 n^2}{4\alpha} V(2k)(1-R_0)^2 \sigma_\varepsilon^2 \tag{6.2.11}$$

for both polarizations, whence we obtain the dependence of brightness temperature on the wavelength in the form

$$T_\text{b} = T_0(1-R_o)\left[1 - \frac{\pi k_0^2 n^2}{4\alpha} V(2k)(1-R_o)\sigma_\varepsilon^2\right].$$

These simple calculations helped explain the frequency profile of the brightness temperature for glacials for various hypothesis about ice parameters (Gurvich at al., 1973).

The above effect of cooling of a scattering medium with small-scale inhomogeneities does not contradict the warming effect that caused by large scale fluctuations of the parameters α and σ in the radiative transfer equation (Section 6.1). These effects are simply related to different causes. In order to describe these effects simultaneously, one should consider an optically thin layer of the medium ($\alpha \text{H} \leq 1$) and supplement the formula (6.2.1) with additional terms caused by the propagation of thermal radiation from a warm layer underlying the ice. Large-scale fluctuations of medium parameters result in a higher transparency of the layer with respect to the radiation coming from below.

6.2.3 Thermal Radio Emission of Rough Surfaces

Studies of the brightness temperature of the ocean surface is an important source of data about oceanic processes. Calculations of the brightness temperature may be carried out by the formula (6.2.2) using some approximation for R_eff (Rytov, Kravtsov, and Tatarskii, 1989). The majority of applications use the perturbation method (which describes small-scale irregularities), the Kirchhoff method (which describes large-scale irregularities), and the combined approach that deals with both large- and small-scale irregularities.

The pioneering calculations of thermal radiation (Stogryn, 1967; Wu and Fung, 1972) were predominantly based on the perturbation method. Genchev (1984) and Raizer and Filonovich (1984) calculated thermal radiation with the Kirchhoff method for a sinusoidal surface. Finally, Wentz (1975) analyzed thermal radiation for a surface described by a two-scale model.

Calculations showed that surface roughness may both increase and decrease the brightness temperature. The sign of the increment in the brightness temperature relative to the brightness temperature $T_\text{b0} = T_0(1-R_0)$ of a plane boundary depends on the nature of roughness, observation angle, and polarization. The temperature contrast ΔT_b may be as large as 5 - 10° K.

The foam on the ocean surface decreases R_eff and, hence, acts toward "warming" of the ocean surface. Bubbles in the foam act as matching elements between the air and water producing the effect of an oil film (Raizer and Sharkov, 1981).

On the contrary, splashes over the ocean surface cause a "cooling" of the ocean (Raizer, 1992; Dombrovskii and Raizer, 1992) because splashes give rise to an additional scattering and may be described by a term similar to Q in formula (6.2.5).

Quasi-periodic disturbances on the surface of a fluid whose characteristics scales are about the wavelength may result in a decrease of the relative temperature in some definite, critical directions θ_{cr} dictated by the condition

$$K = k_0(1 \pm \sin\theta_{cr}). \tag{6.2.12}$$

Here, $K = 2\pi / \Lambda$ is the wave number for a sinusoidal disturbance with period Λ and k_0 is the wave number of the electromagnetic wave.

When the condition (6.2.12) is satisfied, one of the diffraction spectra due to the diffraction at a sinusoidal surface, propagates along the surface and generates (if polarized vertically) a surface wave. Such waves are absorbed effectively in the fluid and, consequently, they reduce the factor λ_{eff} in formula (6.2.1). As a result, the brightness temperature increases in the critical directions θ_{cr} defined by eqn (6.2.12) (Kravtsov et al., 1978).

Critical phenomena are described also by Kong et al. (1984) and Etkin et al. (1992). Allowance for these phenomena markedly increases the accuracy of the evaluation of the parameters of the wavy ocean surface, specifically the surface ocean temperature and the near surface wind speed and direction.

6.3 ACOUSTICAL PROBLEMS

6.3.1 General Wave Nature of Radiative Transfer Theory

The concept of radiance defined as the angular spectrum of a wave field can be applied not only to electromagnetic waves but also to all the other wave fields encountered in acoustics, hydrodynamics, and plasma theory. The behavior of all these fields is described by a transfer equation of the same structure. It states that the radiance decreases along the ray because of absorption and scattering and increases owing to the scattering from other directions and the external sources of the field.

It goes without saying that each kind of waves adds its salient features to the transfer equation. While polarization effects for longitudinal waves are absent in acoustics, these effects are of significance for elastic waves in solids. Specific scattering cross sections, extinction coefficients, and dispersion equations also present their specific features.

Thus, we may treat transfer theory as a general wave method that acquires ever more applications. In what follows, we consider a new application to ocean acoustics.

6.3.2 Ambient Ocean Noise

Transfer theory has proved to be a very efficient tool in calculating acoustic noise fields, namely, the energy and correlation characteristics of noise. Wilson and Tappert (1979) tackled this problem and calculated the field of a point source. Transfer theory, that does not include the interference of rays arriving at the same point, yielded only a smoothed pattern of the field intensity corresponding to quasi-uniform part of the angular spectrum. Voronovich (1987) also attacked this problem, and included the scattering of rays from the rough ocean surface into consideration.

From the standpoint of transfer theory, the calculation of noise intensity is rather simple. The sources of noise, collapsing bubbles, are specified on the surface, and the propagation of noise from the surface into the water is described by the laws of geometrical optics. However simple the formulation of problem may be, the calculations are complicated because of multiple reflections of acoustic waves from the bottom and from the ocean surface, and also because of scattering by random inhomogeneities in the ocean water, on the surface, and in the bottom.

Let us take a brief look at the angular and depth distribution of noise in the ocean using low-frequency acoustic fields as an example. In ocean acoustics, low-frequency fields are usually meant those in the frequency band from 30 to 300 Hz.

Low-frequency sound is only weakly attenuated in ocean water. Therefore, the low-frequency noise field at a point is formed by noise coming from vast surfaces that may be as large as several thousand square kilometers. Low-frequency noise is produced mainly by surface wind and, to some extent, by surface waving.

Methods developed in radiative transfer theory help us to partially overcome these difficulties. Kuryanov and Klyachin (1981) suggested a simple but rather effective technique to calculate the intensity of noise in the ocean. It is based on the invariance principle and on the assumption that the noise field is uniform in the horizontal plane.

This assumption is supported by the approximately layered structure of the ocean. In the model of a strictly layered ocean, the velocity of sound depends only on the depth (z-coordinate) but not upon the horizontal x and y coordinates. We assume that the parameters of noise are independent of x and y, i.e. uniform over large areas of the ocean surface. Under these conditions, the characteristics of a noise field will also be uniform in the horizontal plane, but will vary with depth z.

6.3.3 Acoustic Radiance of a Horizontally Uniform Field

We calculate first the acoustic radiance of noise field in a particular case of no noise scattering from inhomogeneities of the surface, bottom, and bulk of the ocean. Let $I(z,\theta)$ be the acoustic radiance at a depth z in a direction θ measured from the vertical (Fig. 6.3). Consider points A and B on the ocean surface $z = 0$ that are connected by a ray of length l. The length l is called the length of a ray cycle. It depends on the angle θ at which the ray leaves the surface $z = 0 : l = l(\theta)$.

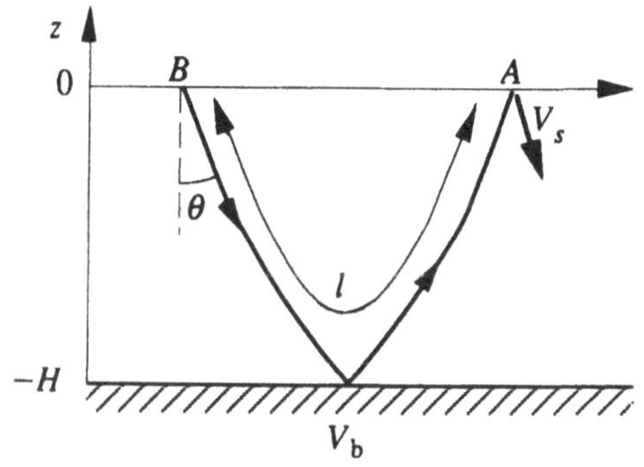

Figure 6.3. In a horizontally homogeneous ocean, the acoustic radiances of noise are identical at points A and B separated by one cycle length l.

The radiances I_A and I_B are related to the radiance of sources $I^0(\theta)$ by

$$I_A(\theta) = I^0(\theta) + V_b V_s I_B(\theta)\exp(-\alpha l), \tag{6.3.1}$$

where V_b and V_s are the energy reflection factors of sound from the bottom and the surface, respectively, and α is the absorption coefficient of sound. Taking into account the dipole nature of sources, we may put $I^0(\theta) = D\cos\theta$, where D is independent of the angle.

The relation (6.3.1) represents the condition of energy balance. In agreement with this condition, the radiance at point A is the sum of the intensity $I^0(\theta)$ of the noise source at point A and the intensity arriving at point A from point B after reflections from the bottom and from the surface (factors V_b and V_s) with the corresponding losses in the water [factor $\exp(-\alpha l)$].

If the condition of uniformity in the horizontal plane is satisfied, the radiances at points A and B must be identical and equal to the intensity $I(0,\theta)$ on the ocean surface:

$$I_A(\theta) = I_B(\theta) = I(0,\theta).$$

This result converts the formula (6.3.1) into an equation for $I(0,\theta)$:

$$I(0,\theta) = I^0(\theta) + V_b V_s I(0,\theta)\exp(-\alpha l). \tag{6.3.2}$$

This equation represents in effect an acoustic analog of the invariance principle suggested by Ambartsumyan to solve astrophysical problems; it is easily solved to yield

$$I(0,\theta)=\frac{I^0(\theta)}{[1-V_bV_s\exp(-\alpha l)]}. \tag{6.3.3}$$

According to this formula, the noise intensity in the ocean grows infinitely when the absorption α tends to zero and the reflection factors V_s and V_b tend to unity. This infinite growth occurs because all noise energy of the sources remains in the oceanic waveguide if losses are absent.

To switch from the radiance at the surface $I(0,\theta)$ to the radiance at an arbitrary depth $I(z,\theta)$, we should multiply $I(0,\theta)$ by the factor $\exp(-\alpha l')$ that describes the attenuation of sound dumping over the ray l' from the surface to a depth z and by the factor $(c_0/c_z)^2$ that allows for the variation of sound velocity with depth. In acoustics, the last factor is an analog of the permittivity ε in the Clausius law. Thus, we have

$$I(z,\theta_z)=\frac{I^0(\theta)\exp(-\alpha l')(c_0/c_z)^2}{1-V_bV_s\exp(-\alpha l)}, \tag{6.3.4}$$

where θ_z is the angle between the ray and z axis at depth z, c_0 and c_z are the sound velocities near the surface and at depth z, respectively.

The distribution of acoustical noise over the depth z may be found by integrating the radiance (6.3.4) over the angle θ_z, or more precisely, over the solid angle $d\Omega=2\pi\sin\theta d\theta$:

$$\mathcal{I}(z)=2\pi\int I(z,\theta_z)\sin\theta_z d\theta_z. \tag{6.3.5}$$

Here, the integration is performed over the sector that contains all the rays leaving the surface, i.e. the rays connecting the surface sources of noise with the observation point.

Calculating the intensity with eqn (6.3.5), we must take into account that both water rays, that do not reach the bottom, and bottom rays that are reflected from the bottom, contribute to the noise field. In a deep ocean, the profile of sound velocity has a minimum at a depth of about 1 km (Fig. 6.4). It is this minimum that gives rise to the far (up to several thousand kilometers) propagation of low frequency sound. In the range below this minimum, the sound velocity increases and at some depth $z=z^*$ it coincides with the velocity at the surface: $c(z^*)=c(0)$. The main energy of the noise field is due to rays that propagate between the surface $z=0$ and the depth $z=z^*$, whereas the contribution of bottom rays may usually be neglected.

Figure 6.4 presents the characteristic curve of the noise intensity versus depth. The maxima of intensity are achieved near the surface and at depths about z^*, because water rays, carrying the main part of noise energy, propagate almost horizontally in

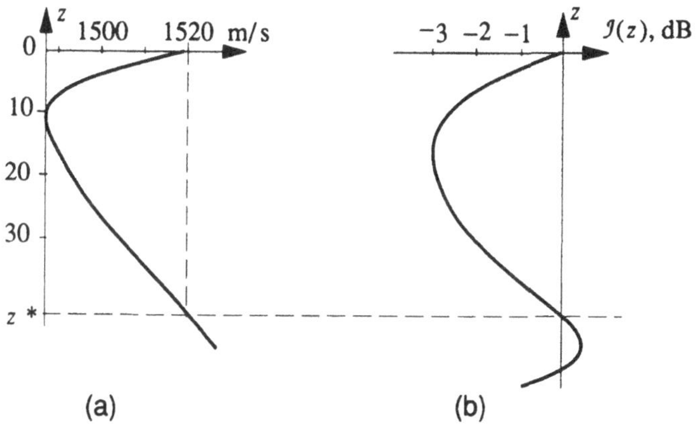

Figure 6.4. (a) Profile of the velocity of sound and (b) the characteristic distribution of noise intensity in a deep ocean (after Kuryanov and Klyachin, 1981)

these regions. This fact results in a concentration of noise energy at the depth $z = z^*$ because the noise flux in the vertical direction is minimal at this depth.

6.3.4 Scattering from the Ocean Surface

Consider now a more complicated problem that includes the multiple scattering of noise signals from surface waving (Klyachin, 1981). If the properties of scattering are uniform in horizontal directions then the noise field also remains uniform in these directions. This means that the radiance will only depend on z and θ as before. However, the acoustic radiance of noise at point A in the direction θ will be composed now of three components rather than two [eqn (6.3.2)]: the intensity of the noise source $I^0(\theta)$, the noise arriving at point A from point B, and an additional noise scattered from the ocean surface and arrived at point A over different rays (Fig. 6.5). As a result, the equation for $I(0, \theta)$ takes the form:

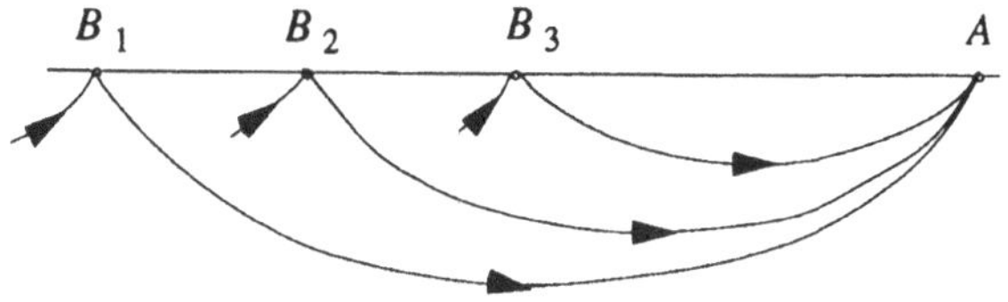

Figure 6.5. All the points on a rough surface contribute to the radiance at point A when the scattering by ocean surface is included.

$$I(0,\theta) = I^o(\theta) + V_b V_s I(0,\theta)\exp(-\alpha l) + \int_0^{\pi/2} m(\theta_1,\theta) I(0,\theta_1)\exp(-\alpha l) 2\pi \sin\theta_1 d\theta_1, \quad (6.3.6)$$

where $m(\theta_1, 0)$ is the scattering coefficient for the ocean surface from the direction θ_1 into the direction θ, and $2\pi \sin\theta_1 d\theta_1$ is the elementary solid angle. This equation can be rewritten as

$$I(0,\theta)(1 - V_b V_s \exp(-\alpha l)) = I^o(0) + 2\pi \int_0^{\pi/2} m(\theta_1,\theta) I(0,\theta_1)\exp(-\alpha l) V_b \sin\theta_1 d\theta_1. \quad (6.3.7)$$

We assume that the roughness height σ_n is small compared to the wavelength. Then we may use a perturbation technique to calculate the scattering coefficient. In the case of an perfectly soft surface of the ocean, the scattering coefficient is (Rytov, Kravtsov, and Tatarskii, 1989b)

$$m(\theta_1,\theta) = m_0 \cos\theta_1,$$

where the quantity m_0 is proportional to the mean square roughness height σ_n^2 and depends on the correlation function of irregularities. The scattering angle cosine is included in this formula because we consider the flux of scattered power over an elementary surface perpendicular to scattering direction rather than over the area parallel to the scattering surface as is common practice in the theory of scattering at a rough surface. The energy reflection coefficient of such a surface may be found by extracting the scattering power flux from the incident power flux:

$$V_s(\theta) = (1 - \frac{2\pi}{3} m_0 \cos\theta).$$

For a given scattering factor $m(\theta, \theta_1)$ depending on the angles of incidence and scattering, the integral in eqn (6.3.7) may be represented as a product of a constant by a factor depending on the angle θ. The unknown constant may be easily found by substituting again the relation for $I(0,\theta)$ into the integral (6.3.7). These calculations lead us to

$$I(0,\theta) = \left(\frac{8}{3}\frac{D}{m_0 \pi}\right) \frac{1}{1 - \sqrt{\left(1 - (c_0 / c_{\min})^2\right)^3}},$$

where $D = I^0(\theta) / \cos\theta$ is a constant that characterizes the intensity of surface dipole sources and $c_{\min}$ is the minimal velocity of sound.

Deriving this formula, we neglected the contribution of rays reflected from the bottom in comparison with the contributions of other rays that were not reflected in such a manner. We neglected also the attenuation of sound. These considerations may be used to calculate the noise intensity with allowance for scattering not only on the surface but in the bulk of the medium and on the bottom of the ocean (Klyachin, 1981).

6.4 TRANSFER EQUATIONS FOR NORMAL WAVES IN MULTIMODE WAVEGUIDES

Examples considered above fall within the realm of propagation in an extended medium, where the boundary effects are of no significance. Nevertheless, the general principles of transfer theory are widely used to describe the behavior of electromagnetic waves in waveguide systems, where the effects of boundaries need be included.

Two alternative methods have been developed to describe the radiative transfer in multimode waveguides — by means of rays and by means of modes. The ray approach, younger of the two, consists in the direct use of the radiative transfer equation written for rays in a waveguide with boundary conditions imposed on its walls. The ray approach was developed by Shatrov (1978) on the basis of statistical wave theory using the small-angle approximation. In wave (mode) approach, normal modes of the waveguide are considered rather than quasi-plane waves. In this case, the system of transfer equations is formulated for the intensities of normal modes rather than for the radiance (Marcuse, 1974; Unger, 1977). The waveguide plays the role of a channel in which normal modes are scattered by random inhomogeneities and are converted to one another. The ray and wave approaches are related by the quantization conditions like that of Bohr–Sommerfeld, where a self-consistent ray congruence corresponds to an individual mode [see Kravtsov and Orlov (1990) and references cited therein].

Methodically, the mode approach looks simpler than the ray approach, and the structure of the system of transfer equations for mode intensities also looks simpler than the structure of transfer equation in three dimensions: a waveguide offers only two directions of mode propagation — forward and backward; moreover, in the case of lightguides, where inhomogeneities are mainly large-scale, backscattering may be neglected. However in multimode case, which may be described formally as the limit $N \to \infty$ (N is the number of modes), both approaches are finally equivalent and led to the same equations of diffusion type. Nevertheless, the ray approach offers definite advantages in the multimode case because it allows sequential results that need cumbersome calculations and additional heuristic considerations in the mode approach.

To illustrate the mode approach, we derive a waveguide set of transfer equations using phenomenological considerations. For simplicity, we limit ourselves to the stationary case and disregard the dependence on time. Also we digress from the possibility of an infinite number of attenuating modes that may bring about nonzero mean energy fluxes because of inhomogeneities (Apresyan, 1978).

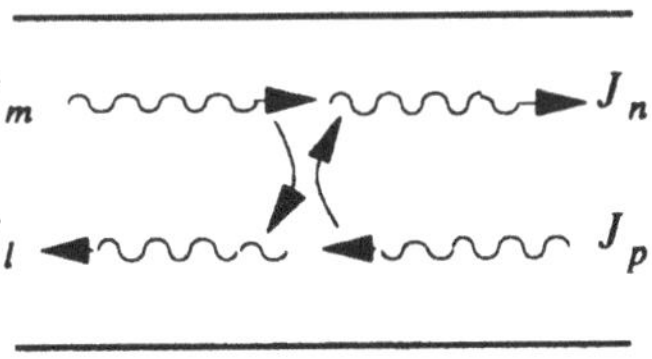

Figure 6.6. Illustration of wave scattering in waveguides.

Let the modes propagating along the z axis of the waveguide be numbered by indexes $m = -N, -(n-1), \ldots, -1, 1, \ldots, N$ and J_m be the intensities of these waves. The average energy flux associated with the mth mode is defined as $(\text{sign}\, m)J_m$ ($\text{sign}\, m = 1$, $m > 0$; $\text{sign}\, m = -1$, $m < 0$) so that the waves with indexes m and $-m$ propagate in the directions $+z$ and $-z$, respectively (Fig. 6.6). In the absence of losses, that the total energy flux will be conserved:

$$\mathrm{d}_z\left(\sum_{m=-N}^{N} J_m \text{sign}\ m\right) = d_z\left(\sum_{m=1}^{N}(J_m - J_{-m})\right) = 0. \tag{6.4.1}$$

The energy flux of the mth mode varies from one point to another because of scattering and transformations into other modes and because of absorption. Whence we may directly come to the system of transfer equations

$$(\text{sign}\ m)d_z J_m + (\alpha_m^{\text{sc}} + \alpha_m^{\text{abs}})J_m = \sum_l \sigma_{m\leftarrow l} J_l. \tag{6.4.2}$$

Here, α_m^{sc} is the total transformation cross section of mth mode into all the other modes, $\sigma_{m\leftarrow m}$ is the scattering cross section without transformation in other modes, $\sigma_{m\leftarrow l}$ is the cross section of transformation of the lth mode into the mth mode, and α_m^{abs} is the energy absorption coefficient for mth mode. According to eqn (6.4.1), in the absence of absorption (i.e. for $\alpha_m^{\text{abs}} = 0$), the optical theorem must be satisfied:

$$\alpha_m^{\text{sc}} = \sum_l \sigma_{l\leftarrow m}.$$

Thus, the system (6.4.2) may be written also in the form

$$(\text{sign}\, m)\text{d}_z J_m + \alpha_m^{\text{abs}} J_m = \sum_l \sigma_{m\leftarrow l} J_l - J_m \sum_l \sigma_{l\leftarrow m}. \tag{6.4.3}$$

If backscattering may be neglected, i.e. if the cross sections $\sigma_{m\leftarrow l}$ let small enough for $(\text{sign m})(\text{sign } l) < 0$, the system (6.4.3) splits into two independent sets of equations, namely, for waves with $m > 0$ and for waves with $m < 0$. The system of transfer equations for the forward traveling waves $(m > 0)$ has the form

$$\text{d}_z J_m + \alpha_m^{\text{abs}} J_m = \sum_{l=1}^{N} \sigma_{m\leftarrow l} J_l - \sum_{l=1}^{N} \sigma_{l\leftarrow m} J_m. \tag{6.4.4}$$

In the conservative case ($\alpha_m^{\text{abs}} = 0$), the law of conservation of the total energy flux for the system (6.4.4) takes the form

$$d_z \sum_{m=1}^{N} J_m = 0.$$

The quantities J_m in (6.4.4) are nonnegative, and may be treated as the probabilities of states of the N-dimensional Markov chain, $\sigma_{m\leftarrow l}$ being the corresponding transient probabilities. In the absence of absorption, the system (6.4.4) takes the form of the Smoluchowski equation for a probability distribution, so that the well developed mathematical formalism of Markov chains (Rytov, Kravtsov, and Tatarskii, 1987) may be invoked to treat this set.

This phenomenological derivation of the transfer equations (6.4.2) and (6.4.4) does not allow the parameters α_m^a and $\sigma_{m\leftarrow l}$ to be expressed in terms of waveguide parameters. Nor does it allow the applicability conditions be deduced for these equations. In the simplest case of weak inhomogeneities, these parameters may be estimated with the Born approximation (see, e.g., Bass and Fuks, 1979). Further refinements of these expressions, as well as a derivation of applicability conditions, are only possible on the basis of the statistical wave approach.

In waveguides with large-scale inhomogeneities, where forward scattering predominates, it is expedient to consider only equations (6.4.4) that describe the interaction between the waves propagating in the forward direction rather than equations (6.4.2) that include backscattering. While the system (6.4.2) requires that the conditions on the ends of the waveguide be imposed and is a boundary-value problem, the system (6.4.4) is uniquely solvable by virtue of the given initial values of the amplitudes. The latter corresponds to a casual problem with initial data and simplifies considerably analysis of its solution.

When backscattering may be neglected, the system (6.4.4) for intensities of parallel waves may be rather rigorously deduced from the wave equations using some asymptotic procedures. The physical meaning of those procedures consists in passing to the limit of long propagation paths with averaging over the wavelength and is similar to the isolation of quasi-uniform part of the spectrum considered in Section 3.5 during the statistical wave substantiation of the radiative transfer equation.

6.5 MISCELLANEOUS PROBLEMS

6.5.1 Nonlinear Problems of Radiative Transfer Theory

In Chapter 3 we constructed a radiative transfer theory as a corollary of the general theory for a statistically quasi-uniform wave field in a scattering medium. Everywhere up to this point we have restricted our consideration to linear theory assuming that the properties of the medium are given and independent of the characteristics of the field. This assumption may be justified only for a weak radiation because, for strong intensities, nonlinear effects come into play. Relatively strong intensities are characteristic of the theory of a weakly turbulent plasma that deals with a homogeneous but nonlinear medium and will be discussed in the next section. Hear touch on the other type of nonlinearity that corresponds to the classic radiative transfer theory and is encountered in astrophysics.

In describing the radiation of stars, the basic equation is the radiation transfer equation of the form

$$d_s I = -(\alpha_{\text{abs}} + \alpha_{\text{sc}})I + \hat{\sigma}I + \varepsilon_q. \tag{6.5.1}$$

In linear theory, the source function ε_q is assumed to be a given characteristic of the medium independent of the properties of radiation. In astrophysics, however, the medium is considered to be radiating only the energy that it has previously absorbed. Then, the stationarity condition leads to the energy balance:

$$\int \varepsilon_q d\omega d\Omega_{\mathbf{n}} = \int \alpha_{\text{abs}} I d\omega d\Omega_{\mathbf{n}}. \tag{6.5.2}$$

This condition relates the source function, i.e. a parameter of the medium, with radiance by a nonlinear relationship in general case. However, the system (6.5.1), (6.5.2) is not closed because these equations do not constitute a complete set to determine I and ε_q. This is why this set is to be augmented by equations based on physical considerations concerning details of the scattering mechanism. For example, one may use the assumption on the so-called local thermodynamic balance. The reader can find the details of that and other approaches in treatises on astrophysics (Chandrasekhar, 1960; Ivanov, 1969; Sobolev, 1975).

In this case, the nonlinearity of the problem is associated with characteristic long observation times, which prevent us from using a linear approximation but, instead, allow us to use the assumption of stationarity. During these times, the "memory" about the initial state dies out, so that an essentially nonlinear regime, described by equations (6.5.1) and (6.5.2), sets up.

The phenomenological approach to make the set (6.5.1) - (6.5.2) closed is, on the one hand, not rigorous and is only allowable for sparse media, where parameters of the transfer equation can be found using simple considerations. On the other hand, the phenomenological approach in no way touches the "non-classic", i.e. wave and correlation, aspects of the transfer theory and deals only with its energy contents. To derive the system (6.5.1) - (6.5.2) more rigorously, one should start with a nonlinear set of stochastic equations that includes the action of radiation on the medium parameters. However, this approach immediately produces complications associated with nonlinearity. We will consider this problem in ensuing section using a simpler example of a homogeneous nonlinear medium.

6.5.2 Radiative Transfer Equation and Kinetic Wave Equations in Weak Turbulence Theory

The theory of weak turbulence studies the behavior of stochastic nonlinear waves in media with both dispersion and nonlinearity. Dispersion prevents the formation of discontinuous (shock) waves, therefore we may limit our consideration with resonant (three- or four-wave) interactions. The theory of weak turbulence was primarily developed for plasma (Tsytovich, 1971), however it was later used to describe the waves on the surface of fluids (see, e.g. Hasselman, 1967).
The theory of weak turbulence is built on the basis of nonlinear equations of the form

$$Lu = j(u,u) + q, \tag{6.5.3}$$

where L is the operator corresponding to a linear medium that is assumed homogeneous and stationary, $j(u,u)$ is the nonlinear current (we take it bilinear for simplicity), and q is the source function. The nonlinear interaction is assumed to be small enough for the field u to approximately satisfy the dispersion equation of linear theory and, consequently, may be represented in the form

$$u = \int A_{\mathbf{k}}(x) e^{i(\mathbf{k}r - \omega(\mathbf{k})t)} d^3k, \tag{6.5.4}$$

where $\omega(\mathbf{k})$ is the solution to the dispersion equation and $A_{\mathbf{k}}(x)$ are the slow amplitudes. For the mean intensities of these amplitudes $\langle |A_{\mathbf{k}}(x)|^2 \rangle \infty N_{\mathbf{k}}(x)$ ($N_{\mathbf{k}}(x)$ is the number of quanta), the theory yields transfer equations similar to eqn (6.5.1) with the difference that they include nonlinear terms describing the interactions of waves. These nonlinear equations are known as kinetic equations for waves. In the simplest case of three-wave interactions, the kinetic equations have the form

$$(d_t + \mathbf{v}_g \nabla) N_{\mathbf{k}}$$
$$= \int W_{\mathbf{k}\mathbf{k}'}(N_{\mathbf{k}'} N_{\mathbf{k}''} - N_{\mathbf{k}} N_{\mathbf{k}'} - N_{\mathbf{k}} N_{\mathbf{k}''}) \delta(\omega - \omega(\mathbf{k}') - \omega(\mathbf{k}'')) d^3k', \tag{6.5.5}$$

where $W_{\mathbf{kk}'}$ is the transition probability, and $\mathbf{k}'' = \mathbf{k} - \mathbf{k}'$. The left-hand side of eqn (6.5.5) corresponds to the radiative transfer to the linear approximation, and the right-hand side describes the nonlinear interactions that may be treated as the scattering of each wave by other waves.

The number of quanta $N_{\mathbf{k}}$ allows us to represent the coherence function of a wave field in the validity region of kinetic equations by relations

$$\Gamma_{12} = \left\langle u(R + \rho / 2) u^{*}(R - \rho / 2) \right\rangle \sim \int N_{\mathbf{k}}(R) e^{i(\mathbf{k}\rho - \omega(\mathbf{k})\tau)} d_3 k, \tag{6.5.6}$$

because the kinetic equations assume the nonlinearity to be small in the sense that the relationships of linear theory hold to the zeroth approximation.

This relation is of fundamental importance. Although the relation (6.5.6) is not directly used in the theory of weak turbulence, it provides information on the coherence function, i.e. on the correlation properties of the field, through the solution of the balance equation (6.5.5).

There is a fundamental difference between this formulation of the problem and that of in the linear case considered above. In contrast to the linear theory that starts from the stochastic wave equation for a field in a fluctuating medium, equation (6.5.3) is written for a nonfluctuating medium. Randomness may formally be included in this equation whether through the source function q or through additional initial (or boundary) conditions. Because of its nonlinearity, equation (6.5.3) allows no derivation of closed equations for second-order moments of the type of Bethe–Salpeter equations of linear theory. Instead, we arrive at an infinite chain of equations for all moments of the field, which is characteristic of nonlinear problems. Now we see that while in the linear case, to derive a transfer equation it was sufficient to assume that the one-group approximation of quasi-uniformity of the field is applicable, in the nonlinear theory, the problem reduces to closing the chain of equations for moments (i.e. to the derivation of a closed set of equations for a finite number of moments).

In pioneering works on the theory of weak turbulence, this problem was solved by invoking the heuristic *postulate of chaotic phases* that is reduced roughly to the assumptions that $A_{\mathbf{k}}$ are Gaussian and $A_{\mathbf{k}}$ and $A_{\mathbf{k}}^{*}$ are uncorrelated for $\mathbf{k} \neq \mathbf{k}'$, so that $\langle A_{\mathbf{k}} A_{\mathbf{k}'}^{*} \rangle = N_{\mathbf{k}} \delta(\mathbf{k} - \mathbf{k}')$. The postulate of chaotic phases contains, but is not reduced to, the condition of quasi-uniformity that was used in linear theory. It requires much stronger limitations connected with the assumptions of a Gaussian distribution for $A_{\mathbf{k}}$. This assumption allows us to reduce the solution for all moments of the field to calculating only the second moments.

The derivation of kinetic equations from nonlinear wave equations has been considerably advanced recently in the theory of chaos. This theory treats the problem, in which the behavior of nonlinear dynamic systems turns out to be nearly random, so that the randomness is inherent in the behavior of the nonlinear system itself rather then introduced in the description of the system "from the outside" by assuming the randomness of medium parameters or initial conditions. This approach permits us to do away with the use of the likelihood hypotheses and to come to a better understanding of the nature of turbulence.

Thus, we can conclude that the methods of linear theory considered above are inapplicable in principle to a nonlinear dynamic equation of the form (6.5.3) [few exceptions like those in Apresyan and Kravtsov (1972) merely confirm this conclusion]. It should be mentioned, however, that some cases allow transition from the nonlinear problem to equations for a distribution function or a characteristic functional that admit description with the linear theory formalism (Klyatskin, 1975, 1985).

6.5.3 Neutron Transfer Theory

The class of problems devoted to the transport of particles, particularly the problems of neutron transfer theory, is of immense importance for applications. The neutron transfer theory is built around a linear transfer equation (a linearized Boltzmann equation) identical in form to the radiative transfer equation if the radiance I is substituted with the distribution of neutrons over the coordinates and velocities. This equation is commonly derived with help of phenomenological considerations about the balance of the number of neutrons in a physically infinitely small volume of a scattering medium (Case and Zweifel, 1967). Such an approach is very similar to the phenomenological transfer theory for the electromagnetic field presented in Chapter 2.

The strategy of substantiation of radiative transfer theory outlined in Chapter 3 can be used, without substantial alterations, to derive the linear neutron-transfer equation. All one has to do is to consider the wave function of the neutron in the given random external field of scatterers instead of the field u in a scattering medium. Then the Wigner function becomes the quasi-probability for a joint distribution over coordinates and momenta, and the conditions of the quasi-classic region, which transform the quasi-probability function into an ordinary distribution function, are an analog of the conditions of quasi-uniformity that allowed for the transfer to the classic transfer theory. Applications of the quasi-probability function to problems of quantum mechanics have been discussed,e. g., by Mori et al. (1962).

When the number density of particles increases, the consideration of their collisions may become necessary. This implication results in a nonlinear theory closely connected with the substantiation of statistical mechanics derived on the ground of the kinetic Boltzmann equation, with all the difficulties inherent in this theory, as mentioned in the foregoing section (Uhlenbeck and Ford, 1963).

APPENDICES

Appendix A: The Method of Geometrical Optics

The method of geometrical optics yields a short-wave asymptotic of the field in smoothly inhomogeneous, slowly nonstationary and weakly nonconserving media. It is assumed that the characteristic scales of medium inhomogeneities and nonstationarities are far greater than the characteristic wavelengths and period of radiation, so that locally the field has the character of traveling plane waves such as in the case of a homogeneous and stationary medium. Without posing on details (for them, see Kravtsov and Orlov, 1990) we describe the scheme of deriving equations of the method of geometrical optics in the very general form.

Let the field u satisfy the homogeneous equation

$$\hat{L}u = \int L(x,x')u(x')d^4x' \equiv \left(\hat{L}_0 + \mu\hat{L}'\right)u = 0, \tag{A1}$$

where $L(x,x')$ is the kernel of the operator $\hat{L}$, and the operator $\hat{L}_0$ corresponds to a conserving medium, whereas the "small" operator $\mu\hat{L}'$ stands for the terms describing nonconserving properties of the medium and other terms that on some grounds are deemed to be "small". The auxiliary small parameter μ serves to express all the assumption on the "slowness" and "smallness" and is put equal to unity in final results.[1]

In agreement with the main assumption of the method of geometrical optics, the field u is represented as a product of a "fast" phase factor and a "slow" amplitude factor. Mathematically, this assumption can be expressed by representing u in the form of an asymptotic expansion:

$$\begin{aligned} u &= \exp\left(i\frac{\psi(\mu x)}{\mu}\right)A(\mu x,\mu) \\ &= \exp\left(i\frac{\psi(\mu x)}{\mu}\right)\left[A_0(\mu x) + \mu A_1(\mu x) + \ldots\right], \end{aligned} \tag{A2}$$

where ψ is the phase or eikonal, and A is the amplitude.

Owing to the presence of a large parameter $1/\mu$ in eqn (A2), small variations in ψ bring about large changes in u, so that the phase factor $exp(i\psi/\mu)$ is a rapidly oscillating quantity compared to the slowly varying amplitude A.

[1] A less formal approach to the construction of a geometrical optics asymptotic is associated with the so-called Debye procedure where the wavelength $\lambda \to 0$ plays the role of the small dimensionless parameter μ.

We will deem that the operator $\hat{L}$ is close to a differential operator by assuming that the kernel $L(x,x')$ is nonzero only at small $\rho' = x' - x$. Then we may substitute in eqn (A2) the formal expansion in μ

$$\begin{aligned} u(x') &= \exp\left(\frac{i}{\mu}\psi(x+\rho')\right)A(\mu(x+\rho'),\mu) \\ &= \exp\left\{\frac{i}{\mu}\left[\psi(x_1)+\mu\rho'\partial_{x_1}\psi(x_1)+\frac{\mu^2}{2}\left(\rho\partial_{x_1}\right)^2\psi(x_1)+O(\mu^3)\right]\right\} \\ &\quad\times\left[1+\mu\rho'\partial_{x_1}+O\left(\mu^2\right)\right]A(x_1) \\ &= \exp\left(i\frac{\psi(x_1)}{\mu}+iK\rho'\right)\left\{1+\mu\left[\rho'\partial_{x_1}+\frac{i}{2}\left(\rho'\partial_{x_1}\right)^2\psi(x_1)\right]+O\left(\mu^2\right)\right\}A(x_1), \end{aligned} \tag{A3}$$

where $x_1 = \mu x$ is the "slow" argument, and $K = K(x_1) = \partial_{x_1}\psi(x_1) = (\mathbf{k}, -\omega)$ is the four-dimensional wave vector. Using (A3), after simple algebra we obtain

$$\begin{aligned} \hat{L}u &= \int L(x,x')u(x')d^4x' \\ &= \exp\left(i\frac{\psi}{\mu}\right)\left\{1-\mu i\left[\partial_K\partial_{x_1}+\frac{1}{2}\left(\partial_K\partial_y\right)^2\psi(y)\right]_{y=x_1}+O\left(\mu^2\right)\right\}L(K)A(x_1), \end{aligned} \tag{A4}$$

where

$$L(K) = \int L(x,x')\exp(iK(x'-x))d^4x' = e^{-iKx}\hat{L}e^{iKx}. \tag{A5}$$

In eqn (A4) we introduced an auxiliary argument refining the action of the operator with the second derivatives.

We expand the amplitude A in powers of μ, as we did in (A2), write $L = L_0 + \mu L'$, and equate the terms at μ^n, $n = 0,1,2,\ldots$ to zero and $\mu = 1$ to obtain a chain of equations:

$$L_0(K)A_0 = 0, \tag{A6}$$

$$L_0(K)A_1 = \left\{i\left[\partial_K\partial_x+\frac{1}{2}\left(\partial_K\partial_y\right)^2\psi(y)\right]_{y=x}L_0(K)-L'(K)\right\}A_0, \tag{A7}$$

and so forth.

In the case of a multicomponent field u, amplitudes A_n may be considered as vectors, and $L_0(K)$ and $L'(K)$ as matrices. Then, for nontrivial resolvability of the zero approximation equation (A6), the determinant of matrix $L_0(K)$ must be zero. This condition yields the *dispersion equation*

$$|L_0| \equiv \det L_0(K) = 0. \tag{A8}$$

Since, by assumption, the matrix $L_0(K)$ corresponds to a conservative system, it may be deemed Hermitian and represented in the canonical form

$$L_0(K) = \sum_i \mathbf{e}_i \otimes \mathbf{e}_i^* \lambda_i, \tag{A9}$$

where $\mathbf{e}_i$ are the orthonormal eigenvectors $\left(\mathbf{e}_i^* \mathbf{e}_j = \delta_{ij}\right)$, and λ_i are the real-valued eigenvalues of $L_0(K)$. In view of eqn (A9), the determinant of $L_0(K)$ equals the product of eigenvalues: $|L_0| = \prod_i \lambda_i$, so that the dispersion equation (A8) splits into several equations

$$\lambda_i \equiv \lambda_i(K,x) = 0\,, \quad i = 1,2,\ldots. \tag{A10}$$

In the general case, these equations correspond to independent normal modes.

If we take into account that the wave vector K is the gradient of the phase ψ, $K = \partial_x \psi$, then eqn (A10) should be viewed as a first-order partial derivative equation for ψ. Equation (A10) may be solved by the method of characteristics (Korn and Korn, 1968). In this case the equations of characteristics play the role of the *ray equations*. It is convenient to write them in the Hamiltonian form

$$\frac{dx}{d\sigma} \equiv d_\sigma x = \partial_K \lambda_i\,, \quad d_\sigma K = -\partial_x \lambda_i\,, \tag{A11}$$

where σ is a parameter varying along the ray.

Equations (A11) for various eigenvalues describe the rays corresponding to various types of rays. We assume that the polarization degeneracy is absent: $\lambda_i \neq \lambda_j$ for $i \neq j$, and consider equation $\lambda_1 = 0$ for $i = 1$. Equations (A6) and (A9) define the direction of the zero approximation vector

$$\mathbf{A}_0 = \mathbf{e}_1 a_1\,, \tag{A12}$$

where $\mathbf{e}_1$ is the unit vector of polarization associated with the dispersion surface $\lambda_1 = 0$, and a_1 is the unknown amplitude factor. The behavior of a_1 is governed by the consistency of the first approximation equations (A7). This condition will be satisfied if we project both sides of eqn (A7) on the direction of $\mathbf{e}_1$. In view of (A9) and (A10) the left-hand side of eqn (A7)vanishes and we arrive at the equation for a_1

$$\mathbf{e}_1^* \cdot \left\{ \left[\partial_K \partial_x + \frac{1}{2}\left(\partial_K \partial_y\right)^2 \psi(y) \right]_{y=x} L_0(K) + iL'(K) \right\}\Big|_{\lambda_1 = 0} \mathbf{e}_1(x) a_1(x) = 0. \tag{A13}$$

Illustrative example: *A scalar wave field.* To illustrate general relations we consider the scalar wave equation $\hat{L} = \partial_{\mathbf{r}}^2 - \hat{\varepsilon}\partial_{ct}^2$ with the permittivity operator $\hat{\varepsilon}$. In this case, unit polarization vectors may be omitted and $\lambda(k)$ coincides with

$$L(K) = e^{-iKx}\hat{L}e^{iKx} = -\mathbf{k}^2 + \left(\frac{\omega}{c}\right)^2 \varepsilon(K,x), \tag{A14}$$

where

$$\varepsilon(K,x) = e^{-iKx}\hat{\varepsilon}e^{iKx}. \tag{A15}$$

The dependence $\varepsilon(K,x)$ on $K = (\mathbf{k}, -\omega)$ describes the spatial and temporal dispersion of the medium.

By writing $\varepsilon(K,x) = \varepsilon' + i\varepsilon''$, where ε' and ε'' are the real and imaginary parts of $\varepsilon(K,x)$ and assuming the absorption to be weak ($\varepsilon'' << \varepsilon'$) we relate ε' to L_0 so that $\mu L' = (\omega / c)^2 i\varepsilon''$.

The dispersion equation $\lambda(k) \equiv L_0(K) = 0$ may be represented in various forms. If we distract from the effects of dispersion and neglect the dependence $\varepsilon(K,x)$upon K, then the dispersion equation may be written as $k - (\omega / c)\sqrt{\varepsilon'} = 0$ or as

$$\tilde{\lambda} \equiv \omega - \omega(\mathbf{k},x) = 0, \tag{A16}$$

where $\omega(\mathbf{k},x) = kc / \sqrt{\varepsilon'}$. Using eqn (A16), we rewrite eqn (A11) by substituting in place of $\lambda(K)$ a proportional quantity $\tilde{\lambda} = \omega - \omega(\mathbf{k},x)$. Changing from the ray parameter σ to the parameter

$$s = \int |d_{\mathbf{k}}\omega(\mathbf{k},x)| d\sigma = \int v_g d\sigma, \tag{A17}$$

that has the sense of the arclength of the spatial ray projection, we write out the four-dimensional dependence to obtain

$$\begin{aligned} d_s\mathbf{r} &\equiv \dot{\mathbf{r}} = \mathbf{n}, \\ \dot{t} &= \frac{1}{v_g}, \\ \dot{\mathbf{k}} &= -\frac{1}{v_g}\partial_{\mathbf{r}}\omega(\mathbf{k},x), \\ \dot{\omega} &= \frac{1}{v_g}\partial_t\omega(\mathbf{k},x). \end{aligned} \tag{A18}$$

Here, $\mathbf{v}_g = d_{\mathbf{k}}\omega(K,x)$ is the group velocity, and $\mathbf{n} = \mathbf{v}_g / v_g$ is the unit vector coinciding with $\mathbf{k} / k$ in view of the isotropic property of the medium.

In this case, eqn (A13) for the amplitude a_1 may be written as

$$\left\{(\partial_{\mathbf{k}}L_0)\partial_{\mathbf{r}} - (\partial_\omega L_0)\partial_t + \frac{1}{2}(\partial_{\mathbf{k}}\partial_{\mathbf{r}} - \partial_\omega\partial_t)^2 L_0(K)\dot{\psi}(x) + iL'\right\}a_1 = 0. \tag{A19}$$

In view of

$$\mathcal{L}_0 = -k^2 + (\omega / c)^2 \varepsilon',$$
$$\partial_{\mathbf{k}} \mathcal{L}_0 = -2\mathbf{k}, \quad \partial_\omega \mathcal{L}_0 = 2\omega\varepsilon' / c^2,$$
$$\partial_{\mathbf{k}} \otimes \partial_{\mathbf{k}} \mathcal{L}_0 = -2\hat{1}, \quad \partial_\omega \partial_\omega \mathcal{L}_0 = 2\varepsilon' / c^2, \quad \partial_\omega \partial_{\mathbf{k}} \mathcal{L}_0 = 0,$$

where $\hat{1}$ is the unit matrix, after a simple algebra we have from (A19)

$$\dot{a}_1 + \frac{1}{2k}\left[\left(\Delta\psi - \frac{1}{v_g}\partial_t^2 \psi\right) + \left(\frac{\omega}{c}\right)^2 \varepsilon''\right] a_1 = 0, \tag{A20}$$

where $\dot{a}_1 = d_s a_1$ is the derivative of a_1 along the ray.

The ray equations (A18) specify the ray structure of the field. Together with eqn (A20) for the field amplitude a_1 they constitute the main equations of the method of geometrical optics.

Appendix B: The Quasi-Isotropic Approximation

In the case of polarization degeneracy (isotropic medium), two eigenvalues vanish, $\lambda_1 = \lambda_2 = 0$, and from the zero approximation equation it follows that the vector $\mathbf{A}_0$ can have two nonzero components

$$\mathbf{A}_0 = a_1 \mathbf{e}_1 + a_2 \mathbf{e}_2, \tag{B1}$$

where $\mathbf{e}_1$ and $\mathbf{e}_2$ are the different eigenvectors corresponding to the eigenvalues λ_1 and λ_2. Let us consider the effects associated with the polarization degeneracy by way of example of the electric field strength:

$$L\mathbf{E} = \left(-\nabla \times \nabla \times - c^2 \partial_t^2 \hat{\varepsilon}\right)\mathbf{E} = 0, \tag{B2}$$

where $\hat{\varepsilon}$ is the medium's permittivity tensor. In contrast to the scalar example considered in Appendix A, eqn (B2) contains the operator $\partial_t^2 \hat{\varepsilon}$ rather than $\hat{\varepsilon}\partial_t^2$. If the medium is stationary, this difference is insignificant. For a nonstationary medium, this operators differ by small addends containing derivatives of $\hat{\varepsilon}$.

In this case,

$$L(K) = e^{-iKx} \hat{L} e^{iKx} = \mathbf{k} \times \mathbf{k} \times + \left(\frac{\omega}{c}\right)^2 \hat{\varepsilon}. \tag{B3}$$

For a weakly anisotropic and weakly nonconserving medium, the permittivity tensor $\hat{\varepsilon}$ may be written as

$$\hat{\varepsilon} = \varepsilon_0 \hat{1} + \hat{\chi}, \tag{B4}$$

where ε_0 is the real quantity, and $\hat{\chi}$ is the tensor corresponding to small corrections due to the anisotropy and nonconserving property of the medium. To ensure the unique representation of eqn (B4) it is convenient to impose the additional condition

$$\mathrm{Re}\,\mathrm{tr}\,\hat{\chi} = 0 \tag{B5}$$

from which it follows $\varepsilon_0 = (1/3)\mathrm{Re}\,\mathrm{tr}\,\hat{\varepsilon}$. Then, in the zero approximation,

$$L_0 = \mathbf{k}\times\mathbf{k}\times + \left(\frac{\omega}{c}\right)^2 \varepsilon_0 \hat{1}, \tag{B6}$$

whereas

$$\mu L' = \left(\frac{\omega}{c}\right)^2 \hat{\chi} = \left(\frac{\omega}{c}\right)^2 \left(\hat{\varepsilon} - \frac{\hat{1}}{3}\mathrm{Re}\,\mathrm{tr}\,\hat{\varepsilon}\right), \tag{B7}$$

In the canonical representation (A9),

$$L_0(K) = \left(\mathbf{e}_1 \otimes \mathbf{e}_1^* + \mathbf{e}_2 \otimes \mathbf{e}_2^*\right)\left[\left(\frac{\omega}{c}\right)^2 \varepsilon_0 - k^2\right] + \mathbf{n}\otimes\mathbf{n}\left(\frac{\omega}{c}\right)^2 \varepsilon_0, \tag{B8}$$

where $\mathbf{n} = \mathbf{k}/k$ is the unit vector tangent to the ray, and $\mathbf{e}_1$ and $\mathbf{e}_2$ are arbitrary unit vectors completing $\mathbf{n} \equiv \mathbf{e}_3$ to an orthonormal basis. Here, the eigenvalues λ_1 and λ_2 coincide: $\lambda_1 = \lambda_2 = (\omega/c)^2 \varepsilon_0 - k^2$, which corresponds to a polarization degeneracy of transverse electromagnetic waves.

The ray equations (A10) for these waves coincide with eqns (A18) for the scalar problem. Let us consider how the polarization vector varies along the ray. In the zero approximation, from eqn (B1) we have $\mathbf{A}_0 = a_1\mathbf{e}_1 + a_2\mathbf{e}_2$. The amplitudes a_1 and a_2 are defined by the system of equations of quasi-isotropic approximation (Kravtsov, 1968). These equations can be obtained by projecting eqn (A7) on a plane with the normal $\mathbf{n} = \mathbf{k}/k$:

$$\hat{\rho}\left[d_s + \frac{1}{2k}\left(\Delta\psi - \frac{1}{v_g^2}\partial_t^2\psi\right)\right]\mathbf{E} = \frac{i\omega}{2c\sqrt{\varepsilon_0}}\hat{\chi}\mathbf{E} \equiv \mathbf{A}\mathbf{E}. \tag{B9}$$

Here, $\hat{\rho} = \hat{1} - \mathbf{n}\otimes\mathbf{n}$ is the projector on the plane with normal $\mathbf{n}$, $\hat{\chi} = \hat{\varepsilon} - (\hat{1}/3)\mathrm{Re}\,\mathrm{tr}\,\hat{\varepsilon}$, and the derivative is taken along the nonstationary geometrical optics ray:

$$d_s = \dot{\mathbf{r}}\partial_{\mathbf{r}} + \dot{t}\partial_t = \mathbf{n}\nabla + \frac{1}{v_g}\partial_t. \tag{B10}$$

In the stationary case, the terms with derivatives ∂_t vanish. In the absence of anisotropy and absorption, when $\hat{\varepsilon} = \varepsilon \hat{1}$ and $\operatorname{Im} \varepsilon = 0$, eqn (B9) takes an especially simple form

$$\hat{\rho}[d_s + \gamma]\mathbf{E} = 0 \tag{B11}$$

where

$$\gamma = \frac{1}{2k}\left(\Delta\psi - \frac{1}{v_g^2}\partial_t^2\psi\right).$$

Substituting (B1) in this expression we readily obtain equations for amplitudes

$$(d_s + \gamma)a_\alpha + \sum_{\beta=1,2} \mathbf{e}_2^* \dot{\mathbf{e}}_\beta a_\beta = 0, \tag{B12}$$

where the derivatives $\dot{\mathbf{e}}_\beta = d_s \mathbf{e}_\beta$ describe rotation of the basis vectors along the ray.

In the case of a curvilinear ray, it is convenient to choose as the basis vectors $\mathbf{e}_1$ and $\mathbf{e}_2$ the normal $\mathrm{v} = \ddot{\mathbf{r}}/|\ddot{\mathbf{r}}|$ and binormal $\mathbf{b} = -\mathrm{v} \times \dot{\mathbf{r}}$ to the spatial projection of the ray. Recalling the Frenêt formulas we have

$$\begin{aligned} \dot{\mathrm{v}} &= -\dot{\mathbf{r}} / R + \mathbf{b} / T \\ \dot{\mathbf{b}} &= -\mathrm{v} / T \end{aligned} \tag{B13}$$

where $R^{-1} = |\ddot{\mathbf{r}}|$ is the curvature, T^{-1} is the twisting, and R and T are the respective radii, we can recast eqns (B12) into the form

$$(d_s + \gamma)a_1 = a_2 / T, \quad (d_s + \gamma)a_2 = -a_1 / T. \tag{B14}$$

From these expressions we immediately obtain the law of rotation of polarization vectors (Rytov, 1938)

$$d_s\theta = T^{-1}, \tag{B15}$$

where $\theta = \arctan(a_1 / a_2)$ is the angle between $\mathbf{E}_0$ and the binormal $\mathbf{b} = \mathbf{e}_2$.

Let us consider simple corollaries of the equations of quasi-isotropic approximation (B12). By putting $\mathbf{E}_0$ in the form

$$\mathbf{E}_0 = \mathbf{e}_0 E_0, \tag{B16}$$

where $E_0 = |\mathbf{E}_0|$, it is not hard to obtain from eqn (B12) the nonlinear equation for $\mathbf{e}_0$

$$\hat{\rho} d_s \mathbf{e}_0 = \left\{\hat{A} - \mathbf{e}_0 \otimes \mathbf{e}_0^* \frac{1}{2}\left(\hat{A} + \hat{A}^+\right)\right\}\mathbf{e}_0. \tag{B17}$$

In the absence of absorption, $\hat{\varepsilon} = \hat{\varepsilon}^+$, and, as can be easily seen from eqn (A3), $\hat{A} = -\hat{A}^+$, so that the nonlinear equation (B17) reduces to the linear equation

$$\hat{p} d_s \mathbf{e}_0 = \hat{A} \mathbf{e}_0. \tag{B18}$$

This equation is used in Section 2.3.3. Other forms of the equation for polarization of the electromagnetic field in weakly anisotropic media may be obtained in the pioneering paper of Kravtsov (1968) and in the book of Kravtsov and Orlov (1990).

Appendix C: The Ray Refractive Index

Here we describe a transition from eqn (2.2.3) to eqn (2.2.8). Changing in eqn (2.2.3) to the integration over the initial volume of the wave train in the k-space, $\Delta\mathbf{k}_0$, we obtain

$$\int_{\Delta\mathbf{k}_0} \left[\partial_t \left(v_g^{-1} I \mathcal{J}\right) + \nabla(\mathbf{n} I \mathcal{J})\right] d^3k_0 = \int_{\Delta\mathbf{k}_0} \left[v_g^{-1}\partial_t + \partial_t v_g^{-1} + \mathbf{n}\nabla + (\nabla\mathbf{n})\right] I \mathcal{J} d^3k_0, \tag{C1}$$

where $\mathcal{J} = |d\omega d\Omega_\mathbf{n} / d^3k_0|$ is the Jacobian of this transition, which, in the general case, is dependent on $\mathbf{r}$ and t. We now take into account the relations

$$\partial_t v_g^{-1} = v_g^{-1}\partial_t \ln v_g^{-1}, \quad \mathbf{n} = \mathbf{v}_g / v_g, \quad \nabla\mathbf{n} = v_g^{-1}\nabla\mathbf{v}_g + \mathbf{n}\nabla \ln v_g^{-1}, \tag{C2}$$

$$d_s = v_g^{-1}\partial_t + \mathbf{n}\nabla, \tag{C3}$$

$$v_g^{-1}\nabla\mathbf{v}_g = d_s \ln\delta = d_s \ln\left|d^3r / d^3r_0\right|, \tag{C4}$$

where d_s is the derivative along the spatial projection of the ray, and $\delta = |d^3r / d^3r_0|$ is the Jacobian of the transition from the initial volume element d^3r_0 of the wave train in the r-space to the element d^3r of the volume moving along the ray. The relations (C2) and (C3) are evident, and eqn (C3) can be proved by differentiating δ along the ray and using the known properties of the Jacobian .

Substituting eqns (C2) – (C4) in (C1) yields the expression

$$\begin{aligned}
&\int_{\Delta\mathbf{k}_0} \left[d_s + \left(d_s \ln v_g^{-1}\right) + (d_s \ln\delta) + d_s \ln(c/\omega)^2\right] I \mathcal{J} d^3k \\
&\qquad = \int_{\Delta\mathbf{k}_0} \left[\mathcal{J} d_s + d_s \ln\left((c/\omega)^2 v_g^{-1} \mathcal{J}\delta\right)\right] I d^3k,
\end{aligned} \tag{C5}$$

in which we added the summand $d_s \ln(c/\omega)^2$ equal to zero (in a stationary medium, the frequency along the ray remains invariable). Equation (C5) coincides with eqn

(2.2.8) since

$$\mathcal{J}\delta = \left|\frac{d\omega d\Omega_{\mathbf{n}}}{d^3k_0}\right| \left|\frac{d^3r}{d^3r_0}\right| = \left|\frac{d\omega d\Omega_{\mathbf{n}}}{d^3k}\right| \left|\frac{d^3k}{d^3k_0}\right| \left|\frac{d^3r}{d^3r_0}\right| = \left|\frac{d\omega d\Omega_{\mathbf{n}}}{d^3k}\right| \tag{C6}$$

in view of the invariance of the phase volume element $d^3kd^3r = d^3k_0d^3r_0$ along the ray.

The derivative along the ray (C3) contains the full derivative with respect to $\mathbf{r}$ and t. If the quantity to be differentiated contains also an implicit dependence on $x = (\mathbf{r}, t)$ by virtue of the wave vector $\mathbf{k} = \mathbf{k}(x)$, then eqn (C3) takes the form

$$d_s = v_g^{-1}\partial_t + \mathbf{n}\nabla + \dot{\mathbf{k}}\nabla_k, \tag{C7}$$

where $\dot{\mathbf{k}} = d_s\mathbf{k}$, and the differentiation with respect to $\mathbf{r}$ and t is performed at a fixed $\mathbf{k}$. If in place of the wave vector one has $\mathbf{n} = \mathbf{v}_g / v_g, \omega$, then this expression changes to

$$d_s = v_g^{-1}\partial_t + \mathbf{n}\nabla + \dot{\mathbf{n}}\nabla_n + \dot{\omega}\partial_\omega. \tag{C8}$$

Expressions for the derivatives along the ray $\dot{\mathbf{n}} = d_s\mathbf{n}$ and $\dot{\omega} = d_s\omega$ can be easily derived with the aid of the ray equations.

References[1]

A

Acquista, C. and J.L.Anderson, 1977, A derivation of the radiative transfer equation for partially polarized light from quantum electrodynamics, *Ann. Phys.* **106**(2), 435–443.

Abrahams, E., P.W.Anderson, D.C.Liccardello and T.V.Ramakrishnan, 1979, Scaling theory of localization: absence of quantum diffusion in two dimensions, *Phys. Rev. Lett.* **42**(10), 673–676.

Akkerman, E., P.E.Wolf and R.Maynard, 1986, Coherent backscattering of light by disordered media: Analysis of the peak line shape, *Phys. Rev. Lett.* **56**(14), 1471–1474.

Allis, W.P., S.J.Buchsbaum and A.Bers, 1963, *Waves in Anisotropic Plasmas* (MIT Press, Cambridge, Mass).

Altshuler, B.L., A.G.Aronov, D.E.Khmelnitskii and A.I.Larkin, 1982, Coherent effects in disordered conductors. In: *Quantum Theory of Solids*, ed. by E.M.Lifshits (Mir, Moscow) pp. 130–237.

Altshuler, B.L., V.E.Kravtsov and I.V.Lerner, 1986, Statistics of mesoscopic fluctuations and single-parameter scaling instability, *Zh. Eksp. Teor. Fiz.* **91**(6), 2276–2302.

Ambartsumyan, V.A., 1943, Diffuse scattering of light by a turbid medium, *Dokl. Akad. Nauk SSSR* **38**(8), 85–91.

Ambartsumyan, V.A., 1944, On a one-dimensional case of a scattering and absorbing medium of finite optical thickness, *Izv. Akad. Nauk Armyansk. SSR* N 1–2, 83–101.

Anderson, P.W., 1958, Absence of diffusion in certain random lattices, *Phys. Rev.* **109**(5), 1492–1505.

Apresyan, L.A., 1973, Radiative transfer equation with allowance for longitudinal waves, *Izv. VUZ Radiofiz.* **16**(3), 461–472.

---

[1] The titles of Russian papers have been translated for convenience. Notice that some Soviet (Russian) journals have been translated into English on a cover-to-cover basis, e.g., Akust. Zh. [Sov. Phys.-Acoust., now Acoust. Phys.], Dokl. Akad. Nauk SSSR [Sov. Phys.-Dokl.], Izv. VUZ Radiofiz. [Radiophys. Quantum Electron.], Radiotekh. Electron. [Radio Eng. Electron. Phys. till 1986 and Sov. J. Commun. Electronics Technology after 1986], Kvant. Elektron. [Sov. J. Quant. Electronics], Zh. Eksp. Teor. Fiz. [Sov. Phys. - JETP], Kratk. Soobsh. Fiz. [Sov. Phys. Lebedev Inst. Reports], Zh. Tekhnich. Fiz. [Sov. Phys. - JTP], Izv. Akad. Nauk SSSR, Fiz. Atm. Okeana [Izv. Acad. Sci. USSR, Atm. Ocean Phys.], Astronom. Zh. [Astrnom.J], Opt. Spektrosk.[Opt. Spectr.]

Apresyan, L.A., 1974, Methods of statistical perturbation theory, *Izv. VUZ Radiofiz.* **17**(2), 165–184.

Apresyan, L.A., 1975, Application of the radiative transfer equation to a free electromagnetic field, *Izv. VUZ Radiofiz.* **18**(12), 1870–1873.

Apresyan, L.A., 1976, Limiting polarization of a wave passing through a thick weakly anisotropic slab with random inhomogeneities, *Astronom. Zh.* **53**(1), 53–62.

Apresyan, L.A., 1978, Allowance for inhomogeneous waves and backscattering in the theory of multimode waveguides with random inhomogeneities, *Izv. VUZ Radiofiz.* **21**(12), 1868–1869.

Apresyan, L.A., 1981a, The Wave content of the radiative transfer equation, *Volny i Diffractsia* (Waves and Diffraction) (VIII All-Union Symposium on diffraction and propagation), Moscow.

Apresyan, L.A., 1981b, Condition of conservation of the adiabatic invariant in radiative transfer theory, *Izv. VUZ Radiofiz.* **24**(3), 308–313.

Apresyan, L.A., 1987, Equations of invariant embedding for linear scattering problems, *Izv. VUZ Radiofiz.* **30**(8), 998–1006.

Apresyan, L.A., 1989, Nonlinear radiative transport equation and Anderson localization in scattering media. Available from VINITI, Moscow, no. 3504–B89. Summary in: Izv. VUZ Radiofiz. 32(6), 721.

Apresyan, L.A., 1990a, Invariant imbedding method for the transfer equation and a description of scattering of bounded beams, *Izv. VUZ Radiofiz.* 33(9), 1047–1054.

Apresyan, L.A., 1990b, Nonlinear Bethe-Salpeter equation: diagrammatic approach, *Izv. VUZ Radiofiz.* **33**(7), 82–884.

Apresyan, L.A., 1991, Markovian description of scattering in a turbulent medium with discrete inhomogeneities, *Opt. Spektrosk.* **71**(4), 643–648.

Apresyan, L.A., 1993, Classification of the mechanisms of backward scattering, *Izv. VUZ Radiofiz.* **34**(10–12), 1125–1134.

Apresyan, L.A. and D.V.Vlasov, 1988, Role of large-scale focusing inhomogeneities in remote sounding experiments, *Izv. VUZ Radiofiz.* **21**(7), 823–833.

Apresyan, L.A. and Yu.A.Kravtsov, 1972, Application of multiple scattering theory to the derivation of kinetic equations for waves in a weakly turbulent plasma, *Izv. VUZ Radiofiz.* **15**(8), 1260–1262.

Apresyan, L.A. and Yu.A.Kravtsov, 1984, Photometry and coherence: wave aspects of the theory of radiation transport, *Sov. Phys. Usp.* **27**(4), 301–313; reprinted in: Friborg, A.T. (Ed), 1993, *Selected Papers on Coherence and Radiometry, SPIE Milestone Ser.*, **69**, (SPIE Optical Engineering Press, Bellingham).

Azzam, R.M.A. and N.M.Bashara, 1977, *Ellipsometry and Polarized Light* (North-Holland, Amsterdam).

B

Badruzzman, A., 1990, Nodal methods in transport theory, *Advances in Nucl. Sci. and Technology* **21**, 1–119 (Plenum Press, New York).

Balescu, R., 1963, *Statistical Mechanics of Charged Particles* (Wiley Interscience, New York).

Baltes, H.P. (ed.), 1978, *Inverse Source Problems in Optics* (Springer, Berlin).

Banakh, V.A. and V.L.Mironov, 1987, *Laser Lidar Ranging in Turbulent*

Atmosphere (Artech House, Boston).
Barabanenkov, Yu.N., 1967, Radiative transfer equation in the model of isotropic point scatterers, *Dokl. Akad. Nauk SSSR* **174**(1), 53–55.
Barabanenkov, Yu.N., 1969a, Correlation function of a random field at a large optical depth, *Izv. VUZ Radiofiz.* **12**(6), 894–899.
Barabanenkov, Yu.N., 1969b, Spectral theory of the transport equation, *Zh. Eksp. Teor. Fiz.* **56**(4), 1262–1272.
Barabanenkov, Yu.N., 1971a, Fraunhofer approximation in the theory of multiple scattering of waves, *Izv. VUZ Radiofiz.* **14**(2), 234–243.
Barabanenkov, Yu.N., 1971b, The depth regime of an electromagnetic field in a scattering medium, Izv. VUZ Radiofiz. **14**(6), 887–891.
Barabanenkov, Yu.N., 1973, Wave corrections to the transport equation for the backscattering direction, *Izv. VUZ Radiofiz.* **16**(1), 88–96.
Barabanenkov, Yu.N., 1974, Energy equivalence of applicability of the Dayson and Bethe equations, *Izv. VUZ Radiofiz.* **17**(1), 113–120.
Barabanenkov, Yu.N., 1975a, Multiple scattering of waves by ensembles of particles and the theory of radiation transport, *Sov. Phys. Uspekhi* **18**(5), 673–689.
Barabanenkov, Yu.N., 1975b, Resolvent method of error estimation of the Bethe-Salpeter equation in the Fraunhofer approximation, *Izv. VUZ Radiofiz.* **18**(5), 716–723.
Barabanenkov, Yu.N. and V.M.Finkelberg, 1967, Radiative transfer equation for correlated scatterers, *Zh. Eksp.Teor. Fiz.* **53**(3), 978–985.
Barabanenkov, Yu.N. and V.M.Finkelberg, 1968, Optical theorem in the theory of multiple scattering, *Izv. VUZ Radiofiz.* **11**(5), 719–725.
Barabanenkov, Yu.N., Yu.A.Kravtsov, S.M.Rytov, and V.I.Tatarskii, 1971, Status of the theory of propagation of waves in a randomly inhomogeneous medium, *Sov. Phys. Uspekhi* **13**(5), 551.
Barabanenkov, Yu.N., Yu.A.Kravtsov, V.D.Ozrin, and A.I.Saichev, 1991, Enhanced backscattering in optics, *Progr. in Optics* **29** ed. by E.Wolf (North-Holland, Amsterdam) pp. 65–197.
Barabanenkov, Yu.N. and V.D.Ozrin, 1991, Diffusion approximation in the theory of weak localization of radiation in a discrete random medium, *Radio Sci.* **26**(3), 747–750.
Barabanenkov, Yu.N., A.G.Vinogradov, Yu.A.Kravtsov, and V.I.Tatarskii, 1972, Application of the theory of multiple scattering of waves to the derivation of the radiation transfer equation for a statistically inhomogeneous medium, *Radiophys. Quantum Electron.* **15**(12), 1420–1425.
Barabashev, N.P., 1922, Bestimmung der Erdalbedo und des Reflexionsgezetzes fur die Oberflache der Mondmeere Theorie der Rillen, *Astron. Nachr* .**217**, 445–452.
Bartelt, H.O., K.N.Brenner, and A.W.Lohmann, 1980, The Wigner distribution function and its optical production, *Opt. Commun.* **32**(1), 32–38.
Bass, F.G. and I.M.Fuks, 1979, *Wave Scattering from a Statistically Rough Surface* (Pergamon Press, Oxford).
Bass, L.P., A.M.Voloshchenko, and T.A.Germogenova, 1986, Method of discrete ordinates in radiation transport problems, Preprint of Inst. Appl. Math., Russ. Acad. Sci. [in Russian].
Beichman, C.A. and S.Ridgway, 1991, Adaptive optics and interferometry, *Phys.*

Today **44**(4), 48–51.

Bekefi, G., 1966, *Radiation Processes in Plasmas* (Wiley, New York).

Belenkii, M.S. and V.L.Mironov, 1972, Diffraction of optical radiation at a Mirror disc in a turbulent atmosphere, *Kvant. Elektron.*, **5**(11), 38–45.

Bell, G.L. and S.Glasstone, 1970, *Nuclear Reactor Theory* (Van Nostrand-Reinhold, Princeton).

Bellman, R. and G.M.Wing, 1975, *An Introduction to Invariant Imbedding* (Wiley, New York).

Bergman, D.J., 1978, The dielectric constant of a composite material. A problem in classical physics, *Phys. Rep.* **43C**, 377.

Bergman, D.J., 1982, Resonances in the bulk properties of composite media–Theory and applications. *Lecture Notes in Physics*, 154 (Springer, New York), pp. 10–37.

Berkovitz, R., 1991, Sensitivity of multiple-scattering speckle pattern to the motion of a single scatterer, *Phys. Rev.* **B43**(10), 8638–8640.

Berkovitz, R., M.Kaveh, and S.Feng, 1989, Memory effect of waves in disordered systems: A real-space approach, *Phys. Rev.* **B40**(1), 737–739.

Berkovitz, R. and M.Kaveh, 1990a, Time reversed memory effects, *Phys. Rev.* **B41**(4), 2635–2638.

Berkovitz, R. and M.Kaveh, 1990b, The vector memory effect for waves, *Europhys.Lett.* **13**(2), 97–101.

Berkovitz, R. and M.Kaveh, 1990c, Theory of speckle-pattern tomography in multiple-scattering media, *Phys. Rev. Lett.* **65**(25), 3120–3123.

Berry, M., 1990, Anticipations of the geometric phase, *Phys. Today* **43**(2), 34–40.

Biryukov, S.V., Yu.V.Gulyaev, V.V.Krylov, and V.P.Plessky, 1995, *Surface Acoustic Waves in Inhomogeneous Media*, (Springer, Berlin).

Bohren, C.F., 1987, Multiple scattering of light and some of it observable consequences, *Am. J. Phys.* **55**(6), 524–533.

Bohren, C.F., 1988, Scattering by a sphere and reflection by a slab: some notable similarities, *Appl. Opt.* **27**(2), 205–206.

Bohren, G. and L.J.Battan, 1980, Radar backscattering by inhomogeneous precipitation particles, *J. Atmos. Sci.* **37**(6), 1821–1827.

Bohren, G. and D.R.Huffman, 1983, *Absorption and Scattering of Light by Small Particles* (Wiley, New York).

Born, M. and E.Wolf, 1980, *Principles of Optics*, 6th ed. (Pergamon Press, New York).

Borovoi, A.G., 1966, Iteration technique in multiple scattering. The transport equation, *Izv. VUZ Fizika* No 6, 50–54.

Borovoi, A.G., 1984, Radiative transfer in inhomogeneous media, *Dokl. Akad. Nauk SSSR* **276**(6), 1374–76.

Borovoi, A.G., G.Ya.Patrushev and A.I.Petrov, 1988, Laser beam propogation through the turbulent atmosphere with precipitation, *Appl. Opt.* **27**(17), 3704–3714.

Bouger, P., 1729, *Essai d'optique sar la qradation de la lumiere*, Paris.

Bourret, R.C., 1960, Coherence properties of blackbody radiation, *Nuovo Cimento* **18**(2), 347–356.

Bremmer, H., 1964, Random volume scattering, *J. Res. NBS* **D68**(3), 967–981.

Bruggeman, D.A.G., 1935, Berechnung verchiedener physikalischer Konstanten von heterogenen Substanzen. I. Dielektrizitatskonstanten und Leitfahigkeiten der Mischkorper aus isotropen Substanzen, *Ann. Phys.* **24**(3), 636–679.

Bugnolo, D., 1960, Transport equation for the spectral density of a multiple–scattered electromagnetic field, *Appl. Phys.* **31**(7), 1176–1182.

Burridge, R., S.Childress, and E.Papanicolaou (eds.), 1982, *Macroscopic Properties of Disordered Media, Lecture Notes in Physics,* vol 154 (Springer, Berlin).

C

Calogero, F., 1967, *Variable Phase Approach in Potential Scattering* (Academic Press, New York).

Carter, W.H. and E.Wolf, 1975, Coherence properties of lambertian and nonlambertian sources, *J. Opt. Soc. Am.* **65**(9), 1067–1071.

Carter, W.H. and E.Wolf, 1977, Coherence and radiometry with quasi-homogeneous planar sources, *J.Opt.Soc.Am.* **67**(4), 785–796.

Case, K.M. and P.F.Zweifel, 1967, *Linear Transport Theory* (Addison–Wesley, Reading, MA).

Chandrasekhar, S., 1960, *Radiative Transfer* (Dover, New York).

Cohen, L., 1989, Time-Frequency Distributions–A Review, *Proc. IEEE* **77**(7), 941–981.

Collin, R.E., 1986, The dyadic Green's functions as an inverse operator, *Radio Sci.* **21**(6), 883–890.

Cornette, W.M. and J.G.Shanks, 1992, Physically reasonable expression for the single-scattering phase function, *Appl. Opt.*, **31**(16), 3152–3160.

Crosignani, B., P.Di Porto, and M.Bertolotty, 1975, *Statistical Properties of Scattered Light* (Academic Press, New York).

Current Science 1990 vol. 59, n 21. Papers presented in the Raman Centrary Symp. "Waves and Symmetry", Bangalore, 1988.

D

Dainty, J.C. (ed.), 1984, *Laser Speckle and Related Phenomena*, 2nd ed.(Berlin: Springer Verlag).

Davison, B., 1957, *Neutron Transport Theory*, (Oxford, New York).

De Wolf, D.A., 1971, Electromagnetic reflection from an extended turbulent medium: cumulative forward-scatter single back-scatter approximation, *IEEE Trans.* **AP-19**(2), 254–262.

Diener, G., 1982, Effective scattering in strongly heterogeneous media, *Physica* **115A**(2), 259–274.

Dolghinov, A.Z., Yu.N.Gnedin, and N.A.Silantiev, 1970, Photon polarization and frequency change in multiple scattering, *J. Quant. Spectrosc. Radiat. Transfer* **10**(7), 707–754.

Dolghinov, A.Z., Yu.N.Gnedin, and N.A.Silant'ev, 1995, *Propagation and Polarization of Radiation in Cosmic Media* (Gordon and Breach, Reading).

Dolin, L.S., 1964a, Ray description of weakly inhomogeneous wave fields, *Izv. VUZ Radiofiz.* **7**(3), 559–562.

Dolin, L.S., 1964b, Scattering of light in a layer of turbid medium, *Izv. VUZ Radiofiz.* **7**(2), 380–382.

Dolin, L.S., 1966, Propagation of a narrow light beam in a medium with strongly anisotropic scattering, *Radiophys. Quantum Electron.* **9**(1), 61–68.

Dolin, L.S., 1968, Equations for correlation functions of a wave beam in a random inhomogeneous medium, *Izv. VUZ Radiofiz.* **11**(6), 840–849.

Dolin, L.S., 1984, On the randomization of a radiation field in a medium with fluctuating extinction coefficient, *Dokl. Akad. Nauk SSSR* **227**(1), 77–80.

Dombrovskii, L.A. and V.Yu.Raizer, 1992, Microwave model of two-phase media in near-water ocean layer, *Izv. Russ. Acad. Sci.* **28**(2), 863–872.

E

Eberly, J.H. and K.Wodkiewicz, 1977, The time dependent physical spectrum of light, *J. Opt. Soc. Am.* **67**(9), 1252–1261.

Edrei, I. and M.Kaveh, 1987, Weak localization of photons and backscattering from finite systems, *Phys. Rev.* **B35**(12), 6461–6463.

Egan, W.G. and T.Hilgeman, 1977, Retroreflectance measurements of photometric standards and coatings. Pt. 2, *Appl. Opt.* **16**(10), 2861–2864.

Eliyahu, D., R.Berkovitz, and M.Kaveh, 1991, Long-range angular correlations of waves in a tube geometry, Phys. Rev. B43(16), 13501–13505.

England, A.W., 1974, Thermal microwave emission from a halfspace containing scatterers, *Rad. Sci.* **9**(4), 447–454.

Erdos, P. and R.C.Herndon, 1982, Theories of electrons in one-dimensional disordered systems, *Adv. Phys.* **31**(2), 65–163.

Erukhimov, L.M. and P.I.Kirsh, 1973, Transport of polarization in a statistically inhomogeneous interstellar plasma, *Izv. VUZ Radiofiz.* **16**(12), 1783–1788.

Erukhimov, L.M., I.G.Zarnitsyna, and P.I.Kirsh, 1973, Discrimination properties and shape of pulse signals crossing a statistically inhomogeneous layer of arbitrary thickness, *Izv. VUZ Radiofiz.* **16**(4), 573–580.

Etemad, S., R.Thompson, and M.J.Andrejco, 1986, Weak localization of photons: universal fluctuations and ensemble averaging, *Phys. Rev. Lett.* **57**(5), 575–578.

Etkin, V.S., V.G.Irisov, and Yu.G.Trokhimovkii, 1992, Resonant radiothermal emission of water surface with non-small periodic roughness, *Int. Geosci. and Remote Sensing Symposium IGARSS'92* (Houston, Texas, May 26–29, 1992) vol. 2, pp. 1457–1459.

F

Felderhov, B.V., G.W.Ford and E.G.D.Cohen, 1983, The Clausius-Mossotti formula and its nonlocal generalization for a dilute suspension of spherical inclusions, *J. Stat. Phys.* **33**(2), 241–260.

Feng, S., C.Kane, P.A.Lee and A.D.Stone, 1988, Correlations and fluctuations of coherent wave transmission through disordered media, *Phys. Rev. Lett.* **61**(7), 834–837.

Finkelberg, V.M., 1964, Dielectric permittivity of mixtures, *Zh. Tekh. Fiz.* **34**(3), 503–518.

Finkelberg, V.M., 1967, Propagation of waves in a random inhomogeneous medium. The method of correlation groups, *Zh. Eksp. Teor. Fiz.* **53**(17), 401–415.

Foldy, L.O., 1945, The multiple scattering of waves. I. General theory of isotropic

scattering by randomly distributed scatterers, *Phys. Rev.* **67**(3, 4), 107–119.
Freund, I., M.Rosenbluh and S.Feng, 1988, Memory effects in propagation of optical waves through disordered media, *Phys.Rev. Lett.* **61**(20), 2328–2331.
Freund, I., 1990a, Stokes-vector reconstruction, *Opt. Lett.* **15**(24), 1425–1427.
Freund, I., 1990b, Correlation imaging through multiply scattering media, *Phys. Lett.A* **147**(8, 9), 502–506.
Freund, I., 1991, Optical Intensity fluctuations in multiply scattering media, *Opt. Commun.* **81**(3, 4), 251–258.
Freund, I. and R.Berkovitz, 1990, Surface reflection and transport through random media: coherent backscattering, optical memory effect, frequency and dynamical correlations, *Phys. Rev.B* **41**(1), 496–502.
Freund, I. M.Kaveh, R.Berkovits and M.Rosenbluh, 1990, Universal polarization correlations and microstatistics of optical waves in random media, *Phys. Rev.*, **B42**(4), 2613–2616.
Friberg, A.T. (Ed), 1993, *Selected papers on Coherence and Radiometry*, SPIE Milestone Ser., **69** (SPIE Optical Eng. Press, Bellingham).
Friberg, A.T. and E.Wolf, 1983, Reciprocity relations with partially coherent sources, *Opt. Acta* 30(10), 1417–1435.
Frisch, U., 1968, Wave propagation in random media. *Probabilistic Methods in Applied Mathematics,* ed by. A.T.Bharucha-Reid (Academic Press, New York) pp. 76–198.
Frisch, U., A.Pouquet, and A.Bourret, 1973, Brownian motion of harmonic oscillator with stochastic frequency, *Physica* **65**(2), 303–320.
Fuks, I.M., 1976, The effect of enhanced backscattering from a shadowed rough surface, *Radiotekh. Elektron.* **21**(3), 625–628.
Furutsu, K., 1975, Multiple scattering of waves in a medium of randomly distributed particles and derivation of transport equation, *Rad. Sci.* **10**(1), 29–44.
Furutsu, K., 1979, An analytical theory of pulse wave propagation in turbulent media, *J. Math. Phys.* **20**(4), 617–628.
Furutsu, K., 1980a, Diffusion equation derived from space–time transport equation in anisotropic media, *J. Math. Phys.* **21**(4), 765–777.
Furutsu, K., 1980b, Diffusion equation derived from space–time transport equation, *J. Opt. Soc. Am.* **70**(4), 360–366.

G

Galinas, R.J. and R.L.Ott, 1970, Quantum theoretical deduction of radiative transfer equations in the spectral line regime, *Ann. Phys.* **59**(2), 323–374.
Gantmacher, F.R., 1959, *The Theory of Matrices*, vols. 1, 2 (Chelsea, New York).
Gaponov, A.V. and M.I.Rabinovich, 1979, L.I.Mandelstam and modern theory of nonlinear oscillations and waves, *Usp. Fiz. Nauk* **128**(4), 579–624.
Garland, J.C. and D.B.Tanner (eds.), 1978, Electrical Transport and Optical Properties of Inhomogeneous Media, *Am. Inst. Phys. Conf. Proc.* **40**(4).
Gazaryan, Yu.L., 1969, One-dimensional propagation of a wave in a medium with random inhomogeneities, *Zh. Eksp. Teor. Fiz.* **56**(6), 1856–1878.
Genchev, J.D., 1984, Electromagnetic wave scattering on a surface with small and smooth inhomogeneities, *Izv. VUZ Radiofiz.* **27**(1), 48–55.

Ginzburg, V.L., 1970, *The Propagation of Electromagnetic Waves in Plasmas*, 2nd ed. (Oxford, Pergamon).
Gnedin, Yu.N. and A.Z.Dolginov, 1963, Theory of multiple scattering, *Zh. Eksp. Teor. Fiz.* **45**(4), 1136–1149.
Golden, K. and G.Papanicolaou, 1983, Bounds for effective parameters of heterogeneous media by analytic continuation, *Commun. Math. Phys.* **90**, 473.
Golubentsev, A.A., 1984a, Suppression of interference effects in multiple scattering, *Zh. Eksp. Teor. Fiz.* **86**(1), 47–59(transl.).
Golubentsev, A.A., 1984b, Contribution of interference into the albedo of a strongly gyrotropic medium with random inhomogeneities, *Izv. VUZ Radiofiz.* **27**(6), 734–745.
Goodman, Y.W., 1985, *Statistical Optics* (Wiley-Interscience, New York).
Gorodnichev, E.E., S.L.Dudarev, and D.B.Rogozkin, 1989, Coherent enhancement of backscattering under weak localization of waves in disordered two- and three-dimensional systems, *Zh. Eksp. Teor. Phys.* **96**(3), 847–864.
Goetze, W., 1978, An elementary approach towards the Anderson transition, *Solid State Comm.* **27**(12), 1393–1395.
Greenspan, (ed.), 1968, *Computing Methods in Reactor Physics* (Gordon and Breach, New York).
Gurvich, A.S., V.I.Kalinin, and D.H.Matveev, 1973, Influence of the internal structure of glaciers on their thermal radio emission, *Izv. Acad. Sci. USSR, Atm. Ocean Phys.* **9**(12), 713.
Gurvich, A.S. and T.G.Krasilnikova, 1977, The polarization of the thermal radiation of permanent snow fields from measurements obtained from the Meteor satellite, *Radio Eng. Electron. Phys.* **22**(9), 88.
Gutnic, M., 1965, *Handbook of Geophysics and Space Environment* (McGraw-Hill, New York).

H

Habetler, G.J. and B.J.Matkovsky, 1975, Uniform asymptotic expansion in transport theory with small mean free paths and the diffusion approximation, *J. Math. Phys.* **16**(4), 846–854.
Halpern, L. and L.N.Trefethen, 1988, Wide-angle one-wave equations, J. *Acoust. Soc. Am.* **84**(4), 1397–1404.
Hapke, B.W., 1963, A theoretical photometric function of Lunar surface, *J. Geoph. Res.* **86**(B4), 3039–3054.
Hapke, B.W., 1981, Bidirectional reflectance spectroscopy, I. Theory, *J. Geophys. Res.*, **86**, 3039–3054.
Hashin, Z. and S.Shtrikman, 1962, A variational approach to the theory of the effective magnetic permeability of multiphase materials, *J. Appl. Phys.* **33**, 3125.
Hasselman, K., 1967, Nonlinear interactions treated by methods of theoretical physics, *Proc. Roy. Soc. Lond.* **A299**, 77–100.
Henyey, L.C. and J.L.Greenstein, 1941, Diffuse radiation in the galaxy, *Atrohpys. J.*, **93**, 73–83.
Hilborn, R.C., 1984, A book review, *Am. J. Phys.* **52**(7), 668–669.
Holland, O.W. and J.H.Barrett, 1987, New aspects of enhanced ion scattering near 180°, *Phys. Rev.* **B35**(13), 6495–6503.

Hori, H., 1977, Theory of effective dielectric, thermal and magnetic properties of random heterogeneous materials. VIII Comparison of different approaches, *J. Math. Phys.* **18**(3), 487–501.

Howe, M.S., 1973, On the kinetic theory of wave propagation in random media, *Philos. Trans. R. Soc. London* **274**(1242), 523–549.

I

Ishimaru, A., 1978a, *Wave Propagation and Scattering in Random Media* (Academic, New York), Vols. 1 and 2.

Ishimaru, A., 1978b, Diffusion of a pulse in densely distributed scatterers, *J. Opt. Soc. Am.* **68**(8), 1045–1050.

Ishimaru, A., 1984, Difference between Ishimaru's and Furutsu's theories on pulse propagation in discrete random media, *J. Opt. Soc. Am.* **A1**(5), 506–509.

Ito, S., 1980, On the theory of pulse wave propagation in media of discrete random scatterers, *Rad. Sci.* **15**(5), 893–901.

Ito, S. and K.Furutsu, 1980, Theory of light pulse propagation through thick clouds, *J. Opt. Soc. Am.* **70**(4), 366–374.

Ivanov, V.V., 1969, Radiative Transfer and the Spectra of Celestial Bodies (Nauka, Moscow, in Russian) [Translation: 1972, *Transfer in Spectral Lines* (NBS Spec. Publ. 385, V.S. 9001, US Govt. Printing office, Washington, D.C.)].

J

Jackson, J.D., 1975, *Classical Electrodynamics*, 2nd ed. (Wiley, New York).

Jakeman, E., 1988, Enhanced backscattering through a deep random phase screen, *J. Opt. Soc. Am.* **5**(10), 1638–1647.

Jakeman, E. and R.J.Tough, 1988, Non-Gaussian models for the statistics of scattering waves, *Adv. Phys.* **37**(5), 471–529.

James, D.F.V. and E.Wolf, 1990, Doppler-like frequency shifts generated by dynamic scattering, *Phys. Lett.* **A146**(4), 167–171.

K

Kadomtsev, B.B. and V.I.Karpman, 1971, Nonlinear waves, *Usp. Fiz. Nauk* **103**(2), 193–232.

Kano, Y. and E.Wolf, 1962, Temporal coherence of blackbody radiation, *Proc. Phys. Soc.* **80**(6), 1273–1276.

Kashkarov, S.S. and A.S.Gurvich, 1977, On the enhanced scattering in a turbulent medium, *Izv. VUZ Radiofiz.* **20**(5), 794–800.

Kato, T., 1966, *Perturbation Theory for Linear Operators* (Springer, Berlin).

Kaveh, M., M.Rosenbluth, I.Edrei, and I.Freund, 1986, Weak localization and light scattering from disordered solids, *Phys. Rev. Lett.* **57**(16), 2049–2052.

Kaveh, M., 1987, Localization of photons in disordered systems, *Philos. Mag.* **56**(6), 693–703.

Kaveh, M., 1991, New phenomena in the propagation of optical waves through random media, *Waves in Random Media* **3**(1), S121–S128.

Khvolson, O.D., 1890, Grundzuge einer mathematischen Theorie der inneren

Diffusion des Lichtes, *Bulletin St.Petersburg Acad. Sci.* **33**(2), 221–256.

Kirkpatrick, T.R., 1973, Percolation and conduction, *Rev. Mod. Phys.* **45**(4), 574–588.

Kirkpatrick, T.R., 1985, Localization of acoustic waves, *Phys. Rev.* **B31**(9), 5746–5755.

Kirkpatrick, T.R. and J.R.Dorfman, 1985, Aspects of localization, *Fundamental Problems in Statistical Mechanics*, vol. VI, ed. by E.G.D.Cohen, (Elsevier North Holland, Amsterdam), 365–398.

Kirkwood, J.G., 1936, On the theory of dielectric polarization, *J. Chem. Phys.* **4**(9), 592–601.

Klauder, J.R. and E.C.G.Sudarshan, 1968, Fundamentals of Quantum Optics (Benjamin, New York).

Kleeorin, N.I., Yu.A.Kravtsov, A.E.Mereminskii and V.G.Mirovskii, 1989, Effect of translucence and radiative transfer in a media with large-scale fluctuation of scatterers concentration, *Izv. VUZ Radiofiz.* **32**(9), 1072–1080.

Klyatskin, V.I., 1975, *Statistical Description of Dynamical Systems with Fluctuating Parameters* (Nauka, Moscow) [in Russian].

Klyatskin, V.I., 1985, *Ondes et Equations Stochastiques dens les Milieus Allatoirement Non-homogenes* (Editons de Physique, Besancon).

Klyatskin, V.I., 1986, *Embedding Method in Wave Propagation Theory* (Nauka, Moscow) [in Russian].

Klyachin, B.I., 1981, Influence of scattering on the anisotropy of the ocean noise field, *Sov. Phys. Acoust* **27**(4), 293–296.

Kohler, W. and G.C.Papanicolaou, 1982, Bounds for effective conductivity of random media. In: *Macroscopic Properties of Disordered Media*, ed by R. Burridge, S. Childress, and G.Papanicolaou (Springer Verlag, Berlin), p.111.

Kong, J.A., S.L.Lin, and S.L.Chuang, 1984, Microwave thermal emission from periodic surface, *IEEE Trans.* **GE-22**(4), 377–381.

Korn, G.A. and T.M.Korn, 1968, *Mathematical Handbook for Scientists and Engineers* (McGraw-Hill, New York).

Krein, S.G., ed., 1964, *Functional Analysis* (Nauka, Moscow) [in Russian].

Kraichnan, R.H., 1961, Dynamics of nonlinear stochastic systems, *J. Math. Phys.* **2**(1), 124–134.

Kramers, B., C.Bergman, and Y.Bruynseraede (eds.), 1985, *Localization, Interaction and Transport Phenomena*, vol. 61 (Springer Verlag, Berlin).

Kravtsov, Yu.A., 1968, Quasi-isotropic approximation of geometric optics, *Dokl. Akad. Nauk SSSR* **183**(1), 74–76.

Kravtsov, Yu.A., E.A.Mirovskaya, A.E.Popov, I.A.Troitskii and V.S.Etkin, 1978, Critical phenomena in thermal radiation of periodically uneven water surface, *Izv. Akad. Nauk SSSR, Fiz. Atm. Ocean* **14**(7), 733–739.

Kravtsov, Yu.A. and Yu.I.Orlov, 1990, *Geometrical Optics of Inhomogeneous Media* (Springer Verlag, Berlin).

Kravtsov, Yu.A. and Orlov, Yu.I., 1993 *Caustics, Catastrophes and Wave Fields* (Springer Verlag, Berlin).

Kravtsov, Yu.A. and A.I.Saichev, 1982, Effects of double passage of waves in randomly inhomogeneous media, *Sov. Phys. Uspekhi* **25**(7), 494–508.

Kravtsov, Yu.A., M.V.Tinin and Yu.N.Cherkashin, 1979, On possible mechanisms of excitation of the ionospheric wave ducts, *Geomagnetizm and Aeronomia* **19**(5), 769–787.

Kuga, Y. and A.Ishimaru, 1984, Retroreflectance from a dense distribution of spherical particles, *J. Opt. Soc. Am.* **A1**(8), 831–839.

Kuga, Y., L.Tsang and A.Ishimaru, 1985, Depolarization effects of the enhanced retroreflectance from a dense distribution of spherical particles, *J. Opt. Soc. Am.* **A2**(6), 616–618.

Kumar, M.S. and R.Simon, 1992, Characterization of Mueller matrices in polarization optics, *Opt. Commun.* **88**(4–6), 464–470.

Kur'yanov, B.F. and B.I.Klyachin, 1981, A theory of the depth dependence of low-frequency noise of the ocean, *Dokl. Akad. Nauk SSSR* **259**(6), 1483–1487.

L

Lambert, J.H., 1760, Photometria, sive de mensura et gradibus luminis, colorum et umbral, Augsburg.

Lampard, D.G., 1954, Generalization of the Wiener-Khinchin theorem to nonstationary processes, *J. Appl. Phys.* **25**(6), 802–803.

Landau, L.D. and E.M.Lifshitz, 1975, *The Classical Theory of Fields*, 4th edn. (Pergamon, Oxford).

Landau, L.D. and E.M.Lifshitz, 1976, *Mechanics*, 3rd edn. (Pergamon, Oxford).

Landauer, R., 1978, Electrical conductivity in inhomogeneous media. In: *Electrical Transport and Optical Properties of Inhomogeneous Media*, p.2, ed. by J.C. Garland and D.B Tanner (Amer. Inst. Phys., New York).

Larsen, E.W. and J.B.Keller, 1974, Asymptotic solution of neutron transport problems for small mean free paths, *J. Math. Phys.* **15**(1), 75–81.

Larsen, E.W. and J.B.Keller, 1976, Asymptotic theory of the linear transport equation for small mean free paths, *Phys. Rev.* **A13**(5), 1933–1939.

Law, C.W. and K.M.Watson, 1970, Radiation transport along curved ray paths, *J. Math. Phys.* **11**(11), 3125–3137.

Lax, M., 1951, Multiple scattering of waves, *Rev. Mod. Phys.* **23**(4), 287–310.

Lax, M., 1952, Multiple scattering of waves, II. The effective field in dense systems, *Phys. Rev.* **85**(3), 621–629.

Lee, A.P. and T.V.Ramakrishnan, 1985, Disordered electronic systems, *Rev. Mod. Phys.* **57**(2), 287–338.

Levin, M.L. and S.M.Rytov, 1967, *Theory of Equilibrium Thermal Fluctuations in Electrodynamics* (Nauka, Moscow) [in Russian].

Lewis, E.E. and W.F.Miller, 1984, *Computational Methods of Neutron Transport* (John Wiley, New York).

M

Ma, S.-K., 1976, *Modern Theory of Critical Phenomena* (Benjamin-Cummings, Reading, Mass.).

Maheu, B., J.N.Letoulouzan and G.Gouesbet, 1984, Four-flux models to solve the scattering transfer equation in terms of Lorenz-Mie parameters, *Appl. Opt.* **23**(14), 3353.

Maheu, B., J.P.Briton and G.Gouesbet, 1989, Four-flux model and Monte-Carlo code: comparisons between two simple, complementary tools for multiple scattering calculations, *Appl. Opt.* **28**(1), 22–24.

Malakhov, A.N., 1978, Cumulant Analysis of Random Non-Gaussian Processes and Their Transformations (Sov. Radio, Moscow) [in Russian].

Malakhov, A.N. and A.I.Saichev, 1979, Representation of a wave reflected from a random inhomogeneous layer as a series satisfying the causality condition, *Izv. VUZ Radiofiz.* **22**(7), 1324–1330.

Mandelstam, L.I., 1972, *Lectures on Theory of Oscillations* (Nauka, Moscow) [in Russian].

Manning, R.M., 1989, Radiative transfer for inhomogeneous media with random extinction and scattering coefficient, *J. Math. Phys.* **30**(8), 2432–2440.

Marc, W.D., 1970, Spectral analysis of the convolution and filtering of non-stationary stochastic processes, *J. Sound. Vibr.* **11**(1), 19–63.

Marcuse, D., 1974, *Theory of Dielectric Optical Wavequides* (Academic Press, New York).

Marchuk, G.I. and V.I.Lebedev, 1986, *Numerical Methods in the Theory of the Neutron Transport* (Harwood Acad. Publishers, Churchhill?).

Marton, J.P. and J.R.Lemon, 1971, Optical properties of aggregated metal systems, I. Theory. *Phys. Rev.* **4**(2), 271–280.

Maxwell Garnett, J.C., 1904, Colors in metal glasses and in metallic films, *Phil. Trans. Roy. Soc. London* **A203**, 385–420.

Mayes, T.W., 1976, Polar intensity profile of elliptically polarized light, *Am. J. Phys.* **44**(11), 1101–1103.

McCartney, E.J., 1976, *Optics of the Atmosphere* (Wiley, New York).

Metha, C.L. and E.Wolf, 1964, Coherence properties of blackbody radiation, *Phys. Rev.* **134A**(5), 1143–1149.

Milton, G.W., 1981, Bounds on the complex permittivity of a two-component composite material, *J. Appl. Phys.* **52**(18), 5286–5293.

Montgomery, W.D., 1967, Self-imaging objects of infinite aperture, *J. Opt. Soc. Am.* **57**(6), 772–778.

Mori, H., I.Oppenheim, and J.Ross, 1962, Wigner function and transport theory studies, in: *Studies in Statistical Mechanics*, 1, (North Holland, Amsterdam), pp. 217–302.

Mott, N.F., 1974, *Metal-Insulator Transition* (Taylor & Francis, London).

N

Nagaoka, Y. and H.Fukuyama (eds), 1982, *Anderson Localization* (Springer Verlag, Berlin).

Nayfeh, A.H., 1972, *Perturbation Methods* (Wiley, New York).

Newton, R.G., 1982, *Scattering Theory of Waves and Particles*, 2nd ed. (Springer-Verlag, New York).

O

Odelevskii, V.I., 1951, Calculation of generalized conductivity of heterogeneous systems, 2. Statistical combinations of nonextended particles, *Zh. Tekh. Fiz.* **21**(5), 667–678.

Oetking, P., 1966, Photometric studies of diffusely reflecting surface with applications to the brightness of the Moon., *J. Geoph. Res.* **71**(10), 2505–2513.

Ojeda-Castaneda, J. and E.E.Sicre, 1985, Quasi ray-optical approach to longitudinal periodicities of free and bounded wavefields, *Opt. Acta* **32**(1), 17–26.

Ovchinnikov, G.I., 1973, Equation for the field spectral density in medium with random inhomogeneities, *Radiotekh. Elektron.* **18**(10), 2044–2049.

Ovchinnikov, G.I. and V.I.Tatarskii, 1972, On the relation between coherence theory and radiative transfer equation, *Izv. VUZ Radiofiz.* **15**(9), 1419–1421.

Ozrin, V.D., 1992, Exact solution for coherent backscattering from a semi-infinite random medium of anisotropic scattering, *Phys. Lett.* **A162**, 341–345.

P

Pacholczyk, A.G., 1970, *Radio Astrophysics* (Freeman, San Francisco).

Page, C.H., 1952, Instantaneous power spectra, *J. Appl. Phys.* **23**(1), 103–106.

Pancharatnam, S., 1956, Generalized theory of interference and its application. Pt.1. Coherent pencils, *Proc. Ind. Acad. Sci.* **A44**(5), 247–262 [Reprinted in: A.Shapere and F.Wilczek (eds), 1989, *Geometric Phase in Physics* (World Scientific, Singapore)].

Peacher, J.I. and K.M.Watson, 1970, Doppler shift in frequency in the transport of electromagnetic wave through an underdense plasmas, *J. Math. Phys.* **11**(4), 1496–1504.

Perina, J., 1972, *Coherence of Light* (Van Nostrand Reinhold, London).

Picibono, B., 1977, Statistical error due to finite-time averaging, *Phys. Rev.A* **16**(5), 2174–2177.

Pnini, R. and B.Shapiro, 1989, Fluctuations in transmission of waves through disordered slabs, *Phys. Rev.B* **39**(10), 6986–6994.

Polder, D. and J.H.van Santen, 1946, The effective permeability of mixture of solids, *Physica* **12**, 257–271.

Pomraning, G.C., 1968, Radiative transfer in dispersive media, *Astrophys. J.* **153**(1), 321–324.

Pomraning, G.C., 1990, Near-infinite-medium solutions of the equation of transfer, *J. Quant. Spectrosc. Radiat. Transfer.* **44**(3), 317–338.

Pronko, P.P., B.R.Appleton, O.W.Holland and S.R.Wilson, 1979, Anomalous field enhancement for highly collimated 180° scattering of He in amorphous and polycrystalline materials, *Phys. Rev. Lett.* **43**(11), 779–782.

R

Raizer, V.Yu., 1992, Two-phase ocean-surface structures and microwave remote sensing, *Int.Geosci. and Remote Sensing Simp. IGARSS '92* (Houston, Texas, May 26–29, 1992), 1460–1462.

Raizer, V.Yu. and E.A.Sharkov, 1981, Electrodynamic description of densely packed dispersed systems, *Radiophys. Quant. Electron* **24**(7), 553–560.

Raizer, V.Yu. and S.R. Filonovich, Radiative characteristics of sinusoidal surfaces in the Kirchhoff approximation, *Izv. VUZ Radiofiz.*, **27**(7), 940–942.

Ramshow, J.D., 1984, Dielectric polarization in random media, *J. Stat. Phys.* **35**(1/2), 49–75.

Rayleigh, 1881, On copying diffraction-gratings, and on some phenomenon connected therewith, *Philos. Mag.* **11**(67), 196–205.

Richtmyer, R.D., 1978, *Principles of Advanced Mathematical Physics* (Springer, New York).

Rochon, P. and D.Bissonnette, 1990, Lensless imaging due to backscattering, *Nature* **348**(6303), 708–710.

Roth, L.M., 1974, Effective medium approximation for liquid metals, *Phys. Rev.B* **9**(6), 2476–2484.

Rozenberg, G.V., 1955, Stokes vector-parameter, *Usp. Fiz. Nauk* **56**(1), 77–110.

Rozenberg, G.V., 1970, Statistical and electrodynamical content of the photometric quantities and basic concepts of radiative transfer theory, *Opt. Spektrosk.* **28**(2), 392–398.

Rozenberg, G.V., 1977, The light ray (contribution to theory of light field), *Sov. Phys. Usp.* **20**, 55–79.

Ruelle, D., 1969, *Statistical Mechanics. Rigorous Results* (Bejamin, New York).

Rye, B.J., 1981, Refractive-turbulence contribution to incoherent backscatter heterodyne lidar returns *J. Opt. Soc. Am.* **71**(6), 687–691.

Rytov, S.M., 1938, On the transition from wave optics to geometrical optics, *Dokl. Akad. Nauk SSSR* **18**(2), 263–268.

Rytov, S.M., 1953, *Theory of Electric Fluctuations and Thermal Radiation* (Izd. Akad. Nauk SSSR, Moscow) [in Russian].

Rytov, S.M., Yu.A.Kravtsov, and V.I.Tatarskii, 1987a, *Principles of Statistical Radiophysics*, vol. 1: *Random Processes* (Springer Verlag, Berlin).

Rytov, S.M., Yu.A.Kravtsov, and V.I.Tatarskii, 1987b, *Principles of Statistical Radiophysics*, vol. 2: *Correlation Theory of Random Processes* (Springer Verlag, Berlin).

Rytov, S.M., Yu.A.Kravtsov and V.I.Tatarskii, 1989a, *Principles of Statistical Radiophysics*, vol. 3: *Random Fields* (Springer Verlag, Berlin).

Rytov, S.M., Yu.A.Kravtsov, and V.I.Tatarskii, 1989b, *Principles of Statistical Radiophysics*, vol. 4: *Wave Propagation Through Random Media* (Springer Verlag, Berlin).

Ryzhov, Y.A., V.V.Tamoikin and V.I.Tatarskii, 1965, Spatial dispersion of inhomogenoeus media, *Zh. Eksp. Teor. Fiz.*, **48**(2), 656–665.

Ryzhov, Y.A. and V.V.Tamoikin, 1970, Radiation and propagation of electromagnetic waves in randomly inhomogeneous media, *Radiophys. Quantum Electron.* **13**(3), 273–300.

S

Sazonov, V.N. and V.N.Tsytovich, 1968, Polarization effects in the generation and transfer of radiation of relativistic electrons in a magnetoactive plasma, *Izv. VUZ Radiofiz.* 11(9), 1287–1299.

Schuster, A., 1905, Radiation through a foggy atmosphere, *Astrophys. J.* **21**(1), 1–22. [Reprinted in: *Selected Papers on the Transfer of Radiation*, ed. by D. H. Menzel, 1966 (Dover, New York)].

Seeliger, H., 1895, Theorie der beleuchtung staubformiger kosmishen Masses insbesondere des Saturnringes, *Abh. Bayer. Akad. Wiss. Math.-Natur. Wiss* **18**(1), 1–72.

Sen, P.N., C.Scala and M.H.Cohen, 1981, A self-similar model for sedimentary rocks with application to the dielectric constant of fused glass beads, *Geophysics* **46**(5), 781–795.

Seraphin, B.O., ed., 1979, *Solar Energy Conversion*. Topics in Applied Physics, vol. 31 (Springer Verlag, Berlin). Shapere, A. and F. Wilczek, (eds), 1989, *Geometric Phase in Physics* (World Scientific, Singapore).

Shapere, A, and Wilczek, F. (eds), 1989, *Geometric Phase in Physics* (World Scientific, Singapore)

Shapiro, B., 1986, Large intensity fluctuations for wave propagation in random media, *Phys. Rev. Lett.* **57**(17), 2168–2171.

Shatrov, A.D., 1978, Energy description of radiative transfer through a multimode optical fiber. *Radiotekh. Elektron.* **23**(10), 2084–2092.

Sheffield, J., 1975, *Plasma Scattering of Electromagnetic Radiation* (Academic Press, New York).

Sheng, P., 1980, Theory of the dielectric function of granular composite media, *Phys. Rev. Lett.* **45**(1), 60–63.

Sheng, P., 1990, *Scattering and Localization of Classical Waves in Random Media* (World Science, Singapore).

Sherman, G.C., J.J.Stamnes and E.Lalor, 1976, Asymptotic approximations to angular spectrum representations, *J. Math. Phys.* **17**(5), 760–776.

Shnerb, N. and M.Kaveh, 1991, Non-Rayleigh statistics of waves in random systems, *Phys. Rev.B* **43**(4), 1279–1282.

Shurcliff, W.A., 1962, *Polirized Light: Production and Use* (Harvard Univ. Press, Cambridge, Mass.).

Siegman, A.E., 1966, The antenna properties of optical heterodyne receivers, *Appl. Opt.* **5**(10), 1588–1594.

Sihvola, A., 1989, Macroscopic permittivity of dielectric mixtures with application to microwave attenuation of rain and hail, *IEE Proc.* **136,** Pt. H, no. 1, 24–28.

Skolnic, H. (ed), 1970, *Radar Handbook*, vol. 1 (McGraw-Hill, New York).

Sobolev, V.V., 1949, On the scattered light polarization, *Uchen. Zapiski LGU* **16**(1), 1–28.

Sobolev, V.V., 1956, *Light Scattering in Planetary Atmospheres* (Nauka, Moscow). [English transl.: 1975, Pergamon, Oxford].

Sobolev, V.V., 1963, *A Treatise on Radiative Transfer* (Van Nostrand-Reinhold, Princeton, N.J.).

Sobolev, V.V., 1975, *A Course in Theoretical Astrophysics* (Nauka, Moscow) [in Russian].

Sornette, D., 1989a, Anderson localization and wave absorption, *J. Stat. Phys.* **56**(15, 6), 669–680.

Sornette, D., 1989b, Acoustic waves in random media II. Coherent effects and strong disorder regime, *Acustica* **67**(4), 251–265.

Sornette, D., 1989c, Acoustical wave in random media III. Experimental situations, *Acustica* **68**(1), 15–25.

Stephen, M.J. and G.Cwilich, 1987, Intensity correlation functions and fluctuations in light scattered from a random media, *Phys. Rev. Lett.* **59**(3), 285–287.

Stott, P.E., 1968, A transport equation for the multiple scattering of electromagnetic waves by a turbulent plasma, *J. Phys.* **A1**(6), 675–689.

Stratonovich, R.L., 1967, *Topics in the Theory of Random Noise*, vol. 1 (Gordon and Breach, New York).

Strogryn, A., 1967, The apparent temperature of sea at microwave frequencies, *IEEE Trans.* **AP-15**(2), 278–286.

Stroud, D., 1975, Generalized effective-medium approach to the conductivity of an inhomogeneous medium, *Phys. Rev.B* **12**(11), 3368–3373.

T

Talbot, M.F., 1836, Facts relating to optical science. IV. *Philos. Mag.* **9**, 401.

Tatarskii, V.I., 1961, *Wave Propagation in a Turbulent Medium* (McGraw-Hill, New York).

Tatarskii, V.I. and M.E.Gertsenshtein, 1963, Propagation of waves in a medium with strong fluctuations of the refractive index, *Sov. Phys. JETP* **17**(3), 458 – 463.

Tatarskii, V.I., 1983, Wigner representation in quantum mechanics, *Usp. Fiz. Nauk* **139**(4), 587–620.

Taylor, J.R., 1972, *Scattering Theory. The Quantum Theory of Nonrelativistic Collisions* (Wiley, New York).

Thiele, E., 1963, Equation of state for hard spheres, *J. Chem. Phys.* **39**(2), 474–479.

Thouless, D.J., 1974, Electrons in disordered systems and the theory of localization, *Phys. Rep.* **13**(3), 94 –142.

Tsang, L. and J.A.Kong, 1975, The brightness temperature of a half space random medium with nonuniform temperature profile, *Rad. Sci.* **10**(12), 1025–1033.

Tsolakis, A.I., I.M.Besieris and W.E.Kohler, 1985, Two-frequency radiative transfer equation for scalar waves in a random distribution of discrete scatterers with pair correlations, *Radio Sci.* **20**(12), 1037–1052.

Tsytovich, V.N., 1971, *Nonlinear Effects in Plasma* (Plenum Press, New York).

Twersky, V., 1962a, On scattering of waves by random distributions.I. Free space scatterer formalism, *J. Math. Phys.* **3**(4), 700–715.

Twersky, V., 1962b, On scattering of waves by random distributions.II. Two space scattering formalism, *J. Math. Phys.* **3**(4), 724–734.

U

Uhlenbeck, G.E, and G.W. Ford, 1963, *Lectures in Statistical Mechanics* (Amer. Math. Soc., Providcence, R.I.).

Unger, H.G., 1977, *Planar Optical Waveguides and Fibres* (Oxford Univ. Press, Clarendon).

V

Vaganov, R.B., R.F.Matveev and V.V.Meriakri, 1972, *Multiple Mode Waveguides with Random Irregularities* (Sov. Radio, Moscow) [in Russian].

Van Albada, M.P. and A.Lagendijk, 1985, Observation of weak localization of light in a random medium, *Phys. Rev. Lett.* **55**(24), 2692–2695.

Van de Huslt, H.C., 1957, *Light Scatering by Small Particles* (Wiley, New York)

Van de Hulst, H.C., 1980, *Multiple Light Scattering*, vols. 1, 2 (Academic, New York).

Vardanyan, R.C., 1988, Radiation transfer in stochastic media, *Astrophys. Spac. Sci.* **141**(2), 375–383.

Vinogradov, A.G. and Yu.A.Kravtsov, 1973, Hybrid method for the calculation of field fluctiations in a medium with large and small random inhomogeneities, *Izv. VUZ Radiofiz.* **16**(7), 1055–1063.

Vinogradov, A.G., Yu.A.Kravtsov and V.I.Tatarskii, 1973, Enhancement of backscatter from bodies immersed in a medium with random inhomogeneities, *Izv. VUZ Radiofiz.* **16**(7), 1064–1070.

Vlasov, D.V., 1985, Laser sounding of an upper layer of the ocean, *Izv. Akad. Nauk SSSR, Ser. Fiz.* **49**(3), 463–472.

Vollhardt, D. and P.Wölfle, 1980, Diagrammatic self-consistent treatment of the Anderson localization problem, *Phys. Rev.B* **22**(10), 4666–4679.

Vollhardt, D. and P.Wölfle, 1982, Scaling equations from a self-consistent theory of Anderson localization, *Phys, Rev. Lett.* **48**(10), 699–702.

Voronovich, A, G., 1987, An approximation of uncorrelated reflections in propagation of sound through a waveguide with statistically rough boundary, *Akust. Zh.* **33**(1), 19–30.

W

Walther, A., 1968, Radiometry and coherence, *J. Opt. Soc. Am.* **58**(9), 1256–1259.

Walther, A., 1973, Radiometry and coherence, *J. Opt. Soc. Am.* **63**(12), 1622–1623.

Wang, L. and S.Feng, 1989, Correlations and fluctuations in reflection coefficients for coherent wave propagation in disordered scattering media, *Phys. Rev.B* **40**(12), 8284–8289.

Watson, K.M., 1960, Quantum mechanical transport theory.I. Incoherent processes, *Phys. Rev.* **118**(4), 886–898.

Watson, K.M., 1969, Multiple scattering of electromagnetic waves in an underdense plasmas, *J. Math. Phys.* **10**(4), 688–702.

Waterman, P.C. and R.Truell, 1961, Multiple scattering of waves, *J. Math. Phys.* **2**(4), 512–537.

Wells, A.W., 1988, *Fourier-Transform Spectroscopy*, 2nd ed. (McGraw-Hill, New York).

Wen, B., L.Tsang, D.R.Winebrenner, and A.Ishimaru, 1990, Dense medium radiative transfer theory: comparison with experiment and application to microwave remote sensing and polarimetry, *IEEE Trans., Geosci. Remote Sensing* **28**(1), 46–58.

Wentz, F.J., 1975, A two-scale scattering model for foam-free sea microwave brightness temperatures, *J. Geophys. Res.*?? **20**(24), 3441–3446.

Wertheim, M.S., 1963, Exact solution of the Percus-Yevick integral equation for hard spheres, *Phys. Rev. Lett.* **10**(2), 321–323.

Wigner, E., 1932, On quantum corrections for thermodynamic equilibrium, *Phys. Rev.* **40**(6), 749–759.

Wilson, H.L. and F.D.Tappert, 1979, Acoustical propagation ocean using the radiation transport equation, *J. Acoust. Soc. Am.* **66**(1), 256–274.

Wolf, E., 1978, Coherence and radiometry, *J. Opt. Soc. Am.* **68**(1), 6–16.

Wolf, E., 1986, Invariance of spectrum of light on propagation, *Phys. Rev. Lett.* **56**(12), 1370.

Wolf, E., 1987, Redshifts and blueshifts of spectral lines emitted by two correlated sources, *Phys. Rev. Lett.* **58**(24), 2646.

Wolf, E., 1991, Influence of source correlations on spectra of radiated fields. In: *International Trends in Optics,* ed by J.W.Goodman (Academic Press, Boston), p. 221.

Wolf, E. and W.H.Carter, 1977, Coherence and radiometry with quasihomogneous planar sources, *J. Opt. Soc. Am.*, **67**(6), 785–796.

Wolf, E. and J.R.Fienup, 1991, Changes in the spectrum of light arising on propagation through a linear, time-invariant system, *Opt. Commun.* **82**(3, 4), 209–212.

Wolf, P.E. and G.Maret, 1985, Weak localization and coherent backscattering of photons in disordered media, *Phys. Rev. Lett.* **55**(24), 2696–2699.

Wolfe, W.L., 1980, Radiometry, in: *Applied Optics and Optical Engineering*, vol 8, ed. by R.R. Shannon and J.C. Wyant (Academic Press, New York), pp. 119–171.

Wu, S.T. and A.K.Fung, 1972, A noncoherent model for microwave emissions and backscattering from the sea surface, *J. Geophys. Res.* **77**(30), 5917–5928.

Y

Yaghjian, A.D., 1980, Electric dyadic Green's functions in the source region, *Proc. IEEE* **68**(2), 248–262.

Yaglom, A.M., 1981, Correlation Theory of Stationary Random Functions (Gidrometeoizdat, Leningrad) [in Russian].

Yakushkin, I.G., 1985, Intensity fluctuations in small-angle scatering of wave fields, *Izv. VUZ Radiofiz.* **28**(5), 555–565.

Yvon, J., 1937, *Recherches sur la Theorie Cinetique des Liquïdes* (Paris, Hermann).

Z

Zakhar-Itkin, M.H., 1973, Matrix Riccati differential equation and the semigroup of linear fractional transformations, *Usp. Mat. Nauk* **28**(3), 83–120.

Zhang, Z.-Q., Q.-J.Chu, W.Xue, and P.Sheng, 1990, Anderson localization in anisotropic random media, *Phys.Rev.B* **42**(7), 4613–4630.

Zheleznyakov, V.V., 1968, Radiation polarization transfer in magneto-active space plasma, *Astr. Sp. Sci.* **2**(4), 403–416.

Zheleznyakov, V.V., 1970, *Radioemission of Sun and Planets* (Pergamon, Oxford).

Zuzin, A.Yu. and B.Z.Spivak, 1987, Langevine description of mesoscopic fluctuations in disordered media, *Zh. Eksp. Teor. Fiz.* **93**(3), 994–1006.

Notation

Dependence on time for monochromatic radiation	$e^{-i\omega t}$
Linear operator, tensor or matrix	$\hat{a}$ or a
Tensor product of vectors	$\mathbf{a} \otimes \mathbf{b} = \|a_i b_j\|$
Tensor products of matrices	$\hat{a} \otimes \hat{b} = \|a_{ij} b_{kl}\|$
Partial derivative	$\partial_t = \partial / \partial t$
Full derivative	$d_t = d / dt$
Transverse (with respect to z) component of **a**	$\mathbf{a}_\perp = (a_x, a_y, 0)$
Longitudinal component of vector **a**	$\mathbf{a}_\parallel = (0, 0, a_z)$
Spatial arguments	$\mathbf{r} = (x, y, z)$, $\mathbf{R} = (R_x, R_y, R_z)$
Temporal arguments	t, T, τ
Space-time arguments	$x = (\mathbf{r}, t)$, $R = (\mathbf{R}, T)$, $\rho = (\mathrm{\rho}, \tau)$
Four-dimensional wave vector	$K = (\mathbf{k}, \omega)$
Scalar product of 3-vectors	$\mathbf{r} \cdot \mathbf{n} = xn_x + yn_y + zn_z$
Scalar product of 4-tuples	$Kx = \mathbf{kr} - \omega t$,
	$\partial_x \partial_K = \partial_\mathbf{r} \partial_\mathbf{k} - \partial_t \omega$
Four-dimensional differentials	$d^4K = d^3k d\omega$, $d^4x = d^3r dt$

Index

K

L

M

O

P

Q

R

S

T

V

W

Printed in the United States
by Baker & Taylor Publisher Services